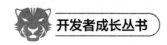开发者成长丛书

Flink原理深入与编程实战
Scala+Java 微课视频版

辛立伟 编著

清华大学出版社
北京

内 容 简 介

本书讲述 Apache Flink 大数据框架的原理，以及如何将 Apache Flink 应用于大数据的实时流处理、批处理、批流一体分析等各个场景。通过对原理的深入学习和对实践示例、案例的学习应用，使读者了解并掌握 Apache Flink 流处理的基本原理和技能，拉近理论与实践的距离。

全书共分为 8 章，主要内容包括 Flink 架构与集群安装、Flink 开发环境准备（基于 IntelliJ IDEA 和 Maven）、开发 Flink 实时数据处理程序、开发 Flink 批数据处理程序、使用 Table API 进行数据处理、Flink on YARN、基于 Flink 构建流批一体数仓、基于 Flink 和 Iceberg 数据湖构建实时数仓。本书源码全部在 Apache Flink 1.13.2 上调试成功，所有示例和案例均提供 Scala 语言和 Java 语言两套 API 的实现（第 8 章除外），供读者参考。

本书内容全面、实例丰富、可操作性强，做到了理论与实践相结合。本书适合大数据学习爱好者、想要入门 Apache Flink 的读者作为入门和提高的技术参考书，也适合用作高等院校大数据专业相关的学生和老师的教材或教学参考书。

本书封面贴有清华大学出版社防伪标签，无标签者不得销售。
版权所有，侵权必究。举报：010-62782989，beiqinquan@tup.tsinghua.edu.cn。

图书在版编目(CIP)数据

Flink 原理深入与编程实战：Scala＋Java：微课视频版/辛立伟编著. —北京：清华大学出版社，2023.2
（开发者成长丛书）
ISBN 978-7-302-62694-7

Ⅰ. ①F… Ⅱ. ①辛… Ⅲ. ①数据处理软件 Ⅳ. ①TP274

中国国家版本馆 CIP 数据核字(2023)第 023858 号

责任编辑：赵佳霓
封面设计：刘　键
责任校对：徐俊伟
责任印制：丛怀宇

出版发行：清华大学出版社
　　　　　网　　址：http://www.tup.com.cn，http://www.wqbook.com
　　　　　地　　址：北京清华大学学研大厦 A 座　　　邮　编：100084
　　　　　社 总 机：010-83470000　　　邮　购：010-62786544
　　　　　投稿与读者服务：010-62776969，c-service@tup.tsinghua.edu.cn
　　　　　质量反馈：010-62772015，zhiliang@tup.tsinghua.edu.cn
　　　　　课件下载：http://www.tup.com.cn，010-83470236
印 装 者：三河市人民印务有限公司
经　　销：全国新华书店
开　　本：186mm×240mm　　　印　张：42.25　　　字　数：1055 千字
版　　次：2023 年 2 月第 1 版　　　印　次：2023 年 2 月第 1 次印刷
印　　数：1～2000
定　　价：159.00 元

产品编号：097800-01

前言
PREFACE

实时数据分析一直是一个热门话题,需要实时数据分析的场景也越来越多,如金融支付中的风控、基础运维中的监控告警、实时大盘等,此外,AI模型也需要依据更为实时的聚合结果来达到很好的预测效果。

Apache Flink是下一代开源大数据处理引擎。它是一个分布式大数据处理引擎,可对有限数据流和无限数据流进行有状态计算;可部署在各种集群环境中,对各种大小规模的数据进行快速计算。

Apache Flink已经被证明可以扩展到数千个内核和TB级的应用程序状态,提供高吞吐量和低时延,并支持世界上一些要求很高的流处理应用程序。例如,Apache Flink在2019年阿里巴巴"双11"场景中突破实时计算消息处理峰值,达到25亿条/秒;2020年"双11"当时的实时计算峰值达到了破纪录的40亿条/秒,数据量也达到了惊人的7TB/s,相当于一秒读完500万本《新华字典》!随着2020年"双11"阿里巴巴基于Flink实时计算场景的成功,毋庸置疑,Flink将会加速成为大厂主流的数据处理框架,最终化身为下一代大数据处理标准。

Apache Flink作为当前热门的实时计算框架之一,是从业人员及希望进入大数据行业的人员必须学习和掌握的大数据技术之一,但是作为大数据的初学者,在学习Flink时通常会遇到以下几个难题:

(1) 缺少面向零基础读者的Flink入门教程。
(2) 缺少系统化的Flink大数据教程。
(3) 现有的Flink资料、教程或图书较陈旧。
(4) 官方全英文文档难以阅读和理解。
(5) 缺少必要的数据集、可运行的实践案例及学习平台。

特别是Apache Flink从2019年被阿里巴巴收购以后,进入快速版本迭代期,不但版本更新快,而且API变化频繁,笔者在应用和研究Apache Flink时,每当遇到Flink版本更新,就不得不花费大量精力重构已经完成的代码。

为此,一方面是为了笔者自己能更系统、更及时地跟进Flink的演进和迭代;另一方面也是为了解决面向零基础读者学习Flink(及其他大数据技术)的入门难度,编写了《Flink原理深入与编程实战——Scala+Java(微课视频版)》。笔者以为,本书具有以下几个特点:

(1) 面向零基础读者,知识点深浅适当,代码完整易懂。

(2) 内容全面系统,包括架构原理、开发环境及程序部署、流和批计算等,并特别包含了第 7 章"基于 Flink 构建批流一体数仓"和第 8 章"基于 Flink 和 Iceberg 数据湖构建实时数仓"内容。

(3) 所有代码均基于 Flink 1.13.2。

(4) 双语实现,大部分示例、案例包含 Scala 和 Java 两种语言版本的实现。

为降低读者学习大数据技术的门槛,本书除提供了丰富的上机实践操作和详细的范例程序讲解之外,作者还为购买和使用本书的读者提供了搭建好的 Hadoop 和 Flink 大数据开发和学习环境。读者既可以参照本书的讲解自行搭建 Hadoop 和 Flink 环境,也可直接使用作者提供的开发和学习环境,快速开始对大数据和 Flink 的学习。

本书特别适合想要入门并深入掌握 Apache Flink、流计算的读者,需要大数据系统参考教材的老师及想要了解最新 Flink 版本应用的从业人员。

当然,由于笔者水平所限,书中难免存在疏漏,敬请读者批评指正。

辛立伟

2022 年 10 月

教学课件(PPT)

本书源代码

目录
CONTENTS

第1章　Flink 架构与集群安装(▶81min)······1
- 1.1　Flink 简介······1
  - 1.1.1　Flink 发展历程······2
  - 1.1.2　Flink 特性······2
- 1.2　Flink 应用场景······3
  - 1.2.1　事件驱动应用程序······3
  - 1.2.2　数据分析应用程序······5
  - 1.2.3　数据管道应用程序······6
- 1.3　Flink 体系架构······6
  - 1.3.1　Flink 系统架构······7
  - 1.3.2　Flink 运行时架构······7
  - 1.3.3　Flink 资源管理······10
  - 1.3.4　Flink 作业调度······15
  - 1.3.5　Flink 故障恢复······18
  - 1.3.6　Flink 程序执行模式······19
- 1.4　Flink 集群安装······23
  - 1.4.1　Flink 独立集群安装和测试······23
  - 1.4.2　Flink 完全分布式集群安装······29
  - 1.4.3　Flink 常用命令······36

第2章　Flink 开发环境准备(▶37min)······39
- 2.1　安装和配置······39
  - 2.1.1　安装和配置 Maven······39
  - 2.1.2　修改 Maven 本地仓库默认位置······41
  - 2.1.3　创建初始模板项目······43
  - 2.1.4　构建打包项目······45
- 2.2　使用 IntelliJ IDEA＋Maven 开发 Flink 项目······46
  - 2.2.1　在 IntelliJ IDEA 中创建 Flink 项目······46
  - 2.2.2　设置项目基本依赖······50
  - 2.2.3　用于构建具有依赖项的 JAR 的模板······51

2.2.4　编写批处理代码并测试执行 ·· 52
2.2.5　项目打包并提交 Flink 集群执行 ·· 55
2.3　Flink 相关概念 ··· 57
2.3.1　Flink 数据流 ·· 57
2.3.2　Flink 分层 API ··· 58

第 3 章　开发 Flink 实时数据处理程序（▷ 376min） ··· 60

3.1　Flink 流处理程序编程模型 ·· 60
3.1.1　流数据类型 ··· 60
3.1.2　流应用程序实现 ··· 64
3.1.3　流应用程序剖析 ··· 67
3.2　Flink 支持的数据源 ··· 72
3.2.1　基于 Socket 的数据源 ·· 73
3.2.2　基于文件的数据源 ··· 77
3.2.3　基于集合的数据源 ··· 80
3.2.4　自定义数据源 ··· 82
3.3　Flink 数据转换 ·· 102
3.3.1　map 转换 ·· 102
3.3.2　flatMap 转换 ··· 103
3.3.3　filter 转换 ·· 106
3.3.4　keyBy 转换 ··· 109
3.3.5　reduce 转换 ·· 116
3.3.6　聚合转换 ·· 118
3.3.7　union 转换 ·· 120
3.3.8　connect 转换 ··· 123
3.3.9　coMap 及 coFlatMap 转换 ·· 128
3.3.10　iterate 转换 ··· 132
3.3.11　project 转换 ·· 136
3.4　Flink 流数据分区 ·· 137
3.4.1　流数据分发模式 ··· 137
3.4.2　数据分区方法 ·· 138
3.4.3　数据分区示例 ·· 155
3.4.4　理解操作符链 ·· 170
3.5　Flink 数据接收器 ·· 172
3.5.1　内置数据接收器 ··· 172
3.5.2　使用流文件连接器 ··· 176
3.5.3　自定义数据接收器 ··· 183
3.6　时间和水印概念 ·· 189
3.6.1　时间概念 ·· 189
3.6.2　事件时间和水印 ··· 189

- 3.6.3 水印策略 …… 191
- 3.6.4 处理空闲数据源 …… 195
- 3.6.5 编写水印生成器 …… 196
- 3.6.6 内置水印生成器 …… 202
- 3.6.7 分配时间戳和水印示例 …… 204

3.7 窗口操作 …… 214
- 3.7.1 理解 Flink 窗口概念 …… 214
- 3.7.2 窗口分配器 …… 217
- 3.7.3 窗口函数 …… 232
- 3.7.4 触发器 …… 260
- 3.7.5 清除器 …… 262
- 3.7.6 处理迟到数据 …… 263
- 3.7.7 处理窗口结果 …… 265
- 3.7.8 窗口连接 …… 266

3.8 低级操作 …… 272
- 3.8.1 ProcessFunction …… 273
- 3.8.2 KeyedProcessFunction 示例 …… 274
- 3.8.3 案例：服务器故障检测报警程序 …… 284

3.9 状态和容错 …… 290
- 3.9.1 状态运算 …… 290
- 3.9.2 状态的类型 …… 291
- 3.9.3 使用托管的 Keyed State …… 296
- 3.9.4 使用托管 Operator State …… 307
- 3.9.5 广播状态 …… 313
- 3.9.6 状态后端 …… 318
- 3.9.7 检查点机制 …… 321
- 3.9.8 状态快照 …… 327

3.10 侧输出流 …… 330
- 3.10.1 什么是侧输出流 …… 330
- 3.10.2 侧输出流应用示例 …… 332

3.11 Flink 流连接器 …… 338
- 3.11.1 Kafka 连接器 …… 339
- 3.11.2 JDBC 连接器 …… 359

3.12 其他 …… 364
- 3.12.1 在应用程序中使用日志 …… 364
- 3.12.2 使用 ParameterTool …… 365
- 3.12.3 命名大型 TupleX 类型 …… 366

3.13 Flink 流处理案例 …… 366
- 3.13.1 处理 IoT 事件流 …… 367

3.13.2 运输公司车辆超速实时监测 ... 375

第4章 开发Flink批数据处理程序 ▶64min ... 390

4.1 Flink批处理程序编程模型 ... 390
4.1.1 批应用程序实现 ... 390
4.1.2 批应用程序剖析 ... 392

4.2 数据源 ... 395
4.2.1 基于文件的数据源 ... 395
4.2.2 基于集合的数据源 ... 402
4.2.3 通用的数据源 ... 406
4.2.4 压缩文件 ... 407

4.3 数据转换 ... 407
4.3.1 map转换 ... 407
4.3.2 flatMap转换 ... 408
4.3.3 mapPartition转换 ... 410
4.3.4 filter转换 ... 412
4.3.5 reduce转换 ... 413
4.3.6 在分组数据集上的reduce转换 ... 414
4.3.7 在分组数据集上的GroupReduce转换 ... 416
4.3.8 在分组数据集上的GroupCombine转换 ... 419
4.3.9 在分组元组数据集上执行聚合 ... 422
4.3.10 在分组元组数据集上执行minBy转换 ... 424
4.3.11 在分组元组数据集上执行maxBy转换 ... 425
4.3.12 在全部元组数据集上执行聚合操作 ... 427
4.3.13 distinct转换 ... 428
4.3.14 join连接转换 ... 430
4.3.15 union转换 ... 432
4.3.16 project转换 ... 434
4.3.17 first-n转换 ... 435

4.4 数据接收器 ... 436
4.4.1 将计算结果保存到文本文件 ... 436
4.4.2 将计算结果保存到JDBC ... 439
4.4.3 标准DataSink方法 ... 441
4.4.4 本地排序输出 ... 442

4.5 广播变量 ... 444
4.6 分布式缓存 ... 446
4.7 参数传递 ... 451
4.7.1 通过构造函数传参 ... 451
4.7.2 通过withParameters(Configuration)传参 ... 453
4.7.3 通过ExecutionConfig传递全局参数 ... 455

4.8 数据集中的拉链操作 457
    4.8.1 密集索引 458
    4.8.2 唯一索引 459
4.9 Flink 批处理示例 462
    4.9.1 分析豆瓣热门电影数据集——Scala 实现 463
    4.9.2 分析豆瓣热门电影数据集——Java 实现 468

第 5 章 使用 Table API 进行数据处理 ▷136min 476

5.1 依赖 476
5.2 Table API 与 SQL 编程模式 477
    5.2.1 TableEnvironment 477
    5.2.2 Table API 与 SQL 程序的结构 480
    5.2.3 在 Catalog 中创建表 481
    5.2.4 查询表 485
    5.2.5 向下游发送表 490
    5.2.6 翻译并执行查询 491
5.3 Table API 498
    5.3.1 关系运算 498
    5.3.2 窗口运算 511
    5.3.3 基于行的操作 518
5.4 Table API 与 DataStream API 集成 525
    5.4.1 依赖 525
    5.4.2 在 DataStream 和 Table 之间转换 526
    5.4.3 处理 insert-only 流 532
    5.4.4 处理变更日志流 542
5.5 Table API 实时流处理案例 551
    5.5.1 传感器温度实时统计 551
    5.5.2 车辆超速实时监测 556
    5.5.3 电商用户行为实时分析 562

第 6 章 Flink on YARN ▷13min 573

6.1 Flink on YARN session 573
    6.1.1 下载 Flink 集成 Hadoop 依赖包 573
    6.1.2 运行 Flink on YARN session 574
    6.1.3 提交 Flink 作业 577
    6.1.4 停止 Flink on YARN session 579
6.2 Flink on YARN 支持的部署模式 580
    6.2.1 Application 模式 580
    6.2.2 Per-Job 集群模式 581
    6.2.3 session 模式 582

第 7 章 基于 Flink 构建流批一体数仓 ▷27min 583

7.1 Flink 集成 Hive 数仓 583

- 7.1.1 Flink 集成 Hive 的方式 583
- 7.1.2 Flink 集成 Hive 的步骤 584
- 7.1.3 Flink 连接 Hive 模板代码 586
- 7.2 批流一体数仓构建实例 587
  - 7.2.1 数据集说明 588
  - 7.2.2 创建 Flink 项目 589
  - 7.2.3 创建执行环境 595
  - 7.2.4 注册 HiveCatalog 596
  - 7.2.5 创建 Kafka 流表 597
  - 7.2.6 创建 Hive 表 599
  - 7.2.7 流写 Hive 表 602
  - 7.2.8 动态读取 Hive 流表 605
  - 7.2.9 完整示例代码 606
  - 7.2.10 执行步骤 615
- 7.3 纯 SQL 构建批流一体数仓 616
  - 7.3.1 使用 Flink SQL 客户端 616
  - 7.3.2 集成 Flink SQL CLI 和 Hive 619
  - 7.3.3 注册 HiveCatalog 621
  - 7.3.4 使用 SQL Client 提交作业 625
  - 7.3.5 构建批流一体数仓完整过程 629

第 8 章 基于 Flink 和 Iceberg 数据湖构建实时数仓（▶23min） 637
- 8.1 现代数据湖概述 637
  - 8.1.1 什么是数据湖 637
  - 8.1.2 数据湖架构 639
  - 8.1.3 开源数据湖框架 639
- 8.2 基于 Flink＋Iceberg 构建企业数据湖 641
  - 8.2.1 Apache Iceberg 的优势 641
  - 8.2.2 Apache Iceberg 经典业务场景 642
  - 8.2.3 应用 Apache Iceberg 的准备工作 645
  - 8.2.4 创建和使用 Catalog 647
  - 8.2.5 Iceberg DDL 命令 649
  - 8.2.6 Iceberg SQL 查询 650
  - 8.2.7 Iceberg SQL 写入 651
  - 8.2.8 使用 DataStream 读取 652
  - 8.2.9 使用 DataStream 写入 653
  - 8.2.10 重写文件操作 654
  - 8.2.11 未来改进 654
- 8.3 基于 Flink＋Iceberg 构建准实时数仓 654
  - 8.3.1 实时数仓构建 655
  - 8.3.2 执行 OLAP 联机分析 659

图书推荐 662

第 1 章 Flink 架构与集群安装

CHAPTER 1

现实世界中，许多系统数据是作为连续的事件流进行的，例如汽车 GPS 定位信号、金融交易记录、手机信号塔与智能手机用户之间的信号交换、网络流量、服务器日志、工业传感器和可穿戴设备的测量等。如果用户能够及时有效地大规模分析这些流数据，就能够更好地理解这些系统，并及时地进行分析。

实时数据分析一直是个热门话题，需要实时数据分析的场景也越来越多，如金融支付中的风控、基础运维中的监控告警、实时大盘等，此外，AI 模型也需要依据更为实时的聚合结果来达到很好的预测效果。

Apache Flink 是下一代开源大数据处理引擎。它是一个分布式大数据处理引擎，可对有限数据流和无限数据流进行有状态计算；可部署在各种集群环境，对各种大小的数据规模进行快速计算，如图 1-1 所示。

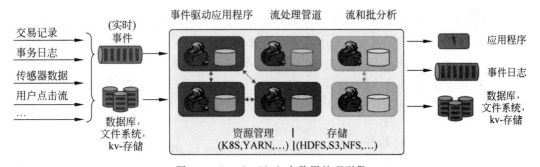

图 1-1　Apache Flink 大数据处理引擎

1.1　Flink 简介

作为下一代开源大数据处理引擎，首先要了解它的发展历程的特性。

1.1.1　Flink 发展历程

Flink 起源于 Stratosphere 项目，这是 2010—2014 年由三所柏林大学和其他欧洲大学共同开展的一项研究项目。2014 年 4 月，Stratosphere 代码的一个分支被捐赠给了 Apache 软件基金会作为一个孵化项目，其初始提交者由系统的核心开发人员组成。此后不久，许多创始人离开大学，创办了一家名叫 Data Artisans 的公司，用于将 Flink 商业化。在孵化期间，为了防止与其他不相关的项目混淆，对项目名称进行了更改，选择 Flink 作为该项目的新名称。

注意：Data Artisans 公司于 2019 年 1 月被阿里巴巴以 9000 万欧元收购。

在德语中，Flink 一词的意思是快速或敏捷，它代表该项目所具有的流和批处理程序的风格。因为松鼠速度快、敏捷，所以 Flink 选择柏林郊外的一种红棕色松鼠作为 Logo。在图 1-2 中，左图为柏林郊外的红棕色松鼠，右图为 Flink 的 Logo。

图 1-2　Apache Flink 名称的由来和 Logo

项目快速完成孵化，2014 年 12 月，Flink 成为 Apache 软件基金会的顶级项目。

Flink 是 Apache 软件基金会最大的 5 个大数据项目之一，在全球拥有超过 200 名开发人员的社区。作为公认的新一代大数据计算引擎，Flink 已成为阿里巴巴、腾讯、滴滴、美团、字节跳动、Netflix、Lyft 等国内外知名公司建设流计算平台的首选。部分使用 Apache Flink 的企业如图 1-3 所示。

图 1-3　部分使用 Apache Flink 的成功企业

1.1.2　Flink 特性

Flink 支持流和批处理、复杂的状态管理、事件时间处理语义，以及对状态的一次一致性保证。此外,Flink 可以部署在各种资源提供者(如 YARN、Apache Mesos 和 Kubernetes)上,也可以作为独立集群部署在裸机硬件上。可以将 Flink 集群配置为高可用的以避免单点故障。

Flink 设计用于在任何规模上运行有状态流应用程序。应用程序可能被并行化为数千个任务,这些任务分布在集群中并且并行执行,因此,一个应用程序可以利用几乎无限数量的 CPU、主内存、磁盘和网络 IO。此外,Flink 很容易维护非常大的应用程序状态。它的异步和增量检查点算法在保证精确一次性的状态一致性的同时,确保对处理时延的影响最小。

Apache Flink 为用户提供了更强大的计算能力和更易用的编程接口:

(1)批流统一。Flink 在 Runtime 和 SQL 层批流统一,提供高吞吐低延时计算能力和更强大的 SQL 支持。

(2)生态兼容。Flink 能与 Hadoop YARN/Apache Mesos/Kubernetes 集成,并且支持单机模式运行。

(3)性能卓越。Flink 提供了性能卓越的批处理与流处理支持。

(4)规模计算。Flink 的作业可被分解成上千个任务,分布在集群中并行执行。

Flink 已经被证明可以扩展到数千个内核和 TB 级的应用程序状态,提供高吞吐量和低时延,并支持世界上一些要求最高的流处理应用程序。例如,Apache Flink 在 2019 年阿里巴巴"双 11"场景中突破实时计算消息处理峰值达到 25 亿条/秒,2020 年"双 11"当时的实时计算峰值达到了破纪录的每秒 40 亿条记录,数据量也达到了惊人的每秒 7TB,相当于一秒需要读完 500 万本《新华字典》!随着 2020 年"双 11"阿里巴巴基于 Flink 实时计算场景的成功,毋庸置疑,Flink 将会加速成为大厂主流的数据处理框架,最终化身为下一代大数据处理标准。

1.2 Flink 应用场景

Apache Flink 是一款非常适合做流批处理的计算框架,适用于以下场景:

(1)事件驱动类型,例如信用卡交易、刷单、监控等。

(2)数据分析类型,例如库存分析、"双 11"数据分析等。

(3)数据管道类型,也就是 ETL 场景,例如一些日志的解析等。

1.2.1 事件驱动应用程序

任何类型的数据都是作为事件流产生的。信用卡交易、传感器测量、机器日志、网站或移动应用程序上的用户交互,所有这些数据都以流的形式生成。

事件驱动的应用程序是一个有状态的应用程序,它从一个或多个事件流中摄取事件,并通过触发计算、状态更新或外部操作来响应传入的事件。

事件驱动的应用程序是传统应用程序(具有独立的计算层和数据存储层)设计的演化。

在这种传统体系结构中,应用程序从远程事务数据库读取数据并将数据持久存储。

而事件驱动的应用程序则基于有状态流来处理应用程序。在这种设计中,数据和计算是共存的,从而产生本地(内存或磁盘)数据访问。通过定期将检查点写入远程持久存储,可以实现容错。传统事务型应用程序和事件驱动应用程序体系结构之间的区别如图1-4所示。

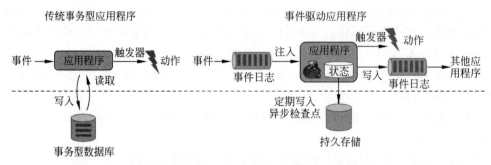

图1-4 传统事务型应用程序和事件驱动应用程序体系结构之间的区别

事件驱动的应用程序访问本地数据,而不是查询远程数据库,从而在吞吐量和时延方面获得更好的性能。远程持久存储的定期检查点可以异步和增量地完成,因此,检查点对常规事件处理的影响非常小。事件驱动的应用程序设计提供的好处不仅是本地数据访问。在分层体系结构中,多个应用程序共享同一个数据库是很常见的,因此,需要协调数据库的任何更改,例如由于应用程序更新或扩展服务而更改数据布局。由于每个事件驱动的应用程序都负责自己的数据,因此对数据表示的更改或应用程序的扩展需要较少的协调。

事件驱动应用程序的限制由流处理器处理时间和状态的能力来定义,Flink的许多突出特性都围绕这些概念。Flink提供了一组丰富的状态原语,这些原语可以管理非常大的数据量(最多可达几TB),并且具有严格的精确一次性的一致性保证。此外,Flink支持事件时间、高度可定制的窗口逻辑及ProcessFunction提供的对时间的细粒度控制,从而支持高级业务逻辑的实现。此外,Flink还提供了一个用于复杂事件处理(Complex Event Processing,CEP)的库,用于检测数据流中的模式。

对于事件驱动的应用程序,Flink的突出特性是保存点(savepoint)。保存点是一个一致的状态镜像,可以用作兼容应用程序的起点。给定一个保存点,应用程序可以更新或调整其规模,或者可以启动应用程序的多个版本进行A/B测试。

典型的事件驱动程序包括:
(1) 欺诈检测。
(2) 异常检测。
(3) 基于规则的提醒。
(4) 业务流程监控。
(5) Web应用程序(例如,社交网络)。

1.2.2 数据分析应用程序

分析工作从原始数据中提取信息。传统上，分析是作为对记录事件的有界数据集的批处理查询或应用程序执行的。为了将最新的数据合并到分析结果中，必须将其添加到待分析的数据集中，然后重新运行查询或应用程序。结果被写入存储系统或作为报告发出。

使用复杂的流处理引擎，还可以实时执行分析。流查询或应用程序不是读取有限的数据集，而是摄入实时事件流，并随着事件被消费而不断生成和更新结果。结果要么写入外部数据库，要么作为内部状态维护。例如，Dashboard 应用程序可以从外部数据库读取最新结果，也可以直接查询应用程序的内部状态。

Apache Flink 支持流及批处理分析应用程序，如图 1-5 所示。

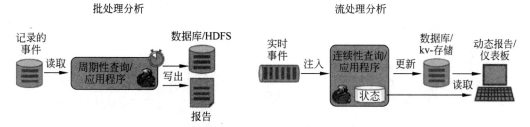

图 1-5　Apache Flink 支持流及批处理分析应用程序

与批处理分析相比，连续流分析的优势并不仅在于低时延（由于消除了周期性导入和查询执行，从事件到洞察的时延要低得多），另一个方面是更简单的应用程序架构。批处理分析管道由几个独立组件组成，用于定期调度数据摄入和查询执行。可靠地运行这样的管道并不简单，因为一个组件的故障会影响管道的后续步骤。相比之下，流分析应用程序运行在复杂的流处理器（如 Flink）上，它包含从数据摄取到连续结果计算的所有步骤，因此，它可以依靠引擎的故障恢复机制。

Flink 为连续流和批处理分析提供了非常好的支持。具体来讲，它具有一个符合 ANSI 的 SQL 接口，具有用于批处理和流查询的统一语义。无论 SQL 查询是在记录事件的静态数据集上运行，还是在实时事件流上运行，它们都会计算出相同的结果。对用户定义函数的丰富支持确保可以在 SQL 查询中执行定制代码。如果需要更多的定制逻辑，Flink 的 DataStream API 或 DataSet API 提供了更多的底层控制。此外，Flink 的 Gelly 库为批量数据集的大规模和高性能图分析提供了算法和构建块。

典型的数据分析应用程序包括：

（1）电信网络质量监控。

（2）移动应用程序中的产品更新及用户体验分析。

（3）消费者技术中实时数据的即时（Ad Hoc）分析。

（4）大规模图分析。

1.2.3 数据管道应用程序

提取-转换-加载(Extract-Transform-Load,ETL)是在存储系统之间转换和移动数据的常用方法。ETL作业通常定期触发，以便将数据从事务型数据库系统复制到分析数据库或数据仓库。

数据管道的作用类似于ETL作业。它们转换和丰富数据，并能将数据从一个存储系统移动到另一个存储系统，然而，它们以连续流模式运行，而不是周期性地触发，因此，它们能够从不断产生数据的源读取记录，并以较低的时延将其移动到目标。例如，数据管道可以监视文件系统目录中的新文件，并将其数据写入事件日志。另一个应用程序可能将事件流物化到数据库，或者增量地构建和细化搜索索引。

Apache Flink 周期性 ETL 作业和连续数据管道之间的区别如图 1-6 所示。

图 1-6 Apache Flink 周期性 ETL 作业和连续数据管道之间的区别

与定期 ETL 作业相比，连续数据管道的明显优势是减少了将数据移动到其目的地的时延。此外，数据管道更加通用，可以用于更多的用例，因为它们能够持续地消费和发出数据。

Flink 的 SQL 接口(或 Table API)及其对用户定义函数(UDF)的支持可以解决许多常见的数据转换问题。使用更加通用的 DataStream API 可以实现具有更高要求的数据管道。Flink 为各种存储系统(如 Kafka、Kinesis、Elasticsearch 和 JDBC 数据库系统)提供了一组丰富的连接器。它还为文件系统提供了连续源，用于监视以时间间隔方式写文件的目录和接收器。

典型的数据管道应用程序包括:
(1) 电子商务中的实时搜索索引构建。
(2) 电子商务中的持续 ETL。

1.3 Flink 体系架构

在大数据领域，有许多流计算框架，但是通常很难兼顾时延性和吞吐量。Apache Storm 支持低时延，但目前不支持高吞吐量，也不支持在发生故障时正确处理状态。Apache Spark Streaming 的微批处理方法实现了高吞吐量的容错性，但是难以实现真正的低延时和实时处理，并且表达能力方面也不是特别丰富；而 Apache Flink 兼顾了低时延和高吞吐量，是企业部署流计算时的首选。对 Storm、Spark Streaming 和 Flink 这 3 种流计算框架的比较，见表 1-1。

表 1-1　3 种流计算框架比较

流处理框架	高吞吐量	低时延	易于使用和表达	正确的时间/窗口语义	压力下保持正确性
Storm	×	√	×	×	×
Spark Streaming	√	×	×	×	√
Flink	√	√	√	√	√

1.3.1　Flink 系统架构

Flink 可以运行在多种不同的环境中，例如，它可以通过单进程多线程的方式直接运行，从而提供调试的能力。它也可以运行在 YARN 或者 K8S 这种资源管理系统上，也可以在各种云环境中执行。

Flink 的整体系统架构如图 1-7 所示。

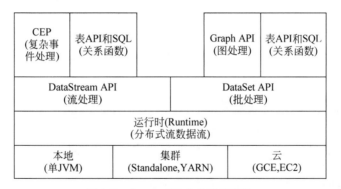

图 1-7　Apache Flink 的系统架构

针对不同的执行环境，Flink 提供了一套统一的分布式作业执行引擎，也就是 Flink 运行时层。Flink 在运行时层之上提供了 DataStream 和 DataSet 两套 API，分别用来编写流作业与批作业，以及一组更高级的 API 库来简化特定作业的编写（例如用于复杂事件处理的 CEP 库）。

1.3.2　Flink 运行时架构

Flink 运行时是 Flink 的核心计算结构，这是一个分布式系统，它接受流数据处理程序，并在一台或多台机器上以容错的方式执行这些数据流程序。这个运行时可以作为 YARN 的应用程序在集群中运行，也可以在 Mesos 集群中运行，或者在一台机器中运行（通常用于调试 Flink 应用程序）。

Flink 运行时层的整个架构采用了标准 Master-Slave 的结构，即总是由一个 Flink Master(JobManager)和一个或多个 Flink Slave(TaskManager)组成。Flink 运行时层的主

要架构如图 1-8 所示。

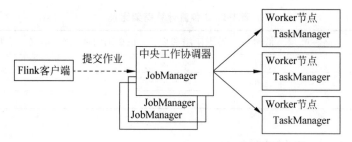

图 1-8　Apache Flink 运行时层的主要架构

图 1-8 展示了一个 Flink 集群的基本结构。在部署 Flink 时,每个组件通常有多个可用选项,见表 1-2。

表 1-2　Flink 程序部署选项

组件	作　　用	实　　现
Flink Client	将批处理或流应用程序编译成数据流图,然后提交给 JobManager	命令行接口 REST Endpoint SQL 客户端 Python REPL Scala REPL
JobManager	JobManager 是 Flink 的中心工作协调组件的名称。它有针对不同资源提供者的实现,这些提供者在高可用性、资源分配行为和支持的作业提交模式上有所不同。用于作业提交的 JobManager 模式有: Application Mode:仅为一个应用程序运行集群。作业的 main 方法(或 client)在 JobManager 上执行。支持在应用程序中多次调用 'execute'/'executeAsync'; Per-Job Mode:仅为一个作业运行集群。作业的 main 方法(或 client)仅在创建集群之前运行; Session Mode:一个 JobManager 实例管理共享同一个 TaskManager 集群的多个作业	Standalone Kubernetes YARN Mesos
TaskManager	TaskManager 是实际执行 Flink Job 工作的服务	

Flink 运行时由两种类型的进程组成:一个 JobManager 和一个或多个 TaskManager。客户端不是运行时和程序执行的一部分,而是用于准备数据流并将数据流发送到 JobManager。在此之后,客户端可以断开连接(分离模式),或者保持连接以接收进度报告(附加模式)。客户端可以作为触发执行的 Java/Scala 程序的一部分运行,也可以在命令行进程(./bin/flink run)中运行。

对 Flink 运行时架构更为详细的描述如图 1-9 所示。

JobManager 和 TaskManager 可以通过多种方式启动,例如,可以直接在机器上作为独

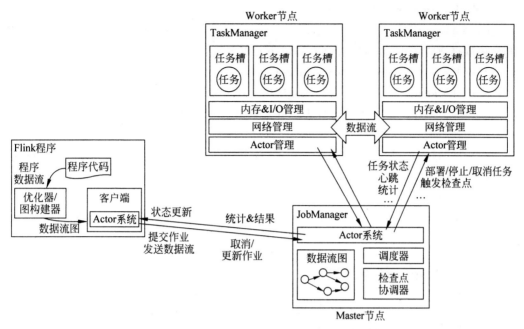

图 1-9　Apache Flink 运行时架构详细描述

立集群启动，也可以在容器中启动，或者由 YARN 或 Mesos 等资源框架管理。TaskManager 连接到 JobManager，宣布自己可用，并被分配工作。

1）JobManager

JobManager 有许多与协调 Flink 应用程序的分布式执行相关的职责：它决定什么时候安排下一个任务（或一组任务），对完成的任务或执行失败做出反应，协调检查点，协调故障恢复等。始终至少要有一个 JobManager。高可用性集群可能有多个 JobManager，其中一个始终是 leader，其他的都是 standby。

JobManager 是 Flink 集群的主进程，它由 3 个不同的组件组成：

（1）ResourceManager。ResourceManager 负责 Flink 集群中的资源分配和释放。它管理任务槽（Task Slots），任务槽是 Flink 集群中的资源调度单元。Flink 为不同的环境和资源提供者（如 YARN、Mesos、Kubernetes 和独立部署）实现了多个 ResourceManager。在 Standalone 独立集群中，ResourceManager 只能分发可用的 TaskManager 的任务槽，不能自己启动新的 TaskManager。

（2）Dispatcher。Dispatcher 提供了一个 REST 接口来提交 Flink 应用程序以供执行，并为每个提交的作业（通过 WebUI 或命令行）启动一个新的 JobMaster。它还运行 Flink WebUI 来提供关于作业执行的信息。

（3）JobMaster。JobMaster 负责管理单个 JobGraph 的执行。多个作业可以在一个 Flink 集群中同时运行，每个作业都有自己的 JobMaster。

2) TaskManager

TaskManagers 是一个 Flink 集群的工作（worker）进程，负责执行数据流的任务，并缓冲和交换数据流。关于 TaskManager，有以下几点要求：

(1) 必须始终至少有一个 TaskManager。

(2) TaskManager 中最小的资源调度单位是任务槽。

(3) TaskManager 的任务槽数量表示并发处理的任务数。注意，多个操作符可以在一个任务槽中执行。

(4) TaskManager 启动后，TaskManager 将其槽位注册到 ResourceManager。当得到 ResourceManager 的指示时，TaskManager 会将一个或多个它的槽位提供给 JobMaster。

(5) JobMaster 可以将任务分配给这些槽以执行它们。

(6) 在执行过程中，一个 TaskManager 与运行同一应用程序任务的其他 TaskManager 交换数据。

1.3.3　Flink 资源管理

Apache Flink 是一个分布式系统，需要计算资源才能执行应用程序。Flink 集成了所有常见的集群资源管理器，如 Hadoop YARN、Apache Mesos 和 Kubernetes，不过也可以设置作为 Standalone 独立集群运行。

注意：在部署 Flink 应用程序时，Flink 根据应用程序配置的并行性自动标识所需的资源，并从资源管理器中请求这些资源。如果发生故障，Flink 则可通过请求新的资源来替换失败的容器。所有提交或控制应用程序的通信都是通过 REST 调用进行的，因此简化了 Flink 在许多环境中的集成。

在 Flink 中，资源是由 TaskManager 上的 Slot 来表示的，每个 Slot 可以用来执行不同的任务（Task），而 Job 中实际的 Task 包含了待执行的用户逻辑代码。作业调度的主要目的就是给 Task 找到匹配的 Slot。实际上，Flink 作业调度可以看作对资源和任务进行匹配的过程。

注意：从逻辑上来讲，每个 Slot 都应该有一个向量来描述它所能提供的各种资源的量，每个 Task 也需要相应地说明它所需要的各种资源的量，但是实际上在 Flink 1.9 之前，Flink 是不支持细粒度的资源描述的，而是统一地认为每个 Slot 提供的资源和 Task 需要的资源都是相同的。从 Flink 1.9 开始，Flink 开始增加对细粒度资源匹配的支持和实现，但这部分功能目前仍在完善中。

1. Task 执行

这是很难解释和理解的部分。在深入讲解之前，读者应该记住什么是 Flink 中的操作符和任务。一般情况下,需要记住:Operator=(Task1+Task2+…+TaskN)。

(1) Operator in Flink =>将一个数据流转换为另一个数据流(可以是相同类型的数据流,也可以是不同类型的数据流)。Operator(操作符)是逻辑数据流图(也称为 JobGraph)的节点。

(2) Task in Flink =>是由 Flink 的运行时执行的基本工作单元。任务是物理数据流图(也称为 Execution Graph)的节点。

(3) 任务是操作符或操作符链的一个并行实例(两个或多个连续操作符之间没有任何重分区)。

(4) 在 Apache Flink 中,每个 Task 只有一个线程,Task 之间没有共享知识。例如,Task1 不知道另一个 Task 正在发生什么。这意味着 API 可以从一个 Task 中访问每种状态段,但没有办法访问其他线程中的状态。

(5) 还有子任务(Sub-Task)的概念。子任务是在数据流的一部分上工作的任务。子任务指的是同一个操作符或操作符链有多个并行任务(这实际上意味着存在数据并行性)。

了解了术语操作符(Operator)、任务(Task)和子任务(Sub-Task)之后,接下来了解任务执行部分,包括:

(1) TaskManager(worker/slave 进程)可以同时执行多个任务,并发数通常与 TaskManager 所在机器的 CPU 数有关。例如,如果机器的 CPU 数是 16,则一个 TaskManager 可以同时运行 16 个任务。这是 Apache Flink 提供的最佳解决方案,可以同时运行 16 个任务,但这可能会导致一些问题,如 Flink 集群可能会经常重新启动自己等。

(2) 一个 TaskManager 在同一个 JVM 进程中以多线程方式执行它的任务。

(3) 任务可以是相同操作符(记住数据并行性)的子任务,也可以是不同操作符(记住任务并行性)的子任务,甚至可以是来自不同应用程序的子任务。

(4) TaskManager 提供一定数量的槽位(与机器的 CPU 数量相关)来控制它能够并发执行的任务。换句话说,如果机器有 16 个 CPU,则 TaskManager 可以有 16 个槽,每个槽用于处理一个任务。

注意:一个 Slot 槽可以包含一个特定的任务或多个关联的任务。

2. 操作符链

对于分布式执行,Flink 将操作子任务连成 Tasks。每个 Task 由一个线程执行。将操作符连接到 Tasks 中是一种优化行为,它减少了线程到线程切换和缓冲的开销,并在降低时延的同时增加了总体吞吐量。

例如,一个数据流使用 5 个子任务执行,因此使用 5 个并行线程,如图 1-10 所示。

操作符链允许非 shuffle 操作在同一个线程中共存,完全避免了序列化和反序列化。

3. 任务槽

每个 TaskManager(Worker 进程)都是一个 JVM 进程,可以在单独的线程中执行一个

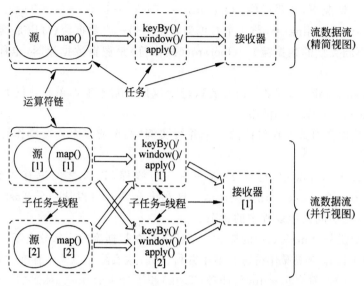

图 1-10 Apache Flink 使用操作符链来优化执行

或多个子任务。为了控制一个 worker 接受多少任务,一个 worker 具有至少一个"任务插槽"。

每个 Task Slot 表示 TaskManager 资源的一个固定子集。例如,一个 TaskManager 有 3 个插槽,它会将其 1/3 的托管内存分配给每个插槽。对资源进行插槽化意味着子任务不会与来自其他作业的子任务争夺托管内存,而是拥有一定数量的预留托管内存。注意,这里没有发生 CPU 隔离;当前插槽只分隔任务的托管内存,如图 1-11 所示。

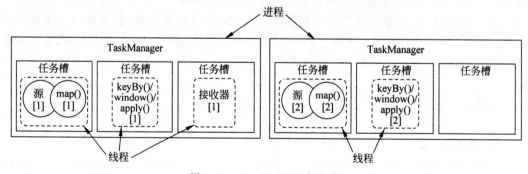

图 1-11 Apache Flink 任务槽

通过调整任务槽的数量,用户可以定义子任务如何彼此隔离。每个 TaskManager 有一个插槽(Slot)意味着每个任务组运行在各自的 JVM 中(例如,可以在单独的容器中启动 JVM)。拥有多个插槽意味着更多的子任务共享同一个 JVM。相同 JVM 中的任务共享 TCP 连接(通过多路复用)和心跳消息。它们还可以共享数据集和数据结构,从而减少每个任务的开销。

默认情况下，Flink 允许子任务共享槽位，即使它们是不同任务的子任务，只要它们来自相同的作业。结果是一个槽位可以容纳作业的整个管道，如图 1-12 所示。

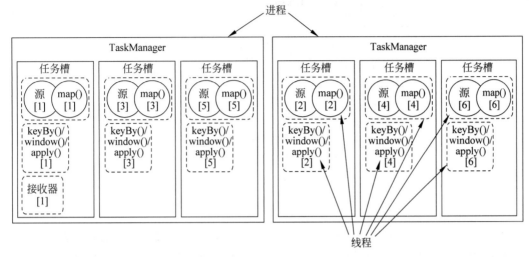

图 1-12　Apache Flink 允许子任务共享槽位

这种槽位共享有以下两个主要好处：

（1）Flink 集群需要的任务插槽与作业中使用的最高并行度一样多。不需要计算一个程序总共包含多少任务（具有不同的并行度）。

（2）更容易得到更好的资源利用。如果没有槽位共享，则非密集型 source/map() 子任务将阻塞与资源密集型窗口子任务一样多的资源。使用插槽共享，将示例中的基本并行度从 2 提高到 6，可以充分利用插槽资源，同时确保繁重的子任务在 TaskManager 中得到公平分配。

API 还包括一个资源组（Resource Group）机制，可用于防止不需要的槽位共享。

根据经验，恰当的默认任务槽位数应该是 CPU 核的数量。使用超线程，每个槽位将接受 2 个或更多的硬件线程上下文。

4. 资源申请

在 ResourceManager 中，有一个子组件叫作 SlotManager，SlotManager 用于维护当前集群中所有 TaskManager 上的 Slot 的信息与状态，例如该 Slot 在哪个 TaskManager 中，该 Slot 当前是否空闲等，如图 1-13 所示。

当 JobMaster 为特定 Task 申请资源时，根据当前作业部署模式（关于作业部署模式，可参阅 1.5 节）的区别，TaskManager 可能已经启动或者尚未启动。如果 TaskManager 尚未启动，则 ResourceManager 会去申请资源来启动新的 TaskManager。当 TaskManager 启动之后，它会通过服务找到当前活跃的 ResourceManager 并进行注册。在注册信息中，会包含该 TaskManager 中所有 Slot 的信息。ResourceManager 收到注册信息后，其中的 SlotManager 就会记录下相应的 Slot 信息。当 JobMaster 为某个 Task 来申请资源时，

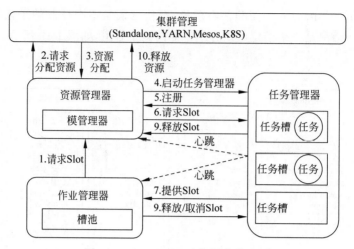

图 1-13　Apache Flink 资源申请流程

SlotManager 就会从当前空闲的 Slot 中按一定规则选择一个空闲的 Slot 进行分配。当分配完成后，ResourceManager 会首先向 TaskManager 发送 RPC 要求将选定的 Slot 分配给特定的 JobManager。TaskManager 如果还没有执行过该 JobMaster 的 Task，则它首先需要与相应的 JobMaster 建立连接，然后发送提供 Slot 的 RPC 请求。在 JobMaster 中，所有 Task 的请求会缓存到 SlotPool 中。当有 Slot 被提供之后，SlotPool 会从缓存的请求中选择相应的请求并结束相应的请求过程。

当 Task 结束之后，无论是正常结束还是异常结束，都会通知 JobMaster 相应的结束状态，然后在 TaskManager 端将 Slot 标记为已占用但未执行任务的状态。JobMaster 会首先将相应的 Slot 缓存到 SlotPool 中，但不会立即释放。这种方式避免了如果将 Slot 直接还给 ResourceManager，在任务异常结束之后需要重启时，则需要立刻重新申请 Slot 的问题。通过延时释放，容错的 Task 可以尽快调度回原来的 TaskManager，从而加快故障切换的速度。当 SlotPool 中缓存的 Slot 超过指定的时间仍未使用时，SlotPool 就会发起释放该 Slot 的过程。与申请 Slot 的过程对应，SlotPool 会首先通知 TaskManager 来释放该 Slot，然后 TaskManager 通知 ResourceManager 该 Slot 已经被释放，从而最终完成释放的逻辑。

5．心跳报告

除了正常的通信逻辑外，在 ResourceManager 和 TaskManager 之间还存在定时的心跳消息同步 Slot 的状态。在分布式系统中，消息的丢失、错乱不可避免，这些问题会在分布式系统的组件中引入不一致状态，如果没有定时消息，则组件无法从这些不一致状态中恢复。此外，当组件之间长时间未收到对方的心跳时，就会认为对应的组件已经失效，并进入容错的流程。

6．共享槽位

在 Slot 管理基础上，Flink 可以将 Task 调度到相应的 Slot 当中。如上文所述，Flink 尚未完全引入细粒度的资源匹配，默认情况下，每个 Slot 可以分配给一个 Task，但是，这种方式在某些情况下会导致资源利用率不高，如图 1-14 所示，假如 A、B、C 依次执行计算逻辑，那么给 A、B、C 分配单独的 Slot 就会导致资源利用率不高。为了解决这一问题，Flink 提供了共享槽位（Share Slot）的机制，如图 1-14 所示。基于共享槽位，每个 Slot 中可以部署来自不同 JobVertex（作业向量）的多个任务，但是不能部署来自同一个 JobVertex 的 Task，如图中所示，每个 Slot 中最多可以部署同一个 A、B 或 C 的 Task，但是可以同时部署 A、B 和 C 的各一个 Task。当单个 Task 占用资源较少时，共享槽位可以提高资源利用率。此外，共享槽位也提供了一种简单的保持负载均衡的方式。

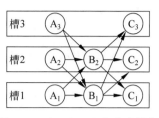

图 1-14　Apache Flink 共享槽位

1.3.4　Flink 作业调度

参与 Flink 程序执行的有多个进程，包括 JobManager、TaskManager 及 JobClient，其中 JobManager 充当 Master 角色，TaskManager 充当 Worker 角色。JobClient 不是 Flink 任务执行过程的内部组件，而是执行过程的起始点。JobClient 负责接收用户提交的应用程序，创建对应的数据流，然后将数据流提交到 JobManager 上执行。一旦执行完毕，JobClient 将执行结果发回给用户。

Flink 程序的执行过程如图 1-15 所示。

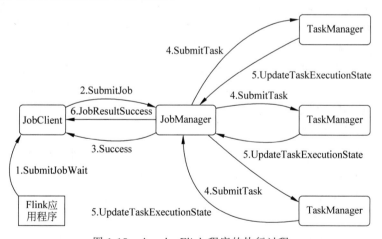

图 1-15　Apache Flink 程序的执行过程

Flink 应用程序首先被提交到 JobClient 上，随后 JobClient 将它提交到 JobManager 上，JobManager 负责安排资源的分配和作业的执行。首先是资源的分配，然后将作业划分为若干任务后提交到对应的 TaskManager 上。TaskManager 在接收到任务后，初始化一个线程并开始执行程序。执行过程中 TaskManager 持续地将状态的变化情况报告给

JobManager，这些状态包括开始执行(starting the execution)、正在执行(in progress)及完成(finished)。一旦作业的执行彻底完成，JobManager 就将结果发回 JobClient 端。

Flink 的分布式作业执行过程包含两个重要的角色：Master 和 Worker，如图 1-16 所示。

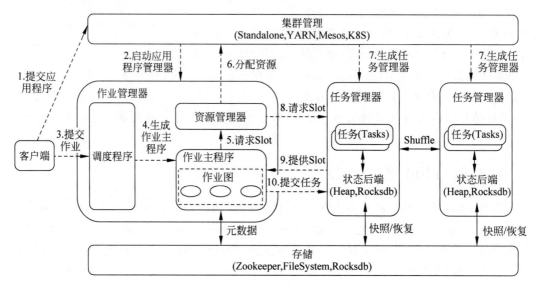

图 1-16 Apache Flink 分布式作业执行过程中的两个重要角色：Master 和 Worker

其中左侧的 JobManager 部分即是 Master，它负责管理整个集群中的资源并处理作业提交、作业监督；右侧的两个 TaskManager 则是 Worker，这是工作进程，负责提供具体的资源并实际执行作业。

当用户提交作业时，提交脚本会先启动一个 JobClient 进程负责作业的编译与提交。JobClient 首先将用户编写的代码编译为一个 JobGraph，在这个过程中，它还会进行一些检查或优化等工作，例如判断哪些运算符(算子)可以连接到同一个任务中，然后，JobClient 将产生的 JobGraph 提交到集群中执行。

当作业提交到 Dispatcher 后，Dispatcher 会首先启动一个 JobMaster 组件，然后 JobMaster 会向 ResourceManager 申请资源来启动作业中具体的任务。这时根据作业部署模式(关于作业部署模式，可参阅 1.3.6 节)的区别，TaskManager 可能已经启动或者尚未启动。

(1) 如果是前者，此时 ResourceManager 中已有记录了 TaskManager 注册的资源，可以直接选取空闲资源进行分配。

(2) 否则 ResourceManager 首先需要向外部资源管理系统申请资源来启动 TaskManager，然后等待 TaskManager 注册相应资源后再继续选择空闲资源进程分配。

目前 Flink 中 TaskManager 的资源是通过 Slot 来描述的，一个 Slot 一般可以执行一个具体的任务，但在一些情况下也可以执行多个相关联的任务。ResourceManager 选择到空闲的 Slot 之后，就会通知相应的 TaskManager"将该 Slot 分配给 JobMaster XX"，然后 TaskManager 进行相应的记录后会向 JobManager 进行注册。JobManager 收到 TaskManager 注册的 Slot 后，就可以实际提交任务了。

TaskManager 收到 JobManager 提交的任务之后会启动一个新的线程来执行该任务。该任务启动后就会开始进行预先指定的计算，并通过数据 shuffle 模块互相交换数据。

基于 1.3.3 节中 Slot 管理和分配的逻辑，JobMaster 负责维护作业中任务执行的状态。如上文所述，客户端会向 JobMaster 提交一个 JobGraph，它代表了作业的逻辑结构。JobMaster 会根据 JobGraph 按并发展开，从而得到 JobMaster 中关键的 ExecutionGraph。ExecutionGraph 的结构如图 1-17 所示。

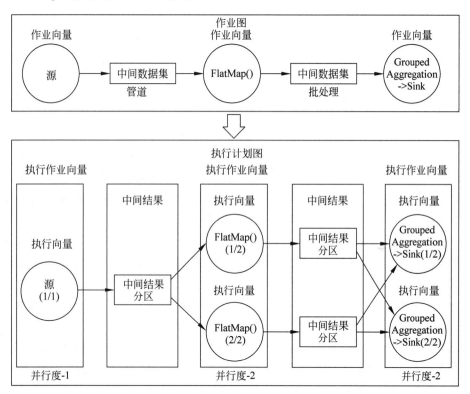

图 1-17　ExecutionGraph 是 JobMaster 中的核心数据结构

与 JobGraph 相比，ExecutionGraph 中对于每个 Task 和中间结果等均创建了对应的对象，从而可以维护这些实体的信息与状态。

在一个 Flink Job 中包含多个 Task，因此另一个关键的问题是在 Flink 中按什么顺序来调度这些 Task。目前 Flink 提供了两种基本的调度策略，即时延调度和即时调度，如图 1-18 所示。

即时调度会在作业启动时申请资源将所有的任务调度起来。这种调度算法主要用来调度可能没有终止的流作业。与之对应，时延调度则是从源开始，按拓扑顺序进行调度。简单

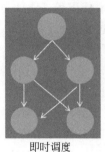 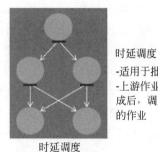

图 1-18　Flink 中两种基本的调度策略

来讲,时延调度会先调度没有上游任务的源任务,当这些任务执行完成时,它会将输出数据缓存到内存或者写入磁盘中,然后,对于后续的任务,当它的前驱任务全部执行完成后,Flink 就会将这些任务调度起来。这些任务会从读取上游缓存的输出数据进行自己的计算。这一过程继续进行直到所有的任务完成计算。

1.3.5　Flink 故障恢复

在 Flink 作业的执行过程中,除正常执行的流程外,还有可能由于环境等原因导致各种类型的错误。整体来讲,错误可分为两大类:Task 执行出现错误或 Flink 集群的 Master 出现错误。由于错误不可避免,为了提高可用性,Flink 需要提供自动错误恢复机制进行重试。

对于第一类 Task 执行错误,Flink 提供了多种不同的错误恢复策略。

(1) 第一种错误恢复策略是 restart-all,即直接重启所有的 Task。对于 Flink 的流任务,由于 Flink 提供了 checkpoint 机制,因此当任务重启后可以直接从上次的 checkpoint 开始继续执行,因此这种方式更适合于流作业。

(2) 第二种错误恢复策略是 restart-individual,它只适用于 Task 之间没有数据传输的情况。在这种情况下,可以直接重启出错的 Task。

由于 Flink 的批作业没有 checkpoint 机制,因此对于需要数据传输的作业,直接重启所有的 Task 会导致作业从头计算,从而导致一定的性能问题。为了增强对批作业的处理,Flink 在 1.9 版本中引入了一种新的基于分区的错误恢复策略。

在一个 Flink 的批作业中,Task 之间存在两种数据传输方式,一种是管道(pipeline)类型的方式,这种方式上下游 Task 之间直接通过网络传输数据,因此需要上下游同时运行;另外一种是 blocking 类型的方式,这种方式下,上游的 Task 会首先将数据进行缓存,因此上下游的 Task 可以单独执行。基于这两种类型的传输,Flink 将 ExecutionGraph 中使用管道方式传输数据的 Task 的子图叫作分区(region),从而将整个 ExecutionGraph 划分为多个子图。分区内的 Task 必须同时重启,而不同分区的 Task 由于在分区边界存在 blocking 的边,因此,可以单独重启下游分区中的 Task。

基于这一思路,如果某个分区中的某个 Task 执行出现错误,可以分两种情况进行考虑。

(1) 如果是由于 Task 本身的问题发生错误，则可以只重启该 Task 所属的分区中的所有 Task，这些 Task 重启之后，可以直接拉取上游分区缓存的输出结果继续进行计算，这个过程如图 1-19 所示。

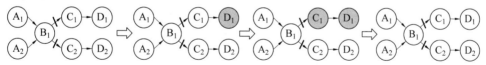

图 1-19　只重启出错 Task 所属的分区中的所有 Task

(2) 另一方面，如果错误是由于读取上游结果出现问题，如网络连接中断、缓存上游输出数据的 TaskManager 异常退出等，则还需要重启上游分区来重新产生相应的数据。在这种情况下，如果上游分区输出的数据分发方式不是确定性的（如 keyBy、broadcast 是确定性的分发方式，而 rebalance、random 则不是，因为每次执行会产生不同的分发结果），为了保证结果的正确性，还需要同时重启上游分区所有的下游分区，这个过程如图 1-20 所示。

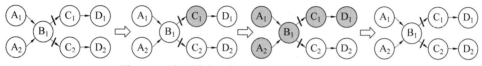

图 1-20　需要同时重启上游分区的故障恢复过程

另一类异常是 Flink 集群的 Master 进程发生异常。目前 Flink 支持启动多个 Master 作为备份，这些 Master 可以通过 ZooKeeper 进行选主 Master，从而保证某一时刻只有一个 Master 在运行。当前活跃的 Master 发生异常时，某个备份的 Master 可以接管协调的工作。为了保证 Master 可以准确维护作业的状态，Flink 目前采用了一种最简单的实现方式，即直接重启整个作业。实际上，由于作业本身可能仍在正常运行，因此这种方式存在一定的改进空间。

1.3.6　Flink 程序执行模式

Flink 可以通过以下 3 种方式来执行应用程序：①以 Application 模式；②以 Per-Job 模式；③以 Session 模式。这 3 种方式如图 1-21 所示。

以上 3 种模式的区别在于：

(1) 集群生命周期和资源隔离保证。

(2) 应用程序的 main() 方法是在客户端上执行还是在集群上执行。

1. Per-Job 模式

为了提供更好的资源隔离保证，Per-Job 模式使用可用的资源提供者框架（例如 YARN、Kubernetes）为每个提交的作业启动一个集群。该集群仅对该作业可用。当作业完成时，将关闭集群，并清除所有滞留的资源（文件等）。

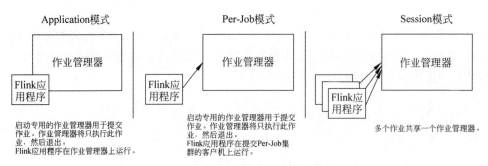

图 1-21 Flink 程序的 3 种执行模式

(1) 集群生命周期：在 Flink Job Cluster 中，可用的集群管理器（如 YARN）用于为每个提交的作业启动集群，该集群仅对该作业可用。在这里，客户端首先从集群管理器请求资源以启动 JobManager，并将作业提交给运行在此进程中的 Dispatcher，然后根据作业的资源需求惰性地分配 TaskManager。一旦作业完成，Flink 作业集群就会被拆除。集群的生命周期与作业的生命周期绑定在一起。

(2) 资源隔离：更好的隔离保证，因为资源不会跨作业共享。JobManager 中的致命错误只会影响 Flink Job Cluster 中运行的一个作业。此外，它将负载分散到多个 JobManager，因为每个作业有一个 JobManager。

由于 ResourceManager 需要申请并等待外部资源管理组件启动 TaskManager 进程并分配资源，所以 Flink Job Clusters 更适合长时间运行、对稳定性要求高、对启动时间不敏感的大型任务。Per-Job 资源分配模型是许多生产环境的首选模式（注：Kubernetes 和 Standalone 集群不支持此模式）。

采用这种模式的有 Flink On YARN，JobManager 不会预先启动，此时 Client 将首先向资源管理系统（如 YARN、K8S）申请资源来启动 JobManager，然后向 JobManager 中的 Dispatcher 提交作业。

例如，可以在 YARN 上使用 Per-Job 模式来部署并运行 Flink 自带的示例程序，命令如下：

```
$ /bin/flink run -t yarn-per-job
--detached ./examples/streaming/TopSpeedWindowing.jar
```

这将在 YARN 上启动一个 Flink 集群，然后在本地运行提供的应用程序 JAR，最后将 JobGraph 提交给 YARN 上的 JobManager。如果传递--detached 参数，则客户端将在提交被接受后停止。一旦作业停止，YARN 集群就会停止 Flink 集群。

一旦部署了 Per-Job Cluster，就可以与它进行交互了，以执行取消或获取保存点等操作。常用命令如下：

```
#列出集群上正在运行的作业
$ ./bin/flink list -t yarn-per-job
-Dyarn.application.id=application_XXXX_YY
```

```
#撤销正在运行的作业
$ ./bin/flink cancel -t yarn-per-job
-Dyarn.application.id=application_XXXX_YY <jobId>
```

需要注意,取消 Per-Job Cluster 上的作业将停止集群。

2. Session 模式

Session 模式假设有一个已经运行的集群(称为 Flink Session Cluster),并使用该集群的资源来执行任何提交的应用程序。在 Session Mode 中,集群的生命周期独立于集群上运行的任何作业的生命周期,并且资源在所有作业之间共享。

(1) 集群生命周期:在 Session 模式中,客户端连接到一个预先存在的、长时间运行的集群,该集群可以接受多个作业提交,即使在所有作业完成后,集群(和 JobManager)仍将继续运行,直到手动停止会话,因此,Flink Session Cluster 的生命周期不与任何 Flink Job 的生命周期绑定。

(2) 资源隔离:TaskManager 槽在作业提交时由 ResourceManager 分配,作业完成后释放。因为所有作业都共享同一个集群,所以在提交作业阶段存在一些对集群资源的竞争,例如网络带宽。这种共享设置的一个限制是,如果一个 TaskManager 崩溃了,则所有在这个 TaskManager 上运行的任务都会失败;以类似的方式,如果 JobManager 上发生一些致命错误,则它将影响集群中运行的所有作业。除了对导致失败的作业产生负面影响外,这还意味着一个潜在的大规模恢复过程,所有重新启动的作业都同时访问文件系统,并使其他服务无法使用它。此外,让一个集群运行多个作业意味着 JobManager 需要承担更多的负载,因为需要 JobManager 负责对集群中的所有作业进行记账。

使用预先存在的集群可以节省大量申请资源和启动 TaskManager 的时间。当作业的执行时间非常短,而高启动时间会对端到端用户体验产生负面影响时,这一点非常重要,就像短查询的交互分析那样,作业可以使用现有资源快速执行计算。

采用这种模式的有 Standalone,JobManager 会预先启动,此时 Client 直接与 Dispatcher 建立连接并提交作业即可。

例如,使用 Session Mode 部署 Flink 自带的 WordCount 程序,命令如下:

```
#假设是在 Flink 发行版的根目录下

#(1)启动 Flink Session 集群
$ ./bin/start-cluster.sh

#(2)现在可以访问 http://localhost:8081 的 Flink Web 界面

#(3)提交作业执行
$ ./bin/flink run ./examples/batch/WordCount.jar

#(4)重新停止集群
$ ./bin/stop-cluster.sh
```

3. Application 模式

在 Per-Job 模式和 Session 模式中，应用程序的 main() 方法都在客户端执行。这个过程包括在本地下载应用程序的依赖项，执行 main() 来提取 Flink 的运行时可以理解的应用程序的表示 (JobGraph)，并将依赖项和 JobGraph 发送到集群。这使客户端成为一个沉重的资源消耗者，因为它可能需要大量的网络带宽来下载依赖项并将二进制文件发送到集群，并需要 CPU 周期性地执行 main()。当客户端在用户之间共享时，这个问题会更加明显。

在此观察的基础上，Application 模式为每个提交的应用程序创建一个集群，并且应用程序的 main() 方法在 JobManager 上执行。在 JobManager 上执行 main() 可以节省所需的 CPU 周期，还可以节省本地下载依赖项所需的带宽。此外，为了下载集群中应用程序的依赖关系，它允许更均匀地分散网络负载，因为每个应用程序有一个 JobManager。

(1) 集群生命周期：一个 Flink Application Cluster 是一个专用的 Flink 集群，它只执行来自一个 Flink 应用程序的作业，其中 main() 方法运行在集群而不是客户机上。作业提交是 one-step 过程：不首先需要启动一个 Flink 集群，然后将作业提交到现有的集群会话；相反，可以将应用程序逻辑和依赖打包到一个可执行的作业 JAR 中，并且集群入口点 (ApplicationClusterEntryPoint) 负责调用 main() 方法来提取 JobGraph。这允许用户像在 Kubernetes 上部署其他应用程序一样部署 Flink 应用程序，因此，Flink Application Cluster 集群的生命周期与 Flink 应用程序的生命周期绑定在一起。

(2) 资源隔离：在 Flink Application Cluster 中，ResourceManager 和 Dispatcher 被限定为单个 Flink 应用程序，这比 Flink Session Cluster 提供了更好的关注点分离。

与 Per-Job 模式相比，Application 模式允许提交由多个作业组成的应用程序。作业执行的顺序不受部署模式的影响，而是受用于启动作业的调用的影响。使用 execute() 是阻塞的，它会建立一个顺序，并导致"下一个"作业的执行被推迟到"这个"作业完成。使用 executeAsync() 方法，它是非阻塞的，将导致"下一个"任务在"这个"任务完成之前开始。

Application 模式相当于为每个应用程序创建一个会话集群，并在集群上执行应用程序的 main() 方法。

在这种部署模式下，应用程序 JAR 文件需要在类路径中可用。最简单的方法就是把 JAR 放到 lib/文件夹中。例如，使用 Application 模式部署 Flink 自带的 TopSpeedWindowing 程序，命令如下：

```
#将 JAR 包复制到 lib/文件夹下
$ cp ./examples/streaming/TopSpeedWindowing.jar lib/

#先启动一个 TaskManager(如果应用程序需要更多资源,则可以启动多个 TaskManager)
$ ./bin/taskmanager.sh start

#启动 JobManager
$ ./bin/standalone-job.sh start --job-classname org.apache.flink.streaming.examples.windowing.TopSpeedWindowing
```

该脚本还支持停止服务。如果想停止多个实例，则可多次调用它们，或者使用 stop-all，命令如下：

```
$ ./bin/taskmanager.sh stop
$ ./bin/standalone-job.sh stop
```

1.4 Flink 集群安装

Flink 运行在 Linux、Mac OS X 和 Windows 上。本教程中将 Flink 集群搭建在 Linux 系统上。由于 Flink 是用 Java 和 Scala 实现的，所以所有组件都运行在 JVM 上。

使用 Flink 需要满足以下先决条件：
（1）需要安装 Java 1.8/Java 11 来运行 Flink 作业/应用程序。
（2）Scala API（可选）依赖于 Scala 2.11。
（3）如果配置为高可用（没有单点故障），则需要 Apache ZooKeeper。
（4）如果配置为高可用（可以从故障中恢复）的流处理，Flink 需要某种形式的检查点分布式存储（HDFS/S3/NFS/SAN/GFS/Kosmos/Ceph/…）。

1.4.1 Flink 独立集群安装和测试

Flink 集群可以运行在单节点上，这称为 Standalone Cluster 模式（在一台机器上，但在不同的进程中）。独立模式是部署 Flink 最简单的方式。

1. Standalone 集群安装

Standalone 集群安装步骤如下。
（1）要运行 Flink，要求必须安装好 Java 1.8。检查 Java 是否已经正确安装，命令如下：

```
$ java -version
```

如果已经正确地安装了 Java 1.8，则输出内容如图 1-22 所示。

图 1-22　在安装 Flink 之前，验证是否已经安装了 JDK

（2）下载 Flink 安装包。下载网址为 https://archive.apache.org/dist/flink/flink-1.13.2/。可以选择任何喜欢的 Hadoop/Scala 组合。本书使用的是 1.13.2 版本，基于 Scala 2.12。因为 Flink 版本更新迭代比较快，并且每次版本升级都有许多 API 变动，因此建议读者学习时也安装与本书相同的版本，如图 1-23 所示。
（3）解压缩安装包。将下载的安装包放在～/software/目录下，然后将其解压缩到指定

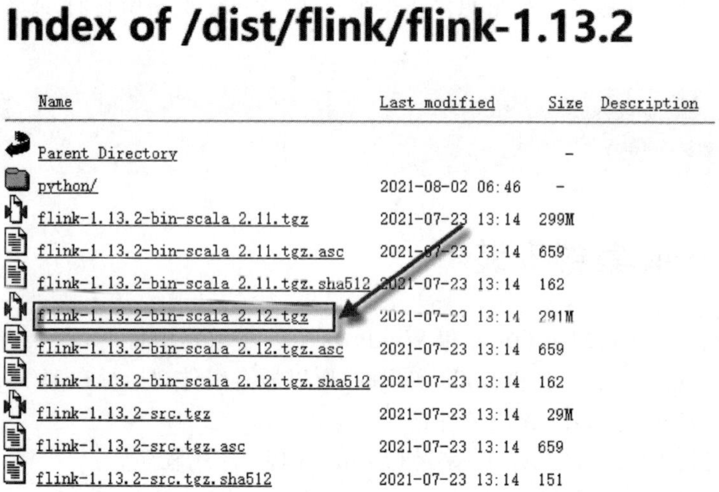

```
Index of /dist/flink/flink-1.13.2

Name                                    Last modified      Size  Description
Parent Directory                                             -
python/                                 2021-08-02 06:46     -
flink-1.13.2-bin-scala_2.11.tgz         2021-07-23 13:14   299M
flink-1.13.2-bin-scala_2.11.tgz.asc     2021-07-23 13:14   659
flink-1.13.2-bin-scala_2.11.tgz.sha512  2021-07-23 13:14   162
flink-1.13.2-bin-scala_2.12.tgz         2021-07-23 13:14   291M
flink-1.13.2-bin-scala_2.12.tgz.asc     2021-07-23 13:14   659
flink-1.13.2-bin-scala_2.12.tgz.sha512  2021-07-23 13:14   162
flink-1.13.2-src.tgz                    2021-07-23 13:14    29M
flink-1.13.2-src.tgz.asc                2021-07-23 13:14   659
flink-1.13.2-src.tgz.sha512             2021-07-23 13:14   151
```

图 1-23　建议与本书保持一致，安装 Flink 1.13.2 版本

的位置(例如，~/bigdata/目录下)。在终端执行的命令如下：

```
$ cd ~/bigdata
$ tar -zxvf ~/software/flink-1.13.2-bin-scala_2.12.tgz
```

(4) 启动一个本地 Flink 集群。

对于单节点设置，Flink 是开箱即用的，即不需要更改默认配置，直接启动即可，命令如下：

```
$ cd flink-1.13.2
$ ./bin/start-cluster.sh
```

使用 jps 命令查看进程，可以看到启动了以下两个进程：

```
2569 StandaloneSessionClusterEntrypoint
2844 TaskManagerRunner
```

打开浏览器，输入地址 http://localhost:8081，可查看调度程序的 Web 前端。Web 前端应该报告有单个可用的 TaskManager 实例，如图 1-24 所示。

还可以通过检查 log 目录中的日志文件来验证系统是否正在运行，命令如下：

```
$ tail log/flink-*-standalonesession-*.log
```

(5) 要关闭 Flink 集群，使用的命令如下：

```
$ ./bin/stop-cluster.sh
```

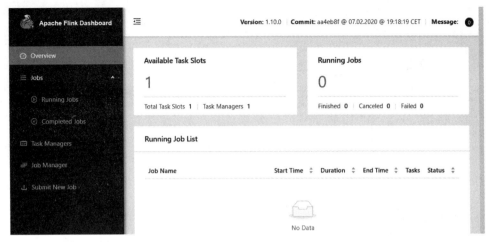

图 1-24 Flink 调度程序的 Web 前端

2. 运行 Flink 自带的实时单词计数流程序

Flink 安装包自带了一个以 Socket 作为数据源的实时统计单词计数的流程序,位于 Flink 下的 example/streaming/SocketWindowWordCount.jar 包中。可以通过运行这个流程序来测试 Flink 集群的使用方法。建议按以下步骤执行。

(1) 首先,启动 netcat 服务器,运行在 9000 端口,使用的命令如下:

```
$ nc -l 9000
```

(2) 打开另一个终端,启动 Flink 集群,执行的命令如下:

```
$ cd ~/bigdata/flink-1.13.2
$ ./bin/start-cluster.sh
```

(3) 启动 Flink 示例程序,监听 netcat 服务器的输入,命令如下:

```
$ ./bin/flink run examples/streaming/SocketWindowWordCount.jar --hostname localhost --port 9000
```

这个实时单词计数流程序将从 Socket 套接字中读取输入的文本内容,并每 5s 打印前 5s 内每个不同单词出现的次数,即处理时间的滚动窗口。

(4) 在 netcat 控制台,键入一些单词,Flink 将会处理这些单词。例如,输入以下内容:

```
good good study
day day up
```

(5) 启动第 3 个终端窗口,并在该窗口中执行以下命令,查看日志中的输出:

```
$ cd ~/bigdata/flink-1.13.2
$ tail -f log/flink-*-taskexecutor-*.out
```

可以看到输出结果如下：

```
good : 2
study : 1
day : 2
up : 1
```

（6）还可以检查 Flink Web UI 来查看 Job 是怎样执行的，如图 1-25 所示。

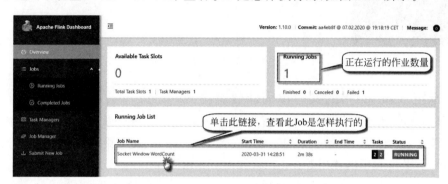

图 1-25　在 Flink Web UI 中查看作业的执行

单击图中的 Running Job List 下正在运行的作业列表，查看某个正在运行的作业执行情况，如图 1-26 所示。

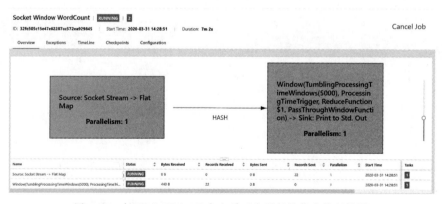

图 1-26　在 Flink Web UI 中查看正在运行的作业执行情况

（7）最后停止 Flink 集群，使用的命令如下：

```
$ ./bin/stop-cluster.sh
```

3. 运行 Flink 自带的单词计数批处理程序

Flink 安装包自带了一个以文本文件作为数据源的单词计数批处理程序，位于 Flink 下的 example/batch/目录下的 WordCount.jar 包中。下面演示如何在 Flink 集群上执行该程序，读取 HDFS 上的输入数据文件进行处理，并将计算结果输出到 HDFS 上。

建议按以下步骤执行。

（1）集成 Hadoop。因为要读取 HDFS 上的源数据文件，所以需要在 Flink 中集成 Hadoop 包。从 Flink 1.8 开始，Hadoop 不再包含在 Flink 的安装包中，所以需要单独下载并复制到 Flink 的 lib 目录下。

如果使用的是 Hadoop 2，则可从 Flink 官网下载 flink-shaded-hadoop2-uber-2.7.5-1.10.0.jar，如图 1-27 所示。

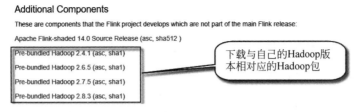

图 1-27　从 Flink 官网下载兼容 Hadoop 2 的集成包

如果使用的是 Hadoop 3，则需要从 Maven 官方资源库下载。下载网址为 https://mvnrepository.com/artifact/org.apache.flink/flink-shaded-hadoop-3-uber，如图 1-28 所示。

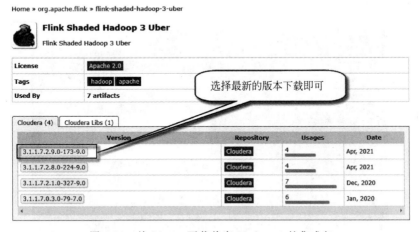

图 1-28　从 Maven 下载兼容 Hadoop 3 的集成包

本书中使用的是 Hadoop-3.2.2 这个版本，所以需要从 Maven 官方资库下载 flink-shaded-hadoop-3-uber-3.1.1.7.2.9.0-173-9.0.jar 包。将下载的 JAR 包复制到 Flink 的 lib 目录下。

（2）准备环境。先启动 Flink 集群，再启动 HDFS 集群。在终端窗口中，执行的命令如下：

```
#首先启动Flink集群
$ ./bin/start-cluster.sh

#再启动HDFS集群
$ start-dfs.sh
```

(3) 准备数据文件。编辑数据文件 wc.txt，内容如下：

```
good good study
day day up
```

然后将该文件上传到 HDFS 的/data/flink/目录下，使用的命令如下：

```
$ hdfs dfs -put wc.txt /data/flink/
```

(4) 执行 Flink 的批处理程序，命令如下：

```
#执行Flink单词计数程序
$ ./bin/flink run ./examples/batch/WordCount.jar \
-- input hdfs://xueai8:8020/data/flink/wc.txt \
-- output hdfs://xueai8:8020/data/flink/wc-result
```

执行过程如图 1-29 所示。

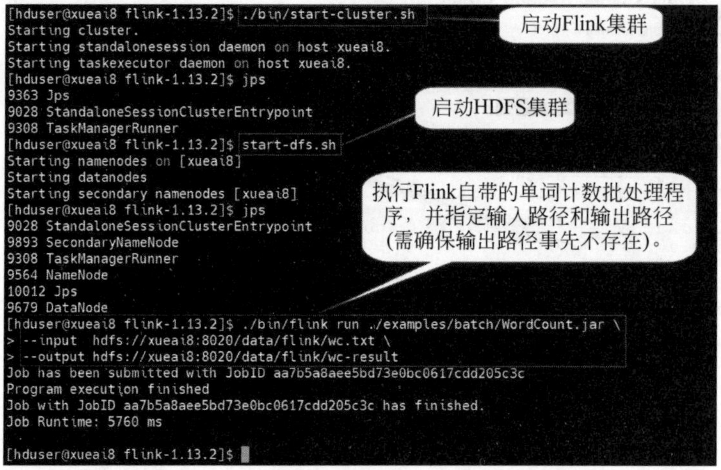

图 1-29　Flink 单词计数批处理程序执行过程

上面的命令是在运行 WordCount 时读写 HDFS 中的文件的，其中--input 参数用于指定要处理的输入文件，--output 参数用于指定将计算结果输出到哪个文件（注：如果不加

hdfs://前缀,则默认使用本地文件系统)。

在执行此程序时,很可能会出现错误信息,信息如下:

```
Java.lang.NoSuchMethodError: org.apache.commons.cli.Option.builder(LJava/lang/String;)
Lorg/apache/commons/cli/Option$Builder;
    at org.apache.flink.client.cli.DynamicPropertiesUtil.<clinit>(DynamicPropertiesUtil.
java:39)
    at org.apache.flink.client.cli.GenericCLI.addGeneralOptions(GenericCLI.java:108)
    at org.apache.flink.client.cli.CliFrontend.<init>(CliFrontend.java:136)
    at org.apache.flink.client.cli.CliFrontend.<init>(CliFrontend.java:119)
    at org.apache.flink.client.cli.CliFrontend.main(CliFrontend.java:1129)
```

原因是 Flink 缺少了 commons-cli 的相关 JAR 包。从 Maven 仓库下载该 JAR 包,复制到 Flink 的 lib 目录下即可。

(5)查询输出结果。在终端窗口中,执行的命令如下:

```
$ hdfs dfs -cat /data/flink/wc-result/*
```

可以看到以下计算结果:

```
day 2
good 2
study 1
up 1
```

1.4.2　Flink 完全分布式集群安装

Flink 支持完全分布式安装模式,这时它由一个 master 节点和多个 worker 节点构成。本节将演示如何搭建一个三节点的 Flink 集群,如图 1-30 所示。

13min

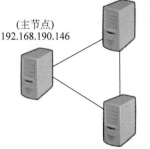

图 1-30　Flink 完全分布式集群设计

1. 完全分布式集群安装

Flink 完全分布式集群搭建步骤如下：

（1）配置从 master（主节点）到 worker（工作节点）的 SSH 无密登录，并保持节点上相同的目录结构。

注意：SSH 原理：master 作为客户端，要实现无密码公钥认证，连接到服务器端 worker 上时，需要在 master 上生成一个密钥对，包括一个公钥和一个私钥，而后将公钥复制到 worker 上。当 master 通过 SSH 连接 worker 时，worker 就会生成一个随机数并用 master 的公钥对随机数进行加密，并发送给 master。master 收到加密数之后再用私钥进行解密，并将解密数回传给 worker，worker 确认解密数无误之后就允许 master 进行连接了。这就是一个公钥认证过程，其间不需要用户手工输入密码。重要过程是将客户端 master 公钥复制到 worker 上。

① 在每台机器上，执行的命令如下：

```
$ ssh localhost
$ ssh exit          #记得最后通过这个命令退出 SSH 连接
```

② 在 master 节点上，生成公私钥，命令如下：

```
$ cd .ssh
$ ssh-keygen -t rsa
```

然后一路按 Enter 键，在.ssh 目录下生成公私钥。

③ 将 master 上的公钥分别加入 master、worker1 和 worker2 机器的授权文件中。

在 master 机器上，执行的命令如下：

```
$ ssh-copy-id hduser@master
$ ssh-copy-id hduser@worker1
$ ssh-copy-id hduser@worker2
```

④ 测试。在 master 机器上，使用 SSH 分别连接 master、worker1 和 worker2，命令如下：

```
$ ssh master
$ ssh worker1
$ ssh worker2
```

这时会发现不需要输入密码，直接就连接上了这两台机器。

（2）Flink 要求在主节点和所有工作节点上设置 Java_HOME 环境变量，并指向 Java

安装的目录。检查 Java 的安装和版本信息,使用的命令如下:

```
$ Java -version
```

(3)下载 Flink 安装包。下载网址为 https://flink.apache.org/downloads.html。可以选择任何喜欢的 Hadoop/Scala 组合。因为 Flink 版本更新迭代比较快,并且每次版本升级都有许多 API 变动,因此建议读者学习时也安装与本书相同的版本。本教程选择使用的是 1.13.2 版本,基于 Scala 2.12。

(4)将下载的最新版本的 Flink 压缩包复制到 master 节点的~/software/目录下,并解压缩到~/bigdata/目录下,命令如下:

```
$ cd ~/bigdata/
$ tar xzf ~/software/flink-1.13.2-bin-scala_2.12.tgz
$ cd flink-1.13.2
```

(5)在 master 节点上配置 Flink。

所有的配置都在 conf/flink-conf.yaml 文件中。在实际应用中,以下几个配置项是非常重要的。

① jobmanager.heap.mb:每个 JobManager 的可用内存量,以 MB 为单位。
② taskmanager.heap.mb:每个 TaskManager 的可用内存量,以 MB 为单位。
③ taskmanager.numberOfTaskSlots:每台机器上可用的 CPU 数量,默认为 1。
④ parallelism.default:集群中 CPU 的总数。
⑤ io.tmp.dirs:临时目录。

首先用编辑器 nano 打开该配置文件(读者也可以用任何自己喜欢的编辑器,如 vim),命令如下:

```
$ nano conf/flink-conf.yaml
```

编辑以下内容(注意,冒号后面一定要有一个空格):

```
jobmanager.rpc.address: master          //指向 master 节点
jobmanager.rpc.port: 6123
jobmanager.heap.size: 1024m             //定义允许 JVM 在每个节点上分配的最大主内存量
taskmanager.memory.process.size: 1024m
taskmanager.numberOfTaskSlots: 2
parallelism.default: 6
```

(6)每个节点下的 Flink 必须保持相同的目录内容,因此将配置好的 Flink 复制到集群中的另外两个节点 worker01 和 worker02,使用的命令如下:

```
$ scp -r ~/bigdata/flink-1.11.1 hduser@worker01:~/bigdata/
$ scp -r ~/bigdata/flink-1.11.1 hduser@worker02:~/bigdata/
```

(7) 最后,必须提供集群中所有用作 worker 节点的列表,每个 worker 节点稍后将运行一个 TaskManager。在 conf/slaves 文件中添加每个 slave 节点信息(IP 或 hostname 均可),每个节点一行,节点信息如下:

```
master
worker1
worker2
```

(8) 启动 Flink 集群,命令如下:

```
$ cd ~/bigdata/flink-1.13.2
$ ./bin/start-cluster.sh
```

这个脚本会在本地节点启动一个 JobManager 并通过 SSH 连接到所有的 worker 节点(在 slaves 文件中列出的)以启动每个节点上的 TaskManager。注意观察启动过程中的输出信息,输出信息如下:

```
Starting cluster.
Starting standalonesession daemon on host master.
Starting taskexecutor daemon on host master.
Starting taskexecutor daemon on host worker1.
Starting taskexecutor daemon on host worker2.
```

可以看出,Flink 先在 master 上启动 standalonesession 进程,然后依次在 master、worker1 和 worker2 上启动 taskexecutor 进程。

启动以后,可以分别在 master、worker1 和 worker2 节点上执行 jps 命令,查看各节点上的进程是否正常启动了。

(9) 关闭集群。执行的命令如下:

```
$ cd ~/bigdata/flink-1.13.2
$ ./bin/stop-cluster.sh
```

也可以分别停止 JobManager 和 TaskManager,命令如下:

```
# 执行以下命令,停止单个的 Job Manager
$ ./bin/jobmanager.sh stop
# 执行以下命令,停止单个的 Task Manager
$ ./bin/taskmanager.sh stop
```

2. 运行 Flink 自带的实时单词计数流程序

Flink 安装包自带了一个以 Socket 作为数据源的实时统计单词计数的流程序,位于 Flink 下的 example/streaming/SocketWindowWordCount.jar 包中。可以通过运行这个流程序来测试 Flink 集群的使用方法。建议按以下步骤执行。

(1) 首先,启动 netcat 服务器,运行在 9000 端口,使用的命令如下:

```
$ nc -l 9000
```

(2) 打开另一个终端启动 Flink 集群,执行的命令如下:

```
$ cd ~/bigdata/flink-1.13.2
$ ./bin/start-cluster.sh
```

(3) 在另一个终端,启动 Flink 示例程序,监听 netcat 服务器。它将从套接字中读取文本,并每 5s 打印前 5s 内每个不同单词出现的次数,即处理时间的滚动窗口,命令如下:

```
$ ./bin/flink run examples/streaming/SocketWindowWordCount.jar --hostname master --port 9000
```

(4) 回到第 1 个正在运行 netcat 的终端窗口,随意输入一些单词,单词之间用空格分隔,Flink 将会处理这些单词。例如,输入以下文本内容:

```
good good study
day day up
```

(5) 分别使用 SSH 登录 master、worker01 和 worker02 节点,并执行以下命令,查看日志中的输出:

```
$ cd ~/bigdata/flink-1.13.2
$ tail -f log/flink-*-taskexecutor-*.out
```

可以看到输出结果如下:

```
good : 2
study : 1
day : 2
up : 1
```

(6) 还可以检查 Flink Web UI 来查看 Job 是怎样执行的。

打开浏览器,输入地址 http://localhost:8081,可查看调度程序的 Web 前端。Web 前端应该报告有 3 个可用的 TaskManager 实例,以及正在执行的作业。Flink Web UI 包含许多关于 Flink 集群及其作业（JobGraph、指标、检查点统计、TaskManager 状态等）的有用且有趣的信息，如图 1-31 所示。

图 1-31　在 Flink Web UI 中查看作业的执行

单击图中的 Running Job List 下正在运行的作业列表，查看某个正在运行的作业执行情况，如图 1-32 所示。

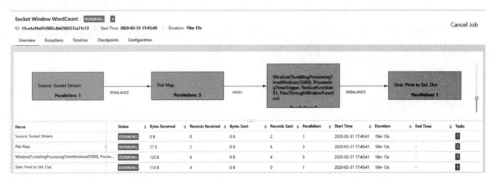

图 1-32　在 Flink Web UI 中查看正在运行的作业执行情况

（7）最后停止 Flink 集群，使用的命令如下：

```
$ ./bin/stop-cluster.sh
```

3. 运行 Flink 自带的单词计数批处理程序

Flink 安装包自带了一个以文本文件作为数据源的单词计数批处理程序，位于 Flink 下的 example/batch/ 目录下的 WordCount.jar 包中。下面演示如何在 Flink 集群上执行该程序，读取 HDFS 上的输入数据文件进行处理，并将计算结果输出到 HDFS 上。

建议按以下步骤执行。

（1）集成 Hadoop。因为要读取 HDFS 上的源数据文件，所以需要在 Flink 中集成 Hadoop 包。从 Flink 1.8 开始，Hadoop 不再包含在 Flink 的安装包中，所以需要单独下载并复制到 Flink 的 lib 目录下。集成过程可参见 1.4.1 节中的部分内容。

（2）准备环境。在终端窗口中，先启动 Flink 集群，再启动 HDFS 集群，执行的命令如下：

```
# 首先启动 Flink 集群
$ ./bin/start-cluster.sh

# 再启动 HDFS 集群
$ start-dfs.sh
```

（3）准备数据文件。编辑数据文件 wc.txt，内容如下：

```
good good study
day day up
```

然后将该文件上传到 HDFS 的 /data/flink/ 目录下，使用的命令如下：

```
$ hdfs dfs -put wc.txt /data/flink/
```

（4）执行 Flink 的批处理程序，命令如下：

```
# 执行 Flink 单词计数程序
$ ./bin/flink run ./examples/batch/WordCount.jar \
--input hdfs://xueai8:8020/data/flink/wc.txt \
--output hdfs://xueai8:8020/data/flink/wc-result
```

执行过程如图 1-33 所示。

上面的命令是在运行 WordCount 时读写 HDFS 中的文件的，其中 --input 参数用于指定要处理的输入文件，--output 参数用于指定将计算结果输出到哪个文件（注：如果不加 hdfs:// 前缀，则默认使用本地文件系统）。

（5）查询输出结果。在终端窗口中，执行的命令如下：

```
$ hdfs dfs -cat /data/flink/wc-result/*
```

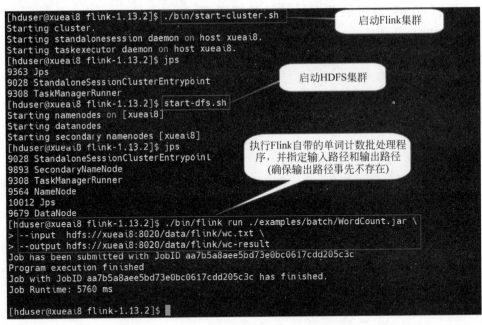

图 1-33　Flink 单词计数批处理程序执行过程

可以看到计算结果如下：

```
day 2
good 2
study 1
up 1
```

1.4.3　Flink 常用命令

Flink 以 CLI 的方式提供了一些常见的操作任务。下面是常用的 CLI 命令。
(1) 查看正在运行的作业，有两种方式。
第 1 种，CLI 方式，命令如下：

```
$ flink list
```

第 2 种，可用 REST API 方式查看正在运行的作业，命令如下：

```
$ curl localhost:8081/jobs
```

(2) 运行示例程序，不带参数，命令如下：

```
$ ./bin/flink run ./examples/batch/WordCount.jar
```

（3）运行示例程序，带有 input 和 output 参数，命令如下：

```
$ ./bin/flink run ./examples/batch/WordCount.jar \
    -- input file:///home/user/hamlet.txt \
    -- output file:///home/user/wordcount_out
```

（4）运行示例程序，带有 input 和 output 参数，并且将并行度指定为 16，命令如下：

```
$ ./bin/flink run -p 16 ./examples/batch/WordCount.jar \
    -- input file:///home/user/hamlet.txt \
    -- output file:///home/user/wordcount_out
```

（5）运行示例程序，禁用 Flink 日志输出，命令如下：

```
$ ./bin/flink run -q ./examples/batch/WordCount.jar
```

（6）运行示例程序，在分离模式（Detached Mode）下，命令如下：

```
$ ./bin/flink run -d ./examples/batch/WordCount.jar
```

（7）运行示例程序，在特定的 JobManager 上，命令如下：

```
$ ./bin/flink run -m myJMHost:8081 ./examples/batch/WordCount.jar \
    -- input file:///home/user/hamlet.txt \
    -- output file:///home/user/wordcount_out
```

（8）运行示例程序，以特定的类作为入口点，命令如下：

```
$ ./bin/flink run -c org.apache.flink.examples.java.wordcount.WordCount ./examples/batch/WordCount.jar \
    -- input file:///home/user/hamlet.txt \
    -- output file:///home/user/wordcount_out
```

（9）运行示例程序，使用带有两个 TaskManager 的一个 Per-Job YARN 集群，命令如下：

```
$ ./bin/flink run -m yarn-cluster -yn 2 ./examples/batch/WordCount.jar \
    -- input hdfs:///user/hamlet.txt \
    -- output hdfs:///user/wordcount_out
```

（10）以 JSON 格式显示 WordCount 示例程序的优化执行计划，命令如下：

```
$ ./bin/flink info ./examples/batch/WordCount.jar \
    -- input file:///home/user/hamlet.txt \
    -- output file:///home/user/wordcount_out
```

(11) 列出已调度和正在运行的作业（包括它们的JobIDs），命令如下：

```
$ ./bin/flink list
```

(12) 列出已调度的作业（包括它们的JobIDs），命令如下：

```
$ ./bin/flink list -s
```

(13) 列出正在运行的作业（包括它们的JobIDs），命令如下：

```
$ ./bin/flink list -r
```

(14) 列出所有现有的作业(包括它们的JobIDs)，命令如下：

```
$ ./bin/flink list -a
```

(15) 列出在Flink YARN session 内正在运行的Flink作业，命令如下：

```
$ ./bin/flink list -m yarn-cluster -yid <yarnApplicationID> -r
```

(16) 取消一个作业，命令如下：

```
$ ./bin/flink cancel <jobID>
```

(17) 取消一个作业，带有savepoint，命令如下：

```
$ ./bin/flink cancel -s [targetDirectory] <jobID>
```

(18) 停止一个作业（仅限于流作业），命令如下：

```
$ ./bin/flink stop <jobID>
```

(19) 修改一个正在运行的作业（仅限于流作业），命令如下：

```
$ ./bin/flink modify <jobID> -p <newParallelism>
```

第 2 章 Flink 开发环境准备

CHAPTER 2

"工欲善其事,必先利其器",一个好的开发环境和工具,能极大地提升开发效率。本章将学习如何准备好 Flink 开发环境,并尝试开发自己的第一个 Flink 程序,然后提交到 Flink 集群上运行。最后,将深入理解一些 Flink 的核心概念。

Flink 支持 Linux、macOS X 和 Windows 作为 Flink 程序和本地执行的开发环境。对于 Flink 开发设置,需要安装以下软件:

(1) JDK 8 或以上版本。
(2) Apache Maven 3.x。
(3) 用于 Java(和/或 Scala)开发的 IDE,建议使用 IntelliJ IDEA。

2.1 安装和配置

作为下一代开源大数据处理引擎,首先要了解它的发展历程及特性。

9min

2.1.1 安装和配置 Maven

在 Windows 系统上安装 Maven 的步骤如下。

(1) 首先下载最新的 Maven 安装包。下载网址为 http://maven.apache.org/download.cgi。找到 bin.zip 压缩包的下载链接,单击即可下载,如图 2-1 所示。

(2) 将 Maven 安装包解压到指定目录。例如,解压到 E:\maven\apache-maven-3.6.3 目录下,该目录下包含 bin、lib、conf 等文件夹,如图 2-2 所示。

(3) 配置 Maven 系统环境变量。依次选择"我的计算机→属性→高级系统设置→环境变量→系统变量→新建变量",新建一个系统变量。将变量名设置为 M2_HOME,变量值为 E:\maven\apache-maven-3.6.3,如图 2-3 所示。

(4) 继续配置 Maven 系统环境变量。在系统变量中找到 Path 环境变量,在变量值尾部加入 %M2_HOME%\bin(注意,如果是 Windows 7 系统,则需要在前面加上一个分号,用来和其他路径值分隔开),如图 2-4 所示。

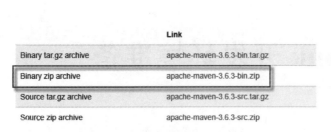
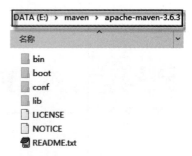

图 2-1　下载最新的 Maven 安装包　　　　图 2-2　Maven 安装目录结构

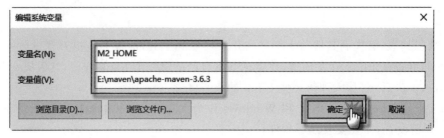

图 2-3　配置 Maven 系统环境变量 M2_HOME

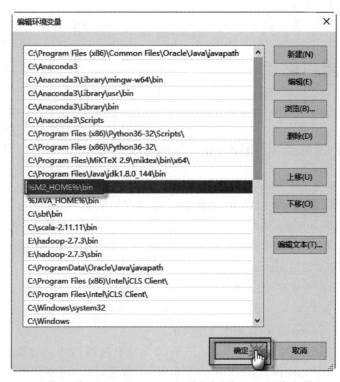

图 2-4　将 Maven 的 bin 目录添加到系统 Path 环境变量

(5) 检查 Maven 的环境变量是否配置成功。打开命令行,执行的命令如下:

```
> mvn - v
```

如果能正确地输出 Maven 的版本信息,则说明配置正确,如图 2-5 所示。

图 2-5　Maven 配置成功信息

2.1.2　修改 Maven 本地仓库默认位置

Maven 会将下载的类库(JAR 包)放置到本地的一个目录下(一般默认情况下 Maven 默认的本地仓库路径为 ${user.home}/.m2/repository,其中 ${user.home}指的是当前用户主目录),如果想重新定义这个仓库目录的位置,则需要修改 Maven 本地仓库的配置。

修改本地仓库默认位置的操作步骤如下:

(1) 在自己喜欢的位置创建文件夹,此处笔者创建的位置是 E:\maven\repository。

(2) 在安装 Maven 的目录下找到 conf/settings.xml 文件,打开编辑,更改默认的仓库位置,并指定仓库的镜像位置。编辑内容,如图 2-6 所示。

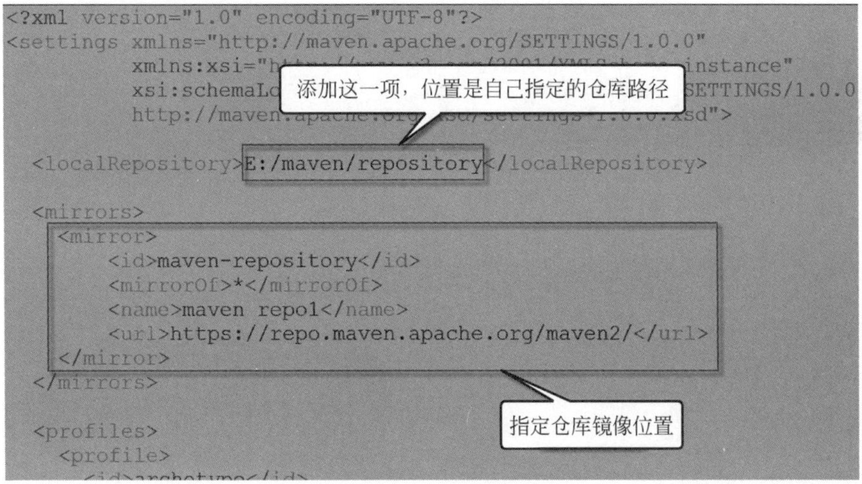

图 2-6　编辑 Maven 配置文件,指定仓库位置和镜像位置

(3) 要基于 Maven 的 Archetype 来快速创建一个新的 Flink 工程,还需要在< mirrors/>标签后加上< profiles >内容和< artiveProfiles >内容(Archetype 是 Maven 工程的模板工具包,会帮助用户创建 Maven 工程模板,并向用户提供生成相关工程模板版本的参数化方法),如

图 2-7 所示。

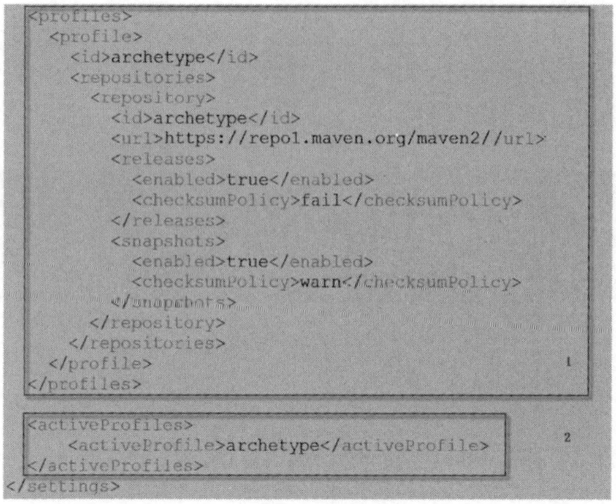

```
<profiles>
  <profile>
    <id>archetype</id>
    <repositories>
      <repository>
        <id>archetype</id>
        <url>https://repo1.maven.org/maven2/</url>
        <releases>
          <enabled>true</enabled>
          <checksumPolicy>fail</checksumPolicy>
        </releases>
        <snapshots>
          <enabled>true</enabled>
          <checksumPolicy>warn</checksumPolicy>
        </snapshots>
      </repository>
    </repositories>
  </profile>
</profiles>                                            1

<activeProfiles>
  <activeProfile>archetype</activeProfile>             2
</activeProfiles>
</settings>
```

图 2-7　编辑 Maven 配置文件，指定工程模板

(4) 最后，settings.xml 文件的完整内容如下：

```xml
<?xml version="1.0" encoding="UTF-8"?>
<settings xmlns="http://maven.apache.org/SETTINGS/1.0.0"
          xmlns:xsi="http://www.w3.org/2001/XMLSchema-instance"
          xsi:schemaLocation="http://maven.apache.org/SETTINGS/1.0.0 http://maven.apache.org/xsd/settings-1.0.0.xsd">

  <localRepository>E:/maven/repository</localRepository>

  <mirrors>
    <mirror>
      <id>maven-repository</id>
      <mirrorOf>*</mirrorOf>
      <name>maven repo1</name>
      <url>https://repo.maven.apache.org/maven2/</url>
    </mirror>
  </mirrors>

  <profiles>
    <profile>
      <id>archetype</id>
      <repositories>
        <repository>
          <id>archetype</id>
          <url>https://repo1.maven.org/maven2/</url>
          <releases>
            <enabled>true</enabled>
```

```xml
          <checksumPolicy>fail</checksumPolicy>
        </releases>
        <snapshots>
          <enabled>true</enabled>
          <checksumPolicy>warn</checksumPolicy>
        </snapshots>
      </repository>
    </repositories>
  </profile>
</profiles>

<activeProfiles>
    <activeProfile>archetype</activeProfile>
</activeProfiles>
</settings>
```

（5）验证修改是否生效。打开命令行，执行的命令如下：

```
> mvn help:system
```

执行完该命令后，在 E:\maven\repository 目录下面会生成很多文件，这些文件就是 Mven 从中央仓库下载到本地仓库的文件。

2.1.3　创建初始模板项目

在命令行中快速创建一个空白 Flink 项目，命令如下：

```
> cd FlinkProjects                     //在指定的工作目录下
> mvn archetype:generate \
-DarchetypeGroupId=org.apache.flink \
-DarchetypeArtifactId=flink-quickstart-scala \
-DarchetypeCatalog=local \
-DarchetypeVersion=1.10.0
```

注意：如果要创建一个 Java 语言的空白 Flink，则只需将上面构建命令中的 flink-quickstart-scala 换成 flink-quickstart-Java，其他操作都相同。

如果创建时很慢，则可以把 http://repo1.maven.org/maven2/archetype-catalog.xml 下载到本地仓库根目录，例如 ~/.m2 下，并在执行 mvn archetype 命令时加上参数 -DarchetypeCatalog=internal。

这会在 FlinkProjects 目录下创建一个 Flink 模板项目。在创建过程中，Maven 会交互式地询问 groupId、artifactId 和 package 名称。这里笔者分别输入以下名称：

```
groupId:com.xueai8
artifactId:FlinkJavaBlank
package:com.xueai8
```

创建过程如图 2-8 所示。

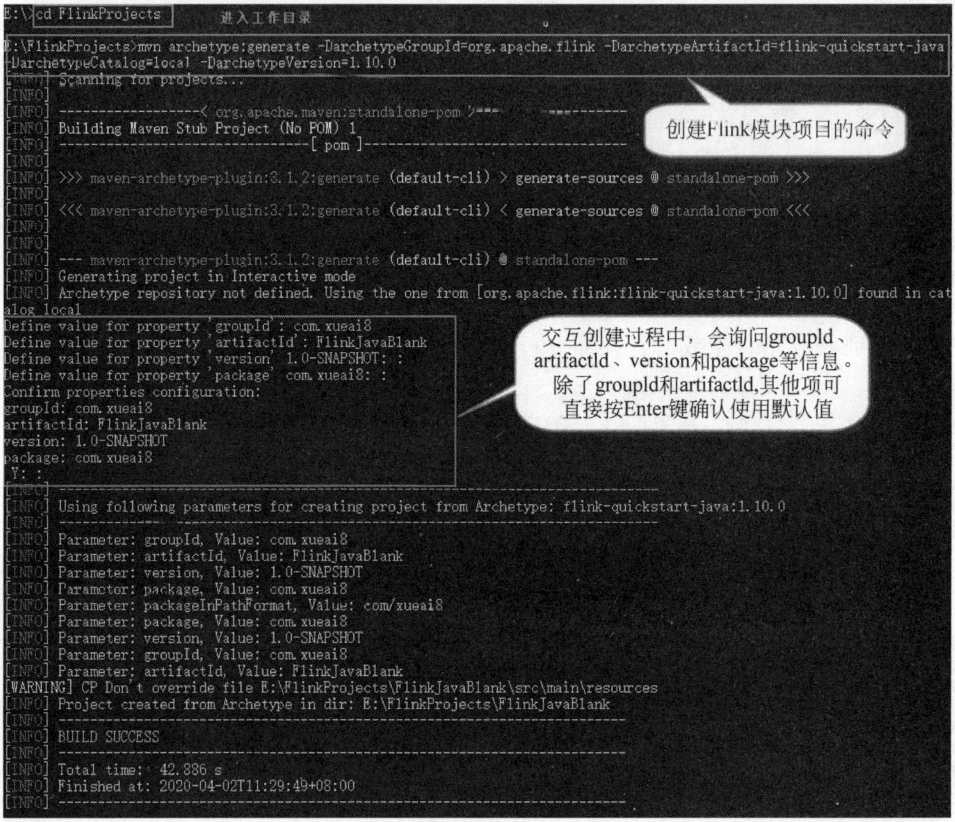

图 2-8 创建初始模板项目过程

工作目录中将增加一个新目录,目录名称是 artifactId 的名称。在命令行中可以使用下面的命令查看项目结构:

```
> tree FlinkJavaBlank
```

项目结构如图 2-9 所示。

示例项目是一个 Maven 项目,它包含两个类:StreamingJob 和 BatchJob,分别是 DataStream 和 DataSet 程序的基本框架程序。类中的 main 方法是程序的入口,用于内部测试/执行和部署。这个基本框架程序可以被导入 IDE 中进行开发。

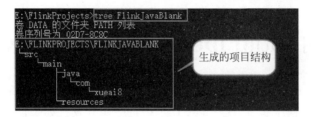

图2-9　Flink Maven 项目结构

2.1.4　构建打包项目

如果想构建/打包 FlinkJavaBlink 项目以便部署，需要运行 mvn clean package 命令，建议按以下步骤操作：

（1）切换到项目目录。在命令行中执行的命令如下：

```
> cd E:/FlinkProjects/FlinkJavaBlank
> mvn clean package
```

（2）在项目目录下会生成一个 target 目录。用户打包的结果文件就放在这个目录中，可以转到相应的文件夹去查看，也可以直接在命令行下执行 dir 命令查看，如图 2-10 所示。

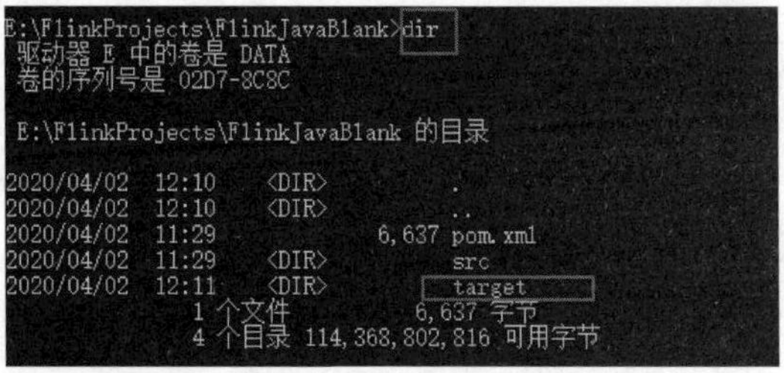

图 2-10　编译打包后的项目目录中会生成一个 target 文件夹

（3）查看 target 文件夹，将会发现一个 JAR 文件，其中包含应用程序及相关的依赖项（如连接器和库），如图 2-11 所示。

图 2-11　编译打包后的 JAR 文件

2.2 使用 IntelliJ IDEA＋Maven 开发 Flink 项目

对于熟悉 Maven 操作的读者来讲,可以像 2.1 节讲的那样,先在命令行用 Maven 命令快速生成 Flink 基本程序框架,然后导入 IDE 中进一步开发。如果读者更熟悉集成开发环境(例如,IntelliJ IDEA),则可以直接在集成开发环境中开发 Maven 程序。下面通过结合使用 IntelliJ IDEA 集成开发工具和 Maven 项目构建工具,讲解如何方便快速地开发 Flink 项目。

2.2.1 在 IntelliJ IDEA 中创建 Flink 项目

要在 IntelliJ IDEA 中创建 Flink Maven 项目,建议按以下步骤操作:
(1) 启动 IntelliJ IDEA,创建一个新的项目,如图 2-12 所示。

图 2-12 在 IntelliJ IDEA 中创建一个新的项目

(2) 选择 Maven 项目,并选择 Create from archetype,如图 2-13 所示。
(3) 因为默认没有 Flink 的 archetype,所以需要自己添加。
添加 flink-quickstart-java 的 archetype,如图 2-14 所示。
添加 flink-quickstart-scala 的 archetype,如图 2-15 所示。

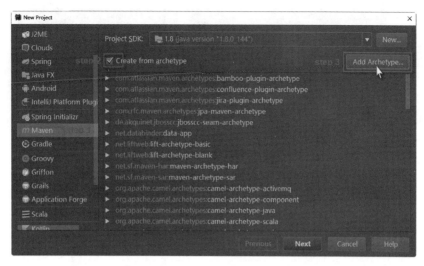

图 2-13　选择 Maven 项目，并勾选 Create from archetype 项

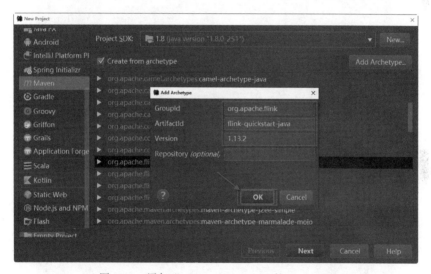

图 2-14　添加 flink-quickstart-java 的 archetype

（4）选择对应的 archetype，例如，这里选择 flink-quickstart-scala，如图 2-16 所示。
（5）指定项目的 groupId、artifactId 名称。这里分别取以下名称：

```
groupId:com.xueai8
artifactId:FlinkScalaDemo
```

如图 2-17 所示。

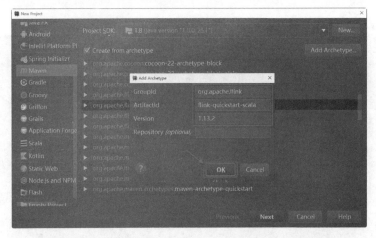

图 2-15　添加 flink-quickstart-scala 的 archetype

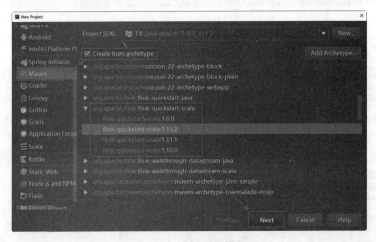

图 2-16　选择项目模板

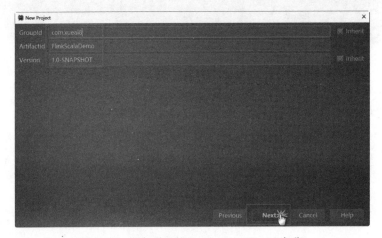

图 2-17　指定项目的 groupId、artifactId 名称

（6）接下来，指定项目的 Maven 配置，此处可采用默认配置，如图 2-18 所示。

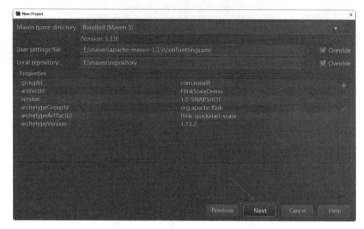

图 2-18　指定项目的 Maven 配置

（7）指定项目的名称和项目文件所在位置。这里保持默认配置即可。单击 Finish 按钮，开始创建项目，如图 2-19 所示。

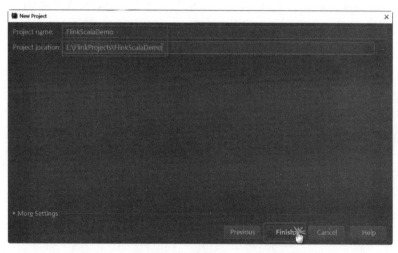

图 2-19　指定项目的名称和项目文件所在位置

（8）Maven 会自动构建项目，最后的项目结构如图 2-20 所示。

可以看出，flink-quickstart-scala 快速地构建了一个基本的 Flink 项目框架，并创建了两个模板程序文件：用于流处理的 StreamingJob 和用于批处理的 BatchJob。

注意：以同样的步骤，选择 flink-quickstart-java，创建一个基于 Java API 的 Flink 项目框架。读者可自行尝试。

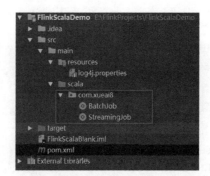

图 2-20　最终生成的项目结构

2.2.2　设置项目基本依赖

每个 Flink 应用程序都依赖于一些核心的 API 库。对于 Maven，可以使用 Java 项目模板（flink-quickstart-java）或 Scala 项目模板（flink-quickstart-scala）创建具有这些初始依赖项的程序框架。

如果没有使用项目模板而手动设置项目，则需要为 Java/Scala API 添加以下依赖项，这里以 Maven 语法表示，但是同样的依赖项也适用于其他构建工具（如 Gradle、SBT 等）。

Scala Maven 依赖内容如下：

```xml
<dependency>
    <groupId>org.apache.flink</groupId>
    <artifactId>flink-scala_2.12</artifactId>
    <version>1.13.2</version>
    <scope>provided</scope>
</dependency>
<dependency>
    <groupId>org.apache.flink</groupId>
    <artifactId>flink-streaming-scala_2.12</artifactId>
    <version>1.13.2</version>
    <scope>provided</scope>
</dependency>
```

Java Maven 依赖内容如下：

```xml
<dependency>
    <groupId>org.apache.flink</groupId>
    <artifactId>flink-Java</artifactId>
    <version>1.13.2</version>
    <scope>provided</scope>
</dependency>
```

```xml
<dependency>
    <groupId>org.apache.flink</groupId>
    <artifactId>flink-streaming-Java_2.12</artifactId>
    <version>1.13.2</version>
    <scope>provided</scope>
</dependency>
```

需要注意，所有这些依赖项的作用域（scope）都被设置为 provided。这意味着需要对它们进行编译，但是不应该将它们打包到项目的最终应用程序 JAR 文件中。这些依赖项是 Flink 核心依赖项，Flink 运行时环境已经包含了这些依赖，因此不需要把它们打包到 JAR 包中。

强烈建议将该依赖关系保持在 provided 范围内。如果没有将它们设置为 provided，则最好的情况是生成的 JAR 包变得非常大，因为它还包含所有 Flink 核心依赖项。最坏的情况是添加到应用程序 JAR 文件中的 Flink 核心依赖项与自己的一些依赖项版本冲突（通常通过反向类加载来避免）。

注意：要使 Flink 应用程序在 IntelliJ IDEA 中运行，需要在 scope compile 中声明 Flink 依赖项，而不是 provided。否则 IntelliJ 将不会将它们添加到类路径中，而 IDE 中的执行将会失败，并抛出 NoClassDefFountError 错误信息。

2.2.3 用于构建具有依赖项的 JAR 的模板

要构建包含声明的连接器和库所需的所有依赖项的应用程序 JAR，可以使用下面的 shade 插件进行定义：

```xml
<build>
    <plugins>
        <plugin>
            <groupId>org.apache.maven.plugins</groupId>
            <artifactId>maven-shade-plugin</artifactId>
            <version>3.1.1</version>
            <executions>
                <execution>
                    <phase>package</phase>
                    <goals>
                        <goal>shade</goal>
                    </goals>
                    <configuration>
                        <artifactSet>
                            <excludes>
```

```xml
            <exclude>com.google.code.findBugs:jsr305</exclude>
                        <exclude>org.slf4j:*</exclude>
                        <exclude>log4j:*</exclude>
                    </excludes>
                </artifactSet>
                <filters>
                    <filter>
                        <artifact>*:*</artifact>
                        <excludes>
                            <exclude>META-INF/*.SF</exclude>
                            <exclude>META-INF/*.DSA</exclude>
                            <exclude>META-INF/*.RSA</exclude>
                        </excludes>
                    </filter>
                </filters>
                <transformers>
                    <transformer implementation="org.apache.maven.plugins.shade.resource.ManifestResourceTransformer">
                        <mainClass>com.xueai8.StreamingJob</mainClass>
                    </transformer>
                </transformers>
            </configuration>
        </execution>
    </executions>
</plugin>
</plugins>
</build>
```

2.2.4 编写批处理代码并测试执行

以2.2.1节所创建项目中的BatchJob源文件为模板,进一步编写一个简单的批处理代码并执行。这里的目的是掌握如何使用IntelliJ IDEA创建Flink Maven项目,所以不必理解代码,在后续的章节中会详细讲解。

在IntelliJ IDEA中打开BatchJob源文件,编辑代码。

Scala代码如下:

```scala
//第2章/BatchJob.scala
import org.apache.flink.api.scala._

object BatchJob {

  def main(args: Array[String]) {
    //设置批执行环境
```

```
    val env = ExecutionEnvironment.getExecutionEnvironment

    //得到输入数据
    val text = env.fromElements("good good study", "day day up")

    //对数据进行转换
    val counts = text
        .flatMap { _.toLowerCase.split("\\W+") }
        .map { (_, 1) }
        .groupBy(0)
        .sum(1)

    //执行并输出结果
    counts.print()
  }
}
```

Java 代码如下：

```
//第 2 章/BatchJob.java
import org.apache.flink.api.common.functions.FlatMapFunction;
import org.apache.flink.api.java.DataSet;
import org.apache.flink.api.java.ExecutionEnvironment;
import org.apache.flink.api.java.tuple.Tuple2;
import org.apache.flink.util.Collector;

/**
 * 要将应用程序打包为 JAR 文件,可在命令行执行如下命令
 * $ mvn clean package
 *
 * 如果改变了主类的名称,则需要在 pom.xml 文件中进行相应修改(可搜索'mainClass')
 */
public class BatchJob {

    public static void main(String[] args) throws Exception {
        //设置批处理执行环境
        final ExecutionEnvironment env =
                ExecutionEnvironment.getExecutionEnvironment();

        //读取数据源
        DataSet<String> text = env.fromElements("Good good study", "Day day up");

        //对数据集进行转换
        DataSet<Tuple2<String,Integer>> result = text
            .flatMap(new FlatMapFunction<String, Tuple2<String,Integer>>() {
```

```
            @Override
            public void flatMap(String s,
                        Collector<Tuple2<String,Integer>> out)
                   throws Exception {
                String line = s.toLowerCase();
                for(String word : line.split(" ")){
                    out.collect(new Tuple2<>(word, 1));
                }
            }
        })
        .groupBy(0)
        .sum(1);

    result.print();
    }
}
```

在文件内任意空白处右击，在弹出的快捷菜单中选择 run BatchJob，执行该程序，在下方的运行窗口可以看到输出结果如下：

```
(up,1)
(day,2)
(good,2)
(study,1)
```

在执行此程序时，有可能会遇到异常信息，内容如下：

```
...
Caused by: Java.lang.ClassNotFoundException:
org.apache.flink.api.scala.typeutils.CaseClassTypeInfo
    at Java.net.urlClassLoader.findClass(URLClassLoader.java:381)
    at Java.lang.ClassLoader.loadClass(ClassLoader.java:424)
    at sun.misc.Launcher$AppClassLoader.loadClass(Launcher.java:335)
    at Java.lang.ClassLoader.loadClass(ClassLoader.java:357)
    ... 13 more
```

异常的原因是 Maven 引入依赖问题。打开项目中的 pom.xml 文件，将其中依赖部分的<scope>provided</scope>注释（或删除）掉，修改后的代码如下：

```
<dependency>
    <groupId>org.apache.flink</groupId>
    <artifactId>flink-scala_2.12</artifactId>
    <version>${flink.version}</version>
    <!--<scope>provided</scope>-->
```

```xml
</dependency>
<dependency>
    <groupId>org.apache.flink</groupId>
    <artifactId>flink-streaming-scala_2.12</artifactId>
    <version>${flink.version}</version>
    <!-- <scope>provided</scope> -->
</dependency>
```

另外,在 IntelliJ IDEA 中运行项目可能会导致 Java.lang.NoClassDefFoundError 异常。这可能是因为没有将所有必需的 Flink 依赖项隐式地加载到类路径中。

解决办法是在 IntelliJ IDEA 中,选择 Run→Edit Configurations→Modify options,然后选择 include dependencies with "Provided" scope。这个运行配置现在将包括从 IDE 中运行应用程序所需的所有类。

2.2.5 项目打包并提交 Flink 集群执行

如果项目想要部署到生产环境并运行,就要先打包成 JAR 包再进行部署。使用 Maven 的 mvn clean package 命令可以很方便地进行打包。具体操作步骤如下:

(1) 打开项目中的 pom.xml 文件,找到以下内容并将 mainClass(可同时按下快捷键 Ctrl+F,查找 mainClass)修改为当前类的全限定名称:

```xml
<transformers>
    <transformer implementation="org.apache.maven.plugins.shade.resource.ManifestResourceTransformer">
        <mainClass>com.xueai8.BatchJob</mainClass>
    </transformer>
</transformers>
```

(2) 在 IntelliJ IDEA 界面最下方左侧,单击 Terminal 选项卡,打开 Terminal 窗口。在 Terminal 窗口中执行命令 mvn clean package 打包项目,如图 2-21 所示。

图 2-21 执行命令 mvn clean package 打包项目

(3) 在编译打包过程中会输出一系列信息,如图 2-22 所示。

(4) 在项目下会生成一个 target 目录,打好的 JAR 包就在这里,如图 2-23 所示。

图 2-22 打包成功提示信息

图 2-23 打包好的 JAR 包文件

（5）启动 Flink 集群。在 Linux 终端中，执行的命令如下：

```
$ cd ~/bigdata/flink-1.13.2/
$ ./bin/start-cluster.sh
```

执行过程如图 2-24 所示。

图 2-24 Flink 集群启动过程

（6）将该 JAR 包复制到 Linux 的指定目录下，例如，~/flinkdemos/目录下，然后提交到 Flink 集群上运行 Job 作业，命令如下：

```
$ cd ~/bigdata/flink-1.13.2/
$ ./bin/flink run -- class com.xueai8.BatchJob ~/flinkdemos/FlinkScalaDemo-1.0-SNAPSHOT.jar
```

执行结果如下：

```
(day,2)
(good,2)
(study,1)
(up,1)
```

2.3 Flink 相关概念

到目前为止,已经准备好了 Flink 应用程序开发环境,并尝试编写了自己的第一个 Flink 应用程序,然后打包为作业提交给 Flink 集群去执行。在进一步深入学习 Flink 技术之前,读者有必要了解一些 Flink 的核心概念。

2.3.1 Flink 数据流

在 Flink 中,应用程序由数据流组成,这些数据流可以由用户定义的运算符(有时将这些运算符称为"算子")进行转换。这些数据流形成有向图,从一个或多个源开始,以一个或多个输出结束,如图 2-25 所示。

图 2-25　Flink 程序流

Flink 支持流处理和批处理,它是一个分布式的流批结合的大数据处理引擎。在 Flink 中,认为所有的数据本质上都是随时间产生的流数据,把批数据看作流数据的特例,只不过流数据是一个无界的数据流,而批数据是一个有界的数据流(例如固定大小的数据集),如图 2-26 所示。

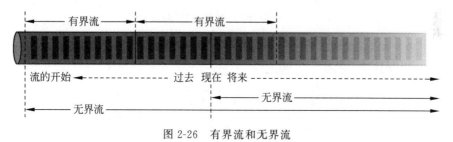

图 2-26　有界流和无界流

因此,Flink 是一个用于在无界和有界数据流上进行有状态计算的通用的处理框架,它既具有处理无界数据流的复杂功能,也具有专门的运算符来高效地处理有界数据流。通常将无界数据称为实时数据,这些数据来自消息队列或分布式日志等流源(如 Apache Kafka 或 Kinesis);而有界数据通常指的是历史数据,这些数据来自各种数据源(如文件、关系型数据库等)。由 Flink 应用程序产生的结果流可以发送到各种各样的系统,并且可以通过 REST API 访问 Flink 中包含的状态,如图 2-27 所示。

当 Flink 处理一个有界的数据流时采用的就是批处理工作模式。在这种操作模式中,可以选择先读取整个数据集,然后对数据进行排序、计算全局统计数据或生成总结所有输入

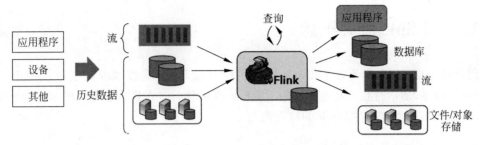

图 2-27　Flink 处理框架

的最终报告。

当 Flink 处理一个无界的数据流时采用的就是流处理工作模式。对于流数据处理，输入可能永远不会结束，因此必须在数据到达时持续不断地对这些数据进行处理。

2.3.2　Flink 分层 API

Flink 提供了开发流/批处理应用程序的不同抽象层次，如图 2-28 所示。

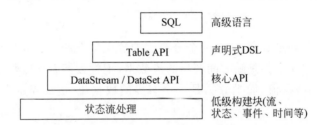

图 2-28　Flink 应用程序的不同抽象层次

Flink 提供了 3 个分层的 API。每个 API 在简洁性和表达性之间提供了不同的权衡，并针对不同的应用场景，如图 2-29 所示。

图 2-29　Flink 分层 AP

最低级别抽象只提供有状态流，它通过 ProcessFunction 嵌入 DataStream API 中。ProcessFunction 是 Flink 提供的最富表现力的函数接口，它允许用户自由地处理来自一个或多个流的事件，并使用一致的容错状态。ProcessFunction 提供了对时间和状态的细粒度控制。ProcessFunction 可以任意修改它的状态并注册定时器，从而在将来触发回调函数，以实现复杂的计算，因此，ProcessFunction 可以根据许多有状态事件驱动的应用程序的需要实现每个事件的复杂业务逻辑。

实际上，大多数应用程序不需要上面描述的最低级别层抽象，而是根据核心 API（DataStream 流处理 API 和 DataSet 批处理 API）进行编程。DataStream API 是数据驱动应用程序和数据管道的主要 API，DataStream API 为许多常见的流处理操作提供基本类型，例如窗口化、record-at-a-time 转换和通过查询外部数据存储来丰富事件。DataStream API 支持 Java 和 Scala 语言，它基于 map()、reduce() 和 aggregate() 等函数。函数可以通过扩展接口或作为 Java 或 Scala lambda 函数来定义。使用核心 API 开发的 Flink 应用程序可以显式地控制时间和状态（状态、触发器、proc、fun 等）。

最低级别层 ProcessFunction 与 DataStream API 集成，使仅对某些操作进行最低级别层抽象成为可能，而 DataSet API 在有界数据集上提供了额外的原语，支持循环/迭代运算。

Flink 有两个关系 API，即 Table API 和 SQL。这两个 API 都是用于批处理和流处理的统一 API，即查询在无界、实时流或有界、记录流上以相同的语义执行，并产生相同的结果。

Table API 是一个以表为中心的声明性 DSL，它可以（在表示流时）动态地更改表。Table API 遵循（扩展的）关系模型：表有一个附加的模式（类似于关系数据库中的表），API 提供了类似的关系操作，例如 select、project、join、group-by、aggregate 等。Table API 程序声明应该执行什么逻辑操作，而不是明确指定操作代码的形式。虽然 Table API 可以通过各种类型的用户定义函数（UDF）进行扩展，使用起来更简洁（编写的代码更少），但是该 API 没有提供对时间和状态的直接访问，与核心 API 相比缺乏表现力。此外，Table API 程序还需要经过一个在执行之前应用优化规则的优化器，从而使应用自动优化。可以在 Table 和 DataStream/DataSet 之间无缝转换，Flink 允许程序编程时混合 Table API 和 DataStream/DataSet API。

Flink 提供的最高级抽象是 SQL。这种抽象在语义和表达性上都类似于 Table API，但将程序表示为 SQL 查询表达式。SQL 抽象与 Table API 紧密交互，SQL 查询可以在 Table API 中定义的表上执行。

Table API 和 SQL 利用 Apache Calcite 进行解析、验证和查询优化。它们可以与 DataStream 和 DataSet API 无缝集成，并支持用户定义的标量、聚合和表值（table-valued）函数。Table API/SQL 正在以跨批处理和流的统一方式成为分析应用场景下的主要 API，旨在简化数据分析、数据管道和 ETL 应用程序的定义。

第 3 章 开发 Flink 实时数据处理程序

CHAPTER 3

实时分析是当前非常重要的一个问题。许多不同的领域需要实时处理数据。例如，来自物联网(IoT)的应用程序需要实时或近实时地存储、处理和分析数据。

为了满足对实时数据处理的需求，Apache Flink 提供了一个名为 DataStream API 的流数据处理 API，用于构建健壮的、有状态的流应用程序。它提供了对状态和时间的细粒度控制，允许实现高级事件驱动系统。

Apache Flink 的流处理 API(DataStream API)位于 org.apache.flink.streaming.api.scala 包(Scala 语言)或 org.apache.flink.streaming.api 包(Java 语言)中。

3.1 Flink 流处理程序编程模型

在深入了解 Flink 实时数据处理程序的开发之前，先通过一个简单示例来了解使用 Flink 的 DataStream API 构建有状态流应用程序的过程。

3.1.1 流数据类型

Flink 以一种独特的方式处理数据类型和序列化，它包含自己的类型描述符、泛型类型提取和类型序列化框架。基于 Java 和 Scala 语言，Flink 实现了一套自己的类型系统，它支持很多类型，包括：

(1) 基本类型。
(2) 数组类型。
(3) 复合类型。
(4) 辅助类型。
(5) 通用类型。

详细的 Flink 类型系统如图 3-1 所示。

Flink 针对 Java 和 Scala 的 DataStream API 要求流数据的内容必须是可序列化的。Flink 内置了以下类型数据的序列化器。

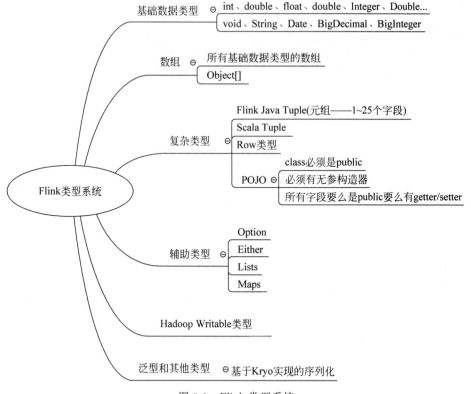

图 3-1　Flink 类型系统

（1）基本数据类型：String、Long、Integer、Boolean、Array。

（2）复合数据类型：Tuple、POJO、Scala case class。

对于其他类型，Flink 会返回 Kryo。也可以在 Flink 中使用其他序列化器。Avro 尤其得到了很好的支持。

1. Java DataStream API 使用的流数据类型

对于 Java API，Flink 定义了自己的 Tuple1～Tuple25 类型来表示元组类型，代码如下：

```
Tuple2 < String, Integer > person = new Tuple2 <>("王老五", 35);

//索引基于 0
String name = person.f0;
Integer age = person.f1;
```

在 Java 中，POJO(Plain Old Java Object)是这样的 Java 类：

（1）有一个无参的默认构造器。

（2）所有的字段要么是 public 的，要么有一个默认的 getter 和 setter。

例如，定义一个名为 Person 的 POJO 类，代码如下：

```java
//第 3 章/Person.java

//定义一个 Person POJO 类
public class Person{
    public String name;
    public Integer age;

    public Person() {};

    public Person(String name, Integer age) {
        this.name = name;
        this.age = age;
    };
}

//创建一个实例
Person person = new Person("王老五", 35);
```

2. Scala DataStream API 使用的流数据类型

对于元组，可以使用 Scala 自己的 Tuple 类型，代码如下：

```scala
val person = ("王老五", 35)

//索引基于 1
val name = person._1
val age = person._2
```

对于对象类型，使用 case class（相当于 Java 中的 JavaBean），代码如下：

```scala
case class Person(name: String, age: Int)

val person = Person("王老五", 35)
```

3. Flink 类型系统

对于创建的任意一个 POJO 类型，看起来它是一个普通的 Java Bean，在 Java 中，可以使用 Class 来描述该类型，但其实在 Flink 引擎中，它被描述为 PojoTypeInfo，而 PojoTypeInfo 是 TypeInformation 的子类。

TypeInformation 是 Flink 类型系统的核心类。Flink 使用 TypeInformation 来描述所有 Flink 支持的数据类型，就像 Java 中的 Class 类型一样。每种 Flink 支持的数据类型都对应于 TypeInformation 的子类。例如 POJO 类型对应的是 PojoTypeInfo、基础数据类型数组对应的是 BasicArrayTypeInfo、Map 类型对应的是 MapTypeInfo、值类型对应的是 ValueTypeInfo。

除了对类型的描述，TypeInformation 还提供了对序列化的支持。在 TypeInformation 中有一种方法：createSerializer 方法，它用来创建序列化器，在序列化器中定义了一系列的方法，其中，通过 serialize 和 deserialize 方法可以将指定类型进行序列化，并且 Flink 的这些序列化器会以稠密的方式将对象写入内存中。Flink 中也提供了非常丰富的序列化器。在基于 Flink 类型系统支持的数据类型进行编程时，Flink 在运行时会推断出数据类型的信息，程序员在基于 Flink 编程时，几乎不需要关心类型和序列化。

4. 类型与 Lambda 表达式支持

在编译时，编译器能够从 Java 源代码中读取完整的类型信息，并强制执行类型的约束，但在生成 class 字节码时会将参数化类型信息删除，这就是类型擦除。类型擦除可以确保不会为泛型创建新的 Java 类，泛型是不会产生额外的开销的。也就是说，泛型只是在编译器编译时能够理解该类型，但编译后在执行时泛型会被擦除。

为了便于说明，参看下面的代码：

```java
public static < T > boolean hasItems(T [ ] items, T item){
    for (T i : items){
        if(i.equals(item)){
            return true;
        }
    }
    return false;
}
```

以上是一段 Java 的泛型方法，但在编译后，编译器会将未绑定类型的 T 擦除，替换为 Object，也就是编译之后的代码如下：

```java
public static Object boolean hasItems(Object [ ] items, Object item){
    for (Object i : items){
        if(i.equals(item)){
            return true;
        }
    }
    return false;
}
```

泛型只能防止在运行时出现类型错误，但运行时会出现以下异常，而且 Flink 以非常友好的方式提示：

```
could not be determined automatically, due to type erasure. You can give type information hints by using the returns(...) method on the result of the transformation call, or by letting your function implement the 'ResultTypeQueryable' interface.
```

因为Java编译器可将类型擦除,所以Flink根本无法推断算子(例如flatMap)要输出的类型是什么,所以在Flink中使用Lambda表达式时,为了防止因类型擦除而出现运行时错误,需要指定TypeInformation或者TypeHint。

创建TypeInformation,代码如下:

```
.returns(TypeInformation.of(String.class))
```

创建TypeHint,代码如下:

```
.returns(new TypeHint<String>() {})
```

3.1.2 流应用程序实现

Flink程序的基本构建块是stream和transformation(流和转换)。从概念上讲,stream是数据记录的流(可能永远不会结束),transformation是一个运算,它接收一个或多个流作为输入,经过处理/计算后生成一个或多个输出流。

下面实现一个完整的可工作的Flink流应用程序示例。

【示例3-1】 将有关人员的记录流作为输入,并从中筛选出未成年人信息。

Scala代码如下:

(1) 在IntelliJ IDEA中创建一个Flink项目,使用flink-quickstart-scala项目模板(Flink项目的创建过程可参考2.2节)。

(2) 设置依赖。在pom.xml文件中添加以下依赖内容:

```xml
<dependency>
    <groupId>org.apache.flink</groupId>
    <artifactId>flink-scala_2.12</artifactId>
    <version>1.13.2</version>
    <scope>provided</scope>
</dependency>
<dependency>
    <groupId>org.apache.flink</groupId>
    <artifactId>flink-streaming-scala_2.12</artifactId>
    <version>1.13.2</version>
    <scope>provided</scope>
</dependency>
```

(3) 创建主程序StreamingJobDemo1,编辑流处理代码如下:

```scala
// 第3章/StreamingJobDemo1.scala

import org.apache.flink.streaming.api.scala._
```

```scala
object StreamingJobDemo1 {
  //定义事件类
  case class Person(name:String, age:Integer)

  def main(args: Array[String]) {

    //设置流执行环境
    val env = StreamExecutionEnvironment.getExecutionEnvironment

    //读取数据源,构造数据流
    val people = env.fromElements(
      Person("张三", 21),
      Person("李四", 16),
      Person("王老五", 35)
    )

    //对数据流执行filter转换
    val adults = people.filter(_.age > 18)

    //输出结果
    adults.print

    //执行
    env.execute("Flink Streaming Job")
  }
}
```

执行以上代码,输出结果如下:

```
7 > Person(张三,21)
1 > Person(王老五,35)
```

Java 代码如下:

(1) 在 IntelliJ IDEA 中创建一个 Flink 项目,使用 flink-quickstart-Java 项目模板(Flink 项目的创建过程可参考 2.2 节)。

(2) 设置依赖。在 pom.xml 文件中添加以下依赖内容:

```xml
<dependency>
    <groupId>org.apache.flink</groupId>
    <artifactId>flink-Java</artifactId>
    <version>1.13.2</version>
    <scope>provided</scope>
</dependency>
<dependency>
```

```xml
    <groupId>org.apache.flink</groupId>
    <artifactId>flink-streaming-Java_2.12</artifactId>
    <version>1.13.2</version>
    <scope>provided</scope>
</dependency>
```

(3) 创建一个POJO类,用来表示流中的数据,代码如下:

```java
//第3章/Person.java

//POJO类,表示人员信息实体
public class Person {
    public String name;                //存储姓名
    public Integer age;                //存储年龄

    //空构造器
    public Person() {};

    //构造器,初始化属性
    public Person(String name, Integer age) {
        this.name = name;
        this.age = age;
    };

    //用于调试时输出信息
    public String toString() {
        return this.name.toString() + ": age " + this.age.toString();
    };
}
```

(4) 打开项目中的StreamingJob对象文件,编辑流处理代码如下:

```java
//第3章/StreamingJobDemo1.java

import org.apache.flink.streaming.api.environment.StreamExecutionEnvironment;
import org.apache.flink.streaming.api.datastream.DataStream;
import org.apache.flink.api.common.functions.FilterFunction;

public class StreamingJobDemo1 {

    public static void main(String[] args) throws Exception {
        //获得流执行环境
        final StreamExecutionEnvironment env =
                StreamExecutionEnvironment.getExecutionEnvironment();
```

```java
//读取数据源,构造 DataStream
DataStream<Person> personDS = env.fromElements(
        new Person("张三", 21),
        new Person("李四", 16),
        new Person("王老五", 35)
);

//执行转换运算(这里是过滤年龄不小于 18 岁的人)
//注意,这里使用了匿名函数
DataStream<Person> adults = personDS.filter(
        new FilterFunction<Person>() {
            @Override
            public boolean filter(Person person) throws Exception {
                return person.age >= 18;
            }
        });

//将结果输出到控制台
adults.print();

//触发流程序开始执行
env.execute("stream demo");
}
}
```

(5) 执行以上程序,输出结果如下。

```
张三: age 21
王老五: age 35
```

注意:Flink 将批处理程序作为流程序的一种特殊情况执行,其中流是有界的(有限数量的元素)。DataSet 在内部被视为数据流,因此,上述概念同样适用于批处理程序,也适用于流程序,只有少数例外:

(1) 批处理程序的容错不使用检查点。错误恢复是通过完全重放流实现的,这使恢复的成本更高,但是因为它避免了检查点,所以使常规处理更轻量。

(2) DataSet API 中的有状态运算使用简化的 in-memory/out-of-核数据结构,而不是 key-value 索引。

(3) DataSet API 引入了特殊的同步(基于 superstep)迭代,这只可能在有界流上实现。

3.1.3 流应用程序剖析

所有的 Flink 应用程序都以特定的步骤来工作,这些工作步骤如图 3-2 所示。

图 3-2 Flink 应用程序工作步骤

也就是说,每个 Flink 程序都由相同的基本部分组成:
(1) 获取一个执行环境。
(2) 加载/创建初始数据。
(3) 指定对该数据的转换。
(4) 指定将计算结果放在哪里。
(5) 触发流程序执行。

1. 获取一个执行环境

Flink 应用程序从其 main() 方法中生成一个或多个 Flink 作业(job)。这些作业可以在本地 JVM(LocalEnvironment)中执行,也可以在具有多台机器的集群的远程设置中执行(RemoteEnvironment)。对于每个程序,ExecutionEnvironment 提供了控制作业执行(例如设置并行性或容错/检查点参数)和与外部环境交互(数据访问)的方法。

每个 Flink 应用程序都需要一个执行环境(本例中为 env)。流应用程序需要的执行环境使用的是 StreamExecutionEnvironment。为了开始编写 Flink 程序,用户首先需要获得一个现有的执行环境,如果没有,就需要先创建一个。根据目的的不同,Flink 支持以下几种方式:
(1) 获得一个已经存在的 Flink 环境。
(2) 创建本地环境。
(3) 创建远程环境。

Flink 流程序的入口是 StreamExecutionEnvironment 类的一个实例,它定义了程序执行的上下文。StreamExecutionEnvironment 是所有 Flink 程序的基础。可以通过一些静态方法获得一个 StreamExecutionEnvironment 的实例,代码如下:

```
StreamExecutionEnvironment.getExecutionEnvironment()
StreamExecutionEnvironment.createLocalEnvironment()
StreamExecutionEnvironment.createRemoteEnvironment(String host, int port, String... jarFiles)
```

要获得执行环境,通常只需调用 getExecutionEnvironment() 方法。这将根据上下文选择正确的执行环境。如果正在 IDE 中的本地环境上执行,则它将启动一个本地执行环境。如果是从程序中创建了一个 JAR 文件,并通过命令行调用它,则 Flink 集群管理器将执行 main() 方法,getExecutionEnvironment() 将返回用于在集群上以分布式方式执行程序的执行环境。

在上面的示例程序中,使用以下语句来获得流程序的执行环境。
Scala 代码如下:

```
//设置流执行环境
val env = StreamExecutionEnvironment.getExecutionEnvironment
```

Java 代码如下:

```
//获得流执行环境
final StreamExecutionEnvironment env =
        StreamExecutionEnvironment.getExecutionEnvironment();
```

StreamExecutionEnvironment 包含 ExecutionConfig,可使用它为运行时设置特定于作业的配置值。例如,如果要设置自动水印发送间隔,则可以像下面这样在代码进行配置。

Scala 代码如下:

```
val env = StreamExecutionEnvironment.getExecutionEnvironment
env.getConfig.setAutoWatermarkInterval(long milliseconds)
```

Java 代码如下:

```
final StreamExecutionEnvironment env =
        StreamExecutionEnvironment.getExecutionEnvironment();
env.getConfig().setAutoWatermarkInterval(long milliseconds);
```

2. 加载/创建初始数据

执行环境可以从多种数据源读取数据,包括文本文件、CSV 文件、Socket 套接字数据等,也可以使用自定义的数据输入格式。例如,要将文本文件读取为行序列,代码如下:

```
final StreamExecutionEnvironment env =
    StreamExecutionEnvironment.getExecutionEnvironment();
DataStream<String> text = env.readTextFile("file://path/to/file");
```

数据被逐行读到内存后,Flink 会将它们组织到 DataStream 中,这是 Flink 中用来表示流数据的特殊类。

在 3.1.2 节的示例程序【示例 3-1】中,使用 fromElements()方法读取集合数据,并将读取的数据存储为 DataStream 类型。

Scala 代码如下:

```
//读取数据源,构造数据流
val personDS = env.fromElements(
    Person("张三", 21),
    Person("李四", 16),
    Person("王老五", 35)
)
```

Java代码如下：

```
//读取数据源,构造 DataStream
DataStream<Person> personDS = env.fromElements(
                                new Person("张三", 21),
                                new Person("李四", 16),
                                new Person("王老五", 35));
```

3. 对数据进行转换

每个 Flink 程序都对分布式数据集合执行转换。Flink 的 DataStream API 提供了多种数据转换功能,包括过滤、映射、连接、分组和聚合。例如,下面是一个 map 转换应用,通过将原始集合中的每个字符串转换为整数来创建一个新的 DataStream,代码如下:

```
DataStream<String> input = env.fromElements("12","3","25","5","32","6");

DataStream<Integer> parsed = input.map(new MapFunction<String, Integer>() {
    @Override
    public Integer map(String value) {
        return Integer.parseInt(value);
    }
});
```

在 3.1.2 节的示例程序【示例 3-1】中使用了 filter 过滤转换,将原始数据集转换为只包含成年人信息的新 DataStream 流。

Scala 代码如下：

```
//对数据流执行 filter 转换
val adults = personDS.filter(_.age > 18)
```

Java 代码如下：

```
//对数据流执行 filter 转换
DataStream<Person> adults = flintstones.filter(
    new FilterFunction<Person>() {
        @Override
        public boolean filter(Person person) throws Exception {
            return person.age >= 18;
        }
    });
```

这里不必了解每个转换的具体含义,后面会详细介绍它们。需要强调的是,Flink 中的转换是惰性的,在调用 sink 操作之前不会真正执行。

4. 指定将计算结果放在哪里

一旦有了包含最终结果的 DataStream，就可以通过创建接收器（sink）将其写入外部系统。例如，将计算结果输出到屏幕上。

Scala 代码如下：

```
//输出结果
adults.print
```

Java 代码如下：

```
//输出结果
adults.print();
```

Flink 中的接收器操作触发流的执行，以生成程序所需的结果，例如将结果保存到文件系统或将其打印到标准输出。上面的示例使用 adults.print() 将结果打印到任务管理器日志中（在 IDE 中运行时，任务管理器日志将显示在 IDE 的控制台中）。这将对流的每个元素调用其 toString() 方法。

5. 触发流程序执行

一旦写好了程序处理逻辑，就需要通过调用 StreamExecutionEnvironment 上的 execute() 来触发程序执行。所有的 Flink 程序都是时延执行的：当程序的主方法执行时，数据加载和转换不会直接发生，而是创建每个运算并添加到程序的执行计划中。当执行环境上的 execute() 调用显式触发执行时，这些操作才真正被执行。程序是在本地执行还是提交到集群中执行取决于 ExecutionEnvironment 的类型。

时延计算可以让用户构建复杂的程序，然后 Flink 将其作为一个整体计划的单元执行。在 3.1.2 节的示例程序【示例 3-1】中，使用如下代码来触发流处理程序的执行。

Scala 代码如下：

```
//触发流程序执行
env.execute("Flink Streaming Job")          //参数是程序名称,会显示在 Web UI 界面上
```

Java 代码如下：

```
//触发流程序执行
env.execute("Flink Streaming Job");          //参数是程序名称,会显示在 Web UI 界面上
```

在应用程序中执行的 DataStream API 调用将构建一个附加到 StreamExecutionEnvironment 的作业图（Job Graph）。调用 env.execute() 时，此图被打包并发送到 Flink Master，该 Master 并行化作业并将其片段分发给 TaskManagers 以供执行。作业的每个并行片段将在一个 Task Slot（任务槽）中执行，如图 3-3 所示。

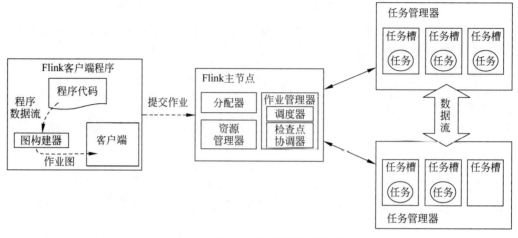

图 3-3　Flink 流应用程序执行原理

这个分布式运行时要求 Flink 应用程序是可序列化的。它还要求集群中的每个节点都可以使用所有依赖项。

StreamExecutionEnvironment 上的 execute() 方法将等待作业完成，然后返回一个 JobExecutionResult，其中包含执行时间和累加器结果。注意，如果不调用 execute()，则应用程序将不会运行。

如果不想等待作业完成，则可以通过调用 StreamExecutionEnvironment 上的 executeAysnc() 来触发异步作业执行。它将返回一个 JobClient，可以使用它与刚才提交的作业进行通信。例如，下面的示例代码演示了如何通过 executeAsync() 实现 execute() 的语义。

Scala 代码如下：

```scala
val jobClient = evn.executeAsync
val jobExecutionResult =
        jobClient.getJobExecutionResult(userClassloader).get
```

Java 代码如下：

```java
final JobClient jobClient = env.executeAsync();
final JobExecutionResult jobExecutionResult =
        jobClient.getJobExecutionResult(userClassloader).get();
```

3.2　Flink 支持的数据源

数据源是 Flink 程序希望从其中获取数据的地方。可以从多个源创建数据集，如 Apache Kafka、CSV、文件或几乎任何其他数据源。Flink 附带了许多预先实现的数据源函数。它还支持编写自定义数据源函数。

首先理解 Flink 内置的数据源函数。可以从 StreamExecutionEnvironment 访问预定义的数据源。

> **注意**：在实际应用程序中，最常用的数据源是那些支持低时延、高吞吐量并行读取及重放和 rewind 的数据源，这是实现高性能和容错的先决条件，例如 Apache Kafka、Kinesis 和各种文件系统。

3.2.1 基于 Socket 的数据源

DataStream API 支持从 Socket 套接字读取数据。只需指定要从其中读取数据的主机和端口号。读取 Socket 套接字的数据源函数的定义如下。

(1) socketTextStream(hostName, port)：指定主机和端口号。

(2) socketTextStream(hostName, port, delimiter)：可指定分隔符。

(3) socketTextStream(hostName, port, delimiter, maxRetry)：可指定 API 应该尝试获取数据的最大次数。

【示例 3-2】 Socket 数据源：流应用程序示例，它接收来自 Web 套接字的单词。

Scala 代码如下：

```scala
//第 3 章/SocketSourceDemo.scala

import org.apache.flink.streaming.api.scala._

object SocketSourceDemo {

  val HOST = "localhost"              //host 主机
  val PORT = 9999                     //端口号

  def main(args: Array[String]): Unit = {
    //设置流执行环境
    val env = StreamExecutionEnvironment.getExecutionEnvironment

    //连接 Socket 数据源
    val input = env.socketTextStream(HOST, PORT)

    //流数据转换,并打印结果
    input
      .map(_.toLowerCase)
      .flatMap(_.split("\\W+"))
      .print
```

```
        //触发流程序执行
        env.execute("Flink Socket Source")
    }
}
```

Java实现：

```java
//第3章/SocketSourceDemo.java

import org.apache.flink.api.common.functions.MapFunction;
import org.apache.flink.api.common.functions.FlatMapFunction;
import org.apache.flink.api.java.tuple.Tuple2;
import org.apache.flink.streaming.api.datastream.DataStream;
import org.apache.flink.streaming.api.environment.StreamExecutionEnvironment;
import org.apache.flink.streaming.api.windowing.time.Time;
import org.apache.flink.util.Collector;

public class SocketSourceDemo {

    public static void main(String[] args) throws Exception {
        //设置流执行环境
        final StreamExecutionEnvironment env =
                StreamExecutionEnvironment.getExecutionEnvironment();

        //源数据流
        String host = "localhost";          //host主机或IP地址
        int port = 9999;                    //端口号
        DataStream<String> input = env.socketTextStream(host, port);

        //对DataStream进行转换,向map()和flatMap()传入匿名内容类
        input.map(new MapFunction<String, String>() {
            @Override
            public String map(String s) throws Exception {
                return s.toLowerCase();     //先转小写
            }
        }).flatMap(new FlatMapFunction<String, String>() {
            @Override
            public void flatMap(String s, Collector<String> collector) throws Exception {
                for(String word : s.split("\\W+")){     //再分词
                    collector.collect(word);            //向下游发送
                }
            }
        }).print();

        //执行流程序
```

```
        env.execute("Flink Socket Source");
    }
}
```

建议按以下步骤执行这个流程序。

(1) 启动一个运行在9999端口的netcat服务器。打开一个终端窗口,执行的命令如下:

```
$ nc -lk 9999
```

(2) 运行上面编写的流应用程序。

(3) 在netcat运行窗口,输入以下内容,并按Enter键:

```
good good study
day day up
```

(4) 在程序执行窗口,可以看到计算后的输出如下:

```
5> good
5> good
5> study
6> day
6> day
6> up
```

在上面的Java示例代码中,也可以将map()的逻辑代码合并到flatMap()函数中,以精简代码。修改后的代码如下:

```java
//第3章/SocketSourceDemo2.java

import org.apache.flink.api.common.functions.FlatMapFunction;
import org.apache.flink.api.java.tuple.Tuple2;
import org.apache.flink.streaming.api.datastream.DataStream;
import org.apache.flink.streaming.api.environment.StreamExecutionEnvironment;
import org.apache.flink.streaming.api.windowing.time.Time;
import org.apache.flink.util.Collector;

public class SocketSourceDemo2 {

    public static void main(String[] args) throws Exception {
        //设置流执行环境
        final StreamExecutionEnvironment env =
                StreamExecutionEnvironment.getExecutionEnvironment();
```

```java
//源数据流
String host = "localhost";           //host 主机或 IP 地址
int port = 9999;                     //端口号
DataStream<String> input = env.socketTextStream(host, port);

//对 DataStream 进行转换,向 map()和 flatMap()传入匿名内容类
input.flatMap(new FlatMapFunction<String, String>() {
    @Override
    public void flatMap(String s, Collector<String> collector) throws Exception {
        String line = s.toLowerCase();              //先转小写
        for(String word : line.split("\\W+")){      //再分词
            collector.collect(word);                //向下游发送
        }
    }
}).print();

//执行流程序
env.execute("Flink Socket Source");
    }
}
```

在前面的 Java 代码示例中,向 map()和 flatMap()传入的都是匿名内部类。也可以代之以 Java 1.8 引入的 Lambda 函数,以简化代码的编写。使用 Lambda 函数重构的代码如下:

```java
//第 3 章/SocketSourceDemo3.java

import org.apache.flink.api.common.functions.FlatMapFunction;
import org.apache.flink.api.java.tuple.Tuple2;
import org.apache.flink.streaming.api.datastream.DataStream;
import org.apache.flink.streaming.api.environment.StreamExecutionEnvironment;
import org.apache.flink.streaming.api.windowing.time.Time;
import org.apache.flink.util.Collector;

public class SocketSourceDemo3 {

    public static void main(String[] args) throws Exception {
        //设置流执行环境
        final StreamExecutionEnvironment env =
            StreamExecutionEnvironment.getExecutionEnvironment();

        //源数据流
        String host = "localhost";           //host 主机或 IP 地址
        int port = 9999;                     //端口号
        DataStream<String> input = env.socketTextStream(host, port);
```

```
        //对 DataStream 进行转换,向 map()和 flatMap()传入 Lambda 函数
        input.flatMap((FlatMapFunction<String, String>) (s, collector) -> {
            String line = s.toLowerCase();              //先转小写
            for(String word : line.split("\\W+")){      //再分词
                collector.collect(word);                //向下游发送
            }
        }).returns(Types.STRING) //因为类型擦除的原因,所以必须提供返回类型
        .print();

        //执行流程序
        env.execute("Flink Socket Source");
    }
}
```

3.2.2 基于文件的数据源

还可以选择使用基于文件的源函数从文件源中传输数据。从文件源读取数据的源函数定义有多个,包括:

(1) readTextFile(String path):逐行读取路径指定的文本文件,即符合 TextInputFormat 规范的文本文件,并以字符串的形式返回。

(2) readFile(FileInputFormat inputFormat, String path):根据指定的文件输入格式读取(一次)文件。

(3) readFile(fileInputFormat, path, watchType, interval, pathFilter):这是前两种方法在内部调用的方法。它根据给定的 fileInputFormat 读取路径中的文件。根据所提供的 watchType,此源可以定期(FileProcessingMode.PROCESS_CONTINUOUSLY,每隔 interval 毫秒)监视新数据的路径,或处理一次(FileProcessingMode.PROCESS_ONCE)当前路径中的数据并退出。使用 pathFilter 进一步排除正在处理的文件。

对于第 3 种方法,在底层,Flink 将文件读取过程分成两个子任务,即目录监视和数据读取。每个子任务都由一个单独的实体实现。监视文件路径的子任务是由单个非并行(并行度=1)任务实现的,而文件读取则由多个并行运行的任务执行,并行度等于作业并行度。单个监视任务的作用是扫描目录(定期或仅扫描一次,这取决于 watchType),查找要处理的文件,将它们划分为分段,并将这些分段分配给下游的读取器,如图 3-4 所示。

图 3-4 文件数据源

读取器将读取实际数据。每个分段只由一个读取器读取,而一个读取器可以逐个读取多个分段。

【示例 3-3】 编写 Flink 流应用程序,读取文件,并实时统计文件内的单词数量。建议按以下步骤实现:

(1) 在 IDEA 中创建 Flink 项目。

(2) 在项目的 src 上右击,创建一个名为 wc.txt 的文本文件,如图 3-5 所示。

图 3-5 创建数据源文件 wc.txt

(3) 编辑 wc.txt 文件,输入以下内容并保存:

```
good good study
day day up
```

(4) 流处理代码实现。

Scala 代码如下:

```scala
/第 3 章/FileSourceDemo.scala

import org.apache.flink.streaming.api.scala._

object FileSourceDemo {
  def main(args: Array[String]): Unit = {
    //设置流执行环境
    val env = StreamExecutionEnvironment.getExecutionEnvironment

    //加载文件数据源,构造 DataStream
    val textPath = "wc.txt"
    val text = env.readTextFile(textPath)

    //对 DataStream 执行转换操作
    text
      .flatMap { _.toLowerCase.split("\\W+").filter( _.nonEmpty ) }
      .map { (_, 1) }
      .print()

    //触发流程序执行
```

```
    env.execute("Simple Flink File Source Demo")
  }
}
```

Java 代码如下：

```java
//第 3 章/FileSourceDemo.java

import org.apache.flink.api.common.functions.FlatMapFunction;
import org.apache.flink.api.java.tuple.Tuple2;
import org.apache.flink.streaming.api.datastream.DataStreamSource;
import org.apache.flink.streaming.api.environment.StreamExecutionEnvironment;
import org.apache.flink.util.Collector;

public class FileSourceDemo {
    public static void main(String[] args) throws Exception {
        //设置流执行环境
        final StreamExecutionEnvironment env =
                StreamExecutionEnvironment.getExecutionEnvironment();

        //加载文件数据源,构造 DataStream
        String textPath = "wc.txt";
        DataStreamSource<String> text = env.readTextFile(textPath);

        //数据流转换
        text.map(String::toLowerCase)          //转小写
            .flatMap(new Splitter())           //flatMap()转换
            .print();

        //触发流程序执行
        env.execute("Simple Flink File Source Demo");
    }

    //实现 FlatMapFunction 接口的函数
    public static class Splitter
            implements FlatMapFunction<String, Tuple2<String, Integer>> {
        @Override
        public void flatMap(String sentence, Collector<Tuple2<String, Integer>> out) throws Exception {
            for (String word: sentence.split(" ")) {    //分词
                out.collect(new Tuple2<>(word, 1));     //构造(word,1)发送给下游算子
            }
        }
    }
}
```

执行以上程序,输出的结果如下:

```
2> (day,1)
6> (good,1)
2> (day,1)
6> (good,1)
2> (up,1)
6> (study,1)
```

3.2.3 基于集合的数据源

也可以直接在内存中将一个数据集合读取为 DataStream。Flink 提供了以下几种方法:

(1) fromCollection(Seq):从 Java Java.util.collection 创建一个数据流。集合中的所有元素必须具有相同的类型。

(2) fromCollection(Iterator):从迭代器创建数据流。该类用于指定迭代器返回的元素的数据类型。

(3) fromElements(elements:_*):根据给定的对象序列创建数据流。所有对象必须具有相同的类型。

(4) fromParallelCollection(SplittableIterator):并行地从迭代器创建数据流。该类用于指定迭代器返回的元素的数据类型。

(5) generateSequence(from,to):并行地生成给定区间内的数字序列。

注意:目前,集合数据源要求数据类型和迭代器实现 Serializable。此外,集合数据源不能并行执行(并行度=1)。

【示例 3-4】 使用集合数据源,实时找出年龄超过 18 岁的人员。
建议按以下步骤操作:

(1) 在 IntelliJ IDEA 中创建一个 Flink 项目,使用 flink-quickstart-scala/flink-quickstart-Java 项目模板。

(2) 打开项目中的 StreamingJob 对象文件,编辑流处理代码。
Scala 代码如下:

```scala
//第3章/CollectionSourceDemo.scala

import org.apache.flink.streaming.api.scala._

object CollectionSourceDemo {
```

```scala
//定义事件类
case class Person(name:String, age:Integer)

def main(args: Array[String]): Unit = {
    //设置流执行环境
    val env = StreamExecutionEnvironment.getExecutionEnvironment

    //首先从环境中获取一些数据,例如
    val people = List(Person("张三", 21),Person("李四", 16),Person("王老五", 35))
    val personStream = env.fromCollection(people)

    //对 DataStream 执行转换操作,并输出计算结果
    personStream.filter(_.age > 18).print()

    //触发流程序执行
    env.execute("Flink Collection Source Demo")
}
}
```

Java 代码如下：

```java
//第 3 章/CollectionSourceDemo.java

import org.apache.flink.api.common.functions.FilterFunction;
import org.apache.flink.streaming.api.datastream.DataStream;
import org.apache.flink.streaming.api.environment.StreamExecutionEnvironment;
import Java.util.ArrayList;
import Java.util.List;

public class CollectionSourceDemo {

    //POJO 类
    public static class Person {
        public String name;
        public Integer age;

        public Person() {}

        public Person(String name, Integer age) {
            this.name = name;
            this.age = age;
        }

        public String toString() {
            return this.name.toString() + ": age " + this.age.toString();
```

```java
        }
    }

    public static void main(String[] args) throws Exception {
        //设置流执行环境
        final StreamExecutionEnvironment env = 
                StreamExecutionEnvironment.getExecutionEnvironment();

        //读取数据源,构造 DataStream
        List<Person> people = new ArrayList<>();
        people.add(new Person("张三", 21));
        people.add(new Person("李四", 16));
        people.add(new Person("王老五", 35));

        DataStream<Person> personStream = env.fromCollection(people);

        //对 DataStream 执行 filter 转换操作
        DataStream<Person> adults = personStream.filter(
          new FilterFunction<Person>() {
            @Override
            public boolean filter(Person person) throws Exception {
                return person.age >= 18;
            }
          });

        //输出流计算结果
        adults.print();

        //触发流程序执行
        env.execute("Flink File Source");
    }
}
```

(3) 执行以上程序,输出的结果如下:

```
张三: age 21
王老五: age 35
```

3.2.4 自定义数据源

除了内置数据源,用户还可以编写自己的定制数据源。对于非并行数据源,实现 SourceFunction;对于并行数据源,实现 ParallelSourceFunction 接口或扩展(继承)自 RichParallelSourceFunction。

1. 了解 SourceFunction 接口

SourceFunction 是 Flink 中所有流数据源的基本接口。SourceFunction 接口的定义如下：

```java
//第3章/SourceFunction.java
package org.apache.flink.streaming.api.functions.source;

import Java.io.Serializable;
import org.apache.flink.annotation.Public;
import org.apache.flink.annotation.PublicEvolving;
import org.apache.flink.api.common.functions.Function;
import org.apache.flink.streaming.api.watermark.Watermark;

@Public
public interface SourceFunction<T> extends Function, Serializable {
    void run(SourceFunction.SourceContext<T> var1) throws Exception;

    void cancel();

    @Public
    public interface SourceContext<T> {
        void collect(T var1);

        @PublicEvolving
        void collectWithTimestamp(T var1, long var2);

        @PublicEvolving
        void emitWatermark(Watermark var1);

        @PublicEvolving
        void markAsTemporarilyIdle();

        Object getCheckpointLock();

        void close();
    }
}
```

从上面的接口定义中可知，在 SourceFunction 接口中定义了 run() 和 cancel() 两种方法以及一个内部接口 SourceContext，其中 run(SourceContex) 方法用来实现数据获取逻辑，并可以通过传入的参数 ctx 向下游节点进行数据转发。cancel() 方法用来取消数据源，一般在 run() 方法中会存在一个循环来持续产生数据，cancel() 方法则可以使该循环终止。SourceContext 内部接口用于发出元素和可能的 watermark 的接口。

2. run()方法

在run()方法中实现了数据源向下游发送数据的主要逻辑。编写模式如下：

(1) 不断调用，以便实现循环发送数据。

(2) 使用一种状态变量控制循环的执行。当cancel()方法执行后必须能够跳出循环，以便停止发送数据。

(3) 使用SourceContext的collect()等方法将元素发送至下游。

(4) 如果使用检查点，在SourceContext收集数据时必须加锁。防止checkpoint操作和发送数据操作同时进行。

3. cancel()方法

在数据源停止时调用cancel()方法。cancel()方法必须能够控制run()方法中的循环，即停止循环的运行，并做一些状态清理操作。

4. SourceContext类

SourceContext在SourceFunction中使用，用于向下游发送数据，或者发送水印。SourceContext的方法包括

(1) collect()方法：向下游发送数据。有以下3种情况：

① 如果使用ProcessingTime，则该元素不携带timestamp。

② 如果使用IngestionTime，则元素使用系统当前时间作为timestamp。

③ 如果使用EventTime，则元素不携带timestamp。需要在数据流后续为元素指定timestamp(assignTimestampAndWatermark)。

(2) collectWithTimestamp()方法：向下游发送带有timestamp的数据。和collect()方法一样也有以下3种情况：

① 如果使用ProcessingTime，则timestamp会被忽略。

② 如果使用IngestionTime，则使用系统时间覆盖timestamp。

③ 如果使用EventTime，则使用指定的timestamp。

(3) emitWatermark()方法：向下游发送watermark。watermark也包含一个timestamp。向下游发送watermark意味着所有在watermark的timestamp之前的数据已经到齐。如果在watermark之后，收到了timestamp比该watermark的timestamp小的元素，该元素会被认为迟到，将会被系统忽略，或者进入侧输出(Side Output)。

(4) markAsTemporarilyIdle()方法：将此数据源暂时标记为闲置。该数据源暂时不会发送任何数据和watermark。仅对IngestionTime和EventTime生效。下游任务前移watermark时将不会再等待被标记为闲置的数据源的watermark。

5. CheckpointedFunction

如果数据源需要保存状态，就需要实现CheckpointedFunction中的相关方法。CheckpointedFunction包含的方法如下。

(1) snapshotState()：保存checkpoint时调用。需要在此方法中编写状态保存逻辑。

(2) initializeState()：在数据源创建或者从 checkpoint 恢复时调用。此方法包含数据源的状态恢复逻辑。

对于自定义的数据源，需要使用 StreamExecutionEnvironment. addSource(sourceFunction) 方法将指定数据源附加到程序中。例如，如果要从 Apache Kafka 中读取数据，则可以使用的模板代码如下：

```
addSource(new FlinkKafkaConsumer<>(…))
```

【示例 3-5】（简单版本）使用自定义数据源，模拟信用卡交易流数据生成器。

建议按以下步骤操作：

(1) 在 IntelliJ IDEA 中创建一个 Flink 项目，使用 flink-quickstart-scala/flink-quickstart-Java 项目模板。

(2) 设置依赖。在 pom.xml 文件中添加如下依赖：

```xml
<dependency>
    <groupId>org.apache.flink</groupId>
    <artifactId>flink-scala_2.12</artifactId>
    <version>1.13.2</version>
    <scope>provided</scope>
</dependency>
<dependency>
    <groupId>org.apache.flink</groupId>
    <artifactId>flink-streaming-scala_2.12</artifactId>
    <version>1.13.2</version>
    <scope>provided</scope>
</dependency>
<dependency>
    <groupId>org.apache.flink</groupId>
    <artifactId>flink-Java</artifactId>
    <version>1.13.2</version>
    <scope>provided</scope>
</dependency>
<dependency>
    <groupId>org.apache.flink</groupId>
    <artifactId>flink-streaming-Java_2.12</artifactId>
    <version>1.13.2</version>
    <scope>provided</scope>
</dependency>
```

(3) 创建 POJO 类，表示信用卡交易数据结构，代码如下：

```
//第 3 章/Transaction.java

import Java.util.Objects;
```

```java
/**
 * 实体类,代表信用卡交易数据
 */
public class Transaction implements Serializable {
    public long accountId;          //交易账户
    public long timestamp;          //交易时间
    public double amount;           //交易金额

    public Transaction() { }

    public Transaction(long accountId, long timestamp, double amount) {
        this.accountId = accountId;
        this.timestamp = timestamp;
        this.amount = amount;
    }

    public long getAccountId() {
        return accountId;
    }

    public void setAccountId(long accountId) {
        this.accountId = accountId;
    }

    public long getTimestamp() {
        return timestamp;
    }

    public void setTimestamp(long timestamp) {
        this.timestamp = timestamp;
    }

    public double getAmount() {
        return amount;
    }

    public void setAmount(double amount) {
        this.amount = amount;
    }

    @Override
    public boolean equals(Object o) {
        if (this == o) {
            return true;
        } else if (o == null || getClass() != o.getClass()) {
            return false;
```

```java
        }
        Transaction that = (Transaction) o;
        return accountId == that.accountId &&
                timestamp == that.timestamp &&
                Double.compare(that.amount, amount) == 0;
    }

    @Override
    public int hashCode() {
        return Objects.hash(accountId, timestamp, amount);
    }

    @Override
    public String toString() {
        return "Transaction{" +
                "accountId=" + accountId +
                ", timestamp=" + timestamp +
                ", amount=" + amount +
                '}';
    }
}
```

(4) 创建自定义的数据源类,继承自 SourceFunction,代码如下:

```java
//第3章/MyTransactionSource.java

import com.xueai8.fraud.entity.Transaction;
import org.apache.flink.streaming.api.functions.source.SourceFunction;

import Java.sql.Timestamp;
import Java.util.Arrays;
import Java.util.List;

/**
 * 自定义数据源,继承自 SourceFunction
 */
public class MyTransactionSource implements SourceFunction<Transaction> {
    private static final long serialVersionUID = 1L;

    private static final Timestamp INITIAL_TIMESTAMP =
                    Timestamp.valueOf("2020-01-01 00:00:00");
    private static final long SIX_MINUTES = 6 * 60 * 1000;

    private final boolean bounded;          //标志变量,指示生成流数据还是批数据
```

```java
    private int index = 0;                    //交易记录的索引
    private long timestamp;                   //交易发生的时间戳
    
    private volatile boolean isRunning = true;
    private List<Transaction> data = null;
    
    public MyTransactionSource(boolean bounded){
        this.bounded = bounded;
        this.timestamp = INITIAL_TIMESTAMP.getTime();
        
        //事先存储的信用卡交易数据,在实际中来自外部数据源系统,如Kafka
        data = Arrays.asList(
                new Transaction(1, 0L, 188.23),
                new Transaction(2, 0L, 374.79),
                new Transaction(3, 0L, 112.15),
                new Transaction(4, 0L, 478.75),
                new Transaction(5, 0L, 208.85),
                new Transaction(1, 0L, 379.64),
                new Transaction(2, 0L, 351.44),
                new Transaction(3, 0L, 320.75),
                new Transaction(4, 0L, 259.42),
                new Transaction(5, 0L, 273.44),
                new Transaction(1, 0L, 267.25),
                new Tran7saction(2, 0L, 397.15),
                new Transaction(3, 0L, 0.219),
                new Transaction(4, 0L, 231.94),
                new Transaction(5, 0L, 384.73),
                new Transaction(1, 0L, 419.62),
                new Transaction(2, 0L, 412.91),
                new Transaction(3, 0L, 0.77),
                new Transaction(4, 0L, 22.10),
                new Transaction(5, 0L, 377.54),
                new Transaction(1, 0L, 375.44),
                new Transaction(2, 0L, 230.18),
                new Transaction(3, 0L, 0.80),
                new Transaction(4, 0L, 350.89),
                new Transaction(5, 0L, 127.55),
                new Transaction(1, 0L, 483.91),
                new Transaction(2, 0L, 228.22),
                new Transaction(3, 0L, 871.15),
                new Transaction(4, 0L, 64.19),
                new Transaction(5, 0L, 79.43),
                new Transaction(1, 0L, 56.12),
                new Transaction(2, 0L, 256.48),
                new Transaction(3, 0L, 148.16),
                new Transaction(4, 0L, 199.95),
```

```java
            new Transaction(5, 0L, 252.37),
            new Transaction(1, 0L, 274.73),
            new Transaction(2, 0L, 473.54),
            new Transaction(3, 0L, 119.92),
            new Transaction(4, 0L, 323.59),
            new Transaction(5, 0L, 353.16),
            new Transaction(1, 0L, 211.90),
            new Transaction(2, 0L, 280.93),
            new Transaction(3, 0L, 347.89),
            new Transaction(4, 0L, 459.86),
            new Transaction(5, 0L, 82.31),
            new Transaction(1, 0L, 373.26),
            new Transaction(2, 0L, 479.83),
            new Transaction(3, 0L, 454.25),
            new Transaction(4, 0L, 83.64),
            new Transaction(5, 0L, 292.44)
    );
}

@Override
public void run(SourceContext<Transaction> sourceContext) throws Exception {
    while(this.isRunning && this.hasNext()) {
        sourceContext.collect(this.next());
    }
}

@Override
public void cancel() {
    this.isRunning = false;
}

private boolean hasNext() {
    //如果还有数据
    if (index < data.size()) {
        return true;
    }
    //如果用于生成批数据
    else if(bounded){
        return false;
    }
    //如果用于生成流数据,则从头循环
    else {
        index = 0;
        return true;
    }
```

```
        }

        //生成下一个交易数据,交易时间相隔为6min
        private Transaction next() {
            try {
                Thread.sleep(100);
            } catch (InterruptedException e) {
                throw new RuntimeException(e);
            }
            Transaction transaction = data.get(index++);
            transaction.setTimestamp(timestamp);
            timestamp += SIX_MINUTES;
            return transaction;
        }
    }
```

(5) 创建一个测试类(带有main方法的主程序),使用addSource()方法添加自定义数据源,编辑代码。

Scala 代码如下:

```
//第 3 章/CustomSourceDemo.scala

import com.xueai8.ch03.source.MyTransactionSource
import org.apache.flink.streaming.api.scala._

/**
 * 自定义数据源的测试类
 */
object CustomSourceDemo {

    def main(args: Array[String]): Unit = {
        //获取流执行环境
        val env = StreamExecutionEnvironment.getExecutionEnvironment

        //添加自定义数据源
        val transactions = env
            .addSource(new MyTransactionSource(false))
            .name("transactions")

        //输出交易记录
        transactions.print

        //触发流程序执行
        env.execute("Flink Custom Source")
    }
}
```

Java 代码如下:

```java
//第 3 章/CustomSourceDemo.java

import com.xueai8.ch03.entity.Transaction;
import com.xueai8.ch03.source.MyTransactionSource;
import org.apache.flink.streaming.api.datastream.DataStream;
import org.apache.flink.streaming.api.environment.StreamExecutionEnvironment;

/**
 * 自定义数据源测试类
 */
public class CustomSourceDemo {

    public static void main(String[] args) throws Exception {
        //设置流执行环境
        final StreamExecutionEnvironment env =
                StreamExecutionEnvironment.getExecutionEnvironment();

        //设置自定义数据源.参数 false 用于指定创建的是流数据源
        DataStream<Transaction> transactions = env
                .addSource(new MyTransactionSource(false))
                .name("transactions");

        //输出查看
        transactions.print();

        //执行流程序
        env.execute("Transaction Stream");
    }
}
```

(6) 执行以上程序, 查看控制台, 输出结果如下(部分结果):

```
1> Transaction{accountId = 1, timestamp = 1577808000000, amount = 188.23}
2> Transaction{accountId = 2, timestamp = 1577808360000, amount = 374.79}
3> Transaction{accountId = 3, timestamp = 1577808720000, amount = 112.15}
4> Transaction{accountId = 4, timestamp = 1577809080000, amount = 478.75}
5> Transaction{accountId = 5, timestamp = 1577809440000, amount = 208.85}
6> Transaction{accountId = 1, timestamp = 1577809800000, amount = 379.64}
7> Transaction{accountId = 2, timestamp = 1577810160000, amount = 351.44}
8> Transaction{accountId = 3, timestamp = 1577810520000, amount = 320.75}
1> Transaction{accountId = 4, timestamp = 1577810880000, amount = 259.42}
2> Transaction{accountId = 5, timestamp = 1577811240000, amount = 273.44}
3> Transaction{accountId = 1, timestamp = 1577811600000, amount = 267.25}
4> Transaction{accountId = 2, timestamp = 1577811960000, amount = 397.15}
5> Transaction{accountId = 3, timestamp = 1577812320000, amount = 0.219}
```

```
6 > Transaction{accountId = 4, timestamp = 1577812680000, amount = 231.94}
7 > Transaction{accountId = 5, timestamp = 1577813040000, amount = 384.73}
8 > Transaction{accountId = 1, timestamp = 1577813400000, amount = 419.62}
......
```

可以对上面的代码加以修改,创建更加通用的自定义数据源类。

【示例 3-6】（复杂版本）使用自定义数据源,模拟信用卡交易流数据生成器:

建议按以下步骤操作:

(1) 在 IntelliJ IDEA 中创建一个 Flink 项目,使用 flink-quickstart-Java 项目模板。

(2) 设置依赖。在 pom.xml 文件中添加的依赖如下:

```xml
<dependency>
    <groupId>org.apache.flink</groupId>
    <artifactId>flink-scala_2.12</artifactId>
    <version>${flink.version}</version>
    <scope>provided</scope>
</dependency>
<dependency>
    <groupId>org.apache.flink</groupId>
    <artifactId>flink-streaming-scala_2.12</artifactId>
    <version>${flink.version}</version>
    <scope>provided</scope>
</dependency>
<dependency>
    <groupId>org.apache.flink</groupId>
    <artifactId>flink-Java</artifactId>
    <version>1.13.2</version>
    <scope>provided</scope>
</dependency>
<dependency>
    <groupId>org.apache.flink</groupId>
    <artifactId>flink-streaming-Java_2.12</artifactId>
    <version>1.13.2</version>
    <scope>provided</scope>
</dependency>
```

(3) 创建 POJO 类,表示信用卡交易数据结构,代码如下:

```java
//第3章/Transaction.java

import Java.util.Objects;

/**
 * 实体类,代表信用卡交易
```

```java
*/
public class Transaction {
    public long accountId;          //交易账户
    public long timestamp;          //交易时间
    public double amount;           //交易金额

    public Transaction() { }

    public Transaction(long accountId, long timestamp, double amount) {
        this.accountId = accountId;
        this.timestamp = timestamp;
        this.amount = amount;
    }

    public long getAccountId() {
        return accountId;
    }

    public void setAccountId(long accountId) {
        this.accountId = accountId;
    }

    public long getTimestamp() {
        return timestamp;
    }

    public void setTimestamp(long timestamp) {
        this.timestamp = timestamp;
    }

    public double getAmount() {
        return amount;
    }

    public void setAmount(double amount) {
        this.amount = amount;
    }

    @Override
    public boolean equals(Object o) {
        if (this == o) {
            return true;
        } else if (o == null || getClass() != o.getClass()) {
            return false;
        }
        Transaction that = (Transaction) o;
```

```java
            return accountId == that.accountId &&
                    timestamp == that.timestamp &&
                    Double.compare(that.amount, amount) == 0;
        }

        @Override
        public int hashCode() {
            return Objects.hash(accountId, timestamp, amount);
        }

        @Override
        public String toString() {
            return "Transaction{" +
                    "accountId = " + accountId +
                    ", timestamp = " + timestamp +
                    ", amount = " + amount +
                    '}';
        }
}
```

(4) 定义一个迭代器类,实现 Iterator 接口,封装了对 Transaction 信用卡交易数据列表的迭代方法。这是一个通用的数据迭代器类,既可以迭代生成批数据,也可以循环迭代生成流数据,由其中的标志变量 bounded 来控制,代码如下:

```java
//第3章/TransactionIterator.java

import com.xueai8.fraud.entity.Transaction;

import Java.io.Serializable;
import Java.sql.Timestamp;
import Java.util.Arrays;
import Java.util.Iterator;
import Java.util.List;

/**
 * 模拟信用卡交易的交易数据源的迭代器
 */
public class TransactionIterator implements Iterator<Transaction>, Serializable {
    private static final long serialVersionUID = 1L;

    private static final Timestamp INITIAL_TIMESTAMP =
            Timestamp.valueOf("2020 - 01 - 01 00:00:00");
    private static final long SIX_MINUTES = 6 * 60 * 1000;

    private final boolean bounded;           //标志变量,指示生成流数据还是批数据
```

```java
    private int index = 0;              //交易记录的索引
    private long timestamp;             //交易发生的时间戳

    static TransactionIterator bounded() {
        return new TransactionIterator(true);
    }

    static TransactionIterator unbounded() {
        return new TransactionIterator(false);
    }

    private TransactionIterator(boolean bounded) {
        this.bounded = bounded;
        this.timestamp = INITIAL_TIMESTAMP.getTime();
    }

    @Override
    public boolean hasNext() {
        //如果还有数据
        if (index < data.size()) {
            return true;
        }
        //如果用于生成流数据,则从头循环
        else if (!bounded) {
            index = 0;
            return true;
        }

        //如果用于生成批数据
        else {
            return false;
        }
    }

    //生成下一个交易数据,交易时间相隔为6min
    @Override
    public Transaction next() {
        Transaction transaction = data.get(index++);
        transaction.setTimestamp(timestamp);
        timestamp += SIX_MINUTES;
        return transaction;
    }

    //事先存储的信用卡交易数据,在实际中来自外部数据源系统,如 Kafka
    private static List<Transaction> data = Arrays.asList(
        new Transaction(1, 0L, 188.23),
```

```
            new Transaction(2, 0L, 374.79),
            new Transaction(3, 0L, 112.15),
            new Transaction(4, 0L, 478.75),
            new Transaction(5, 0L, 208.85),
            new Transaction(1, 0L, 379.64),
            new Transaction(2, 0L, 351.44),
            new Transaction(3, 0L, 320.75),
            new Transaction(4, 0L, 259.42),
            new Transaction(5, 0L, 273.44),
            new Transaction(1, 0L, 267.25),
            new Transaction(2, 0L, 397.15),
            new Transaction(3, 0L, 0.219),
            new Transaction(4, 0L, 231.94),
            new Transaction(5, 0L, 384.73),
            new Transaction(1, 0L, 419.62),
            new Transaction(2, 0L, 412.91),
            new Transaction(3, 0L, 0.77),
            new Transaction(4, 0L, 22.10),
            new Transaction(5, 0L, 377.54),
            new Transaction(1, 0L, 375.44),
            new Transaction(2, 0L, 230.18),
            new Transaction(3, 0L, 0.80),
            new Transaction(4, 0L, 350.89),
            new Transaction(5, 0L, 127.55),
            new Transaction(1, 0L, 483.91),
            new Transaction(2, 0L, 228.22),
            new Transaction(3, 0L, 871.15),
            new Transaction(4, 0L, 64.19),
            new Transaction(5, 0L, 79.43),
            new Transaction(1, 0L, 56.12),
            new Transaction(2, 0L, 256.48),
            new Transaction(3, 0L, 148.16),
            new Transaction(4, 0L, 199.95),
            new Transaction(5, 0L, 252.37),
            new Transaction(1, 0L, 274.73),
            new Transaction(2, 0L, 473.54),
            new Transaction(3, 0L, 119.92),
            new Transaction(4, 0L, 323.59),
            new Transaction(5, 0L, 353.16),
            new Transaction(1, 0L, 211.90),
            new Transaction(2, 0L, 280.93),
            new Transaction(3, 0L, 347.89),
            new Transaction(4, 0L, 459.86),
            new Transaction(5, 0L, 82.31),
            new Transaction(1, 0L, 373.26),
            new Transaction(2, 0L, 479.83),
```

```
            new Transaction(3, 0L, 454.25),
            new Transaction(4, 0L, 83.64),
            new Transaction(5, 0L, 292.44)
    );
}
```

(5) 创建自定义的数据源类,继承自 FromIteratorFunction 类(这是 Flink 自带的一个 SourceFunction 的子类),并封装了一个数据生成器 RateLimitedIterator,代码如下:

```
//第3章/TransactionSource.java

import com.xueai8.fraud.entity.Transaction;
import org.apache.flink.streaming.api.functions.source.FromIteratorFunction;

import Java.io.Serializable;
import Java.util.Iterator;

/**
 * 自定义的数据生成器,运行后生成持续的信用卡交易数据(模拟),QPS 为 0.1s
 *
 * 这个类继承自 Flink,即由 Flink 提供的 org.apache.flink.streaming.api.functions.source.
 * FromIteratorFunction 类
 * 该类中有一个 run 方法,当运行此方法时,会自动连续发送流,以便指定迭代器数据源中的流
 * 数据
 */
public class TransactionSource extends FromIteratorFunction<Transaction> {
    private static final long serialVersionUID = 1L;

    public TransactionSource() {
        //调用超类构造器,默认生成流数据
        super(new RateLimitedIterator<>(TransactionIterator.unbounded()));
    }

    //泛型数据迭代器
    private static class RateLimitedIterator<T> implements Iterator<T>, Serializable {

        private static final long serialVersionUID = 1L;

        private final Iterator<T> inner;         //数据源迭代器

        private RateLimitedIterator(Iterator<T> inner) {
            this.inner = inner;
        }
```

```java
    @Override
    public boolean hasNext() {
        return inner.hasNext();
    }

    @Override
    public T next() {
        try {
            Thread.sleep(100);
        } catch (InterruptedException e) {
            throw new RuntimeException(e);
        }
        return inner.next();
    }
  }
}
```

(6) 创建一个测试类(带有main方法的主程序),使用addSource()方法添加自定义数据源,编辑代码。

Scala 代码如下:

```scala
//第3章/CustomSourceDemo2.scala

import com.xueai8.ch03.source.TransactionSource
import org.apache.flink.streaming.api.scala._

/**
 * 自定义数据源的测试类
 */
object CustomSourceDemo2 {

  def main(args: Array[String]): Unit = {
    //获取流执行环境
    val env = StreamExecutionEnvironment.getExecutionEnvironment

    //添加自定义流数据源
    val transactions = env.addSource(new TransactionSource()).name("transactions")

    //输出交易记录
    transactions.print

    //触发流程序执行
    env.execute("Flink Custom Source")
  }
}
```

Java 代码如下：

```java
import com.xueai8.ch03.entity.Transaction;
import com.xueai8.ch03.source.TransactionSource;
import org.apache.flink.streaming.api.datastream.DataStream;
import org.apache.flink.streaming.api.environment.StreamExecutionEnvironment;

/**
 * 自定义数据源测试类
 */
public class CustomSourceDemo2 {

    public static void main(String[] args) throws Exception {
        //设置流执行环境
        final StreamExecutionEnvironment env =
                StreamExecutionEnvironment.getExecutionEnvironment();

        //设置自定义流数据源
        DataStream<Transaction> transactions = env
            .addSource(new TransactionSource())
            .name("transactions");

        //输出查看
        transactions.print();

        //执行流程序
        env.execute("Transaction Stream");
    }
}
```

（7）执行以上程序，查看控制台，输出结果如下（部分结果）：

```
1> Transaction{accountId = 1, timestamp = 1577808000000, amount = 188.23}
2> Transaction{accountId = 2, timestamp = 1577808360000, amount = 374.79}
3> Transaction{accountId = 3, timestamp = 1577808720000, amount = 112.15}
4> Transaction{accountId = 4, timestamp = 1577809080000, amount = 478.75}
5> Transaction{accountId = 5, timestamp = 1577809440000, amount = 208.85}
6> Transaction{accountId = 1, timestamp = 1577809800000, amount = 379.64}
7> Transaction{accountId = 2, timestamp = 1577810160000, amount = 351.44}
8> Transaction{accountId = 3, timestamp = 1577810520000, amount = 320.75}
1> Transaction{accountId = 4, timestamp = 1577810880000, amount = 259.42}
2> Transaction{accountId = 5, timestamp = 1577811240000, amount = 273.44}
3> Transaction{accountId = 1, timestamp = 1577811600000, amount = 267.25}
4> Transaction{accountId = 2, timestamp = 1577811960000, amount = 397.15}
5> Transaction{accountId = 3, timestamp = 1577812320000, amount = 0.219}
```

```
6> Transaction{accountId = 4, timestamp = 1577812680000, amount = 231.94}
7> Transaction{accountId = 5, timestamp = 1577813040000, amount = 384.73}
8> Transaction{accountId = 1, timestamp = 1577813400000, amount = 419.62}
......
```

在上面的示例代码中，自定义的数据源类TransactionSource并不是直接创建SourceFunction类的子类，而是从FromIteratorFunction类继承的。FromIteratorFunction是Flink的org.apache.flink.streaming.api.functions.source包中自带的一个类，它已经实现了接口SourceFunction<T>，包装了一个具体的泛型迭代器对象属性iterator。通常使用时从它派生一个子类，并在构造时传入一个具体的迭代数据生成器类即可，其源码如下：

```java
//第3章/FromIteratorFunction.java

package org.apache.flink.streaming.api.functions.source;

import java.util.Iterator;
import org.apache.flink.annotation.PublicEvolving;
import org.apache.flink.streaming.api.functions.source.SourceFunction.SourceContext;

@PublicEvolving
public class FromIteratorFunction<T> implements SourceFunction<T> {
    private static final long serialVersionUID = 1L;
    private final Iterator<T> iterator;
    private volatile boolean isRunning = true;

    public FromIteratorFunction(Iterator<T> iterator) {
        this.iterator = iterator;
    }

    public void run(SourceContext<T> ctx) throws Exception {
        while(this.isRunning && this.iterator.hasNext()) {
            ctx.collect(this.iterator.next());
        }
    }

    public void cancel() {
        this.isRunning = false;
    }
}
```

下面这个参考示例是Flink官方给出的自定义数据源样例。这个数据源会将0～999发送到下游系统，代码如下：

```java
//第3章/ExampleCountSource.java

public class ExampleCountSource
        implements SourceFunction<Long>, CheckpointedFunction {

    private long count = 0L;

    //使用一个 volatile 类型变量控制 run 方法内循环的运行
    private volatile boolean isRunning = true;

    //保存数据源状态的变量
    private transient ListState<Long> checkpointedCount;

    public void run(SourceContext<T> ctx) {
        while (isRunning && count < 1000) {
            //这个同步块确保状态检查点,元素的内部状态更新和释放是一个原子操作
            //此处必须加锁,防止在 checkpoint 过程中仍然发送数据
            synchronized (ctx.getCheckpointLock()) {
                ctx.collect(count);
                count++;
            }
        }
    }

    public void cancel() {
        //将 isRunning 设置为 false,以便终止 run 方法内循环的运行
        isRunning = false;
    }

    public void initializeState(FunctionInitializationContext context) {
        //获取存储状态
        this.checkpointedCount = context
            .getOperatorStateStore()
            .getListState(new ListStateDescriptor<>("count", Long.class));

        //如果数据源是从失败中恢复的,则读取 count 的值,恢复数据源 count 状态
        if (context.isRestored()) {
            for (Long count : this.checkpointedCount.get()) {
                this.count = count;
            }
        }
    }

    public void snapshotState(FunctionSnapshotContext context) {
        //将数据保存到状态变量
        this.checkpointedCount.clear();
        this.checkpointedCount.add(count);
    }
}
```

3.3 Flink 数据转换

数据转换使用操作符(operator)将一个或多个数据流转换为新的数据流。转换输入可以是一个或多个数据流，转换输出也可以是零个、一个或多个数据流。程序可以将多个转换组合成复杂的数据流拓扑。

本节将介绍 Flink 提供的这些基本的转换运算。

3.3.1 map 转换

这是最简单的转换之一，其中输入是一个数据流，输出也是一个数据流。关于 map 转换的简单介绍见表 3-1。

表 3-1 map 转换运算

转换运算符	描 述	用 法
map DataStream→DataStream	获取一个元素并生成一个元素	dataStream.map{x=> x * 2}

在下面的示例程序中，应用了 map 转换来处理流数据。

Scala 代码如下：

```scala
//第 3 章/TransformerMap.scala

import org.apache.flink.streaming.api.scala._

object TransformerMap{
  def main(args: Array[String]) {
    //设置流执行环境
    val env = StreamExecutionEnvironment.getExecutionEnvironment

    //首先从环境中获取一些数据，并使用操作符转换 DataStream[String]
    env.fromElements("Good good study", "Day day up")
      .map(_.toLowerCase)
      .print()

    //对于流程序，只有执行了下面这种方法，流程序才真正开始执行
    env.execute("flink map transformatiion")
  }
}
```

Java 代码如下：

```java
//第3章/TransformerMap.java

import org.apache.flink.streaming.api.datastream.DataStream;
import org.apache.flink.streaming.api.environment.StreamExecutionEnvironment;

class TransformerMap{

    public static void main(String[] args) throws Exception {
        //设置流执行环境
        final StreamExecutionEnvironment env =
            StreamExecutionEnvironment.getExecutionEnvironment();

        //首先从环境中获取一些数据
        DataStream<String> ds = env.fromElements("Good good study","Day day up");

        //执行 map 转换
        DataStream<String> ds_map = ds.map(String::toLowerCase);

        //输出结果
        ds_map.print();

        //对于流程序,只有执行了下面这种方法,流程序才真正开始执行
        env.execute("flink map transformatiion");
    }
}
```

执行以上代码,输出结果如下:

```
good good study
day day up
```

3.3.2　flatMap 转换

这也是很常用的一个转换操作。flatMap 接收一条记录并输出 0 条、一条或多条记录。关于 flatMap 转换的简单介绍见表 3-2。

表 3-2　flatMap 转换运算

转换运算符	描　　述	用　　法
flatMap DataStream→DataStream	获取一个元素并生成零个、一个或多个元素	dataStream.flatMap{ str => str.split("") }

在下面的示例中演示了如何应用 flatMap 转换。
Scala 代码如下:

```scala
//第 3 章/TransformerFlatMap.scala

import org.apache.flink.api.scala._
import org.apache.flink.streaming.api.scala.StreamExecutionEnvironment

object TransformerFlatMap{

  def main(args: Array[String]) {
    //设置流处理执行环境
    val env = StreamExecutionEnvironment.getExecutionEnvironment

    //得到输入数据,flatMap 转换
    env.fromElements("Good good study", "Day day up")
      .map(_.toLowerCase)
      .flatMap(_.split("\\W+"))     //相当于先执行 map,再执行 flatten
      .print

    //执行
    env.execute("flink flatmap transformatiion")
  }
}
```

Java 代码如下：

```java
//第 3 章/TransformerFlatMap.java

import org.apache.flink.api.common.functions.FlatMapFunction;
import org.apache.flink.api.common.typeinfo.Types;
import org.apache.flink.streaming.api.environment.StreamExecutionEnvironment;

public class TransformerFlatMap {
    public static void main(String[] args) throws Exception {
        //设置流执行环境
        final StreamExecutionEnvironment env =
            StreamExecutionEnvironment.getExecutionEnvironment();

        //首先从环境中获取一些数据,再执行 map 和 flatMap 转换
        env.fromElements("Good good study","Day day up")
            .map(String::toLowerCase)
            //传入一个匿名函数
            .flatMap((FlatMapFunction<String, String>) (value, out) -> {
                for(String word: value.split("\\W+")){
                    out.collect(word);
                }
            }).returns(Types.STRING)
```

```
            .print();

        //对于流程序,只有执行了下面这种方法,流程序才真正开始执行
        env.execute("flink flatmap transformatiion");
    }
}
```

在上面的代码中,flatMap()函数的传入参数是一个 Lambda 表达式,它对于 flatMap()的支持是无法猜测出来类型的,必须通过 returns(Types.STRING)指定具体的返回值类型。

或者,也可以像下面这样,使用匿名内部类(而不是 Lambda 表达式),代码如下:

```
//第 3 章/TransformerFlatMap2.java

import org.apache.flink.api.common.functions.FlatMapFunction;
import org.apache.flink.streaming.api.datastream.DataStream;
import org.apache.flink.streaming.api.environment.StreamExecutionEnvironment;
import org.apache.flink.util.Collector;

public class TransformerFlatMap2 {
    public static void main(String[] args) throws Exception {
        //设置流执行环境
        final StreamExecutionEnvironment env =
            StreamExecutionEnvironment.getExecutionEnvironment();

        //首先从环境中获取一些数据,再执行 map 和 flatMap 转换
        DataStream<String> ds = env
            .fromElements("Good good study","Day day up")
            .map(String::toLowerCase)
            .flatMap(new FlatMapFunction<String, String>() {
                @Override
                public void flatMap(String value, Collector<String> out) throws Exception {
                    for(String word: value.split("\\W+")){
                        out.collect(word);
                    }
                }
            });

        //输出
        ds.print();

        //对于流程序,只有执行了下面这种方法,流程序才真正开始执行
        env.execute("flink flatmap transformatiion");
    }
}
```

执行以上代码,输出结果如下:

```
good
good
study
day
day
up
```

3.3.3　filter 转换

这也是很常用的一个转换操作。应用 filter()函数对条件进行评估,如果结果为 true,则输出该条数据。filter()函数可以输出 0 个记录。关于 filter 转换的简单介绍见表 3-3。

表 3-3　filter 转换运算

转换运算符	描　　述	用　　法
filter DataStream→DataStream	对每个元素求布尔函数的值,并保留函数的返回值为 true 的元素	dataStream.filter{_!=0}

在下面的示例中演示了 filter 转换运算,代码如下:
Scala 代码如下:

```scala
//第3章/TransformerFilter.scala

import org.apache.flink.api.scala._
import org.apache.flink.streaming.api.scala.StreamExecutionEnvironment

object TransformerFilter{

  def main(args: Array[String]) {
    //设置批处理执行环境
    val env = StreamExecutionEnvironment.getExecutionEnvironment

    //得到输入数据,然后执行filter转换
    env.fromElements("Good good study", "Day day up")
      .map(_.toLowerCase)
      .filter(_.contains("study"))
      .print()

    //执行
    env.execute("flink filter transformatiion")
  }
}
```

Java 代码如下：

```java
//第3章/TransformerFilter.java

import org.apache.flink.api.common.functions.FilterFunction;
import org.apache.flink.streaming.api.environment.StreamExecutionEnvironment;

public class TransformerFilter {
    public static void main(String[] args) throws Exception {
        //设置流执行环境
        final StreamExecutionEnvironment env =
            StreamExecutionEnvironment.getExecutionEnvironment();

        //首先从环境中获取一些数据,再执行 map 和 filter 转换
        env.fromElements("Good good study","Day day up")
            .map(String::toLowerCase)
            .filter((FilterFunction<String>) s -> s.contains("study"))
            .print();

        //对于流程序,只有执行了下面这种方法,流程序才真正开始执行
        env.execute("flink map transformatiion");
    }
}
```

执行以上代码,输出结果如下：

```
good good study
```

数据流中的数据也可以封装到事件对象中(在 Scala 中是 case class,在 Java 中是 POJO 类中),代码如下。

Scala 代码如下：

```scala
//第3章/TransformerFilter2.scala

import org.apache.flink.streaming.api.scala._

object TransformerFilter2 {

    //case class 类,用来表示事件流中的事件
    case class Person(name:String,age:Int)        //需要定义在 main 方法外部

    def main(args: Array[String]): Unit = {

        //设置批处理执行环境
        val env = StreamExecutionEnvironment.getExecutionEnvironment
```

```scala
        //得到输入数据,然后执行filter转换
        val personDS = env.fromElements(
            Person("张三", 25),
            Person("李四", 18),
            Person("小多米", 2)
        )
        val adults = personDS.filter(_.age >= 18)

        //输出过滤后的数据流
        adults.print

        //执行
        env.execute("flink filter transformatiion")
    }
}
```

Java代码如下:

```java
//第3章/TransformerFilter2.java

import org.apache.flink.api.common.functions.FilterFunction;
import org.apache.flink.streaming.api.datastream.DataStream;
import org.apache.flink.streaming.api.environment.StreamExecutionEnvironment;

public class TransformerFilter2 {

    //POJO类,用来表示事件流中的事件
    public static class Person{
        String name;
        Integer age;

        Person() {}

        Person(String name, Integer age) {
            this.name = name;
            this.age = age;
        }

        public String toString() {
            return this.name + ": 年龄 " + this.age.toString();
        }
    }

    public static void main(String[] args) throws Exception {
        //设置流执行环境
```

```java
        final StreamExecutionEnvironment env =
            StreamExecutionEnvironment.getExecutionEnvironment();

        //加载数据源
        DataStream<Person> persons = env.fromElements(
            new Person("张三", 35),
            new Person("李四", 35),
            new Person("小多米", 2));

        //执行 filter 转换
        DataStream<Person> adults = persons.filter(
          new FilterFunction<Person>() {
            @Override
            public boolean filter(Person person) throws Exception {
                return person.age >= 18;
            }
          });

        //输出结果流
        adults.print();

        //对于流程序,只有执行了下面这种方法,流程序才真正开始执行
        env.execute("flink filter transformatiion");
    }
}
```

执行以上代码,输出结果如下:

```
张三: 年龄 35
李四: 年龄 35
```

3.3.4 keyBy 转换

有一些转换(如 join、coGroup、keyBy、groupBy)要求在元素集合上定义一个 key。还有一些转换(如 reduce、groupReduce、aggregate、windows)可以应用在按 key 分组的数据上。关于 keyBy 转换的简单介绍见表 3-4。

表 3-4 keyBy 转换运算

转换运算符	描述	用法
keyBy DataStream→KeyedStream	逻辑上将一个流划分为不相连的分区,每个分区包含相同 key 的元素。在内部,这是通过哈希分区实现的。这个转换会返回一个 KeyedStream	//按字段 dataStream.keyBy ("someKey") //按元组的第 1 个元素 dataStream.keyBy(0)

Flink 的数据模型不是基于键-值对的，因此，不需要将数据集类型物理打包为 key 和 value。key 是"虚拟的"，它们被定义一个函数，该函数可指定数据流中实际数据的哪个字段（或属性）用作 key。需要注意的是，如果流元素是 POJO 类型，则该 POJO 类必须重写 hashCode()方法。

最简单的情况是对元组的一个或多个字段进行分组，参看下面的示例。

Scala 代码如下：

```scala
//第3章/TransformerKeyBy.scala

import org.apache.flink.streaming.api.scala._

object TransformerKeyBy{
  def main(args: Array[String]): Unit = {
    //设置流执行环境
    val env = StreamExecutionEnvironment.getExecutionEnvironment

    //得到输入数据，进行转换
    env.fromElements("Good good study", "Day day up")
      .map(_.toLowerCase)
      .flatMap(_.split("\\W+"))         //相当于先执行 map,再执行 flatten
      .map((_,1))
      .keyBy(0)                          //按元组索引
      //.keyBy(t => t._1)                //或按元组的第1个字段分区
      .print

    //触发流程序执行
    env.execute("flink keyBy transformatiion")
  }
}
```

Java 代码如下：

```java
//第3章/TransformerKeyBy.java

import org.apache.flink.api.common.functions.FlatMapFunction;
import org.apache.flink.api.common.functions.MapFunction;
import org.apache.flink.api.java.tuple.Tuple;
import org.apache.flink.api.java.tuple.Tuple2;
import org.apache.flink.streaming.api.datastream.DataStream;
import org.apache.flink.streaming.api.datastream.KeyedStream;
import org.apache.flink.streaming.api.environment.StreamExecutionEnvironment;
import org.apache.flink.util.Collector;

public class TransformerKeyBy {
```

```java
public static void main(String[] args) throws Exception {
    //设置流执行环境
    final StreamExecutionEnvironment env =
            StreamExecutionEnvironment.getExecutionEnvironment();

    //加载数据源,并执行flatMap转换
    DataStream<String> ds = env
        .fromElements("good good study","day day up")
        .flatMap(new FlatMapFunction<String, String>() {
            @Override
            public void flatMap(String value, Collector<String> out)
                    throws Exception {
                for(String word: value.split("\\W+")){
                    out.collect(word);
                }
            }
        });

    //通过map转换,将事件流中事件的数据类型变换为(word,1)元组的形式
    DataStream<Tuple2<String,Integer>> dm = ds
        .map(new MapFunction<String, Tuple2<String,Integer>>() {
            @Override
            public Tuple2<String, Integer> map(String s) throws Exception {
                return new Tuple2<>(s,1);
            }
        });

    //keyBy转换,按key重分区
    KeyedStream<Tuple2<String,Integer>,Tuple> ds_keyed = dm.keyBy(0);

    //输出
    ds_keyed.print();

    //执行
    env.execute("flink keyBy transformatiion");
}
}
```

执行以上代码,输出结果如下:

```
(up,1)
(study,1)
(day,1)
(day,1)
(good,1)
(good,1)
```

1. 使用字段表达式来定义 key

在 Flink 1.11 版本之前,也可以使用字段表达式来定义 key(从 Flink 1.11 开始已弃用)。可以使用基于字符串的字段表达式来引用嵌套的字段,并为分组、排序、连接或联合分组定义 key。字段表达式使在(嵌套的)复合类型(如 Tuple 和 POJO 类型)中选择字段变得非常容易。例如,在下面的示例中,按成员性别分区。

Scala 代码如下:

```scala
//第 3 章/TransformerKeyBy2.scala

import org.apache.flink.streaming.api.scala._

object TransformerKeyBy2 {

  //事件流中事件的数据类型采用 case class
  case class Person(name:String, gender:String)

  def main(args: Array[String]): Unit = {
    //设置流执行环境
    val env = StreamExecutionEnvironment.getExecutionEnvironment

    //得到输入数据,进行转换
    val personDS = env.fromElements(
        Person("张三","男"),
        Person("李四","女"),
        Person("小多米","男")
    )

    personDS.keyBy("gender").print

    env.execute("flink keyBy transformatiion")
  }
}
```

Java 代码如下:

```java
//第 3 章/TransformerKeyBy2.java

import org.apache.flink.streaming.api.datastream.DataStream;
import org.apache.flink.streaming.api.environment.StreamExecutionEnvironment;

public class TransformerKeyBy2 {

    //POJO 类,用来表示事件流中的事件
    public static class Person {
```

```java
    public String name;              //姓名
    public String gender;            //性别

    public Person() {}

    public Person(String name, String gender) {
        this.name = name;
        this.gender = gender;
    }

    @Override
    public String toString() {
        return this.name + ": 性别 " + this.gender;
    }
}

public static void main(String[] args) throws Exception {
    //设置流执行环境
    final StreamExecutionEnvironment env =
        StreamExecutionEnvironment.getExecutionEnvironment();

    //加载数据源
    DataStream<Person> personDS = env.fromElements(
        new Person("张三","男"),
        new Person("李四","女"),
        new Person("小多米","男"));

    //执行 keyBy 转换,按性别(gender)分区
    personDS.keyBy("gender").print();

    //执行
    env.execute("flink keyBy transformatiion");
    }
}
```

执行以上程序,输出结果如下:

```
6> 李四: 性别 女
1> 张三: 性别 男
1> 小多米: 性别 男
```

注意到,相同的 key 在同一分区内被计算。另外要特别注意的是,对于要充当 key 的 POJO 类,必须满足以下条件:

(1) 字段名必须声明为 public。
(2) 必须有默认的无参构造器。

(3) 所有构造器必须声明为 public。

2. 使用 key selector 函数来定义 key

定义 key 的另一种方法是使用 key selector 函数。一个 key selector 函数可接收单个元素作为输入,并返回该元素的 key。返回的 key 可以是任何类型的,可以从确定性计算中得到。例如,在下面的示例中,根据成员的年龄,将成员分为两组:成年人和未成年人。

Scala 代码如下:

```scala
//第 3 章/TransformerKeyBy3.scala

import org.apache.flink.streaming.api.scala._

object TransformerKeyBy3 {

    //事件流中事件的数据类型采用 case class
    case class Person(name:String, age:Integer)

    def main(args: Array[String]): Unit = {
        //设置流执行环境
        val env = StreamExecutionEnvironment.getExecutionEnvironment

        //得到输入数据,进行转换
        val personDS = env.fromElements(
            Person("张三", 16),
            Person("李四", 18),
            Person("王老五", 35),
            Person("赵小六", 23),
            Person("小多米", 12))

        personDS.keyBy(person => if (person.age >= 18) "adult" else "young").print

        env.execute("flink keyBy transformatiion")
    }
}
```

Java 代码:

```java
//第 3 章/TransformerKeyBy3.java

import org.apache.flink.api.java.functions.KeySelector;
import org.apache.flink.streaming.api.datastream.DataStream;
import org.apache.flink.streaming.api.datastream.KeyedStream;
import org.apache.flink.streaming.api.environment.StreamExecutionEnvironment;

public class TransformerKeyBy3 {
```

```java
//POJO 类,用来表示事件流中的事件
public static class Person {
    public String name;            //姓名
    public Integer age;            //年龄

    public Person() {}

    public Person(String name, Integer age) {
        this.name = name;
        this.age = age;
    }

    @Override
    public String toString() {
        return this.name + ": 年龄 " + this.age;
    }
}

public static void main(String[] args) throws Exception {
    //设置流执行环境
    final StreamExecutionEnvironment env =
        StreamExecutionEnvironment.getExecutionEnvironment();

    //加载数据源
    DataStream<Person> personDS = env.fromElements(
        new Person("张三", 16),
        new Person("李四", 18),
        new Person("王老五", 35),
        new Person("赵小六", 23),
        new Person("小多米", 12)
    );

    //执行 keyBy 转换,按年龄分区(成年人和未成年人)
    KeyedStream<Person, String> keyed = personDS
      .keyBy(new KeySelector<Person, String>() {
        @Override
        public String getKey(Person person) throws Exception {
            return person.age >= 18 ? "adult" : "young";
        }
    });

    keyed.print();

    //执行
    env.execute("flink keyBy transformatiion");
}
}
```

执行以下代码,输出结果如下:

```
6> 李四：年龄 18
1> 张三：年龄 16
6> 王老五：年龄 35
1> 小多米：年龄 12
6> 赵小六：年龄 23
```

3.3.5 reduce 转换

这个转换操作类似于 Scala 或 Python 语言中的 reduce 函数，它接收一个函数作为累加器，将流中的每个值(从左到右)开始缩减，最终计算为一个值。关于 reduce 转换的简单介绍见表 3-5。

表 3-5 reduce 转换运算

转换运算符	描述	用法
reduce KeyedStream→DataStream	key-value 数据流上的"滚动"缩减。将当前元素与最后一个减少的值组合在一起，并计算出新的值	keyedStream.reduce{_+_}

在下面的示例代码中，演示了使用 reduce 转换操作对 DataStream 中的元素进行求和。Scala 代码如下：

```scala
//第3章/TransformerReduce.scala

import org.apache.flink.api.scala._
import org.apache.flink.streaming.api.scala.StreamExecutionEnvironment

object TransformerReduce {

  def main(args: Array[String]): Unit = {
    //设置批处理执行环境
    val env = StreamExecutionEnvironment.getExecutionEnvironment

    //加载数据源，然后执行 DataStream 转换
    env.fromElements("Good good study", "Day day up")
      .map(_.toLowerCase)                            //转换为小写
      .flatMap(_.split("\\W+"))                      //相当于先执行map,再执行flatten
      .map((_,1))                                    //转换为元组类型
      .keyBy(_._1)                                   //按单词进行分组
      .reduce((a,b) => (a._1, a._2 + b._2))          //reduce 转换
      .print
    //触发执行
```

```
        env.execute("flink reduce transformatiion")
    }
}
```

Java 代码如下:

```java
//第 3 章/TransformerReduce.java

import org.apache.flink.api.common.functions.FlatMapFunction;
import org.apache.flink.api.common.functions.ReduceFunction;
import org.apache.flink.api.java.tuple.Tuple2;
import org.apache.flink.streaming.api.datastream.DataStream;
import org.apache.flink.streaming.api.environment.StreamExecutionEnvironment;
import org.apache.flink.util.Collector;

public class TransformerReduce {
    public static void main(String[] args) throws Exception {
        //设置流执行环境
        final StreamExecutionEnvironment env =
                StreamExecutionEnvironment.getExecutionEnvironment();

        //首先加载数据,然后执行转换
        DataStream < Tuple2 < String, Integer >> ds = env
            .fromElements("Good good study","Day day up")
            .map(String::toLowerCase)
            .flatMap(new FlatMapFunction < String, Tuple2 < String,Integer >>() {
                @Override
                public void flatMap(String s, Collector < Tuple2 < String, Integer >> collector) throws Exception {
                    for(String word : s.split("\\W+")){
                        collector.collect(new Tuple2 <>(word,1));
                    }
                }
            });

        ds
            .keyBy(t -> t.f0)
            .reduce(new ReduceFunction < Tuple2 < String, Integer >>() {
                @Override
                public Tuple2 < String, Integer > reduce(Tuple2 < String, Integer > t1, Tuple2 < String, Integer > t2) throws Exception {
                    return new Tuple2 <>(t1.f0, t1.f1 + t2.f1);
                }
            }).print();
        //对于流程序,只有执行了下面这种方法,流程序才真正开始执行
```

```
            env.execute("flink reduce transformatiion");
        }
    }
```

执行以上程序,输出结果如下:

```
6> (good,1)
4> (up,1)
6> (good,2)
7> (day,1)
7> (day,2)
5> (study,1)
```

3.3.6 聚合转换

DataStream API 支持各种聚合操作,例如 min、max、sum 等。这些聚合函数可以应用于 KeyedDataStream 类型的流上,以获得滚动聚合。关于聚合转换的简单介绍见表 3-6。

表 3-6 map 转换运算

转换运算符	描 述	用 法
Aggregations KeyedStream→DataStream	在 key-value 数据流上滚动聚合。 min 和 minBy 之间的区别是,min 返回最小值,而 minBy 返回该字段中具有最小值的元素(max 和 maxBy 也是如此)	keyedStream.sum(0) keyedStream.sum("key") keyedStream.min(0) keyedStream.min("key") keyedStream.max(0) keyedStream.max("key") keyedStream.minBy(0) keyedStream.minBy("key") keyedStream.maxBy(0) keyedStream.maxBy("key")

下面的示例演示了如何在 Flink 流上执行聚合转换。
Scala 代码如下:

```
//第 3 章/TransformerAgg.scala

import org.apache.flink.streaming.api.scala._

object TransformerAgg{
  def main(args: Array[String]): Unit = {
    //设置流执行环境
    val env = StreamExecutionEnvironment.getExecutionEnvironment
```

```scala
//首先从环境中获取一些数据,并使用操作符转换 DataStream[String]
val ds_keyed = env
    .fromElements(("good",1),("good",2),("study",1))
    .keyBy(_._1)

ds_keyed.sum(1).print
ds_keyed.min(1).print
ds_keyed.max(1).print
ds_keyed.minBy(1).print
ds_keyed.maxBy(1).print

env.execute("flink aggregation transformatiion")
  }
}
```

Java 代码如下:

```java
//第3章/TransformerAgg.java

import org.apache.flink.api.java.tuple.Tuple2;
import org.apache.flink.streaming.api.datastream.DataStream;
import org.apache.flink.streaming.api.environment.StreamExecutionEnvironment;

public class TransformerAgg {
    public static void main(String[] args) throws Exception {
        //设置流执行环境
        final StreamExecutionEnvironment env =
            StreamExecutionEnvironment.getExecutionEnvironment();

        //首先从环境中获取一些数据,再执行 map 和 flatMap 转换
        DataStream<Tuple2<String,Integer>> ds = env.fromElements(
            new Tuple2<>("good",1),
            new Tuple2<>("good",2),
            new Tuple2<>("study",1));

        ds.keyBy(t -> t.f0).sum(1).print();      //参数也可以是字段名
        ds.keyBy(t -> t.f0).min(1).print();
        ds.keyBy(t -> t.f0).max(1).print();
        ds.keyBy(t -> t.f0).minBy(1).print();
        ds.keyBy(t -> t.f0).maxBy(1).print();

        //对于流程序,只有执行了下面这种方法,流程序才真正开始执行
        env.execute("flink aggregation transformatiion");
    }
}
```

3.3.7 union 转换

这个转换执行两个或多个数据流的合并。它对两个或者两个以上的 DataStream 流执行 union 操作,产生一个包含所有 DataStream 元素的新的 DataStream 流。关于 union 转换的简单介绍见表 3-7。

表 3-7 union 转换运算

转换运算符	描述	用法
union DataStream * → DataStream	合并两个或多个数据流(并集),创建包含来自所有流的所有元素的新流。 注意:如果将一个数据流与它自己相合并,则将得到结果流中的每个元素两次	dataStream.union(stream1, stream2, ...)

在 DataStream 上使用 union 操作可以合并多个同类型的数据流,并生成同类型的数据流,即可将多个 DataStream[T]合并为一个新的 DataStream[T]。数据将按照先进先出(First In First Out)的模式合并,并且不去重。例如,执行 union 转换操作对白色和深色两个数据流进行合并,生成一个结果数据流,如图 3-6 所示。

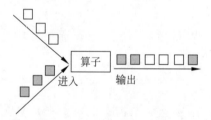

图 3-6 使用 union 转换操作合并数据流

在下面的示例中演示了 union 转换操作的用法。
Scala 代码如下:

```
//第 3 章/TransformerUnion.scala

import org.apache.flink.streaming.api.scala._

object TransformerUnion {

  def main(args: Array[String]): Unit = {
    //设置流执行环境
    val env = StreamExecutionEnvironment.getExecutionEnvironment
```

```scala
//union
//第1个数据集
val ds1 = env
  .fromElements("good good study")
  .flatMap(_.toLowerCase.split("\\W+")).map( (_, 1))

//第2个数据集
val ds2 = env
  .fromElements("day day up")
  .flatMap(_.toLowerCase.split("\\W+"))
  .map { (_, 1) }

//合并两个数据集并输出
ds1.union(ds2).print()

//执行
env.execute("flink union transformatiion")
  }
}
```

Java 代码如下：

```java
//第3章/TransformerUnion.java

import org.apache.flink.api.common.functions.FlatMapFunction;
import org.apache.flink.api.common.functions.MapFunction;
import org.apache.flink.api.java.tuple.Tuple2;
import org.apache.flink.streaming.api.datastream.DataStream;
import org.apache.flink.streaming.api.environment.StreamExecutionEnvironment;
import org.apache.flink.util.Collector;

public class TransformerUnion {
    public static void main(String[] args) throws Exception {
        //设置流执行环境
        final StreamExecutionEnvironment env =
            StreamExecutionEnvironment.getExecutionEnvironment();

        //union 转换
        //第1个数据集
        DataStream<String> ds1 = env.fromElements("good good study")
          .map(String::toLowerCase)
          .flatMap(new FlatMapFunction<String, String>() {
            @Override
            public void flatMap(String value, Collector<String> out) throws Exception {
                for(String word: value.split("\\W+")){
```

```java
                    out.collect(word);
                }
            }
        });
        DataStream<Tuple2<String,Integer>> ds1m = ds1.map(
            new MapFunction<String,Tuple2<String,Integer>>() {
                @Override
                public Tuple2<String,Integer> map(String s) throws Exception {
                    return new Tuple2<>(s, 1);
                }
            });

        //第2个数据集
        DataStream<String> ds2 = env.fromElements("day day up")
            .map(String::toLowerCase)
            .flatMap(new FlatMapFunction<String,String>() {
                @Override
                public void flatMap(String value, Collector<String> out) throws Exception {
                    for(String word: value.split("\\W+")){
                        out.collect(word);
                    }
                }
            });
        DataStream<Tuple2<String,Integer>> ds2m = ds2.map(
            new MapFunction<String,Tuple2<String,Integer>>() {
                @Override
                public Tuple2<String,Integer> map(String s) throws Exception {
                    return new Tuple2<>(s, 1);
                }
            });

        //合并两个数据集并输出
        DataStream<Tuple2<String,Integer>> ds1_and_ds2 = ds1m.union(ds2m);
        ds1_and_ds2.print();

        //执行
        env.execute("flink union transformatiion");
    }
}
```

执行以上代码,输出结果如下:

```
6> (good,1)
7> (day,1)
7> (day,1)
7> (up,1)
6> (good,1)
6> (study,1)
```

3.3.8 connect 转换

DataStream 的 connect 操作用来合并两个数据流，合并结果创建的是 ConnectedStreams 或 BroadcastConnectedStream 类型的流。它用了两个泛型，即不要求两个 DataStream 的元素是同一类型。关于 connect 转换的简单介绍见表 3-8。

表 3-8 connect 转换运算

转换运算符	描 述	用 法
connect DataStream, DataStream → ConnectedStreams	"连接"两个数据流，保持它们的类型，允许在两个流之间共享状态	stream1：DataStream[Int] = ... stream2：DataStream[String] = ... val connectedStreams = stream1.connect（stream2）

虽然前面讲的 union 转换操作也可以合并多个数据流，但 union 有一个限制，即多个数据流的数据类型必须相同；而 connect 转换虽然提供了和 union 类似的功能，也用来连接两个数据流，但它与 union 的区别在于：

（1）connect 只能连接两个数据流，union 可以连接多个数据流。

（2）connect 所连接的两个数据流的数据类型可以不一致，union 所连接的两个数据流的数据类型必须一致。

（3）两个 DataStream 经过 connect 之后被转换为 ConnectedStreams，ConnectedStreams 会对两个流的数据应用不同的处理方法，并且双流之间可以共享状态。

两个输入流经过 connect 合并后，可以进一步使用 CoProcessFunction、CoMap、CoFlatMap、KeyedCoProcessFunction 等 API 对两个流分别处理，如图 3-7 所示。

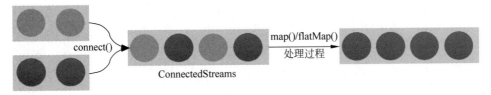

图 3-7 使用 connect 转换操作合并数据流

ConnectedStreams 提供了 keyBy() 方法，用于指定两个流的 keySelector，提供了 map、flatMap、process、transform 操作，其中前 3 个操作最后都调用 transform 操作。transform 操作接收 TwoInputStreamOperator 类型的 operator，然后转换为 SingleOutputStreamOperator。map 操作接收 CoMapFunction，flatMap 操作接收 CoFlatMapFunction，process 操作接收 CoProcessFunction。

下面的示例演示了如何执行 connect 转换。

Scala 代码如下:

```scala
//第 3 章/TransformerConnect.scala

import org.apache.flink.streaming.api.functions.co.{CoMapFunction, CoProcessFunction}
import org.apache.flink.streaming.api.scala._
import org.apache.flink.util.Collector

object TransformerConnect {
  def main(args: Array[String]): Unit = {
    //设置流执行环境
    val env = StreamExecutionEnvironment.getExecutionEnvironment

    //connect
    //第 1 个数据集
    val ds1:DataStream[(String, Int)] = env
      .fromElements("good good study")
      .flatMap(_.split("\\W+"))
      .map((_, 1))

    //第 2 个数据集
    val ds2:DataStream[String] = env
      .fromElements("day day up")
      .flatMap(_.split("\\W+"))

    //连接两个数据集
    val ds:ConnectedStreams[(String, Int),String] = ds1.connect(ds2)

    //调用 process 方法
    //CoProcessFunction 泛型参数:[输入流 1 数据类型,输入流 2 数据类型,输出流数据类型]
    ds.process(new CoProcessFunction[(String, Int), String, (String, Int)]{
      //处理输入流 1 的元素
      override def processElement1(
          in1: (String, Int),
          context: CoProcessFunction[(String, Int),
          String, (String, Int)]#Context,
          out: Collector[(String, Int)]): Unit = {
        out.collect(in1)         //发送给下游算子
      }

      //处理输入流 2 的元素
      override def processElement2(
          in2: String,
          context: CoProcessFunction[(String, Int),
          String, (String, Int)]#Context,
          out: Collector[(String, Int)]): Unit = {
```

```
        out.collect((in2,1)) //将来自输入流2的元素转换为元组,再发送给下游算子
      }
    }).print

    //map
    ds.map(new CoMapFunction[(String, Int), String, (String, Int)] {
      override def map1(in1: (String, Int)): (String, Int) = {
        (in1._1.toUpperCase, in1._2)
      }

      override def map2(in2: String): (String, Int) = {
        (in2, 1)
      }
    }).print

    //执行
    env.execute("flink connect transformatiion")
  }
}
```

Java 代码如下:

```
//第3章/TransformerConnect.java

import org.apache.flink.api.common.functions.FlatMapFunction;
import org.apache.flink.api.java.tuple.Tuple2;
import org.apache.flink.streaming.api.datastream.ConnectedStreams;
import org.apache.flink.streaming.api.datastream.DataStream;
import org.apache.flink.streaming.api.environment.StreamExecutionEnvironment;
import org.apache.flink.streaming.api.functions.co.CoMapFunction;
import org.apache.flink.streaming.api.functions.co.CoProcessFunction;
import org.apache.flink.util.Collector;

public class TransformerConnect {
    public static void main(String[] args) throws Exception {
        //设置流执行环境
        final StreamExecutionEnvironment env =
            StreamExecutionEnvironment.getExecutionEnvironment();

        //connect 转换操作
        //第1个数据集
        DataStream<Tuple2<String, Integer>> ds1 = env
            .fromElements("good good study")
            .map(String::toLowerCase)
            .flatMap(new FlatMapFunction<String, Tuple2<String, Integer>>() {
```

```java
            @Override
            public void flatMap(String value, Collector<Tuple2<String, Integer>> out) 
throws Exception {
                for(String word: value.split("\\W+")){
                    out.collect(new Tuple2<>(word,1));
                }
            }
        });

        //第2个数据集
        DataStream<String> ds2 = env
            .fromElements("day day up")
            .map(String::toLowerCase)
            .flatMap(new FlatMapFunction<String, String>() {
                @Override
                public void flatMap(String value, Collector<String> out) throws Exception {
                    for(String word: value.split("\\W+")){
                        out.collect(word);
                    }
                }
            });

        //连接两个数据集
        ConnectedStreams<Tuple2<String,Integer>,String> ds = ds1
            .connect(ds2);

        ds.process(new CoProcessFunction<Tuple2<String,Integer>, String, Tuple2<String,Integer>>() {
            @Override
            public void processElement1(
                Tuple2<String, Integer> t,
                Context context,
                Collector<Tuple2<String,Integer>> out) throws Exception {
              out.collect(t);
            }

            @Override
            public void processElement2(
                String s,
                Context context,
                Collector<Tuple2<String,Integer>> out) throws Exception {
              out.collect(new Tuple2<>(s,1));
            }
        }).print("process");
```

```java
        ds.map(new CoMapFunction<Tuple2<String,Integer>, String, Tuple2<String,Integer>>() {
            @Override
            public Tuple2<String, Integer> map1(Tuple2<String, Integer> t) throws Exception {
                return new Tuple2<>(t.f0.toUpperCase(),t.f1);
            }

            @Override
            public Tuple2<String, Integer> map2(String s) throws Exception {
                return new Tuple2<>(s, 1);
            }
        }).print("map");

        //执行
        env.execute("flink connect transformatiion");
    }
}
```

执行以上代码,输出结果如下:

```
process:1 > (day,1)
process:1 > (day,1)
process:1 > (up,1)
map:1 > (day,1)
map:1 > (day,1)
map:1 > (up,1)
map:5 > (GOOD,1)
map:5 > (GOOD,1)
map:5 > (STUDY,1)
process:5 > (good,1)
process:5 > (good,1)
process:5 > (study,1)
```

可以看出,connect 和 union 都有一个共同的作用,就是将两个流或多个流合成一个流,但是两者的区别是:union 连接的两个流的类型必须一致,connect 连接的流可以不一致,但是可以统一处理。

另外,在 Scala 实现的 API 中,对于 ConnectedStreams 的 map 转换,还可以分别执行两个不同的函数,这时两个流元素输出的数据结构可以不同,代码如下:

```scala
//第 3 章/TransformerConnect2.scala

import org.apache.flink.streaming.api.scala._

object TransformerConnect2 {
```

```scala
def main(args: Array[String]): Unit = {
  //设置流执行环境
  val env = StreamExecutionEnvironment.getExecutionEnvironment

  //connect
  //第1个数据集
  val ds1:DataStream[(String, Int)] = env
    .fromElements("good good study")
    .flatMap(_.split("\\W+"))
    .map((_, 1))

  //第2个数据集
  val ds2:DataStream[String] = env
    .fromElements("day day up")
    .flatMap(_.split("\\W+"))

  //连接两个数据集
  val ds:ConnectedStreams[(String, Int),String] = ds1.connect(ds2)

  //map,执行两个不同的函数,这时两个流元素输出的数据结构可以不同
  ds.map(
    d1 => {(d1._1.toUpperCase, d1._2)},
    d2 => {d2.toUpperCase}
  ).print

  //执行
  env.execute("flink connect transformatiion")
}
```

执行以上程序,输出结果如下:

```
2 > DAY
1 > (GOOD,1)
2 > DAY
1 > (GOOD,1)
2 > UP
1 > (STUDY,1)
```

3.3.9 coMap 及 coFlatMap 转换

这两个转换操作用来作用于 ConnectedStreams 类型的流上,其功能与 map 和 flatMap 一样,对 ConnectedStreams 中的每个流分别进行 map 和 flatMap 处理。关于这两个转换的简单介绍见表 3-9。

表 3-9 coMap 及 coFlatMap 转换运算

转换运算符	描述	用法

coMap 及 coFlatMap
 	
ConnectedStreams→DataStream
	类似于连接数据流上的 map 和 flatMap	connectedStreams.map(
		  (_ : Int) => true,
		  (_ : String) => false
		)
		connectedStreams.flatMap(
		  (_ : Int) => true,
		  (_ : String) => false
		)

在下面的示例代码中演示了 flatMap 转换的用法。
Scala 代码如下：

```scala
//第3章/TransformerConnect3.scala

import org.apache.flink.streaming.api.functions.co.CoFlatMapFunction
import org.apache.flink.streaming.api.scala._
import org.apache.flink.util.Collector

object TransformerConnect2 {
  def main(args: Array[String]): Unit = {
    //设置流执行环境
    val env = StreamExecutionEnvironment.getExecutionEnvironment

    //union
    //第 1 个数据集
    val ds1:DataStream[String] = env
          .fromElements("good good study","day day up")

    //第 2 个数据集
    val ds2:DataStream[Int] = env.fromElements(1,2,3)

    //连接两个数据集
    val ds:ConnectedStreams[String, Int] = ds1.connect(ds2)

    //flatMap
    ds.flatMap(new CoFlatMapFunction[String,Int,String] {
      override def flatMap1(in1: String, out: Collector[String]): Unit = {
        in1.split(" ").foreach(out.collect)
      }

      override def flatMap2(in2: Int, out: Collector[String]): Unit = {
        1.to(in2).foreach((e) => out.collect(e.toString))
```

```
            }
        }
    ).print

    //执行
    env.execute("flink connect transformatiion")
  }
}
```

Java 代码如下:

```java
//第 3 章/TransformerConnect3.java

import org.apache.flink.streaming.api.datastream.ConnectedStreams;
import org.apache.flink.streaming.api.datastream.DataStream;
import org.apache.flink.streaming.api.environment.StreamExecutionEnvironment;
import org.apache.flink.streaming.api.functions.co.CoFlatMapFunction;
import org.apache.flink.util.Collector;

public class TransformerConnect2 {
    public static void main(String[] args) throws Exception {
        //设置流执行环境
        final StreamExecutionEnvironment env =
                StreamExecutionEnvironment.getExecutionEnvironment();

        //connect 转换操作
        //第 1 个数据集
        DataStream<String> ds1 = env
                .fromElements("good good study","day day up");

        //第 2 个数据集
        DataStream<Integer> ds2 = env.fromElements(1,2,3);

        //连接两个数据集
        ConnectedStreams<String, Integer> ds = ds1.connect(ds2);

        //flatMap
        ds.flatMap(new CoFlatMapFunction<String, Integer, String>() {
            @Override
            public void flatMap1(String s, Collector<String> out) throws Exception {
                for(String word : s.split(" ")){
                    out.collect(word);
                }
            }
```

```java
            @Override
            public void flatMap2(Integer in, Collector<String> out) throws Exception {
                for(int i = 1;i <= in;i++){
                    out.collect(String.valueOf(i));
                }
            }
        }).print();

        //执行
        env.execute("flink connect transformatiion");
    }
}
```

执行以上程序,输出结果如下:

```
4 > 1
2 > good
2 > good
2 > study
3 > day
4 > 2
3 > day
3 > up
4 > 3
3 > 1
3 > 2
2 > 1
```

对于 Scala 实现的 API,因为它支持函数作为参数,所以还可以更加简洁地编码,代码如下:

```scala
//第 3 章/TransformerConnect4.scala

import org.apache.flink.streaming.api.scala._

object TransformerConnect4 {

  def main(args: Array[String]): Unit = {
    //设置流执行环境
    val env = StreamExecutionEnvironment.getExecutionEnvironment

    //union
    //第 1 个数据集
    val ds1:DataStream[String] = env
        .fromElements("good good study","day day up")
```

```
        //第 2 个数据集
        val ds2:DataStream[Int] = env.fromElements(1,2,3)

        //连接两个数据集
        val ds:ConnectedStreams[String, Int] = ds1.connect(ds2)

        //flatMap
        ds.flatMap(
          (e1:String) => e1.split(" "),
          (e2:Int) => 1 to e2
        ).print

        //执行
        env.execute("flink connect transformatiion")
    }
}
```

3.3.10 iterate 转换

这个转换操作通过将一个算子的输出重定向到某个先前的算子在流中创建反馈循环。这对于定义不断更新模型的算法特别有用。关于 iterate 转换的简单介绍见表 3-10。

表 3-10　iterate 转换运算

转换运算符	描　　述	用　　法

iterate
 	
DataStream → IterativeStream →DataStream
	通过将一个操作符的输出重定向到前一个操作符，在流中创建一个"反馈"循环。这对于定义不断更新模型的算法特别有用。右侧的代码从一个流开始，并持续地应用迭代体。大于 0 的元素被发送回反馈通道，其余的元素被转发到下游	initialStream.iterate{
		　　iteration => {
		　　　val iterationBody = iteration.map {/* do something */}
		　　　(iterationBody.filter(_ > 0), iterationBody.filter(_ <= 0))
		　　}
		}

对 DataStream 使用 iterate() 方法创建 IterativeStream，使用 IterativeStream 的 closeWith() 方法来关闭 feedbackStream。

DataStream 提供了两个 iterate() 方法，它们创建并返回 IterativeStream，无参的 iterate() 方法其 maxWaitTimeMillis 为 0。模板代码如下：

```
@Public
public class DataStream<T> {
    ...
```

```java
@PublicEvolving
public IterativeStream<T> iterate() {
        return new IterativeStream<>(this, 0);
}

@PublicEvolving
public IterativeStream<T> iterate(long maxWaitTimeMillis) {
        return new IterativeStream<>(this, maxWaitTimeMillis);
}

...
}
```

IterativeStream 主要提供了两种方法,一个是 closeWith()方法,用于关闭迭代,它主要用于定义要被反馈到 iteration 头部的这部分 iteration(可以理解为回流,或者类似递归的操作,filter 控制的是递归的条件,通过 filter 的 elements 会重新进入 IterativeStream 的头部继续参与后面的运算操作); withFeedbackType()方法创建了 ConnectedIterativeStreams。这个过程如图 3-8 所示。

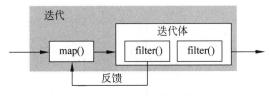

图 3-8　iterate 转换操作

下面通过一个示例来理解这个运算。在下面的示例中,生成"-2,3,4,5"这样的数据流。流程序的目标是只输出非负整数。如果遇到负整数,则被发送回反馈通道,并不断地应用迭代体(每次加 1,直到非负为止);对于遇到的是非负整数,则被向下转发。

Scala 代码如下:

```scala
//第 3 章/TransformerIterate.scala

import org.apache.flink.streaming.api.scala._

object TransformerIterate {
  def main(args: Array[String]): Unit = {
    //设置流执行环境
    val env = StreamExecutionEnvironment.getExecutionEnvironment

    //加载数据集
    val initialStream = env.fromElements(-2,3,4,5)
```

```scala
        //迭代操作
        //使用iterate创建IterativeStream
        initialStream.iterate {
          iteration => {
            //迭代计算部分
            val iterationBody = iteration.map { value =>
                if(value < 0) value + 1 else value
            }
            //下面的元组,第一部分发回到上一步迭代计算,第二部分则继续正常的流
            (iterationBody.filter(_ < 0), iterationBody.filter(_ >= 0))
          }
        }.print
         .setParallelism(1)

        //执行
        env.execute("flink iterate transformatiion")
    }
}
```

Java代码如下:

```java
//第3章/TransformerIterate.java

import org.apache.flink.api.common.functions.FilterFunction;
import org.apache.flink.api.common.functions.MapFunction;
import org.apache.flink.streaming.api.datastream.DataStream;
import org.apache.flink.streaming.api.datastream.IterativeStream;
import org.apache.flink.streaming.api.environment.StreamExecutionEnvironment;

public class TransformerIterate {

    public static void main(String[] args) throws Exception {
        //设置流执行环境
        final StreamExecutionEnvironment env =
            StreamExecutionEnvironment.getExecutionEnvironment();

        //加载数据集
        DataStream<Long> initialStream = env
           .setParallelism(1)
           .fromElements( - 2L, 3L, 4L, 5L);

        //转换为迭代流
        IterativeStream<Long> iteration = initialStream.iterate();

        //迭代流中的前运算符(迭代体)
```

```java
DataStream<Long> iterationBody = iteration
    .map(new MapFunction<Long, Long>() {
        @Override
        public Long map(Long value) throws Exception {
            return value < 0 ? value + 1 : value;
        }
    });

//过滤,凡是小于0的元素都被发回前运算符进行迭代计算
DataStream<Long> feedback = iterationBody
    .filter(new FilterFunction<Long>(){
        @Override
        public boolean filter(Long value) throws Exception {
            return value < 0;
        }
    });

//这里设置feedback这个数据流是被反馈的通道
//只要是value<0的数据都会被重新迭代计算
iteration.closeWith(feedback);

//迭代流中大于0的数会被挑出来作为输出
DataStream<Long> output = iterationBody
    .filter(new FilterFunction<Long>(){
        @Override
        public boolean filter(Long value) throws Exception {
            return value >= 0;
        }
    });

output.print();

//执行
env.execute("flink iterate transformatiion");
    }
}
```

执行以上程序,输出结果如下:

```
3
4
5
0
```

默认情况下,带有迭代的 DataStream 将永远不会终止,但用户可以使用 maxWaitTime

参数设置迭代头的最大等待时间。如果在设置的时间内没有接收到数据，则流将终止，代码如下：

```
public IterativeStream<T> iterate(long maxWaitTimeMillis)
```

3.3.11 project 转换

project 转换操作从事件流中选择一组属性子集，并且只将选中的元素发送到下一个处理流（相当于 SQL 语句中的投影概念）。下面是进行 project 转换的示例代码。

Scala 代码：不支持。

Java 代码如下：

```java
//第3章/BatchJob.java

import org.apache.flink.api.java.tuple.Tuple3;
import org.apache.flink.streaming.api.datastream.DataStream;
import org.apache.flink.streaming.api.environment.StreamExecutionEnvironment;

public class BatchJob {

    public static void main(String[] args) throws Exception {
        //设置批处理执行环境
        final StreamExecutionEnvironment env =
                StreamExecutionEnvironment.getExecutionEnvironment();

        //project 转换
        Tuple3<Integer, String, Double> user01 = new Tuple3<>(1,"张三",12000.00);
        Tuple3<Integer, String, Double> user02 = new Tuple3<>(2,"李四",22000.00);
        Tuple3<Integer, String, Double> user03 = new Tuple3<>(3,"王老五",18000.00);

        DataStream<Tuple3<Integer, String, Double>> ds =
                env.fromElements(user01,user02,user03);

        //选择第3列和第2列
        DataStream<Tuple3<Integer, String, Double>> dsp = ds.project(2,1);

        dsp.print();
    }
}
```

执行以上程序，输出结果如下：

```
(张三,12000.0)
(李四,22000.0)
(王老五,18000.0)
```

3.4 Flink 流数据分区

Flink 应用程序本质上是并行和分布式的。在程序执行期间，一个流被切分为一个或多个流分区，每个运算符有一个或多个运算子任务。运算子任务彼此独立，并在不同的线程中执行，也可能在不同的机器或容器上执行。运算子任务的数量是该运算符的并行度。

有状态运算符的并行实例集实际上是切分的 key-value 存储。每个并行实例负责处理特定 key 组的事件，这些 key 的状态保存在本地。

3.4.1 流数据分发模式

Flink 程序在执行时被映射到流式数据流，由流和转换运算符组成。每个数据流开始于一个或多个源（source），结束于一个或多个接收器（sinks）。数据流类似于任意有向无环图（DAG），它是一个执行计划。

JobClient 将接收的程序转换为对应的数据流的典型过程如图 3-9 所示。

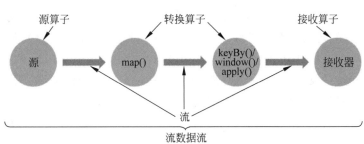

图 3-9 Flink 流程序数据流图转换

通常，程序中的转换与数据流中的运算符之间存在一对一的对应关系，然而，有时一个转换可能包含多个转换运算符。

Flink 数据流默认为并行地分布式地执行,因此实际的转化结果可能更像图 3-10 所示(其中上半部分为执行简图,下半部分为并行执行视图)。

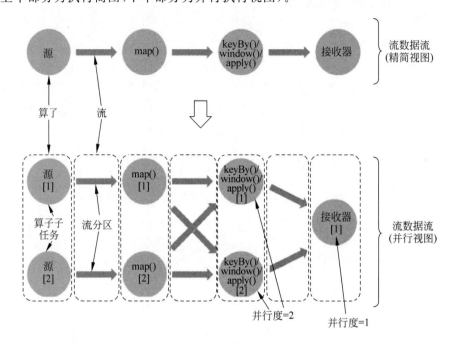

图 3-10 Flink 流程序分布式并行执行

Flink 流中两个运算符之间的数据分发有以下两种模式。

(1) one-to-one 流:一对一分发的模式,例如图 3-10 中的 Source 运算符和 map()运算符之间。这种方式会保持元素的分区和顺序,也就是说,它保证原来的数据分区和排序不会改变。这意味着 map()运算符的子任务[1]将看到与 Source 运算符的子任务[1]生成的元素顺序相同的元素。

(2) redistributing 流:重新分发的模式,就像上面的 map()和 keyBy/window 之间,以及 keyBy/window 和 Sink 之间。这种方式会打乱流数据原有的划分情况和排序情况。每个运算符子任务根据所选的转换向不同的目标子任务发送数据。例如 keyBy()(通过散列 key 重新分区)、broadcast()或 rebalance()(随机重新分区)。在 Redistributing 交换中,元素之间的顺序只保留在每对发送和接收子任务中(例如 map()的子任务[1]和 keyBy/window 的子任务[2])。对于 keyBy 来讲,就是把相同的 key 的数据分发到同一个节点上,而对于最终的 Sink,由于并行执行,它收到的数据可能不是按照原有的排序情况到达的。

3.4.2 数据分区方法

在 Flink 中,用户可以对转换后的流数据执行物理分区,以进行低级控制(如果需要)。Flink 支持 8 种分区方法,见表 3-11。

表 3-11　Flink 支持的 8 种分区方法

分区方法	操作方式
shuffle	随机对数据流进行分区,根据均匀分布随机划分元素
rebalance	round-robin 方式。使用循环分配分区元素的方法,为每个分区创建相等的负载
rescale	根据上下游运算符的数量,对元素进行一个均匀分配
broadcast	将输出元素广播到下一个操作(算子)的每个并行实例
keyBy	按 key 划分数据流,相同 key 的元素划分到一个分区上
forward	将输出元素转发到下一个操作的本地子任务
global	将所有的数据都发送到下游 0 号分区中
自定义分区	使用用户定义的分区程序(Partitioner)为每个元素选择目标任务

不同的分区方法代表了 Flink 中不同的数据分区策略,数据的分区策略决定了数据会分发到下游算子的那个分区。下面详细了解这 8 种分区方法的原理和应用。

1) shuffle()方法

该方法使用 ShufflePartitioner 分区程序设置 DataStream 的分区,通过随机选择一个输出通道平均分配数据。该方法会将输出元素均匀随机地打乱到下一个操作(算子),其分区原理如图 3-11 所示。

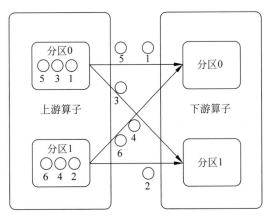

图 3-11　shuffle()方法分区示意图

下面的代码演示了如何使用 shuffle()方法随机进行分区。
Scala 代码如下:

```scala
//第 3 章/PartitionShuffle.scala

import org.apache.flink.streaming.api.scala._

object PartitionShuffle {
  def main(args: Array[String]) {
    //设置流执行环境
```

```scala
val env = StreamExecutionEnvironment.getExecutionEnvironment

//从自定义的集合中读取数据
val stream = env.fromCollection(List(1,2,3,4,5))

//这里只是为了能够将并行度设置为2
val stream2 = stream
  .map(v =>{(v%2,v)})          //偶数 key 为 0,奇数 key 为 1
  .keyBy(0)                     //按奇偶进行分区
  .map(v =>(v._1,v._2))
  .setParallelism(2)
println(stream2.parallelism)    //查看并行度

//查看随机分区的结果
stream2.shuffle.print("shuffle").setParallelism(3)

//触发流程序执行
env.execute("shuffle 分区示例")
}
}
```

Java 代码如下:

```java
//第 3 章/PartitionShuffle.java

import org.apache.flink.api.common.functions.MapFunction;
import org.apache.flink.api.java.tuple.Tuple2;
import org.apache.flink.streaming.api.datastream.DataStream;
import org.apache.flink.streaming.api.environment.StreamExecutionEnvironment;
import Java.util.Arrays;

public class PartitionShuffle {

    public static void main(String[] args) throws Exception {
        //设置流执行环境
        final StreamExecutionEnvironment env =
            StreamExecutionEnvironment.getExecutionEnvironment();

        //从自定义的集合中读取数据
        DataStream<Integer> stream =
            env.fromCollection(Arrays.asList(1,2,3,4,5));

        //这里只是为了能够将并行度设置为2
        DataStream<Tuple2<Integer,Integer>> stream2 = stream
            .map(new MapFunction<Integer, Tuple2<Integer,Integer>>() {
```

```java
            @Override
            public Tuple2<Integer, Integer> map(Integer input) throws Exception {
                return new Tuple2<>(input % 2, input);
            }
        })                                              //偶数key为0,奇数key为1
        .keyBy(t -> t.f0)                               //按奇偶进行分区
        .map(new MapFunction<Tuple2<Integer,Integer>, Tuple2<Integer,Integer>>() {
            @Override
            public Tuple2<Integer, Integer> map(Tuple2<Integer, Integer> t) throws Exception {
                return new Tuple2<>(t.f0, t.f1);
            }
        })
        .setParallelism(2);
        System.out.println(stream2.getParallelism());   //查看并行度

        //查看随机分区的结果
        stream2.shuffle().print("shuffle").setParallelism(3);

        //触发流程序执行
        env.execute("shuffle分区示例");
    }
}
```

执行以上代码,输出结果如下:

```
2
shuffle:1> (0,2)
shuffle:3> (1,5)
shuffle:2> (1,3)
shuffle:1> (1,1)
shuffle:3> (0,4)
```

2) rebalance()方法

该方法使用 RebalancePartitioner 分区程序设置 DataStream 的分区,使用循环通过输出通道平均分配数据。该方法会将输出元素以轮询方式均匀地分布到下一个操作(算子)的实例中,其分区原理如图 3-12 所示。

这种类型的分区有助于均匀地分布数据。它使用循环分配分区元素的方法,为每个分区创建相等的负载。这种类型的分区对于存在数据倾斜的情况下的性能优化非常有用,如图 3-13 所示。

在下面的示例中,演示了如何使用 rebalance()方法。

Scala 代码如下:

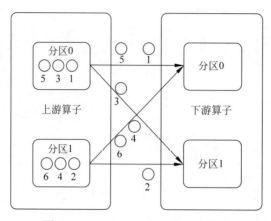

图 3-12 rebalance()方法分区示意图

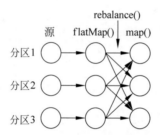

图 3-13 在 flatMap 转换操作之后,均匀分布数据

```
//第3章/RebalancePartitioner.scala

import org.apache.flink.streaming.api.scala._

object RebalancePartitioner extends App{

  //创建执行环境
  val env = StreamExecutionEnvironment.getExecutionEnvironment

  //从自定义的集合中读取数据
  val stream = env.fromCollection(List(1,2,3,4,5,6))

  //直接打印数据
  stream.rebalance.print("rebalance").setParallelism(2)

  env.execute("rebalance 分区示例")
}
```

Java 代码如下:

```java
//第 3 章/RebalancePartitioner.java

import org.apache.flink.api.common.functions.MapFunction;
import org.apache.flink.api.java.tuple.Tuple2;
import org.apache.flink.streaming.api.datastream.DataStream;
import org.apache.flink.streaming.api.environment.StreamExecutionEnvironment;
import Java.util.Arrays;

public class PartitionRebalance {

    public static void main(String[] args) throws Exception {
        //设置流执行环境
        final StreamExecutionEnvironment env = StreamExecutionEnvironment.getExecutionEnvironment();

        //从自定义的集合中读取数据
        DataStream< Integer > stream = env.fromCollection(Arrays.asList(1,2,3,4,5,6));

        //直接打印数据
        stream.rebalance().print("rebalance").setParallelism(2);

        //触发流程序执行
        env.execute("rebalance 分区示例");
    }
}
```

执行以上程序,输出结果如下:

```
rebalance:2 > 1
rebalance:1 > 2
rebalance:2 > 3
rebalance:1 > 4
rebalance:2 > 5
rebalance:1 > 6
```

3) rescale()方法

该方法使用 RescalePartitioner 分区程序设置 DataStream 的分区,使用循环通过输出通道平均分配数据。该方法会将输出元素以轮询方式均匀地分布到下一个操作(算子)的实例子集,其分区原理如图 3-14 所示。

在这种分区方法中,Flink 循环将元素划分为下游操作的子集。上游操作向其发送元素的下游操作子集取决于上游和下游操作的并行度。例如,如果上游操作的并行度为 2,而下游操作的并行度为 4,则一个上游操作将把元素分配给两个下游操作,而另一个上游操作将分配给另外两个下游操作。另一方面,如果下游操作的并行度为 2,而上游操作的并行度为

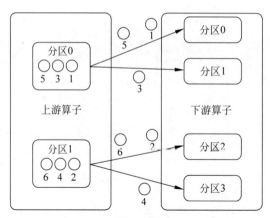

图 3-14 rescale()方法对数据进行重分区

4,则两个上游操作将分配给一个下游操作,而另外两个上游操作将分配给另一个下游操作。

在上下游算子的并行度不是彼此的倍数的情况下,一个或几个下游操作与上游操作的输入数量不同。

如果用户希望有这样的管道,例如,从一个数据源的每个并行实例分散到几个 mapper 的一个子集来分配负载,但又不希望使用 rebalance()导致完全的重新平衡,则这种方法是非常有用的。根据其他配置值,例如 TaskManager 的槽位数,这将只需本地数据传输,而不需要通过网络传输数据。

在下面的示例中,演示了如何使用 rescale()方法。

Scala 代码如下:

```
//第 3 章/PartitionRescale.scala

import org.apache.flink.streaming.api.scala._

object PartitionRescale {
  def main(args: Array[String]) {
    //设置流执行环境
    val env = StreamExecutionEnvironment.getExecutionEnvironment

    //从自定义的集合中读取数据
    val stream = env.fromCollection(List(1,2,3,4,5,6,7,8))
    stream.print("before rescale")

    //直接打印数据
    stream.rescale.print("rescale").setParallelism(2)

    //触发流程序执行
    env.execute("rescale 分区示例")
  }
}
```

Java 代码如下：

```java
//第3章/PartitionRescale.java

import org.apache.flink.streaming.api.datastream.DataStream;
import org.apache.flink.streaming.api.environment.StreamExecutionEnvironment;
import Java.util.Arrays;

public class PartitionRescale {

    public static void main(String[] args) throws Exception {
        //设置流执行环境
        final StreamExecutionEnvironment env =
                StreamExecutionEnvironment.getExecutionEnvironment();

        //从自定义的集合中读取数据
        DataStream < Integer > stream =
                env.fromCollection(Arrays.asList(1,2,3,4,5,6,7,8));
        stream.print("before rescale");

        //直接打印数据
        stream.rescale().print("rescale").setParallelism(2);

        //触发流程序执行
        env.execute("rescale 分区示例");
    }
}
```

执行以上代码，输出结果如下：

```
before rescale:7 > 7
before rescale:3 > 3
before rescale:1 > 1
before rescale:5 > 5
before rescale:8 > 8
before rescale:4 > 4
before rescale:2 > 2
before rescale:6 > 6
rescale:1 > 1
rescale:1 > 3
rescale:2 > 2
rescale:2 > 4
rescale:1 > 5
rescale:2 > 6
rescale:1 > 7
rescale:2 > 8
```

简而言之，rescale()方法用于跨操作(算子)分发数据，对数据子集执行转换并将它们组合在一起。这种重新平衡只发生在单个节点上，因此不需要跨网络进行任何数据传输。

4) broadcast()方法

该方法使用 BroadcastPartitioner 分区程序设置 DataStream 的分区，通过选择所有输出通道将所有记录分发到每个分区。该方法会将输出元素广播到下一个操作(算子)的每个并行实例，其分区原理如图 3-15 所示。

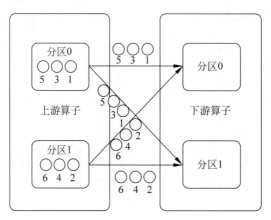

图 3-15   broadcast()方法对数据进行广播

在下面的示例中，演示了如何使用 broadcast()方法。
Scala 代码如下：

```scala
//第 3 章/PartitionBroadcast.scala

import org.apache.flink.streaming.api.scala._

object PartitionBroadcast {
  def main(args: Array[String]) {
    //设置流执行环境
    val env = StreamExecutionEnvironment.getExecutionEnvironment

    //从自定义的集合中读取数据
    val stream = env.fromCollection(List(1,2,3,4,5))

    //直接打印数据
    stream.broadcast.print("broadcast").setParallelism(2)

    //触发流程序执行
    env.execute("broadcast 分区示例")
  }
}
```

Java 代码如下：

```java
//第 3 章/PartitionBroadcast.java

import org.apache.flink.streaming.api.datastream.DataStream;
import org.apache.flink.streaming.api.environment.StreamExecutionEnvironment;

import Java.util.Arrays;

public class PartitionBroadcast {

    public static void main(String[] args) throws Exception {
        //设置流执行环境
        final StreamExecutionEnvironment env =
                StreamExecutionEnvironment.getExecutionEnvironment();

        //从自定义的集合中读取数据
        DataStream < Integer > stream =
                env.fromCollection(Arrays.asList(1,2,3,4,5));

        //直接打印数据
        stream.broadcast().print("broadcast").setParallelism(2);

        //触发流程序执行
        env.execute("broadcast 分区示例");
    }
}
```

执行以上代码，输出结果如下：

```
broadcast:1 > 1
broadcast:2 > 1
broadcast:1 > 2
broadcast:2 > 2
broadcast:1 > 3
broadcast:2 > 3
broadcast:1 > 4
broadcast:2 > 4
broadcast:1 > 5
broadcast:2 > 5
```

5）forward()方法

该方法使用 ForwardPartitioner 分区程序设置 DataStream 的分区，仅将元素转发到本地运行的下游操作（算子）。该方法会将输出元素转发到下一个操作（算子）的本地子任务，其分区原理如图 3-16 所示。

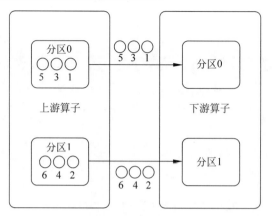

图 3-16　forward()方法仅将元素转发到本地运行的下游操作

在上下游的算子没有指定分区器的情况下，如果上下游的算子并行度一致，则使用 ForwardPartitioner，否则使用 RebalancePartitioner。对于 ForwardPartitioner，必须保证上下游算子并行度一致，即上游算子与下游算子是一对一的关系，否则会抛出异常。

在下面的示例中，演示了如何使用 forward()方法。

Scala 代码如下：

```scala
//第 3 章/PartitionForward.scala
import org.apache.flink.streaming.api.scala._

object PartitionForward {
  def main(args: Array[String]) {
    //设置流执行环境
    val env = StreamExecutionEnvironment.getExecutionEnvironment

    //从自定义的集合中读取数据
    val stream = env.fromCollection(List(1,2,3,4,5))

    //直接打印数据
    Stream
      .map(v =>{v * v})
      .setParallelism(2)
      .forward
      .print()
      .setParallelism(2)

    //触发流程序执行
    env.execute("forward 分区示例")
  }
}
```

Java 代码如下:

```java
//第 3 章/PartitionForward.java

import org.apache.flink.api.common.functions.MapFunction;
import org.apache.flink.streaming.api.datastream.DataStream;
import org.apache.flink.streaming.api.environment.StreamExecutionEnvironment;

import Java.util.Arrays;

public class PartitionForward {

    public static void main(String[] args) throws Exception {
        //设置流执行环境
        final StreamExecutionEnvironment env =
                StreamExecutionEnvironment.getExecutionEnvironment();

        //从自定义的集合中读取数据
        DataStream< Integer > stream =
                env.fromCollection(Arrays.asList(1,2,3,4,5));

        //直接打印数据
        stream
            .map(new MapFunction< Integer, Integer >() {
                @Override
                public Integer map(Integer input) throws Exception {
                    return input * input;
                }
            })
            .setParallelism(2)
            .forward()
            .print("forward")
            .setParallelism(2);

        //触发流程序执行
        env.execute("forward 分区示例");
    }
}
```

执行以上代码,输出结果如下:

```
forward:1 > 1
forward:2 > 4
forward:1 > 9
forward:2 > 16
forward:1 > 25
```

6) keyBy()方法

该方法使用 KeyGroupStreamPartitioner 分区程序设置 DataStream 的分区,根据 key 的分组索引选择目标通道,将输出元素发送到相对应的下游分区。该方法会创建一个新的 KeyedStream,使用提供的 key 来划分其操作符(算子)状态。该方法的分区原理如图 3-17 所示。

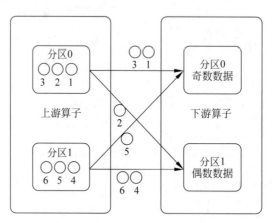

图 3-17　keyBy()方法分区示意图

在下面的示例中,演示了如何使用 keyBy()方法。

Scala 代码如下:

```scala
//第 3 章/PartitionKeyBy.scala
import org.apache.flink.streaming.api.scala._

object PartitionKeyBy {

  def main(args: Array[String]) {
    //设置流执行环境
    val env = StreamExecutionEnvironment.getExecutionEnvironment

    //从自定义的集合中读取数据
    val stream = env.fromCollection(List(1,2,3,4,5,6))

    //先转换为(k,v)对,再执行 keyBy,然后打印数据
    val stream2 = stream.map(v => {(v%3,v)})
    stream2.setParallelism(2).keyBy(0).print("key")

    //触发流程序执行
    env.execute("keyBy 分区示例")
  }
}
```

Java 代码如下：

```java
//第3章/PartitionKeyBy.java

import org.apache.flink.api.common.functions.MapFunction;
import org.apache.flink.api.java.tuple.Tuple2;
import org.apache.flink.streaming.api.datastream.DataStream;
import org.apache.flink.streaming.api.environment.StreamExecutionEnvironment;

import Java.util.Arrays;

public class PartitionKeyBy {

    public static void main(String[] args) throws Exception {
        //设置流执行环境
        final StreamExecutionEnvironment env =
                StreamExecutionEnvironment.getExecutionEnvironment();

        //从自定义的集合中读取数据
        DataStream<Integer> stream =
                env.fromCollection(Arrays.asList(1,2,3,4,5,6));

        //直接打印数据
        DataStream<Tuple2<Integer, Integer>> stream2 = stream
          .map(new MapFunction<Integer, Tuple2<Integer, Integer>>() {
            @Override
            public Tuple2<Integer, Integer> map(Integer input) throws Exception {
                return new Tuple2<>(input % 3, input);
            }
        }).setParallelism(2);

        stream2.keyBy(t -> t.f0).print("key");

        //触发流程序执行
        env.execute("keyBy 分区示例");
    }
}
```

执行以上代码，输出结果如下：

```
key:8 > (2,2)
key:6 > (1,1)
key:6 > (0,3)
key:8 > (2,5)
key:6 > (1,4)
key:6 > (0,6)
```

7）global()方法

该方法使用 GlobalPartitioner 分区程序设置 DataStream 的分区，以便将输出值都转到下一个处理操作符（算子）的第 1 个实例。使用此设置时要小心，因为它可能会在应用程序中造成严重的性能瓶颈。该方法的分区原理如图 3-18 所示。

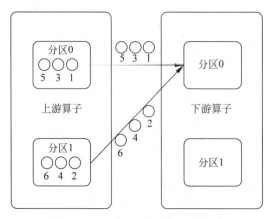

图 3-18 global()方法分区示意图

在下面的示例中，演示了如何使用 global()方法。
Scala 代码如下：

```scala
//第 3 章/PartitionGlobal.scala

import org.apache.flink.streaming.api.scala._

object PartitionGlobal {
  def main(args: Array[String]) {
    //设置流执行环境
    val env = StreamExecutionEnvironment.getExecutionEnvironment

    //从自定义的集合中读取数据
    val stream = env.fromCollection(List(1,2,3,4,5))

    //直接打印数据
    stream.print()

    //使用 GLobalPartitioner 之后打印数据
    stream.global.print("global")

    //触发流程序执行
    env.execute("global 分区示例")
  }
}
```

Java 代码如下：

```java
//第3章/PartitionGlobal.java

import org.apache.flink.streaming.api.datastream.DataStream;
import org.apache.flink.streaming.api.environment.StreamExecutionEnvironment;

import Java.util.Arrays;

public class PartitionGlobal {

    public static void main(String[] args) throws Exception {
        //设置流执行环境
        final StreamExecutionEnvironment env =
                StreamExecutionEnvironment.getExecutionEnvironment();

        //从自定义的集合中读取数据
        DataStream < Integer > stream =
                env.fromCollection(Arrays.asList(1,2,3,4,5));

        //直接打印数据
        stream.print();

        //使用 GLobalPartitioner 之后打印数据
        stream.global().print("global");

        //触发流程序执行
        env.execute("global 分区示例");
    }
}
```

执行以上代码，输出结果如下：

```
global:1 > 2
global:1 > 3
global:1 > 4
global:1 > 5
```

8）自定义分区程序

通过继承 org.apache.flink.api.common.functions.Partitioner，用户也可以自定义分区程序为每个元素选择目标分区。下面的示例提供了一个分区器的自定义实现，将数据流中的整数按奇偶性分发到不同的子分区中。

Scala 代码如下：

```scala
//第3章/PartitionUDF.scala

import org.apache.flink.api.common.functions.Partitioner
import org.apache.flink.streaming.api.scala._

object PartitionUDF {

  //自定义分区程序
  object CustomPartitioner extends Partitioner[String]{
    //重写partition方法
    override def partition(key: String, numPartitions: Int): Int = {
      //根据key值的奇偶性返回不同的分区id
      key.toInt % 2
    }
  }

  def main(args: Array[String]) {
    //设置流执行环境
    val env = StreamExecutionEnvironment.getExecutionEnvironment

    //从自定义的集合中读取数据
    val stream = env.fromCollection(List("1","2","3","4","5"))

    val stream2 = stream.map(value =>{((value.toInt % 2).toString,value)})
    stream2
      .partitionCustom(CustomPartitioner,0)
      .print()
      .setParallelism(2)

    //触发流程序执行
    env.execute("自定义分区示例")
  }
}
```

Java代码如下：

```java
//第3章/PartitionUDF.java

import org.apache.flink.api.common.functions.MapFunction;
import org.apache.flink.api.common.functions.Partitioner;
import org.apache.flink.api.java.tuple.Tuple2;
import org.apache.flink.streaming.api.datastream.DataStream;
import org.apache.flink.streaming.api.environment.StreamExecutionEnvironment;

import Java.util.Arrays;
```

```java
public class PartitionUDF {

    public static void main(String[] args) throws Exception {
        //设置流执行环境
        final StreamExecutionEnvironment env =
                StreamExecutionEnvironment.getExecutionEnvironment();

        //从自定义的集合中读取数据
        DataStream<String> stream =
                env.fromCollection(Arrays.asList("1","2","3","4","5"));

        //先转换为<k,v>
        DataStream<Tuple2<String,String>> stream2 = stream.map(
            new MapFunction<String, Tuple2<String,String>>() {
                @Override
                public Tuple2<String,String> map(String s) throws Exception {
                    return new Tuple2<>(String.valueOf(Integer.parseInt(s) % 2), s);
                }
            });

        //应用自定义分区器
        stream2.partitionCustom(new Partitioner<String>() {
            @Override
            public int partition(String key, int i) {
                return Integer.parseInt(key) % 2;
            }
        }, t -> t.f0).print().setParallelism(2);

        //触发流程序执行
        env.execute("自定义分区示例");
    }
}
```

执行以上代码,输出结果如下:

```
2> (1,1)
1> (0,2)
2> (1,5)
2> (1,3)
1> (0,4)
```

在编写自定义分区程序时,需要确保实现了有效的哈希函数。

3.4.3 数据分区示例

默认情况下,Flink 应用程序会使用 key 的哈希分区或随机分区,但在有些情况下,用户需要自定义分区规则,这就需要自己来定义分区器(分区程序)。

【示例3-7】 (Scala实现)使用自定义分区程序,按年龄对数据流元素进行分区。
建议按以下步骤执行：

(1) 在IntelliJ IDEA中创建一个Flink项目,使用flink-quickstart-scala项目模板(Flink项目的创建过程可参考2.2节)。

(2) 设置依赖。在pom.xml文件中添加如下依赖(根据项目模板创建这个依赖会自动添加,此步可省略)：

```xml
<dependency>
    <groupId>org.apache.flink</groupId>
    <artifactId>flink-scala_2.12</artifactId>
    <version>1.13.2</version>
    <scope>provided</scope>
</dependency>
<dependency>
    <groupId>org.apache.flink</groupId>
    <artifactId>flink-streaming-scala_2.12</artifactId>
    <version>1.13.2</version>
    <scope>provided</scope>
</dependency>
```

(3) 创建一个case class类,用来表示流中的数据类型,代码如下：

```scala
//第3章/PartitionDemo1.scala

import org.apache.flink.streaming.api.scala.StreamExecutionEnvironment

object PartitionDemo1 {

    //定义case class类,表示流数据类型
    case class Person(name:String, age:Int)

    def main(args: Array[String]): Unit = {
        //设置批处理执行环境
        val env = StreamExecutionEnvironment.getExecutionEnvironment

        //触发流程序开始执行
        env.execute("stream demo")
    }
}
```

(4) 自定义分区类,实现了Partitioner接口,按年龄(age字段)划分分区。这里按年龄把流数据划分为3个分区：20岁以下的、20~30岁及30岁以上的,代码如下：

```scala
//第3章/PartitionDemo1.scala

import org.apache.flink.api.common.functions.Partitioner
import org.apache.flink.streaming.api.scala.StreamExecutionEnvironment

object PartitionDemo1 {

  //定义 case class 类,表示流数据类型
  case class Person(name:String, age:Int)

  //自定义分区器,以年龄为 key
  class AgePartitioner extends Partitioner[Int] {
    //重写 partition 方法
    override def partition(key: Int, i: Int): Int = {
      key match {
        case age if age < 20 => 0
        case age if age > 30 => 2
        case _ => 1
      }
    }
  }

  def main(args: Array[String]): Unit = {
    //设置批处理执行环境
    val env = StreamExecutionEnvironment.getExecutionEnvironment

    //触发流程序开始执行
    env.execute("stream demo")
  }
}
```

(5)测试自定义分区逻辑。编辑流处理,代码如下:

```scala
//第3章/PartitionDemo1.scala

import org.apache.flink.api.common.functions.Partitioner
import org.apache.flink.api.scala._
import org.apache.flink.streaming.api.scala.StreamExecutionEnvironment

object PartitionDemo1 {

  //定义 case class 类,表示流数据类型
  case class Person(name:String, age:Int)

  //自定义分区器,以年龄为 key
```

```scala
class AgePartitioner extends Partitioner[Int] {
  //重写 partition 方法
  override def partition(k: Int, i: Int): Int = {
    k match {
      case age if age < 20 => 0
      case age if age > 30 => 2
      case _ => 1
    }
  }
}

def main(args: Array[String]): Unit = {
  //设置批处理执行环境
  val env = StreamExecutionEnvironment.getExecutionEnvironment

  //读取数据源,构造 DataStream
  val people = List(
    Person("张三", 21),
    Person("李四", 16),
    Person("王老五", 35),
    Person("张三 2", 22),
    Person("李四 2", 17),
    Person("王老五 2", 36)
  )
  val personDS = env.fromCollection(people)

  //应用自定义分区器,按年龄字段进行分组
  //注:通过字段位置指定 key 只对元组数据类型有效
  val adults = personDS.partitionCustom(new AgePartitioner, "age")

  //将结果输出到控制台
  adults.print

  //触发流程序开始执行
  env.execute("stream demo")
}
```

(6) 执行以上程序,输出结果如下:

```
分区之前分区数:8
...
2> Person(张三,21)
1> Person(李四,16)
3> Person(王老五,35)
1> Person(李四 2,17)
2> Person(张三 2,22)
3> Person(王老五 2,36)
```

从上面的输出结果可以看到,数据在 3 个分区中分别并行地被处理。

注意:输出结果前面的整数是分区号。

对于简单的自定义分区器,也可以直接使用匿名内部类,以简化代码,代码如下:

```scala
//第 3 章/PartitionDemo2.scala

import org.apache.flink.api.common.functions.Partitioner
import org.apache.flink.api.scala._
import org.apache.flink.streaming.api.scala.StreamExecutionEnvironment

object PartitionDemo2 {

  //定义 case class 类,表示流数据类型
  case class Person(name:String, age:Int)

  def main(args: Array[String]): Unit = {
    //设置批处理执行环境
    val env = StreamExecutionEnvironment.getExecutionEnvironment

    //读取数据源,构造 DataStream
    val people = List(
      Person("张三", 21),
      Person("李四", 16),
      Person("王老五", 35),
      Person("张三 2", 22),
      Person("李四 2", 17),
      Person("王老五 2", 36)
    )
    val personDS = env.fromCollection(people)

    //应用自定义分区器,按年龄字段进行分组
    //注:通过字段位置指定 key 只对元组数据类型有效
    val adults = personDS.partitionCustom(new Partitioner[Int] {
      //重写 partition 方法
      override def partition(k: Int, i: Int): Int = {
        k match {
          case age if age < 20 => 0
          case age if age > 30 => 2
          case _ => 1
        }
      }
    }, "age")    //第 2 个分组字段参数也可以使用_.age 或 person => person.age
```

```
        //将结果输出到控制台
        adults.print

        //触发流程序开始执行
        env.execute("stream demo")
    }
}
```

【示例 3-8】（Java 实现）使用自定义分区程序,按年龄对数据流元素进行分区。

(1) 在 IntelliJ IDEA 中创建一个 Flink 项目,使用 flink-quickstart-Java 项目模板(Flink 项目的创建过程可参考 2.2 节)。

(2) 设置依赖。在 pom.xml 文件中添加如下依赖(根据项目模板创建,这个依赖会自动添加,此步可省略):

```
<dependency>
    <groupId>org.apache.flink</groupId>
    <artifactId>flink-Java</artifactId>
    <version>1.13.2</version>
    <scope>provided</scope>
</dependency>
<dependency>
    <groupId>org.apache.flink</groupId>
    <artifactId>flink-streaming-Java_2.12</artifactId>
    <version>1.13.2</version>
    <scope>provided</scope>
</dependency>
```

(3) 创建一个 POJO 类,用来表示流中的数据,代码如下:

```
//第 3 章/PartitionDemo1.java

import org.apache.flink.api.common.functions.Partitioner;
import org.apache.flink.streaming.api.datastream.DataStream;
import org.apache.flink.streaming.api.environment.StreamExecutionEnvironment;

public class PartitionDemo1 {

    //POJO 类,表示人员信息实体
    public static class Person {
        public String name;           //存储姓名
        public Integer age;           //存储年龄

        //空构造器
        public Person() {}
```

```java
        //构造器,初始化属性
        public Person(String name, Integer age) {
            this.name = name;
            this.age = age;
        }

        //用于调试时输出信息
        public String toString() {
            return this.name.toString() + ": age " + this.age.toString();
        }
    }

    public static void main(String[] args) throws Exception {
        //设置流执行环境
        final StreamExecutionEnvironment env =
            StreamExecutionEnvironment.getExecutionEnvironment();

        //触发流程序开始执行
        env.execute("stream demo");
    }
}
```

(4) 自定义分区类,实现了 Partitioner 接口,按年龄(age 字段)划分分区。这里按年龄把流数据分为 3 个分区:20 岁以下的、20～30 岁及 30 岁以上的,代码如下:

```java
//第 3 章/PartitionDemo1.java
import org.apache.flink.api.common.functions.Partitioner;
import org.apache.flink.streaming.api.datastream.DataStream;
import org.apache.flink.streaming.api.environment.StreamExecutionEnvironment;

public class PartitionDemo1 {

    //POJO 类,表示人员信息实体
    public static class Person {
        public String name;                     //存储姓名
        public Integer age;                     //存储年龄

        //空构造器
        public Person() {};

        //构造器,初始化属性
        public Person(String name, Integer age) {
            this.name = name;
            this.age = age;
```

```java
        };

        //用于调试时输出信息
        public String toString() {
            return this.name.toString() + ": age " + this.age.toString();
        };
    }

    //自定义分区器,以年龄为key
    public static class AgePartitioner implements Partitioner<Integer> {
        @Override
        public int partition(Integer key, int numPartitions) {
            if(key < 20){
                return 0;
            }else if(key > 30){
                return 2;
            }else{
                return 1;
            }
        }
    }

    public static void main(String[] args) throws Exception {
        //设置流执行环境
        final StreamExecutionEnvironment env =
            StreamExecutionEnvironment.getExecutionEnvironment();

        //触发流程序开始执行
        env.execute("stream demo");
    }
}
```

(5) 测试自定义分区逻辑。编辑流处理,代码如下:

```java
//第3章/PartitionDemo1.java
import org.apache.flink.api.common.functions.Partitioner;
import org.apache.flink.streaming.api.datastream.DataStream;
import org.apache.flink.streaming.api.environment.StreamExecutionEnvironment;

public class PartitionDemo1 {

    //POJO类,表示人员信息实体
    public static class Person {
        public String name;         //存储姓名
        public Integer age;         //存储年龄
```

```java
    //空构造器
    public Person() {};

    //构造器,初始化属性
    public Person(String name, Integer age) {
        this.name = name;
        this.age = age;
    };

    //用于调试时输出信息
    public String toString() {
        return this.name.toString() + ": age " + this.age.toString();
    };
}

//自定义分区器,以年龄为 key
public static class AgePartitioner implements Partitioner<Integer> {
    @Override
    public int partition(Integer key, int numPartitions) {
        if(key < 20){
            return 0;
        }else if(key > 30){
            return 2;
        }else{
            return 1;
        }
    }
}

public static void main(String[] args) throws Exception {
    //设置流执行环境
    final StreamExecutionEnvironment env =
        StreamExecutionEnvironment.getExecutionEnvironment();

    //读取数据源,构造 DataStream
    DataStream<Person> personDS = env.fromElements(
        new Person("张三", 21),
        new Person("李四", 16),
        new Person("王老五", 35),
        new Person("张三 2", 22),
        new Person("李四 2", 17),
        new Person("王老五 2", 36));

    //应用自定义分区器,按年龄字段进行分组
    //注:通过字段位置指定 key,只对元组数据类型有效
    DataStream<Person> adults = personDS
```

```java
        .partitionCustom(new AgePartitioner(), p -> p.age);

        //将结果输出到控制台
        adults.print();

        //触发流程序开始执行
        env.execute("stream demo");
    }
}
```

(6) 执行以上程序,输出结果如下:

```
1> 李四: age 16
2> 张三: age 21
3> 王老五: age 35
1> 李四 2: age 17
3> 王老五 2: age 36
2> 张三 2: age 22
```

从上面的输出结果可以看到,数据在 3 个分区中分别并行地被处理。

注意:输出结果前面的整数是分区号。

对于简单的自定义分区器,也可以直接使用匿名内部类,以简化代码,代码如下:

```java
//第 3 章/PartitionDemo2.java

import org.apache.flink.api.common.functions.Partitioner;
import org.apache.flink.streaming.api.datastream.DataStream;
import org.apache.flink.streaming.api.environment.StreamExecutionEnvironment;

public class PartitionDemo2 {

    //POJO 类,表示人员信息实体
    public static class Person {
        public String name;        //存储姓名
        public Integer age;        //存储年龄

        //空构造器
        public Person() {};

        //构造器,初始化属性
        public Person(String name, Integer age) {
            this.name = name;
```

```java
            this.age = age;
        };

        //用于调试时输出信息
        public String toString() {
            return this.name.toString() + ": age " + this.age.toString();
        };
    }

    public static void main(String[] args) throws Exception {
        //设置流执行环境
        final StreamExecutionEnvironment env =
                StreamExecutionEnvironment.getExecutionEnvironment();

        //读取数据源,构造 DataStream
        DataStream<Person> personDS = env.fromElements(
                new Person("张三", 21),
                new Person("李四", 16),
                new Person("王老五", 35),
                new Person("张三 2", 22),
                new Person("李四 2", 17),
                new Person("王老五 2", 36));

        //应用自定义分区器,按年龄字段进行分组
        //注:通过字段位置指定 key,只对元组数据类型有效
        DataStream<Person> adults = personDS.partitionCustom(
                new Partitioner<Integer>(){
                    @Override
                    public int partition(Integer key, int i) {
                        if(key < 20){
                            return 0;
                        }else if(key >= 20 && key < 30){
                            return 1;
                        }else{
                            return 2;
                        }
                    }
                }, person -> person.age);

        //将结果输出到控制台
        adults.print();

        //触发流程序开始执行
        env.execute("stream demo");
    }
}
```

在上面的示例代码中，指定按流数据元素 Person 的字段 age 来作为 key，即将 age 字段作为 partitionCustom 方法的第 2 个参数，代码如下：

```
val adults = personDS
    .partitionCustom(new AgePartitioner, person -> person.age)
```

实际上，其中第 2 个参数是通过 key selector 指定的，这在流数据类型是复杂结构时非常有用。下面的代码是对上一示例的改造。

Scala 代码如下：

```scala
//第 3 章/PartitionDemo3.scala

import org.apache.flink.api.common.functions.Partitioner
import org.apache.flink.api.scala._
import org.apache.flink.streaming.api.scala.StreamExecutionEnvironment

/**
 * 自定义分区，使用 key selector
 */
object PartitionDemo3 {

  //定义 case class 类，表示流数据类型
  case class Person(name:String, age:Int)

  def main(args: Array[String]): Unit = {
    //设置批处理执行环境
    val env = StreamExecutionEnvironment.getExecutionEnvironment

    //读取数据源，构造 DataStream
    val people = List(
      Person("张三", 21),
      Person("李四", 16),
      Person("王老五", 35),
      Person("张三 2", 22),
      Person("李四 2", 17),
      Person("王老五 2", 36)
    )
    val personDS = env.fromCollection(people)

    //应用自定义分区器，按 person.age 字段进行分组
    val adults = personDS.partitionCustom(
      new Partitioner[Int] {
        override def partition(k: Int, i: Int): Int = {
          k match {
            case age if age < 20 => 0
```

```
          case age if age > 30 => 2
          case _ => 1
        }
      }
    },
    person => person.age
  )

  //将结果输出到控制台
  adults.print

  //触发流程序开始执行
  env.execute("stream demo")
  }
}
```

Java 代码如下：

```
//第 3 章/PartitionDemo3.java

import org.apache.flink.api.common.functions.Partitioner;
import org.apache.flink.api.java.functions.KeySelector;
import org.apache.flink.streaming.api.datastream.DataStream;
import org.apache.flink.streaming.api.environment.StreamExecutionEnvironment;

/**
 * 自定义分区
 * 对于简单的自定义分区器,也可以直接使用匿名内部类,以简化代码
 */
public class PartitionDemo3 {

    //POJO 类,表示人员信息实体
    public static class Person {
        public String name;         //存储姓名
        public Integer age;         //存储年龄

        //空构造器
        public Person() {};

        //构造器,初始化属性
        public Person(String name, Integer age) {
            this.name = name;
            this.age = age;
        };
        //用于调试时输出信息
```

```java
        public String toString() {
            return this.name.toString() + ": age " + this.age.toString();
        };
    }

    public static void main(String[] args) throws Exception {
        //设置流执行环境
        final StreamExecutionEnvironment env =
            StreamExecutionEnvironment.getExecutionEnvironment();

        System.out.println("分区之前的分区数:" + env.getParallelism());

        //读取数据源,构造 DataStream
        DataStream< Person > personDS = env.fromElements(
            new Person("张三", 21),
            new Person("李四", 16),
            new Person("王老五", 35),
            new Person("张三 2", 22),
            new Person("李四 2", 17),
            new Person("王老五 2", 36));

        //应用自定义分区器,按年龄字段进行分组
        //注:通过字段位置指定 key,只对元组数据类型有效
        DataStream< Person > adults = personDS.partitionCustom(
            new Partitioner< Integer >() {
                @Override
                public int partition(Integer key, int i) {
                    if (key < 20) {
                        return 0;
                    } else if (key >= 20 && key < 30) {
                        return 1;
                    } else {
                        return 2;
                    }
                }
            }, new KeySelector< Person, Integer >() {
                @Override
                public Integer getKey(Person person) throws Exception {
                    return person.age;
                }
            });

        //将结果输出到控制台
        adults.print();

        //触发流程序开始执行
        env.execute("stream demo");
    }
}
```

1. 事件时间

事件时间(Event Time)是事件在其产生设备上发生的时间。例如在物联网项目中,传感器捕获读数的时间。通常,这些事件时间需要在它们进入 Flink 之前嵌入记录中,由事件中的时间戳描述。在处理时,可以从每个记录中提取这些事件时间戳并考虑用于窗口。Flink 通过时间戳分配器访问事件时间戳。事件时间处理可用于无序事件。

如果要使用事件时间,则还需要提供时间戳提取器和水印生成器,Flink 将使用它们来跟踪事件时间的进展。这种水印机制将在 3.6 节中进行描述。

2. 处理时间

处理时间(Processing Time)是机器执行数据流处理的系统时间,执行基于时间的操作的每个操作符的本地时间。处理时间窗口只考虑事件被处理时的时间戳。处理时间是最简单的时间概念,因为它不需要处理流和生产机器之间的任何同步与协调,然而,在分布式异步环境中,处理时间不提供确定性。

当流程序在处理时间上运行时,所有基于时间的操作(如时间窗口)将使用运行各个操作符的机器的系统时钟。每小时处理时间窗口将包括在系统时钟指示完整小时之间到达特定操作符的所有记录。例如,如果一个应用程序在上午 9:15 开始运行,则第 1 个每小时处理时间窗口将包括上午 9:15 到 10:00 之间处理的事件,下一个窗口将包括上午 10:00 到 11:00 之间处理的事件,以此类推。

3. 注入时间

注入时间(Ingestion Time)是特定事件在源操作符处进入 Flink 数据流的时间。在源操作符中,每个记录以时间戳的形式获取源的当前时间,基于时间的操作(如时间窗口)引用该时间戳。摄入时间概念上位于事件时间和处理时间之间。由于摄取时间使用稳定的时间戳(在源处分配一次),对记录的不同窗口操作将引用相同的时间戳,所以摄取时间虽然是比处理时间更昂贵的操作,但它提供可预测的结果。

与事件时间相比,摄取时间程序不能处理任何无序的事件或时延的数据,因为它只在事件进入 Flink 系统后才分配时间戳,但程序不必指定如何生成水印。在内部,摄取时间处理得很像事件时间,但具有自动时间戳分配和自动水印生成功能。

4. 处理时间 vs. 事件时间

处理时间是通过直接调用本地机器的时间来确定的,而事件时间是由每个处理记录所携带的时间戳确定的。这两种时间在 Flink 的内部处理和用户的实际使用上是不同的。相对来讲,处理时间比较容易处理,而事件时间比较难处理。当使用处理时间时,用户得到的处理结果(或流处理应用程序的内部状态)是不确定的。因为在 Flink 中事件时间是有保证的,所以无论数据重放多少次,使用事件时间都可以获得相对确定且可重现的结果。

因此,在判断应该使用处理时间还是事件时间时,可以遵循一个原则,当应用程序遇到一些问题时,需要从前一个检查点或保存点重放它。希望结果完全一样吗?如果想要结果完全相同,则只能使用事件时间;如果接受不同的结果,则可以使用处理时间。处理时间的一个常见用途是实时计算整个系统的吞吐量。例如,如果要计算实时一小时内有多少条数据被处理,则只能使用处理时间。

5. 设置时间特性

时间特性定义了系统如何为依赖时间的顺序和依赖时间的操作(如时间窗口)确定时间。默认情况下,Flink DataStream 程序将使用 EventTime(事件时间)。如果要改用处理时间,则需要在一开始就设置时间特性。下面的代码演示了如何设置流的处理时间特性。

Scala 代码如下:

```
//获得流执行环境
val env = StreamExecutionEnvironment.getExecutionEnvironment

//设置流的时间特性(这里设置为采用处理时间)
env.setStreamTimeCharacteristic(TimeCharacteristic.ProcessingTime)
```

Java 代码如下:

```
//获得流执行环境
final StreamExecutionEnvironment env = 
    StreamExecutionEnvironment.getExecutionEnvironment();

//设置流的时间特性(这里设置为采用处理时间)
env.setStreamTimeCharacteristic(TimeCharacteristic.ProcessingTime);
```

注意:在 Flink 1.12 之前,Flink DataStream 默认使用的是处理时间。从 Flink 1.12 开始,默认的流时间特性已被更改为 EventTime,因此不再需要调用此方法来启用事件时间支持。

当然也可以选择设置其他类型时间特性。例如,设置注入时间和事件时间特征,使用如下代码。

Scala 代码如下:

```
env.setStreamTimeCharacteristic(TimeCharacteristic.IngestionTime)
env.setStreamTimeCharacteristic(TimeCharacteristic.EventTime)
```

Java 代码如下:

```
env.setStreamTimeCharacteristic(TimeCharacteristic.IngestionTime);
env.setStreamTimeCharacteristic(TimeCharacteristic.EventTime);
```

3.4.4 理解操作符链

在 1.3.3 节中,曾经介绍过操作符链,这是 Flink 采用的一种称为 Operator chain 的优化技术,可以在特定条件下减小本地通信的开销。为了满足操作符链的要求,必须将两个及以上的操作符(算子)设为相同的并行度,并通过本地转发(Local Forward)的方式进行连接。

将操作符(算子)合并为运算符链有以下两个必需条件:

(1) 操作符(算子)的并行度相同。

(2) 合并操作符链的运算符(算子)都是 one-to-one 分发模式。

将操作符(算子)合并为操作符链之后,原来的操作符(算子)成为里面的子任务,如图 3-19 所示。

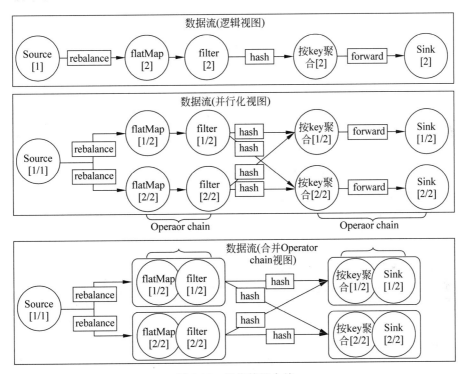

图 3-19　操作符链合并

如果要为流操作符禁用操作符链,则可以使用的方法如下:

```
//禁止全局操作符链合并
env.disableOperatorChaining()

//禁止某个算子的操作符链合并
filter().disableChaining()

//断开前面的操作符链并开始一个新的操作符链
filter().startNewChain()
```

3.5 Flink 数据接收器

在使用 Flink 进行数据处理时,数据经数据源流入,然后通过系列转换,最终可以通过 Data Sink 将计算结果输出,Flink Data Sinks(数据接收器)用于定义数据流最终的输出位置,它消费数据流并将它们转发到文件、套接字、外部系统或打印输出。

3.5.1 内置数据接收器

Flink 提供了多种内置的 Data Sink API,用于日常的开发,具体如下。

1) writeAsText("/path/to/file")

将计算结果以字符串的方式并行地写入指定文件夹下。这些字符串是通过调用每个元素的 toString()方法获得的,使用的输出类是 TextOutputFormat。这种方法除了路径参数是必选外,还可以通过指定第 2 个参数来定义输出模式。输出模式有以下两个可选值。

(1) WriteMode.NO_OVERWRITE:当指定路径上不存在任何文件时才执行写出操作。

(2) WriteMode.OVERWRITE:不论指定路径上是否存在文件,都执行写出操作;如果原来已有文件,则进行覆盖。

(3) 以上的写出是以并行的方式写出到多个文件,如果想要将输出结果全部写出到一个文件,则需要将其并行度设置为 1:

```
streamSource
    .writeAsText(path, FileSystem.WriteMode.OVERWRITE)
    .setParallelism(1);
```

2) writeAsCsv("/path/to/file")

将计算结果以 CSV 文件格式写到指定目录。行和字段分隔符是可配置的。每个字段的值来自对象的 toString()方法。使用的输出类是 CsvOutputFormat。该方法除了路径参数是必选外,还支持传入输出模式、行分隔符和字段分隔符 3 个额外的参数,其方法定义如下:

writeAsCsv(String path, WriteMode writeMode, String rowDelimiter, String fieldDelimiter)

3) print()/printToErr()

在标准输出/标准错误流上打印输出每个元素的 toString()值。可选地,可以提供输出的前缀,这有助于区分不同的 print 调用。如果并行度大于 1,则输出也将以产生输出的任务的 id 作为前缀。print()\ printToErr()是测试当中最常用的方式,用于将计算结果以标准输出流或错误输出流的方式打印到控制台上。

4）writeUsingOutputFormat()

自定义文件输出的基类和方法。支持自定义对象到字节的转换。在定义自定义格式时，需要继承自 FileOutputFormat，它负责序列化和反序列化。上面介绍的 writeAsText 和 writeAsCsv 其底层调用的都是该方法，其方法签名如下：

```
public DataStreamSink<T> writeAsText(String path, WriteMode writeMode) {
    TextOutputFormat<T> tof = new TextOutputFormat<>(new Path(path));
    tof.setWriteMode(writeMode);
    return writeUsingOutputFormat(tof);
}
```

5）writeToSocket(host，port，SerializationSchema)

将计算结果以指定的格式写到指定的 socket 套接字。为了正确地序列化和格式化，需要定义 SerializationSchema。使用示例如下：

```
streamSource.writeToSocket("localhost", 9999, new SimpleStringSchema());
```

6）addSink

调用自定义接收器函数。Flink 与作为接收器函数实现的其他系统（如 Apache Kafka）的连接器捆绑在一起。

注意：在以上方法中，以 writeAs * 开头的方法在 Flink API 文档中已经被标识为 "Deprecated"，即弃用状态，在未来的版本中有可能被删除，因此使用时要慎重。

在生产中，常用的 sinks（接收器）包括 Kafka 及各种数据库和文件系统。下面通过示例来掌握 Flink Data Sink 的常用用法。

【示例 3-9】 分析流数据，并将分析结果写到 CSV 文件中。

建议按以下步骤操作：

（1）在 IntelliJ IDEA 中创建一个 Flink 项目，使用 flink-quickstart-scala/flink-quickstart-Java 项目模板（Flink 项目的创建过程可参考 2.2 节）。

（2）设置依赖。在 pom.xml 文件中添加依赖。

Scala Maven 依赖：

```
<dependency>
    <groupId>org.apache.flink</groupId>
    <artifactId>flink-scala_2.12</artifactId>
    <version>1.13.2</version>
    <scope>provided</scope>
</dependency>
<dependency>
```

```xml
<groupId>org.apache.flink</groupId>
<artifactId>flink-streaming-scala_2.12</artifactId>
<version>1.13.2</version>
<scope>provided</scope>
</dependency>
```

Java Maven 依赖:

```xml
<dependency>
    <groupId>org.apache.flink</groupId>
    <artifactId>flink-Java</artifactId>
    <version>1.13.2</version>
    <scope>provided</scope>
</dependency>
<dependency>
    <groupId>org.apache.flink</groupId>
    <artifactId>flink-streaming-Java_2.12</artifactId>
    <version>1.13.2</version>
    <scope>provided</scope>
</dependency>
```

(3) 创建流应用程序类。

Scala 代码如下:

```scala
import org.apache.flink.streaming.api.scala._

/**
 * Data Sink:writeAsCSV 方法,将结果写入 CSV 文件
 */
object DataSinkAsCSV {

  def main(args: Array[String]): Unit = {
    //设置流执行环境
    val env = StreamExecutionEnvironment.getExecutionEnvironment

    //得到输入数据,进行转换
    env.fromElements("Good good study", "Day day up")
      .map(_.toLowerCase)
      .flatMap(_.split("\\W+"))      //相当于先执行 map,再执行 flatten
      .map((_,1))                     //转换为元组
      .writeAsCsv("result.csv")       //写到结果文件中
      .setParallelism(1)              //将结果写到单个文件中

    env.execute("Data Sink Demo")
  }
}
```

Java 代码如下：

```java
import org.apache.flink.api.common.functions.FlatMapFunction;
import org.apache.flink.api.java.tuple.Tuple2;
import org.apache.flink.streaming.api.environment.StreamExecutionEnvironment;
import org.apache.flink.util.Collector;

/**
 * Data Sink:writeAsCSV 方法,将结果写入 CSV 文件
 */
public class DataSinkDemo1 {
    public static void main(String[] args) throws Exception {
        //设置流执行环境
        final StreamExecutionEnvironment env =
                StreamExecutionEnvironment.getExecutionEnvironment();

        //获得数据,执行 map 和 flatMap 转换
        env.fromElements("Good good study","Day day up")
            .map(String::toLowerCase)
            .flatMap(new FlatMapFunction<String, Tuple2<String,Integer>>() {
                @Override
                public void flatMap(String s, Collector<Tuple2<String, Integer>> out) throws Exception {
                    for(String word : s.split("\\W+")){
                        out.collect(new Tuple2<>(word,1));
                    }
                }
            })
            .writeAsCsv("result.csv")          //写到指定的结果文件中
            .setParallelism(1);                //写到一个结果文件中

        //执行流程序
        env.execute("Data Sink Demo");
    }
}
```

（4）执行以上程序,可以看到,在项目的根目录下生成了一个结果文件 result。查看输出的结果文件 result,内容如下：

```
good,1
good,1
study,1
day,1
day,1
up,1
```

3.5.2 使用流文件连接器

正如3.5.1节所提到的，DataStream上的writeAs*()方法主要用于调试。它们不参与Flink的检查点，这意味着这些函数通常具有at-least-once语义。将数据刷新(flush)到目标系统取决于OutputFormat的实现。这意味着并不是发送到OutputFormat的所有元素都立即出现在目标系统中，而且，在失败的情况下，这些记录可能会丢失。

要可靠、准确地将流一次性地交付到文件系统(精确一次性)，需要使用StreamingFileSink，这是一个流文件连接器，该连接器提供了一个Sink，用于将分区文件写入由Flink FileSystem抽象支持的文件系统。

流文件接收器将传入的数据写入桶(bucket)中。假设传入的流可以不受限制，每个桶中的数据被组织成有限大小的部分文件(Part File)。桶的行为是完全可配置的，默认的基于时间的桶，每小时开始写一个新的桶。这意味着每个结果桶将包含这样的文件：在1h间隔内从流接收的记录。

桶目录本身包含几部分文件，将数据写入该桶的接收器的每个并行子任务至少有一部分文件。这些部分文件包含实际的输出数据。更多的部分文件将根据可配置的滚动策略创建。默认策略根据大小、超时(指定文件可以打开的最大持续时间)和最大不活动超时(之后文件被关闭)来滚动部分文件，如图3-20所示。

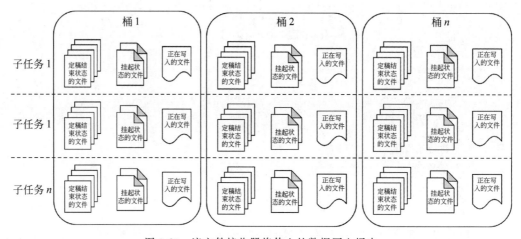

图 3-20 流文件接收器将传入的数据写入桶中

需要注意的是，在流模式下使用FileSink时需要为流作业启用检查点。流数据流的分布式状态将被定期快照。如果出现故障，流数据流将从最近完成的检查点重新启动。部分文件只能在成功的检查点上完成。如果禁用检查点，则部分文件将永远停留在正在写入(in-progress)或挂起(pending)状态，下游系统无法安全地读取。

【示例3-10】 分析流数据，并将分析结果写到文本文件中。

建议按以下步骤操作：

(1) 在 IntelliJ IDEA 中创建一个 Flink 项目,使用 flink-quickstart-scala/flink-quickstart-Java 项目模板(Flink 项目的创建过程可参考 2.2 节)。

(2) 设置依赖。在 pom.xml 文件中添加依赖。

Scala Maven 依赖:

```xml
<dependency>
    <groupId>org.apache.flink</groupId>
    <artifactId>flink-scala_2.12</artifactId>
    <version>1.13.2</version>
    <scope>provided</scope>
</dependency>
<dependency>
    <groupId>org.apache.flink</groupId>
    <artifactId>flink-streaming-scala_2.12</artifactId>
    <version>1.13.2</version>
    <scope>provided</scope>
</dependency>
```

Java Maven 依赖:

```xml
<dependency>
    <groupId>org.apache.flink</groupId>
    <artifactId>flink-Java</artifactId>
    <version>1.13.2</version>
    <scope>provided</scope>
</dependency>
<dependency>
    <groupId>org.apache.flink</groupId>
    <artifactId>flink-streaming-Java_2.12</artifactId>
    <version>1.13.2</version>
    <scope>provided</scope>
</dependency>
```

(3) 创建流应用程序类。

Scala 代码如下:

```scala
import org.apache.flink.api.common.serialization.SimpleStringEncoder
import org.apache.flink.core.fs.Path
import org.apache.flink.streaming.api.functions.sink.filesystem.StreamingFileSink
import org.apache.flink.streaming.api.scala._

/**
 * Data Sink:使用 addSink()方法
 */
```

```scala
object DataSinkWithStreamingFile {
  def main(args: Array[String]): Unit = {
    //设置流执行环境
    val env = StreamExecutionEnvironment.getExecutionEnvironment

    //为流作业启用检查点
    //此方法选择 CheckpointingMode.EXACTLY_ONCE 保证
    env.enableCheckpointing(60000)          //每分钟(60s)执行一次快照

    //自定义的 Sink
    val sink: StreamingFileSink[(String, Int)] = StreamingFileSink
      .forRowFormat(new Path("./result"),
            new SimpleStringEncoder[(String, Int)]("UTF-8"))
      .build()

    //得到输入数据,进行转换
    val input: DataStream[(String, Int)] = env
      .fromElements("Good good study", "Day day up")
      .map(_.toLowerCase)                   //转小写
      .flatMap(_.split("\\W+"))             //相当于先执行 map,再执行 flatten
      .map((_,1))                           //转元组

    //使用自定义的 Sink
    input.addSink(sink)                     //写到结果文件中
      .name("Flink2File")
      .setParallelism(1)                    //将结果写到单个文件中

    //触发流程序执行
    env.execute("Data Sink Demo")
  }
}
```

Java 代码如下:

```java
import org.apache.flink.api.common.functions.FlatMapFunction;
import org.apache.flink.api.common.serialization.SimpleStringEncoder;
import org.apache.flink.api.java.tuple.Tuple2;
import org.apache.flink.core.fs.Path;
import org.apache.flink.streaming.api.datastream.DataStream;
import org.apache.flink.streaming.api.environment.StreamExecutionEnvironment;
import org.apache.flink.streaming.api.functions.sink.filesystem.StreamingFileSink;
import org.apache.flink.util.Collector;

/**
 * Data Sink:使用 addSink()方法
```

```java
 */
public class DataSinkWithStreamingFile {
    public static void main(String[] args) throws Exception {
        //设置流执行环境
        final StreamExecutionEnvironment env =
                StreamExecutionEnvironment.getExecutionEnvironment();

        //为流作业启用检查点
        //此方法选择 CheckpointingMode.EXACTLY_ONCE 保证
        env.enableCheckpointing(60000);           //每分钟(60s)执行一次快照

        //自定义的 Sink
        StreamingFileSink<Tuple2<String,Integer>> sink = StreamingFileSink
            .forRowFormat(new Path("./result"),
                    new SimpleStringEncoder<Tuple2<String,Integer>>())
            .build();

        //获得数据,执行 map 和 flatMap 转换
        DataStream<Tuple2<String,Integer>> input = env
            .fromElements("Good good study","Day day up")
            .map(String::toLowerCase)
            .flatMap(new FlatMapFunction<String, Tuple2<String,Integer>>() {
                @Override
                public void flatMap(String s, Collector<Tuple2<String,Integer>> out) throws Exception {
                    for(String word : s.split("\\W+")){
                        out.collect(new Tuple2<>(word,1));
                    }
                }
            });

        //指定自定义的 Sink,并将并行度设置为 1,以便将结果写入单个文件中
        //Flink 提供了 addSink 方法,用来调用自定义的 Sink 或者第三方的连接器
        input.addSink(sink).name("Flink2File").setParallelism(1);

        //执行流程序
        env.execute("Data Sink Demo");
    }
}
```

（4）执行以上程序,可以看到,在项目的根目录下生成了一个结果文件 result,如图 3-21 所示。

图 3-21　流文件接收器将传入的数据写入结果文件

查看输出的结果文件 result,内容如下:

```
(good,1)
(good,1)
(study,1)
(day,1)
(day,1)
(up,1)
```

关于流文件接收器,还有一些需要了解的概念,解释如下。

1) 文件格式

StreamingFileSink 既支持行编码(row-wise)格式,也支持批量(bulk)编码格式,例如 Apache Parquet。这两个变量都附带了它们各自的构建器,可以使用以下静态方法创建它们。

(1) Row-encoded sink: StreamingFileSink.forRowFormat(basePath,rowEncoder)。

(2) Bulk-encoded sink: StreamingFileSink.forBulkFormat(basePath,bulkWriterFactory)。

在创建行编码的接收器或批编码的接收器时,必须指定存储桶的基本路径和数据的编码逻辑。基本路径中每个桶对应一个目录。

2) 行编码格式

行编码格式需要指定一个 Encoder(编码器),用于将单独的行序列化到 in-progress part file 文件的 OutputStream。

除了桶分配器(Bucket Assigner),RowFormatBuilder 还允许用户指定。

(1) 自定义 RollingPolicy:滚动策略以覆盖 DefaultRollingPolicy。

(2) bucketCheckInterval (default=1min):检查基于时间的滚动策略的间隔(单位为毫秒)。

例如,将字符串流元素写到文件中,使用类似下面这样的代码。

Scala 代码如下:

```scala
import org.apache.flink.api.common.serialization.SimpleStringEncoder
import org.apache.flink.core.fs.Path
import org.apache.flink.streaming.api.functions.sink.filesystem.StreamingFileSink
import org.apache.flink.streaming.api.functions.sink.filesystem.rollingpolicies.DefaultRollingPolicy

val input: DataStream[String] = ...

val sink: StreamingFileSink[String] = StreamingFileSink
    .forRowFormat(new Path(outputPath),
            new SimpleStringEncoder[String]("UTF-8"))
    .withRollingPolicy(
```

```
            DefaultRollingPolicy.builder()
                    .withRolloverInterval(TimeUnit.MINUTES.toMillis(15))
                    .withInactivityInterval(TimeUnit.MINUTES.toMillis(5))
                    .withMaxPartSize(1024 * 1024 * 1024)
                    .build())
        .build()

input.addSink(sink)
```

Java 代码如下:

```
import org.apache.flink.api.common.serialization.SimpleStringEncoder;
import org.apache.flink.core.fs.Path;
import org.apache.flink.streaming.api.functions.sink.filesystem.StreamingFileSink;
import org.apache.flink.streaming.api.functions.sink.filesystem.rollingpolicies.DefaultRollingPolicy;

DataStream<String> input = ...;

final StreamingFileSink<String> sink = StreamingFileSink
    .forRowFormat(new Path(outputPath),
                  new SimpleStringEncoder<String>("UTF-8"))
    .withRollingPolicy(
        DefaultRollingPolicy.builder()
                .withRolloverInterval(TimeUnit.MINUTES.toMillis(15))
                .withInactivityInterval(TimeUnit.MINUTES.toMillis(5))
                .withMaxPartSize(1024 * 1024 * 1024)
                .build())
    .build();

input.addSink(sink);
```

这个示例创建了一个简单的接收器,它将记录分配给默认的一小时时间桶。它还指定了一个滚动策略,在以下 3 种情况下滚动处理 in-progress 状态的 Part File 文件:

(1) 它包含至少 15min 的数据。
(2) 在过去的 5min 里,它没有接收到新的记录。
(3) 文件大小达到了 1GB(在写入最后一条记录之后)。

3) 批编码格式

批编码的接收器与行编码的接收器创建方式类似,但必须指定 BulkWriter.Factory,而不是指定 Encoder(编码器)。BulkWriter 逻辑定义了如何添加和刷新新元素,以及如何为进一步的编码目的完成一批记录。

Flink 有多个内置的 BulkWriter 工厂,包括:

(1) ParquetWriterFactory。

(2) AvroWriterFactory。

(3) SequenceFileWriterFactory。

(4) CompressWriterFactory。

(5) OrcBulkWriterFactory。

Bulk Formats 只能有一个扩展 CheckpointRollingPolicy 的滚动策略，它只在每个检查点上滚动。

4）BucketAssignment

bucket 逻辑定义了如何将数据结构化到基本输出目录的子目录中。

行格式和批量格式都使用 DateTimeBucketAssigner 作为默认分配器。默认情况下，DateTimeBucketAssigner 会根据系统默认时区创建每小时的桶，格式为 yyyy-MM-dd--HH。日期格式（桶大小）和时区都可以手动配置。

可以通过在格式构建器上调用 .withBucketAssigner（assigner）来指定一个自定义的 BucketAssigner。Flink 有两个内置的 BucketAssigners。

(1) DateTimeBucketAssigner：默认的基于时间的分配器。

(2) BasePathBucketAssigner：将所有 Part File 文件存储在基本路径中（单个全局桶）的分配器。

5）RollingPolicy

RollingPolicy 定义了给定的正在写入的 Part File 文件何时被关闭，何时被移到 pending（挂起）状态，何时移到 finished（完成）状态。处于 finished 状态的 Part File 文件是可以查看的，并且保证包含在故障时不会被还原的有效数据。滚动策略与检查点间隔（挂起的文件在下一个检查点完成）相结合，控制 Part File 文件对下游读取器可用的速度，以及这些 Part File 的大小和数量。

Flink 自带两个 RollingPolicies：

(1) DefaultRollingPolicy。

(2) OnCheckpointRollingPolicy。

6）部分文件生命周期

为了在下游系统中使用 StreamingFileSink 的输出，读者需要理解生成的输出文件的命名和生命周期。

部分文件可以处于以下 3 种状态之一。

(1) in-progress：正在写入的 Part File 文件处于 in-progress 状态。

(2) pending：已经关闭（由于指定的滚动策略）正在等待被提交的 in-progress 文件。

(3) finished：在成功的检查点上，挂起的文件转换为 finished。

只有处于 finished 状态的文件才可以被下游系统安全地读取，因为这些文件保证以后不会被修改。

对于任何给定的子任务，部分文件索引都是严格递增的（按照它们创建的顺序），然而，这些索引并不总是顺序的。当作业重新启动时，所有子任务的下一个 part 索引将是 max part index+1,其中 max 是根据所有子任务计算出来的。

对于每个活动 bucket,每个 writer 子任务在任何给定时间都将有单个正在写入的 Part File 文件,但也可能有几个处于 pending 和 finished 状态的文件。

3.5.3 自定义数据接收器

Flink 还支持使用自定义的 Sink 来满足多样化的输出需求。想要实现自定义的 Sink,需要直接或者间接实现 SinkFunction 接口。通常实现其抽象类 RichSinkFunction,相比于 SinkFunction,其提供了更多的与生命周期相关的方法。

在下面的示例中,自定义一个 FlinkToMySQLSink,将流计算结果写到指定的 MySQL 数据库中。

【示例 3-11】 改写上一示例,将流计算结果保存到 MySQL 数据库中。

建议按以下步骤操作:

(1) 在 MySQL 的 xueai8 数据库中,创建一个数据表 wc,用来存储 Flink 的计算结果。创建 wc 表的 SQL 命令如下:

```
mysql> create table wc(word varchar(30), cnt int);
```

(2) 在 IntelliJ IDEA 中创建一个 Flink 项目,使用 flink-quickstart-scala/flink-quickstart-Java 项目模板(Flink 项目的创建过程可参考 2.2 节)。

(3) 设置依赖。在 pom.xml 文件中添加依赖。

```xml
<dependency>
    <groupId>org.apache.flink</groupId>
    <artifactId>flink-scala_2.12</artifactId>
    <version>1.13.2</version>
    <scope>provided</scope>
</dependency>
<dependency>
    <groupId>org.apache.flink</groupId>
    <artifactId>flink-streaming-scala_2.12</artifactId>
    <version>1.13.2</version>
    <scope>provided</scope>
</dependency>

<!-- JDBC 连接器 -->
<dependency>
    <groupId>org.apache.flink</groupId>
    <artifactId>flink-connector-JDBC_2.12</artifactId>
    <version>1.13.2</version>
</dependency>
```

```xml
<!-- MySQL连接器 -->
<dependency>
    <groupId>mysql</groupId>
    <artifactId>mysql-connector-Java</artifactId>
    <version>5.1.48</version>
</dependency>
```

Java Maven依赖：

```xml
<dependency>
    <groupId>org.apache.flink</groupId>
    <artifactId>flink-Java</artifactId>
    <version>1.13.2</version>
    <scope>provided</scope>
</dependency>
<dependency>
    <groupId>org.apache.flink</groupId>
    <artifactId>flink-streaming-Java_2.12</artifactId>
    <version>1.13.2</version>
    <scope>provided</scope>
</dependency>

<!-- JDBC连接器 -->
<dependency>
    <groupId>org.apache.flink</groupId>
    <artifactId>flink-connector-JDBC_2.12</artifactId>
    <version>1.13.2</version>
</dependency>
<dependency>
    <groupId>mysql</groupId>
    <artifactId>mysql-connector-Java</artifactId>
    <version>5.1.48</version>
</dependency>
```

(4) 创建流应用程序类。

Scala代码如下：

```scala
import Java.sql.{Connection, DriverManager, PreparedStatement}

import org.apache.flink.configuration.Configuration
import org.apache.flink.streaming.api.CheckpointingMode
import org.apache.flink.streaming.api.functions.sink.RichSinkFunction
import org.apache.flink.streaming.api.functions.sink.SinkFunction.Context
import org.apache.flink.streaming.api.scala._
```

```scala
/**
 * Data Sink:使用 addSink()方法,指定 MySQL 数据库作为 Data Sink
 */
object DataSinkToMysql {

  //自定义的 Sink 类,继承自 RichSinkFunction
  class FlinkToMySQLSink extends RichSinkFunction[(String, Int)] {
    //定义 MySQL 数据库连接 url 和驱动程序及账号、密码
    val url = "JDBC:mysql://localhost:3306/xueai8?characterEncoding=UTF-8"
    val driver = "com.mysql.JDBC.Driver"
    val username = "root"
    val userpwd = "admin"

    //声明数据库连接对象和 SQL 预编译语句
    var conn: Connection = _
    var stmt: PreparedStatement = _

    //重写 open 方法:加载驱动,建立连接,预编译 SQL 语句
    //先于 invoke 方法调用,仅执行一次
    override def open(parameters: Configuration): Unit = {
      super.open(parameters)
      //加载驱动程序
      Class.forName(driver)
      //连接数据库
      conn = DriverManager.getConnection(url, username, userpwd)
      //执行 SQL 语句
      val sql = "insert into wc(word, cnt) values(?, ?)"
      stmt = conn.prepareStatement(sql)
    }

    //重写事件驱动方法调用
    override def invoke(value: (String, Int), context: Context): Unit = {
      stmt.setString(1, value._1)
      stmt.setInt(2, value._2)
      stmt.executeUpdate
    }

    //关闭资源
    //后于 invoke 方法调用,仅执行一次
    override def close(): Unit = {
      super.close()
      if (stmt != null) stmt.close()
      if (conn != null) conn.close()
    }
  }
```

```scala
//测试
def main(args: Array[String]): Unit = {
    //设置流执行环境
    val env = StreamExecutionEnvironment.getExecutionEnvironment

    //为流数据启用检查点
    env.enableCheckpointing(2000, CheckpointingMode.EXACTLY_ONCE)

    //得到输入数据,进行转换
    val input = env.fromElements("Good good study", "Day day up")
        .map(_.toLowerCase)                    //转小写
        .flatMap(_.split("\\W+"))              //相当于先执行 map,再执行 flatten
        .map((_,1))                            //转换为元组

    //使用自定义的 Sink
    input.addSink(new FlinkToMySQLSink)        //写到数据库中

    //触发流程序执行
    env.execute("Data Sink Demo")
}
}
```

Java 代码如下:

```java
import org.apache.flink.api.common.functions.FlatMapFunction;
import org.apache.flink.api.java.tuple.Tuple2;
import org.apache.flink.configuration.Configuration;
import org.apache.flink.streaming.api.datastream.DataStream;
import org.apache.flink.streaming.api.environment.StreamExecutionEnvironment;
import org.apache.flink.streaming.api.functions.sink.RichSinkFunction;
import org.apache.flink.util.Collector;

import Java.sql.Connection;
import Java.sql.DriverManager;
import Java.sql.PreparedStatement;

/**
 * Data Sink:使用 addSink()方法,指定 MySQL 数据库作为 Data Sink
 */
public class DataSinkToMysql {

    //自定义的 Sink 类,继承自 RichSinkFunction
    public static class FlinkToMySQLSink
                extends RichSinkFunction< Tuple2 < String, Integer >> {
        //定义 MySQL 数据库连接 url 和驱动程序及账号、密码
```

```java
    private static String JDBC_URL = "JDBC:mysql://localhost/xueai8";
    private static String USER_NAME = "root";
    private static String USER_PWD = "admin";

    //声明数据库连接对象和SQL预编译语句
    private PreparedStatement stmt;
    private Connection conn;

    //重写open方法:加载驱动,建立连接,预编译SQL语句
    @Override
    public void open(Configuration parameters) throws Exception {
        Class.forName("com.mysql.JDBC.Driver");
        conn = DriverManager.getConnection(JDBC_URL, USER_NAME, USER_PWD);
        String sql = "insert into wc(word, cnt) values(?, ?)";
        stmt = conn.prepareStatement(sql);
    }

    //重写事件驱动方法调用
    @Override
    public void invoke(Tuple2<String, Integer> t, Context context) throws Exception {
        stmt.setString(1, t.f0);
        stmt.setInt(2, t.f1);
        stmt.executeUpdate();
    }

    //关闭资源
    @Override
    public void close() throws Exception {
        super.close();
        if (stmt != null) {
            stmt.close();
        }
        if (conn != null) {
            conn.close();
        }
    }
}

public static void main(String[] args) throws Exception {
    //设置流执行环境
    final StreamExecutionEnvironment env =
            StreamExecutionEnvironment.getExecutionEnvironment();

    //为流数据启用检查点
    env.enableCheckpointing(2000);
```

```java
//获得数据,执行 map 和 flatMap 转换
DataStream < Tuple2 < String, Integer >> stream = env
    .fromElements("Good good study","Day day up")
    .map(String::toLowerCase)
    .flatMap(new FlatMapFunction < String, Tuple2 < String, Integer >>() {
        @Override
        public void flatMap(String s, Collector < Tuple2 < String, Integer >> out) throws Exception {
            for(String word : s.split("\\W + ")){
                out.collect(new Tuple2 <>(word,1));
            }
        }
    });

//指定使用自定义的 Sink
stream.addSink(new FlinkToMySQLSink());

//执行流程序
env.execute("Data Sink Demo");
```

执行以上程序,然后在 MySQL 中查看表内容,SQL 命令如下:

```
mysql> select * from wc;
```

查询结果如图 3-22 所示。

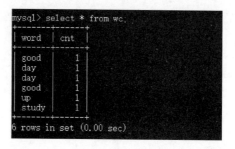

图 3-22 从 MySQL 中查询出的流存储结果

可以看到,流程序计算的结果已经保存到数据库中了。

注意:如果想将流计算结果写到类似 Kafka 这样的第三方消息中间件或者 HDFS 这样的存储系统,则可参考 3.11 节的"流连接器"部分。

在上面的代码中,自定义 Sink 继承自 RichSinkFunction 类,它是 SinkFunction 的 Rich

变体。与 SinkFunction 相比，RichSinkFunction 有一些额外的方法，包括 open(Configuration c)、close() 和 getRuntimeContext()。

在操作符初始化期间调用 Open() 一次。例如，这是加载一些静态数据或打开到外部服务的连接的机会，而 getRuntimeContext() 提供了对底层环境的访问能力，例如创建和访问由 Flink 管理的状态。

3.6　时间和水印概念

时间是流应用程序的另一个重要组成部分。大多数事件流具有固有的时间语义，因为每个事件都在特定的时间点生成。此外，许多常见的流计算都是基于时间的，例如窗口聚合、会话、模式检测和基于时间的连接。

Flink 提供了一组丰富的与时间相关的特性，包括：

（1）事件时间模式：流应用程序（使用事件时间语义处理流）基于事件的时间戳计算结果，因此事件时间处理允许精确和一致的结果。

（2）处理时间模式：除了事件时间模式，Flink 还支持处理时间语义，它执行由处理机器的挂钟时间触发的计算。处理时间模式可以适用于某些具有严格低时延要求的应用程序，这些应用程序可以容忍近似结果。

（3）水印支持：Flink 在事件时间应用程序中使用水印来推断时间。水印是一种灵活的机制，用来平衡结果的时延和完整性。

（4）迟到数据处理：当以事件时间模式处理带有水印的流时，可能会在所有相关事件到达之前完成计算。这样的事件称为迟到事件。Flink 提供了多个处理迟到事件的选项，例如通过侧输出重新路由它们，并更新以前完成的结果。

3.6.1　时间概念

Flink Streaming API 借鉴了谷歌数据流模型（Google Data Flow Model），它的流 API 支持不同的时间概念。

Flink 明确支持以下 3 个不同的时间概念。

（1）事件时间：事件发生的时间，由产生（或存储）事件的设备记录。

（2）接入时间：Flink 在接入事件时记录的时间戳。

（3）处理时间：管道中特定操作符处理事件的时间。

这几种时间概念如图 3-23 所示。

3.6.2　事件时间和水印

支持事件时间的流处理器需要一种方法来度量事件时间的进度。例如，针对事件时间对数据进行窗口或排序的操作符必须缓冲数据，直到它们能够确保已接收到某个时间间隔

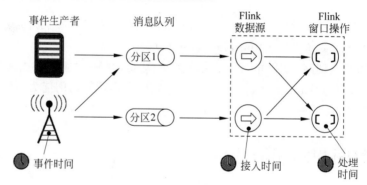

图 3-23 时间概念

的所有时间戳为止。

在Flink中测量事件时间进展的机制是水印(watermark)。水印是一种特殊类型的事件,是告诉系统事件时间进度的一种方式。水印流是数据流的一部分,并带有时间戳t。水印(t)声明事件时间已经到达该流中的时间t,这意味着时间戳t′≤t(时间戳更早或等于水印的事件)的流中不应该有更多的元素,如图3-24所示。

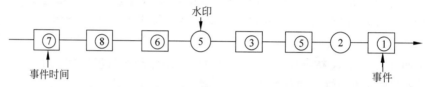

图 3-24 水印流是数据流的一部分,并带有时间戳t

时间t的水印标记了数据流中的一个位置,并断言此时的流在时间t之前已经完成。为了执行基于事件时间的事件处理,Flink需要知道与每个事件相关联的时间,它还需要包含水印的流。水印就是系统事件时间的时钟。水印触发是基于事件时间的计时器的触发。

图3-25显示了带有(逻辑的)时间戳的事件流,以及内联流动的水印。在这个例子中,事件是按顺序排列的(相对于它们的时间戳),这意味着水印只是流中的周期标记。

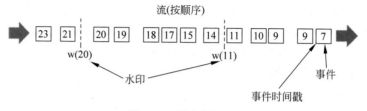

图 3-25 顺序数据流

对于无序流,水印是至关重要的,其中事件不是按照它们的时间戳排序的,如图3-26所示。例如,当操作符接收到w(11)这条水印时,可以认为时间戳小于或等于11的数据已经

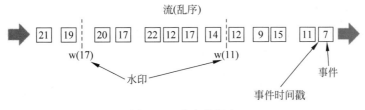

图 3-26 乱序数据流

到达,此时可以触发计算。同样,当接收到 w(17) 这条水印时,可以认为时间戳小于或等于 17 的数据已经到达,此时可以触发计算。

可以看出,水印的时间戳是单调递增的,时间戳为 t 的水印意味着所有后续记录的时间戳将大于 t。一般来讲,水印是一种声明,在流中的那个点之前,即在某个时间戳之前的所有事件都应该已经到达。当水印到达运算符(算子)时,运算符可以将其内部事件时间时钟推进到水印的值。

水印是在源函数处或直接在源函数之后生成的。源函数的每个并行子任务通常可以独立地生成水印。这些水印定义了特定并行源处的事件时间。

当这些水印流经流程序时,它们会将事件时间推进到它们所到达的操作符(算子)处。每当一个操作符推进它的事件时间时,它将为它的后继操作符产生一个新的下游水印。

关于事件和水印在并行流中流动的示例,以及操作符(算子)跟踪事件时间的示例,如图 3-27 所示。

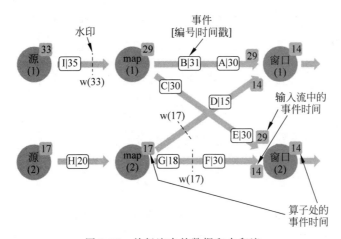

图 3-27 并行流中的数据和水印流

3.6.3 水印策略

Flink 提供了用于处理事件时间、时间戳和水印的 API。

1. 什么是水印策略

为了处理事件时间,Flink 流程序需要知道事件的时间戳,这意味着流中的每个元素都需要分配其事件时间戳。这通常是通过 TimestampAssigner 从元素中的某个字段访问/提取时间戳实现的。

时间戳分配与生成水印密切相关,生成水印是告诉系统事件时间进度的一种方式。可以通过指定 WatermarkGenerator 来配置水印。

TimestampAssigner 和 WatermarkGenerator 形成了一个水印策略,它定义了如何在流源中生成水印。Flink API 需要一个同时包含 TimestampAssigner 和 WatermarkGenerator 的 WatermarkStrategy。许多常用的策略都可以作为 WatermarkStrategy 上的静态方法来使用,但用户也可以在需要时构建自己的策略。

完整的 WatermarkStrategy 接口的定义如下:

```java
public interface WatermarkStrategy<T>
    extends TimestampAssignerSupplier<T>,
        WatermarkGeneratorSupplier<T>{

    /**
     * Instantiates a {@link TimestampAssigner} for assigning timestamps according to this
     * strategy.
     */
    @Override
    TimestampAssigner<T> createTimestampAssigner(TimestampAssignerSupplier.Context context);

    /**
     * Instantiates a WatermarkGenerator that generates watermarks according to this
     * strategy.
     */
    @Override
    WatermarkGenerator<T> createWatermarkGenerator(WatermarkGeneratorSupplier.Context context);
}
```

通常用户不会自己实现这个接口,而是使用 WatermarkStrategy 上的静态辅助方法实现公共水印策略,或者将自定义 TimestampAssigner 与 WatermarkGenerator 捆绑在一起。例如,要使用有界无序水印和 lambda 函数作为时间戳赋值器,可以使用的代码如下。

Scala 代码如下:

```scala
WatermarkStrategy
  .forBoundedOutOfOrderness[(Long, String)](Duration.ofSeconds(20))
  .withTimestampAssigner(
```

```
    new SerializableTimestampAssigner[(Long,String)]{
        override def extractTimestamp(element: (Long, String), recordTimestamp: Long): Long =
    element._1
    })
```

Java 代码如下:

```
WatermarkStrategy
    .<Tuple2<Long, String>> forBoundedOutOfOrderness(Duration.ofSeconds(20))
    .withTimestampAssigner((event, timestamp) -> event.f0);
```

TimestampAssigner 是一个非常简单的函数,它实现的功能仅是从事件中提取字段。指定一个 TimestampAssigner 是可选的,在大多数情况下,实际上用户并不希望指定一个 TimestampAssigner。例如,当使用 Kafka 或 Kinesis 时,用户会直接从 Kafka 或 Kinesis 记录中获得时间戳。

注意:时间戳和水印都指定为自 Java 纪元(1970-01-01 00:00:00)以来的毫秒数。

2. 使用水印策略

在 Flink 应用程序中有以下两个地方可以使用 WatermarkStrategy:

(1)直接在数据源上使用。

(2)在非源操作之后使用。

第 1 种选择更可取,因为它允许数据源利用关于水印逻辑中的分片/分区/分割的知识。数据源通常可以在更细的水平上跟踪水印,数据源产生的整体水印将更准确。直接在数据源上指定 WatermarkStrategy 通常意味着必须使用一个数据源特定的接口。

第 2 个选项(在任意操作后设置 WatermarkStrategy)应该只在不能直接在源上设置策略的情况下使用。

1)直接在数据源上使用

这种方式指的是数据源本身已经嵌入了事件时间,在处理事件时间流时直接提取其中的事件时间的时间戳即可。因为流数据源可以直接将时间戳分配给它们生成的元素,它们还可以发出水印,因此不需要时间戳分配程序。

注意:在源文件中存储记录时,最好存储事件时间。如果使用了时间戳分配器,则会覆盖源提供的任何时间戳和水印。

如果要直接向流数据源中的元素分配时间戳,则流数据源必须使用 SourceContext 上的 collectWithTimestamp()方法。如果要生成水印,则源程序必须调用 emitWatermark(Watermark)函数。

下面是一个简单的(非检查点)流数据源分配时间戳和产生水印的例子。
Scala 代码如下：

```scala
override def run(ctx: SourceContext[MyType]): Unit = {
  while (/* condition */) {
    val next: MyType = getNext()
    ctx.collectWithTimestamp(next, next.eventTimestamp)

    if (next.hasWatermarkTime) {
      ctx.emitWatermark(new Watermark(next.getWatermarkTime))
    }
  }
}
```

Java 代码如下：

```java
@Override
public void run(SourceContext<MyType> ctx) throws Exception {
  while (/* condition */) {
    MyType next = getNext();
    ctx.collectWithTimestamp(next, next.getEventTimestamp());
    if (next.hasWatermarkTime()) {
      ctx.emitWatermark(new Watermark(next.getWatermarkTime()));
    }
  }
}
```

2) 在非源操作之后使用

时间戳分配程序获取一个流并生成一个带有时间戳元素和水印的新数据流。如果原始数据流已经具有时间戳和/或水印，则时间戳分配器将覆盖它们。

时间戳分配程序通常在数据源之后立即指定。例如，一个常见的模式是先对事件进行解析(MapFunction)和筛选(FilterFunction)，然后分配时间戳。在任何情况下，都需要在事件时间的第 1 个操作(例如第 1 个窗口操作)之前指定时间戳分配程序。

注意：作为一种特殊情况，当使用 Kafka 作为流作业的源时，Flink 允许在源(或消费者)内部指定一个时间戳分配者或水印发射器。

下面的代码演示了这种用法。
Scala 代码如下：

```scala
val env = StreamExecutionEnvironment.getExecutionEnvironment

val stream: DataStream[MyEvent] = env.readFile(
```

```
            myFormat,
            myFilePath,
            FileProcessingMode.PROCESS_CONTINUOUSLY,
            100,
            FilePathFilter.createDefaultFilter())

val withTimestampsAndWatermarks: DataStream[MyEvent] = stream
        .filter( _.severity == WARNING )
        .assignTimestampsAndWatermarks(<< watermark strategy >>)

withTimestampsAndWatermarks
        .keyBy( _.getGroup )
        .window(TumblingEventTimeWindows.of(Time.seconds(10)))
        .reduce( (a, b) => a.add(b) )
        .addSink(...)
```

Java 代码如下:

```
final StreamExecutionEnvironment env =
    StreamExecutionEnvironment.getExecutionEnvironment();

DataStream< MyEvent > stream = env.readFile(
            myFormat,
            myFilePath,
            FileProcessingMode.PROCESS_CONTINUOUSLY,
            100,
            FilePathFilter.createDefaultFilter(),
            typeInfo);

DataStream< MyEvent > withTimestampsAndWatermarks = stream
        .filter( event -> event.severity() == WARNING )
        .assignTimestampsAndWatermarks(< watermark strategy >);

withTimestampsAndWatermarks
        .keyBy( (event) -> event.getGroup() )
        .window(TumblingEventTimeWindows.of(Time.seconds(10)))
        .reduce( (a, b) -> a.add(b) )
        .addSink(...);
```

以这种方式使用 WatermarkStrategy 获取一个流，并生成一个带有时间戳的元素和水印的新流。如果原始流已经有时间戳和/或水印，则时间戳分配器将覆盖它们。

3.6.4 处理空闲数据源

如果一个输入 split/partitions/shards 在一段时间内不携带事件，则意味着

WatermarkGenerator 也不会获得任何新信息来作为水印的基础,这种情况称为空闲输入或空闲源。这是一个问题,因为可能会发生一些分区仍然携带事件。在这种情况下,水印将被保留,因为它是作为所有不同的并行水印的最小值计算的。

如果要处理这个问题,则可以使用 WatermarkStrategy 来检测空闲状态并将输入标记为 idle。WatermarkStrategy 为此提供了一个方便的辅助方法。

Scala 代码如下:

```
WatermarkStrategy
  .forBoundedOutOfOrderness[(Long, String)](Duration.ofSeconds(20))
  .withIdleness(Duration.ofMinutes(1))
```

Java 代码如下:

```
WatermarkStrategy
    .<Tuple2<Long, String>>forBoundedOutOfOrderness(Duration.ofSeconds(20))
    .withIdleness(Duration.ofMinutes(1));
```

WatermarkStrategy.withIdleness()方法允许用户在配置的时间内(超时时间内)没有记录到达时将一个流标记为空闲。这样就意味着下游的数据不需要等待水印的到来。目前只有水印生成并发射到下游时,这个数据流才重新变成活跃状态。

3.6.5 编写水印生成器

在 WatermarkStrategy 中,TimestampAssigner 是一个简单的函数,它从事件中提取字段,因此用户不需要过多地关注,但是编写 WatermarkGenerator 稍微复杂一些,所以接下来主要讲解如何实现它。

WatermarkGenerator 接口的声明源码如下:

```
/**
 * The {@code WatermarkGenerator} generates watermarks either based on events or
 * periodically (in a fixed interval).
 *
 * <p><b>Note:</b> This WatermarkGenerator subsumes the previous distinction between the
 * {@code AssignerWithPunctuatedWatermarks} and the {@code AssignerWithPeriodicWatermarks}.
 */
@Public
public interface WatermarkGenerator<T> {

    /**
     * Called for every event, allows the watermark generator to examine and remember the
     * event timestamps, or to emit a watermark based on the event itself.
```

```
     */
    void onEvent(T event, long eventTimestamp, WatermarkOutput output);

    /**
     * Called periodically, and might emit a new watermark, or not.
     *
     * <p> The interval in which this method is called and Watermarks are generated
     * depends on {@link ExecutionConfig#getAutoWatermarkInterval()}.
     */
    void onPeriodicEmit(WatermarkOutput output);
}
```

有两种不同的水印生成方法：periodic（周期型）和 punctuated（标点型）。这两种水印生成方法的区别如下：

（1）周期型水印生成器通常通过 onEvent（）观察传入的事件，然后在框架调用 onPeriodicEmit（）时发出水印。

（2）标点型水印生成器将查看 onEvent（）中的事件，并等待流中携带水印信息的特殊标记事件或标点。当它看到这些事件之一时会立即发出水印。通常，标点生成器不会从 onPeriodicEmit（）发出水印。

1. 编写 Periodic WatermarkGenerator

周期型水印生成器观察流事件并周期性地生成水印（可能取决于流元素，或者纯粹基于处理时间）。生成水印的时间间隔是通过 ExecutionConfig.setAutoWatermarkInterval（）定义的，默认为 200ms。例如，通过下面的方式设置间隔时间：

```
//每3s生成一个水印
env.getConfig().setAutoWatermarkInterval(3000);
```

每次都会调用生成器的 onPeriodicEmit（）方法，如果返回的水印非空且大于前一个水印，则会发出一个新的水印。

下面的代码展示了两个使用周期型水印生成器的简单水印生成示例。

Scala 代码如下：

```
//第 3 章/BoundedOutOfOrdernessGenerator.scala

/**
 * 这个生成器生成的水印,期望元素按顺序到达,但实际上只是在一定程度上可实现
 * 某个时间戳 t 的最新元素最多会在时间戳 t 的最早元素之后 n 毫秒到达
 */
class BoundedOutOfOrdernessGenerator extends WatermarkGenerator[MyEvent] {

    val maxOutOfOrderness = 3500L         //3.5 seconds
```

```
        var currentMaxTimestamp: Long = _

    override def onEvent(element: MyEvent, eventTimestamp: Long, output: WatermarkOutput):
Unit = {
        currentMaxTimestamp = max(eventTimestamp, currentMaxTimestamp)
    }

    override def onPeriodicEmit(output: WatermarkOutput): Unit = {
        //将水印发送为当前最高时间戳减去无序界限(out-of-orderness bound)
        output.emitWatermark(new Watermark(currentMaxTimestamp - maxOutOfOrderness - 1));
    }
}

/**
 * 这个生成器生成的水印比处理时间滞后
 * 它假设元素在有界时延后到达 Flink
 */
class TimeLagWatermarkGenerator extends WatermarkGenerator[MyEvent] {

    val maxTimeLag = 5000L                    //5 seconds

    override def onEvent(element: MyEvent, eventTimestamp: Long, output: WatermarkOutput):
Unit = {
        //不需要做任何事情,因为使用处理时间
    }

    override def onPeriodicEmit(output: WatermarkOutput): Unit = {
        output.emitWatermark(new Watermark(System.currentTimeMillis() - maxTimeLag));
    }
}
```

Java 代码如下:

```
//第 3 章/BoundedOutOfOrdernessGenerator.java

/**
 * 这个生成器生成的水印,期望元素按照一定的顺序到达,但实际上只是在一定程度上可实现
 * 某个时间戳 t 的最新元素最多会在时间戳 t 的最早元素之后 n 毫秒到达
 */
public class BoundedOutOfOrdernessGenerator
                    implements WatermarkGenerator<MyEvent> {

    private final long maxOutOfOrderness = 3500; //3.5 seconds

    private long currentMaxTimestamp;
```

```java
    @Override
    public void onEvent(MyEvent event, long eventTimestamp, WatermarkOutput output) {
        currentMaxTimestamp = Math.max(currentMaxTimestamp, eventTimestamp);
    }

    @Override
    public void onPeriodicEmit(WatermarkOutput output) {
        //将水印发送为当前最高时间戳减去无序界限(out-of-orderness bound)
        output.emitWatermark(new Watermark(currentMaxTimestamp - maxOutOfOrderness - 1));
    }
}

/**
 * 这个生成器生成的水印比处理时间滞后
 * 它假设元素在有界时延后到达Flink
 */
public class TimeLagWatermarkGenerator
            implements WatermarkGenerator<MyEvent> {

    private final long maxTimeLag = 5000;          //5 seconds

    @Override
    public void onEvent(MyEvent event, long eventTimestamp, WatermarkOutput output) {
        //不需要做任何事情,因为在处理时间
    }

    @Override
    public void onPeriodicEmit(WatermarkOutput output) {
        output.emitWatermark(new Watermark(System.currentTimeMillis() - maxTimeLag));
    }
}
```

2. 编写 Punctuated WatermarkGenerator

标点型水印生成器将观察事件流,并在检测到携带水印信息的特殊元素时发出水印。下面的代码用于实现一个标点型水印生成器,每当事件表明它带有某个标记时就会发出水印。

Scala 代码如下:

```scala
//第3章/PunctuatedAssigner.scala

class PunctuatedAssigner extends WatermarkGenerator[MyEvent] {

    override def onEvent(element: MyEvent, eventTimestamp: Long): Unit = {
        if (event.hasWatermarkMarker()) {
```

```
            output.emitWatermark(new Watermark(event.getWatermarkTimestamp()))
        }
    }

    override def onPeriodicEmit(): Unit = {
        //不需要做任何事情,因为我们会对上面的事件做出反应
    }
}
```

Java 代码如下:

```java
//第3章/PunctuatedAssigner.java

public class PunctuatedAssigner implements WatermarkGenerator<MyEvent> {

    @Override
    public void onEvent(MyEvent event, long eventTimestamp, WatermarkOutput output) {
        if (event.hasWatermarkMarker()) {
            output.emitWatermark(new Watermark(event.getWatermarkTimestamp()));
        }
    }

    @Override
    public void onPeriodicEmit(WatermarkOutput output) {
        //不需要做任何事情,因为我们会对上面的事件做出反应
    }
}
```

这种方法简单明了,类中主要包含以下两种方法。

(1) onEvent:每个元素都会调用这种方法,如果想依赖每个元素生成一个水印,然后发送到下游(可选,就是看是否用 output 来收集水印),则可以实现这种方法。

(2) onPeriodicEmit:如果数据量比较大,每条数据都生成一个水印,则会影响性能,所以这里还有一个周期性生成水印的方法。这个水印的生成周期可以这样设置:env.getConfig().setAutoWatermarkInterval(5000L)。

注意:在每个单独的事件上生成水印是可能的,但是,由于每个水印都会引起下游的一些计算,所以过多的水印会降低性能。

3. 水印策略和 Kafka 连接器

当使用 Apache Kafka 作为数据源时,每个 Kafka 分区可能有一个简单的事件时间模式(升序时间戳或有界无序),然而,当使用来自 Kafka 的流时,多个分区经常会被并行地使用,即交叉地使用来自分区的事件并破坏每个分区的模式(这是 Kafka 的消费者客户端固有的

工作方式)。

在这种情况下,可以使用 Flink 的 Kafka-partition-aware 生成水印。使用该特性,可以在 Kafka 消费者内部生成每个 Kafka 分区的水印,每分区(per-partition)的水印合并的方式与在流 shuffle 中合并水印的方式相同。

例如,如果事件时间戳按照 Kafka 分区严格升序,则使用升序时间戳水印生成器生成每分区的水印将得到完美的整体水印。注意,在示例中没有提供 TimestampAssigner,而是使用 Kafka 记录本身的时间戳。

Scala 代码如下:

```
val kafkaSource = new FlinkKafkaConsumer[MyType]("myTopic", schema, props)
kafkaSource.assignTimestampsAndWatermarks(
  WatermarkStrategy.forBoundedOutOfOrderness(Duration.ofSeconds(20)))

val stream: DataStream[MyType] = env.addSource(kafkaSource)
```

Java 代码如下:

```
FlinkKafkaConsumer < MyType > kafkaSource =
            new FlinkKafkaConsumer <>("myTopic", schema, props);
kafkaSource.assignTimestampsAndWatermarks(
WatermarkStrategy.forBoundedOutOfOrderness(Duration.ofSeconds(20)));

DataStream < MyType > stream = env.addSource(kafkaSource);
```

关于如何使用 per-Kafka-partition 生成水印,以及在这种情况下,水印如何通过流数据流传播,如图 3-28 所示。

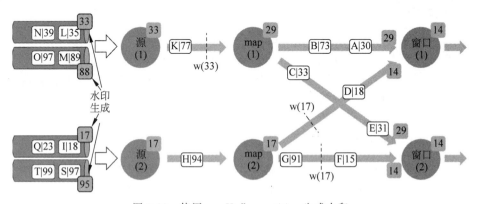

图 3-28　使用 per-Kafka-partition 生成水印

4. 算子(运算符)如何处理水印

作为一般规则,算子在将给定的水印转发到下游之前需要完全处理它。例如,

WindowOperator将首先计算应该被触发的所有窗口，并且只有在生成由水印触发的所有输出之后，水印本身才会被发送到下游。换句话说，由于水印的出现而产生的所有元素都将在水印之前发出。

同样的规则也适用于TwoInputStreamOperator，然而，在这种情况下，算子的当前水印被定义为两个输入的最小值。该行为的细节由OneInputStreamOperator#processWatermark、twinputstreamoperator#processWatermark1和twinputstreamoperator#processWatermark2方法的实现定义。

在注入时间和处理时间的情况下，只需指定时间特性，水印就会生成。可参看下面的代码片段。

Scala代码如下：

```
val env = StreamExecutionEnvironment.getExecutionEnvironment

env.setStreamTimeCharacteristic(TimeCharacteristic.ProcessingTime)
//或
env.setStreamTimeCharacteristic(TimeCharacteristic.IngestionTime)
```

Java代码如下：

```
final StreamExecutionEnvironment env = StreamExecutionEnvironment.getExecutionEnvironment();

env.setStreamTimeCharacteristic(TimeCharacteristic.ProcessingTime);
//或
env.setStreamTimeCharacteristic(TimeCharacteristic.IngestionTime);
```

5. 已弃用的方法

在引入当前使用的WatermarkStrategy、TimestampAssigner和WatermarkGenerator抽象之前，Flink使用的是AssignerWithPeriodicWatermarks和AssignerWithPunctuatedWatermarks。仍然可以在API文档中看到它们，但建议使用新的接口，因为新的接口提供了更清晰的关注点分离，并统一了生成水印的periodic和punctuated样式。

3.6.6 内置水印生成器

正如3.6.5节中所述，Flink提供了抽象，允许程序员分配他们自己的时间戳并发出自己的水印。更具体地说，可以通过实现WatermarkGenerator接口实现。

为了进一步简化此类任务的编程工作，Flink附带了一些预先实现的时间戳赋值器，包括：

(1) 单调递增时间戳。
(2) 固定的迟到时间。

下面详细来了解这两种预实现的时间戳赋值器。

1) 单调递增时间戳

周期性水印生成的最简单特例是给定源任务所看到的时间戳按升序出现的情况。在这种情况下，当前时间戳总是可以作为水印的，因为不会出现更早的时间戳。

需要注意，只有每个并行数据源任务的时间戳是递增的才有必要。例如，如果在一个特定的设置中，一个 Kafka 分区被一个并行数据源实例读取，则在每个 Kafka 分区中只需时间戳升序就可以了。Flink 的水印合并机制将在并行流被 shuffle、union、connect 或 merge 时生成正确的水印。设置单调递增时间戳的代码如下。

Scala 代码如下：

```
WatermarkStrategy.forMonotonousTimestamps()
```

Java 代码如下：

```
WatermarkStrategy.forMonotonousTimestamps();
```

如果时间戳从不乱序，或者用户愿意将所有乱序事件视为迟到，则可在这种情况下使用这种方法为具有单调递增时间戳的情况创建水印策略。

2) 固定的迟到时间

周期性水印生成的另一个例子是当水印滞后于流中看到的最大（事件时间）时间戳时的固定数量的时间。这种情况涵盖在流中可能遇到的最大时延是预先已知的场景，例如，当创建一个包含元素的自定义源时，这些元素的时间戳分布在一个固定的测试时间段内。对于这些情况，Flink 提供了 BoundedOutOfOrderness() 水印生成方法，它将 maxOutOfOrderness 作为参数。这个参数是一个 Duration 类型的时间间隔，也就是用户可以接受的最大的时延时间，即当计算给定窗口的最终结果时，一个元素在被忽略之前允许延迟的最大时间。迟到 lateness 对应于 t- t_w 的结果，其中 t 是元素的（事件时间）时间戳，t_w 是前一个水印的时间戳。如果 lateness>0，则该元素被认为是迟来的，默认情况下，在计算其对应窗口的作业结果时被忽略。

例如，实现一个延迟 10s 的固定时延水印，可以使用下面的代码。

Scala 代码如下：

```
WatermarkStrategy.[T]forBoundedOutOfOrderness(Duration.ofSeconds(10))
```

Java 代码如下：

```
WatermarkStrategy.<T>forBoundedOutOfOrderness(Duration.ofSeconds(10));
```

如果处理的数据流中无序的时间戳是正常的，则可使用 forBoundedOutOfOrderness()，这种方法为记录乱序的情况创建水印策略，但可以设置事件乱序程度的上限。

3.6.7 分配时间戳和水印示例

下面通过几个示例来掌握如何为数据流中的事件分配时间戳和水印。

【示例 3-12】 （Scala 实现）为数据流中的事件分配时间戳和水印。
建议按以下步骤执行：

（1）在 IntelliJ IDEA 中创建一个 Flink 项目，使用 flink-quickstart-scala 项目模板（Flink 项目的创建过程可参考 2.2 节）。

（2）设置依赖。在 pom.xml 文件中添加如下依赖（根据项目模板创建，这个依赖会自动添加，此步可省略）：

```xml
<dependency>
    <groupId>org.apache.flink</groupId>
    <artifactId>flink-scala_2.12</artifactId>
    <version>1.13.2</version>
    <scope>provided</scope>
</dependency>
<dependency>
    <groupId>org.apache.flink</groupId>
    <artifactId>flink-streaming-scala_2.12</artifactId>
    <version>1.13.2</version>
    <scope>provided</scope>
</dependency>
```

（3）创建流应用程序类，代码如下：

```scala
//第 3 章/AssignerDemo2.scala

import org.apache.flink.api.common.eventtime.WatermarkStrategy
import org.apache.flink.streaming.api.scala._

object AssignerDemo2 {

  //case 类,表示流元素数据类型
  case class MessageInfo(hostname: String, status: String)

  //main 方法
  def main(args: Array[String]) {
    //设置流执行环境
    val env = StreamExecutionEnvironment.getExecutionEnvironment

    //数据流源
    val stream = env.fromElements(
```

```
        MessageInfo("host1", "1234"),
        MessageInfo("host2", "2234"),
        MessageInfo("host3", "1234")
    )

    //因为模拟数据没有时间戳,所以可用此方法添加递增时间戳和水印
    //为数据流中的元素分配时间戳,并生成标记以指示事件时间进展
        val withTimestampsAndWatermarks = stream.assignTimestampsAndWatermarks(
          WatermarkStrategy.forMonotonousTimestamps[MessageInfo]()
            .withTimestampAssigner((MessageInfo, ts) => System.currentTimeMillis())
        )

    //将结果输出到控制台
    withTimestampsAndWatermarks.print

    //执行
    env.execute("Flink Watermark Strategy ")
  }
}
```

【示例3-13】 (Java实现)为数据流中的事件分配时间戳和水印。

建议按以下步骤执行:

(1)在IntelliJ IDEA中创建一个Flink项目,使用flink-quickstart-Java项目模板(Flink项目的创建过程可参考2.2节)。

(2)设置依赖。在pom.xml文件中添加如下依赖:

```
<dependency>
    <groupId>org.apache.flink</groupId>
    <artifactId>flink-Java</artifactId>
    <version>1.13.2</version>
    <scope>provided</scope>
</dependency>
<dependency>
    <groupId>org.apache.flink</groupId>
    <artifactId>flink-streaming-Java_2.12</artifactId>
    <version>1.13.2</version>
    <scope>provided</scope>
</dependency>
```

(3)创建事件数据结构,代码如下:

```
//第3章/AssignerDemo.java

import org.apache.flink.streaming.api.environment.StreamExecutionEnvironment;
```

```java
public class AssignerDemo {

    public static class MessageInfo {
        public String hostname;
        public String status;

        public MessageInfo() {}

        public MessageInfo(String hostname, String status){
            this.hostname = hostname;
            this.status = status;
        }

        @Override
        public String toString() {
            return "MessageInfo{" +
                    "hostname = '" + hostname + '\'' +
                    ", status = '" + status + '\'' +
                    '}';
        }
    }

    public static void main(String[] args) throws Exception {
        //设置流执行环境
        final StreamExecutionEnvironment env =
                StreamExecutionEnvironment.getExecutionEnvironment();

        //触发流程序执行
        env.execute("Flink Watermark Strategy");
    }
}
```

(4) 创建流应用程序类,代码如下:

```java
//第 3 章/AssignerDemo2.java

import org.apache.flink.api.common.eventtime.WatermarkStrategy;
import org.apache.flink.streaming.api.datastream.DataStream;
import org.apache.flink.streaming.api.environment.StreamExecutionEnvironment;

public class AssignerDemo {

    public static class MessageInfo {
        public String hostname;
        public String status;
```

```java
    public MessageInfo() {}

    public MessageInfo(String hostname, String status){
        this.hostname = hostname;
        this.status = status;
    }

    @Override
    public String toString() {
        return "MessageInfo{" +
                "hostname = '" + hostname + '\'' +
                ", status = '" + status + '\'' +
                '}';
    }
}

public static void main(String[] args) throws Exception {
    //设置流执行环境
    final StreamExecutionEnvironment env =
        StreamExecutionEnvironment.getExecutionEnvironment();

    //源数据流
    DataStream<MessageInfo> stream = env.fromElements(
        new MessageInfo("host1","1234"),
        new MessageInfo("host2","2234"),
        new MessageInfo("host3","1234")
    );

    //因为模拟数据没有时间戳,所以可用此方法添加单调增加时间戳和水印
    DataStream<MessageInfo> withTimestampsAndWatermarks = stream
        .assignTimestampsAndWatermarks(
            WatermarkStrategy.<MessageInfo>forMonotonousTimestamps()
                .withTimestampAssigner((MessageInfo, ts) -> System.currentTimeMillis())
        );

    //将结果输出到控制台
    withTimestampsAndWatermarks.print();

    //触发流程序执行
    env.execute("Flink Watermark Strategy");
  }
}
```

下面这个示例使用4.2.4节的自定义信用卡交易数据源,指定交易数据的交易时间timestamp字段作为事件时间,实现一个延迟10s的固定时延水印。

【示例3-14】 实现一个延迟10s的固定时延水印。

Scala代码如下：

```scala
//第3章/AssignerDemo2.scala

import java.time.Duration

import com.xueai8.java.ch03.entity.Transaction
import com.xueai8.java.ch03.source.MyTransactionSource
import org.apache.flink.api.common.eventtime.{SerializableTimestampAssigner, WatermarkStrategy}
import org.apache.flink.streaming.api.scala._

/**
 * 功能:为信用卡交易数据流中的事件分配时间戳和水印
 * 使用自定义的 TransactionSource 数据源
 */
object AssignerDemo2 {

  def main(args: Array[String]) {
    //设置流执行环境
    val env = StreamExecutionEnvironment.getExecutionEnvironment

    //设置自定义数据源.参数false用于指定创建的是流数据源
    val stream = env
      .addSource(new MyTransactionSource(false))
      .name("transactions")

    //因为模拟数据没有时间戳,所以可用此方法添加事件时间戳和水印
    /* 使用匿名内部类 */
    val withTimestampsAndWatermarks = stream.assignTimestampsAndWatermarks(
      WatermarkStrategy
        .forBoundedOutOfOrderness[Transaction](Duration.ofMillis(10000))
        .withTimestampAssigner(new SerializableTimestampAssigner[Transaction]() {
          //抽取交易事件中的 timestamp 字段作为事件时间戳
          override def extractTimestamp(t: Transaction, ts: Long): Long = {
            t.timestamp
          }
        })
    )

    //将结果输出到控制台
    stream.print

    //执行
    env.execute("Flink Watermark Strategy")
  }
}
```

Java 代码如下：

```java
//第 3 章/AssignerDemo2.java

import com.xueai8.java.ch03.entity.Transaction;
import com.xueai8.java.ch03.source.MyTransactionSource;
import org.apache.flink.api.common.eventtime.SerializableTimestampAssigner;
import org.apache.flink.api.common.eventtime.WatermarkStrategy;
import org.apache.flink.streaming.api.datastream.DataStream;
import org.apache.flink.streaming.api.environment.StreamExecutionEnvironment;

import Java.time.Duration;

/**
 * 功能:为信用卡交易数据流中的事件分配时间戳和水印
 * 使用自定义的 TransactionSource 数据源
 */
public class AssignerDemo2 {

    public static void main(String[] args) throws Exception {
        //设置流执行环境
        final StreamExecutionEnvironment env =
            StreamExecutionEnvironment.getExecutionEnvironment();

        //设置自定义数据源.参数 false 用于指定创建的是流数据源
        DataStream<Transaction> stream = env
            .addSource(new MyTransactionSource(false))
            .name("transactions");

        //因为模拟数据没有时间戳,所以用此方法添加单调增加时间戳和水印
        /* 使用匿名内部类
        DataStream<Transaction> withTimestampsAndWatermarks = stream
            .assignTimestampsAndWatermarks(WatermarkStrategy
                .<Transaction>forBoundedOutOfOrderness(Duration.ofSeconds(10))
                .withTimestampAssigner(
                    new SerializableTimestampAssigner<Transaction>() {
                        @Override
                        public long extractTimestamp(Transaction transaction,long ts){
                            return transaction.timestamp;
                        }
                    })
            );*/
        //或者使用匿名函数(Lambda 函数)
        DataStream<Transaction> withTimestampsAndWatermarks = stream
            .assignTimestampsAndWatermarks(WatermarkStrategy
                .<Transaction>forBoundedOutOfOrderness(Duration.ofSeconds(10))
```

```
            .withTimestampAssigner((transaction,ts) -> transaction.timestamp)
        );

        //将结果输出到控制台
        withTimestampsAndWatermarks.print();

        //触发流程序执行
        env.execute("Flink Watermark Strategy");
    }
}
```

下面这个示例用于重构 4.2.4 节的自定义信用卡交易数据源,在数据源中分配事件时间戳和生产水印。

【示例 3-15】(Java 实现)在自定义信用卡交易数据源使用水印策略。

```
//第 3 章/MyTransactionSource2.java

import com.xueai8.java.ch03.entity.Transaction;
import org.apache.flink.streaming.api.functions.source.RichSourceFunction;
import org.apache.flink.streaming.api.watermark.Watermark;

import Java.sql.Timestamp;
import Java.util.Arrays;
import Java.util.List;

/**
 * 自定义数据源,继承自 SourceFunction
 * 自带时间戳和水印的数据源
 */
public class MyTransactionSource2 extends RichSourceFunction<Transaction> {
    private static final long serialVersionUID = 1L;

    private static final Timestamp INITIAL_TIMESTAMP =
            Timestamp.valueOf("2020-01-01 00:00:00");
    private static final long SIX_MINUTES = 6 * 60 * 1000;

    private final boolean bounded;              //标志变量,指示生成流数据还是批数据

    private int index = 0;                      //交易记录的索引
    private long timestamp;                     //交易发生的时间戳

    private volatile boolean isRunning = true;
    private List<Transaction> data = null;

    public MyTransactionSource2(boolean bounded){
```

```java
this.bounded = bounded;
this.timestamp = INITIAL_TIMESTAMP.getTime();

//事先存储的信用卡交易数据,在实际中来自外部数据源系统,如 Kafka
data = Arrays.asList(
        new Transaction(1, 0L, 188.23),
        new Transaction(2, 0L, 374.79),
        new Transaction(3, 0L, 112.15),
        new Transaction(4, 0L, 478.75),
        new Transaction(5, 0L, 208.85),
        new Transaction(1, 0L, 379.64),
        new Transaction(2, 0L, 351.44),
        new Transaction(3, 0L, 320.75),
        new Transaction(4, 0L, 259.42),
        new Transaction(5, 0L, 273.44),
        new Transaction(1, 0L, 267.25),
        new Transaction(2, 0L, 397.15),
        new Transaction(3, 0L, 0.219),
        new Transaction(4, 0L, 231.94),
        new Transaction(5, 0L, 384.73),
        new Transaction(1, 0L, 419.62),
        new Transaction(2, 0L, 412.91),
        new Transaction(3, 0L, 0.77),
        new Transaction(4, 0L, 22.10),
        new Transaction(5, 0L, 377.54),
        new Transaction(1, 0L, 375.44),
        new Transaction(2, 0L, 230.18),
        new Transaction(3, 0L, 0.80),
        new Transaction(4, 0L, 350.89),
        new Transaction(5, 0L, 127.55),
        new Transaction(1, 0L, 483.91),
        new Transaction(2, 0L, 228.22),
        new Transaction(3, 0L, 871.15),
        new Transaction(4, 0L, 64.19),
        new Transaction(5, 0L, 79.43),
        new Transaction(1, 0L, 56.12),
        new Transaction(2, 0L, 256.48),
        new Transaction(3, 0L, 148.16),
        new Transaction(4, 0L, 199.95),
        new Transaction(5, 0L, 252.37),
        new Transaction(1, 0L, 274.73),
        new Transaction(2, 0L, 473.54),
        new Transaction(3, 0L, 119.92),
        new Transaction(4, 0L, 323.59),
        new Transaction(5, 0L, 353.16),
        new Transaction(1, 0L, 211.90),
```

```java
                new Transaction(2, 0L, 280.93),
                new Transaction(3, 0L, 347.89),
                new Transaction(4, 0L, 459.86),
                new Transaction(5, 0L, 82.31),
                new Transaction(1, 0L, 373.26),
                new Transaction(2, 0L, 479.83),
                new Transaction(3, 0L, 454.25),
                new Transaction(4, 0L, 83.64),
                new Transaction(5, 0L, 292.44)
        );
    }

    @Override
    public void run(SourceContext<Transaction> sourceContext) throws Exception {
        while(this.isRunning && this.hasNext()) {
            Transaction transaction = this.next();
            sourceContext.collectWithTimestamp(transaction,
                    transaction.timestamp);                    //带时间戳
            //sourceContext.collect(transaction);   //不带时间戳
            sourceContext.emitWatermark(new Watermark(transaction.timestamp + 20000));
                                                    //发出水印

        }

    }

    @Override
    public void cancel() {
        this.isRunning = false;
    }

    private boolean hasNext() {
        //如果还有数据
        if (index < data.size()) {
            return true;
        }
        //如果用于生成批数据
        else if(bounded){
            return false;
        }
        //如果用于生成流数据,则从头循环
        else {
            index = 0;
            return true;
        }

    }
```

```java
//生成下一个交易数据,交易时间相隔为 6min
private Transaction next() {
    try {
        Thread.sleep(500);
    } catch (InterruptedException e) {
        throw new RuntimeException(e);
    }
    Transaction transaction = data.get(index++);
    transaction.timestamp = timestamp;
    timestamp += SIX_MINUTES;          //交易时间间隔为 6min
    return transaction;
}
}
```

接下来实现一个测试程序,代码如下:

```java
//第 3 章/AssignerDemo3.java
import com.xueai8.java.ch03.entity.Transaction;
import com.xueai8.java.ch03.source.MyTransactionSource;
import com.xueai8.java.ch03.source.MyTransactionSource2;
import org.apache.flink.api.common.eventtime.WatermarkStrategy;
import org.apache.flink.streaming.api.datastream.DataStream;
import org.apache.flink.streaming.api.environment.StreamExecutionEnvironment;

import java.time.Duration;

/**
 * 功能:为信用卡交易数据流中的事件分配时间戳和水印
 * 使用自定义带时间戳和水印的 MyTransactionSource2 数据源
 */
public class AssignerDemo3 {

    public static void main(String[] args) throws Exception {
        //设置流执行环境
        final StreamExecutionEnvironment env =
                StreamExecutionEnvironment.getExecutionEnvironment();

        //设置自定义数据源.参数 false 用于指定创建的是流数据源
        DataStream<Transaction> stream = env
                .addSource(new MyTransactionSource2(false))
                .name("transactions");

        //将结果输出到控制台
        stream.print();
```

```
        //触发流程序执行
        env.execute("Flink Watermark Strategy");
    }
}
```

3.7 窗口操作

在进行流处理时,很多时候想要对流的有界子集进行聚合分析。例如回答以下问题:

(1) 每分钟的页面浏览(PV)次数。

(2) 每用户每周的会话次数。

(3) 每分钟每传感器的最高温度。

(4) 当电商发布一个秒杀活动时,想要每隔10min了解流量数据。

显然,要回答这些问题,程序需要处理元素组,而不是单个元素,因此,通常使用窗口来限定在数据流上的聚合(如count、sum等)的范围,例如"过去5min内的计数"或"最后100个元素的总和",所以在处理流数据时,通常更有意义的是考虑有限窗口上的聚合,而不是整个流。

Flink 具有非常富有表现力的窗口语义。

3.7.1 理解 Flink 窗口概念

窗口(window)是处理无限流的核心,使用窗口计算无界流上的聚合。窗口将流分割为有限大小的组,用户可以对这样的组进行计算。窗口可以是由时间驱动的(例如,每30s),也可以是由数据驱动的(例如,每100个元素),如图3-29所示。

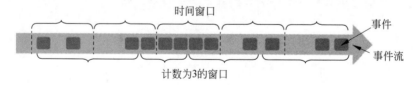

图 3-29 Flink 流的窗口概念

简而言之,流窗口允许用户对流中的元素进行分组,并在每个组上执行用户定义的函数。这个用户定义的函数可以返回0、1或多个元素,通过这种方式,它创建了一个新的流,用户可以在单独的系统中处理或存储这个流,如图3-30所示。

如何对流中的元素进行分组?Flink 支持不同类型的窗口,分别介绍如下。

(1) 滚动窗口:Tumbling Window,是在流中创建不重叠的相邻窗口。它们是固定长度的窗口,没有重叠。可以根据时间对元素进行分组(例如,从 10:00 到 10:05 的所有元素进入一个组),或者根据计数(前 50 个元素进入一个单独的组)对元素进行分组。例如,可以

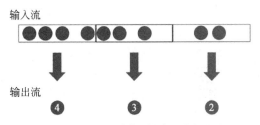

图 3-30　流窗口计算,创建一个新的流

用它来回答这样的问题:"在不重叠的 5min 间隔内计算流中元素的数量"。

(2) 滑动窗口:Sliding Window,类似于滚动窗口,但是窗口可以重叠。滑动窗口是固定长度的窗口,通过用户给定的窗口滑动参数与前面的窗口重叠。例如,如果需要计算最后 5min 的指标,但希望每分钟显示一个输出时。

(3) 会话窗口:Session Window,当对发生的事件进行分组时,将时间接近的分到一组(一个窗口中)。还可以提供会话间隔的配置参数,该参数指示在关闭会话之前需要等待多长时间。

(4) 全局窗口:Global Window,Flink 将所有元素放到一个窗口中。通常在这种情况下,每个元素都被分配给一个单一的 per-key 全局窗口(Global Window)。如果不指定任何触发器,就不会触发任何计算。这只有在定义自定义触发器时才有用,该触发器定义了窗口何时结束。

这几种窗口类型的表示如图 3-31 所示。

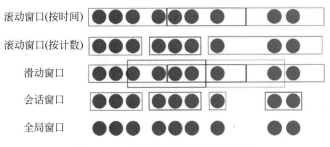

图 3-31　Flink 支持不同类型的窗口

使用 Flink 计算窗口分析依赖于两个主要的抽象:将事件分配给窗口的窗口分配器(必要时创建新的窗口对象),以及应用于分配给窗口的事件的窗口函数。Flink 的窗口 API 还有两个概念:①触发器(Triggers),它决定何时调用窗口函数;②清除器(Evictor),它可以删除窗口中收集的元素。

当应该属于某个窗口的第 1 个元素到达时,就会创建一个窗口,当时间(事件时间或处理时间)超过窗口的结束时间戳加上用户指定的允许时延时,该窗口将被完全删除。Flink 保证只针对基于时间的窗口进行删除,而不针对其他类型,例如全局窗口。例如,一个基于事件时间的窗口策略为每 5min 创建一个非重叠窗口,并且允许迟 1min,那么 Flink 将创建一个12:00-12:05间隔的新窗口(当带有落在该时间窗口的时间戳的第1个元素在这个间隔到来时),当水印通过12:06时间戳时它会删除该窗口。

此外,每个窗口都将有一个触发器(Trigger)和一个附加到该触发器的函数(ProcessWindowFunction、ReduceFunction、AggregateFunction或FoldFunction)。附加函数将包含要应用于窗口内容的计算,而触发器用于指定触发窗口计算(应用函数)的条件。触发策略可能类似于"当窗口中的元素数量超过4个时"或者"当水印经过窗口末端时"。触发器还可以决定在创建和删除窗口之间的任何时间清除窗口的内容。在这种情况下,清除指的仅是窗口中的元素,而不是窗口元数据。这意味着仍然可以向该窗口添加新数据。

除此之外,还可以指定一个清除器(Evictor),它能够在触发器触发之后、函数应用之前和/或之后从窗口中删除元素。

除了选择如何将元素分配给不同的窗口之外,还需要选择一个流类型。Flink的窗口支持以下两种流类型。

(1) Non-keyed stream:在这种情况下,流中的所有元素将一起被处理,我们的用户定义函数将访问流中的所有元素。这种流类型的缺点是它不提供并行性,集群中只有一台机器能够执行我们的代码。

(2) Keyed stream:使用这种流类型,Flink将通过一个key(例如,设备的ID)将单个流划分为多个独立的流。当处理Keyed Stream中的窗口时,定义的函数只能访问具有相同key的项,但是使用多个独立的流可以让Flink并行化工作,如图3-32所示。

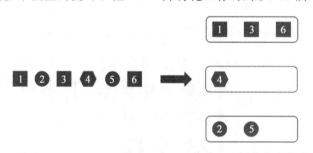

图3-32 Flink Keyed Stream:通过一个key将单个流划分为多个独立的流

必须在定义窗口之前指定要处理的流是否应该被赋予key。使用keyBy()将把无限流分割成逻辑keyed流。如果没有调用keyBy(),则说明流是non-keyed流。

对于keyed流,传入事件的任何属性都可以用作key。这时窗口计算可以由多个任务并行执行,因为每个逻辑keyed流都可以独立于其他流进行处理。所有引用相同key的元素将被发送到相同的并行任务进行处理。

在non-keyed流的情况下,原始流将不会被分割成多个逻辑流,所有的窗口逻辑将由一个任务执行,即并行度为1。

3.7.2 窗口分配器

窗口分配器用于定义如何将元素分配给窗口。这是通过在调用 window()（针对 Keyed Stream）或 windowAll()（针对 non-keyed stream）时指定所选择的 WindowAssigner 实现的。

WindowAssigner 负责将每个传入元素分配给一个或多个窗口。例如，一个乱序的基于事件时间的数据流窗口分配，如图 3-33 所示。

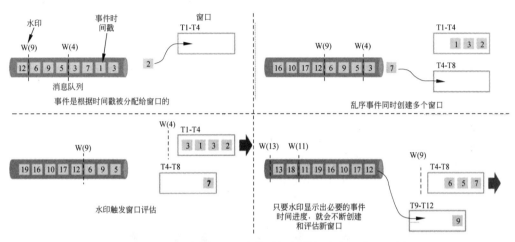

图 3-33　一个乱序的基于事件时间的数据流窗口分配示例

Flink 为最常见的场景（滚动时间窗口、滑动时间窗口、全局窗口和会话窗口）提供了预定义的窗口分配器，它们分别如下。

（1）滚动时间窗口：例如，每分钟 PV 数据（浏览量），代码如下：

```
TumblingEventTimeWindows.of(Time.minutes(1))
```

（2）滑动时间窗口：例如，每 10s 计算一次每分钟的页面浏览量，代码如下：

```
SlidingEventTimeWindows.of(Time.minutes(1), Time.seconds(10))
```

（3）会话窗口：例如，每个会话的 PV 数据，其中会话定义为会话之间至少 30min 的间隔，代码如下：

```
EventTimeSessionWindows.withGap(Time.minutes(30))
```

可以使用 Time.milliseconds(n)、Time.seconds(n)、Time.minutes(n)、Time.hours(n) 和 Time.days(n) 中的一个来指定持续时间。

所有内置的窗口分配器（全局窗口除外）都根据时间向窗口分配元素。基于时间的窗口分配程序(包括会话窗口)有事件时间和处理时间两种形式。在使用这两种时间窗口时需要进行权衡。如果使用处理时间窗口,则必须受到如下限制:

(1) 不能处理历史数据。

(2) 不能正确处理乱序数据。

(3) 结果将是不确定的。

但使用处理时间窗口会具有较低的时延。

基于时间的窗口有一个开始时间戳(包括)和一个结束时间戳(不包括),它们共同描述窗口的大小。在处理基于时间的窗口时,Flink 使用 TimeWindow 类型,该类型具有查询窗口开始和结束时间戳的方法,以及 maxTimestamp() 方法,该方法用于返回给定窗口允许的最大时间戳。

接下来将展示 Flink 预定义的窗口分配程序是如何工作的,以及如何在 DataStream 程序中使用它们。

1. 滚动窗口

滚动窗口分配器将每个元素分配给具有指定窗口大小的窗口。滚动窗口的大小是固定的,事件元素不会重叠参与计算。例如,如果指定一个大小为 5min 的滚动窗口,将计算当前窗口并每 5min 启动一个新窗口,如图 3-34 所示。

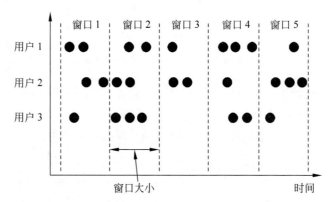

图 3-34　滚动窗口的大小是固定的,事件元素不会重叠参与计算

下面的代码片段展示了如何使用滚动窗口。

Scala 代码如下:

```
val input: DataStream[T] = ...

//滚动事件时间窗口
input
    .keyBy(<key selector>)
    .window(TumblingEventTimeWindows.of(Time.seconds(5)))
```

```
    .<windowed transformation>(<window function>)
//滚动处理时间窗口
input
    .keyBy(<key selector>)
    .window(TumblingProcessingTimeWindows.of(Time.seconds(5)))
    .<windowed transformation>(<window function>)

//每天滚动事件时间窗口,偏移值为-8h
input
    .keyBy(<key selector>)
    .window(TumblingEventTimeWindows.of(Time.days(1), Time.hours(-8)))
    .<windowed transformation>(<window function>)
```

Java 代码如下:

```
DataStream<T> input = ...;

//滚动事件时间窗口-基于事件时间
input
    .keyBy(<key selector>)
    .window(TumblingEventTimeWindows.of(Time.seconds(5)))
    .<windowed transformation>(<window function>);

//滚动处理时间窗口-基于处理时间
input
    .keyBy(<key selector>)
    .window(TumblingProcessingTimeWindows.of(Time.seconds(5)))
    .<windowed transformation>(<window function>);

//每天滚动事件时间窗口,偏移值为-8h
input
    .keyBy(<key selector>)
    .window(TumblingEventTimeWindows.of(Time.days(1), Time.hours(-8)))
    .<windowed transformation>(<window function>);
```

如以上示例所示,滚动窗口分配程序还接受一个可选的偏移参数,该参数可用于更改窗口的对齐方式。例如,如果没有偏移量,每小时滚动窗口都与 epoch 对齐,也就是说用户将得到诸如 1:00:00.000-1:59:59.999、2:00:00.000-2:59:59.999 等窗口。如果想改变,则可以给一个偏移量。例如,如果偏移时间为 15min,则将得到 1:15:00.000-2:14:59.999、2:15:00.000-3:14:59.999 等窗口。偏移量的一个重要用例是将窗口调整到 UTC-0 之外的时区。例如,在中国,必须指定 Time.hours(-8)偏移量。

2. 滑动窗口

滑动窗口分配器用于将元素分配给固定长度的窗口。类似于滚动窗口分配程序,窗口的大小由窗口大小参数配置。附加的窗口滑动参数用于控制滑动窗口启动的频率,因此,如果滑动窗口小于窗口大小,则滑动窗口可以重叠。在这种情况下,元素被分配到多个窗口。

例如,假设有一个10min大小的窗口,它可以滑动5min。这样,每5min就会得到一个包含过去10min内到达的事件的窗口,如图3-35所示。

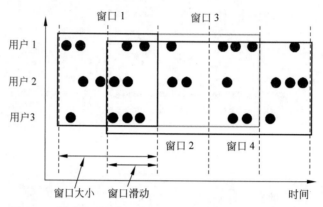

图3-35 滑动窗口包括固定窗口大小和附加的滑动窗口大小

下面的代码片段展示了如何使用滑动窗口。

Scala代码如下:

```
val input: DataStream[T] = ...

//滑动事件时间窗口
input
    .keyBy(<key selector>)
    .window(SlidingEventTimeWindows.of(Time.seconds(10), Time.seconds(5)))
    .<windowed transformation>(<window function>)

//滑动处理时间窗口
input
    .keyBy(<key selector>)
    .window(SlidingProcessingTimeWindows.of(Time.seconds(10), Time.seconds(5)))
    .<windowed transformation>(<window function>)

//滑动处理时间窗口,偏移-8h
input
    .keyBy(<key selector>)
    .window(SlidingProcessingTimeWindows.of(Time.hours(12), Time.hours(1), Time.hours(-8)))
    .<windowed transformation>(<window function>)
```

Java代码如下:

```
DataStream<T> input = ...;

//滑动事件时间窗口-基于事件时间滑动
input
    .keyBy(<key selector>)
    .window(SlidingEventTimeWindows.of(Time.seconds(10), Time.seconds(5)))
    .<windowed transformation>(<window function>);

//滑动处理时间窗口-基于处理时间滑动
input
    .keyBy(<key selector>)
    .window(SlidingProcessingTimeWindows.of(Time.seconds(10),Time.seconds(5)))
    .<windowed transformation>(<window function>);

//滑动处理时间窗口,偏移-8h
input
    .keyBy(<key selector>)
    .window(SlidingProcessingTimeWindows.of(Time.hours(12), Time.hours(1), Time.hours(-8)))
    .<windowed transformation>(<window function>);
```

3. 会话窗口

会话窗口分配器会根据活动会话对元素进行分组。与翻滚窗口和滑动窗口相比,会话窗口不会重叠,也没有固定的开始和结束时间。相反,会话窗口在一段时间内不接收元素时将被关闭,即当一段不活跃的间隔发生时。会话窗口分配程序可以配置为静态会话间隔,也可以配置会话间隔提取器函数,该函数定义不活动的周期有多长。当此期间过期时,当前会话将关闭,随后的元素将分配给新的会话窗口,如图 3-36 所示。

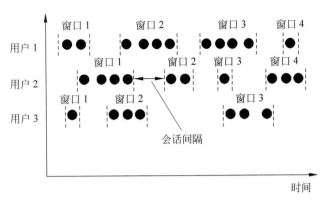

图 3-36 会话窗口没有固定的开始和结束时间

下面的代码片段展示了如何使用会话窗口。
Scala 代码如下:

```
val input: DataStream[T] = ...

//事件时间会话窗口,带有静态间隔
input
    .keyBy(<key selector>)
    .window(EventTimeSessionWindows.withGap(Time.minutes(10)))
    .<windowed transformation>(<window function>)

//事件时间会话窗口,带有动态间隔
input
    .keyBy(<key selector>)
    .window( EventTimeSessionWindows. withDynamicGap ( new  SessionWindowTimeGapExtractor[String] {
        override def extract(element: String): Long = {
            //确定并返回会话间隔
        }
    }))
    .<windowed transformation>(<window function>)

//处理时间会话窗口,带有静态间隔
input
    .keyBy(<key selector>)
    .window(ProcessingTimeSessionWindows.withGap(Time.minutes(10)))
    .<windowed transformation>(<window function>)

//处理时间会话窗口,带有动态间隔
input
    .keyBy(<key selector>)
    .window(DynamicProcessingTimeSessionWindows.withDynamicGap(new SessionWindowTimeGapExtractor[String] {
        override def extract(element: String): Long = {
            //确定并返回会话间隔
        }
    }))
    .<windowed transformation>(<window function>)
```

Java 代码如下:

```
DataStream<T> input = ...;

//事件时间会话窗口,带有静态间隔
input
    .keyBy(<key selector>)
    .window(EventTimeSessionWindows.withGap(Time.minutes(10)))
    .<windowed transformation>(<window function>);
```

```
//事件时间会话窗口,带有动态间隔
input
    .keyBy(< key selector >)
    .window(EventTimeSessionWindows.withDynamicGap((element) -> {
        //确定并返回会话间隔
    }))
    .< windowed transformation >(< window function >);

//处理时间会话窗口,带有静态间隔
input
    .keyBy(< key selector >)
    .window(ProcessingTimeSessionWindows.withGap(Time.minutes(10)))
    .< windowed transformation >(< window function >);

//处理时间会话窗口,带有动态间隔
input
    .keyBy(< key selector >)
    .window(ProcessingTimeSessionWindows.withDynamicGap((element) -> {
        //确定并返回会话间隔
    }))
    .< windowed transformation >(< window function >);
```

可以使用 Time.milliseconds(x)、Time.seconds(x)、Time.minutes(x)、Time.hours(n) 和 Time.days(n) 等指定静态间隔。动态间隔是通过实现 SessionWindowTimeGapExtractor 接口指定的。

4. 全局窗口

全局窗口分配程序将具有相同 key 的所有元素分配给同一个全局窗口,如图 3-37 所示。

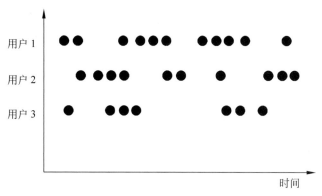

图 3-37 全局窗口没有一个自然的结束

只有在使用自定义触发器执行自定义窗口时,此窗口模式才有用。否则将不执行任何计算,因为全局窗口没有一个自然的结束,不能在此结束位置处理聚合的元素。在很多情况下,最好使用ProcessFunction来代替。

下面的代码片段展示了如何使用全局窗口。

Scala代码如下:

```
val input: DataStream[T] = ...

input
    .keyBy(<key selector>)
    .window(GlobalWindows.create())
    .<windowed transformation>(<window function>)
```

Java代码如下:

```
DataStream<T> input = ...;

input
    .keyBy(<key selector>)
    .window(GlobalWindows.create())
    .<windowed transformation>(<window function>);
```

5. 特别注意

在使用窗口分配器时,需要注意以下几点:

(1) 滑动窗口分配器可以创建许多窗口对象,并将每个事件复制到每个相关窗口。例如,如果每隔15min就有一个24h的滑动窗口,则每个事件都将被复制到$4 \times 24 = 96$个窗口中。

(2) 窗口时间使用的是Epoch时间(纪元时间),指的是一个特定的时间:1970-01-01 00:00:00 UTC。例如,如果使用了一小时的处理时间窗口,并且在12:05启动应用程序,则第1个窗口并不会在1:05关闭。第1个窗口将有55min长,即在3点关闭。

(3) 窗口后面可以再跟窗口,代码如下:

```
stream
  .keyBy(t -> t.key)
  .window(<time specification>)
  .reduce(<reduce function>)
  .windowAll(<same time specification>)
  .reduce(<same reduce function>)
```

(4) 空的时间窗口没有结果。只有当事件被分配给窗口时,才会创建窗口,因此,如果在给定的时间范围内没有事件,则不会报告结果。

(5) 迟来的事件会导致迟来的合并。会话窗口基于可以合并的窗口抽象。每个元素最初被分配给一个新窗口,之后,当窗口之间的间隔足够小时,就会合并窗口。通过这种方式,一个时延的事件可以填补之前两个独立会话之间的空白,从而产生一个时延的合并。

(6) 还可以通过继承 WindowAssigner 类实现自定义窗口分配程序。

6. Flink 窗口分配器示例

一个 Flink 窗口程序的总体结构如图 3-38 所示。

```
Keyed Windows

stream
    .keyBy(...)              <- keyed versus non-keyed windows
    .window(...)             <- required: "assigner"
   [.trigger(...)]           <- optional: "trigger" (else default trigger)
   [.evictor(...)]           <- optional: "evictor" (else no evictor)
   [.allowedLateness(...)]   <- optional: "lateness" (else zero)
   [.sideOutputLateData(...)]<- optional: "output tag" (else no side output for late data)
    .reduce/aggregate/apply()<- required: "function"
   [.getSideOutput(...)]     <- optional: "output tag"

Non-Keyed Windows

stream
    .windowAll(...)          <- required: "assigner"
   [.trigger(...)]           <- optional: "trigger" (else default trigger)
   [.evictor(...)]           <- optional: "evictor" (else no evictor)
   [.allowedLateness(...)]   <- optional: "lateness" (else zero)
   [.sideOutputLateData(...)]<- optional: "output tag" (else no side output for late data)
    .reduce/aggregate/apply()<- required: "function"
   [.getSideOutput(...)]     <- optional: "output tag"
```

图 3-38 Flink 窗口程序的总体结构

其中第 1 个片段代表 Keyed Stream,第 2 个片段代表 non-keyed stream。可以看到,唯一的区别是为 Keyed Stream 调用 keyBy() 函数,而为 non-keyed stream 调用 windowAll() 函数。

在 Keyed Stream 的情况下,可以使用传入事件的任何属性作为 key。在 Keyed Stream 的窗口计算由多个任务并行执行,因为每个逻辑 Keyed Stream 都可以独立于其他流进行处理。所有引用相同 key 的元素将被发送到相同的并行任务。在一个 Keyed Stream 流上应用窗口的基本形式如下:

```
stream.keyBy(< key selector >)
    .window(< window assigner >)
    .reduce|aggregate|process(< window function >)
```

在 non-keyed stream 的情况下,原始流不会被分割成多个逻辑流,所有的窗口逻辑将由一个任务执行,即并行度为 1(在这种情况下,处理将不会并行进行)。在一个 Knon-keyed stream 流上应用窗口的基本形式如下:

```
stream.windowAll(<window assigner>)
       .reduce|aggregate|process(<window function>)
```

下面来看一个滚动窗口示例。

【示例3-16】 计算每5s内来自Web套接字的字数。

建议按以下步骤操作:

(1) 在IntelliJ IDEA中创建一个Flink项目,使用flink-quickstart-scala/flink-quickstart-Java项目模板。

(2) 打开项目中的StreamingJob对象文件,编辑流处理代码。

Scala实现如下:

```
//第3章/WindowOperatorDemo1.scala

import org.apache.flink.streaming.api.scala.StreamExecutionEnvironment
import org.apache.flink.streaming.api.scala._
import org.apache.flink.streaming.api.windowing.assigners.TumblingProcessingTimeWindows
import org.apache.flink.streaming.api.windowing.time.Time

object WindowOperatorDemo1 {
  def main(args: Array[String]): Unit = {
    val host = "localhost"
    val port = 9999

    val env = StreamExecutionEnvironment.getExecutionEnvironment
    val text = env.socketTextStream(host, port)

val counts = text
      .flatMap { _.toLowerCase.split("\\W+").filter{ _.nonEmpty } }
      .map { (_, 1) }
      .keyBy(_._1)              //keyed stream
      .window(TumblingProcessingTimeWindows.of(Time.seconds(5))) //5s滚动窗口
      .sum(1)                   //对元组的第2个字段求和

    counts.print()

    env.execute("Window Stream WordCount")
  }
}
```

Java实现如下:

```
//第3章/WindowOperatorDemo1.java

import org.apache.flink.api.common.functions.FlatMapFunction;
```

```java
import org.apache.flink.api.java.tuple.Tuple2;
import org.apache.flink.streaming.api.datastream.DataStream;
import org.apache.flink.streaming.api.environment.StreamExecutionEnvironment;
import org.apache.flink.streaming.api.windowing.assigners.TumblingProcessingTimeWindows;
import org.apache.flink.streaming.api.windowing.time.Time;
import org.apache.flink.util.Collector;

public class WindowOperatorDemo1 {
    public static void main(String[] args) throws Exception {
        //设置流执行环境
        final StreamExecutionEnvironment env =
            StreamExecutionEnvironment.getExecutionEnvironment();

        //加载数据-转换-执行窗口操作
        //读取 Socket 数据源
        String host = "localhost";
        int port = 9999;

        DataStream< Tuple2 < String, Integer >> dataStream = env
            .socketTextStream(host, port)
            .map(String::toLowerCase)
            .flatMap(new Splitter())
            .keyBy(t -> t.f0)            //keyed stream
            .window(TumblingProcessingTimeWindows.of(Time.seconds(5))) //5s 滚动窗口
            .sum(1);                     //对元组的第 2 个字段求和

        dataStream.print();

        env.execute("Window Stream WordCount");
    }

    //自定义 flatMap 函数
    public static class Splitter implements FlatMapFunction< String, Tuple2 < String, Integer >> {
        @Override
        public void flatMap(String sentence, Collector< Tuple2 < String, Integer >> out) throws Exception {
            for (String word: sentence.split("\\W+")) {
                out.collect(new Tuple2 <>(word, 1));
            }
        }
    }
}
```

(3) 另起一个终端窗口,执行如下命令,启动一个 Web Socket 服务器:

```
$ nc -l 9999
```

(4) 回到 Flink 代码文件,并运行程序。
(5) 在 Socket 运行窗口,随便输入一些内容,以空格分隔。例如

```
good good study
day day up
```

(6) 在 IntelliJ IDEA 的运行窗口,可以观察到输出的统计结果如下:

```
(good,2)
(study,1)
(up,1)
(day,2)
```

下面来看一个滑动窗口示例。

【示例 3-17】 使用滑动窗口,每 5s 计算一次前 10s 内来自 Web 套接字的单词数量。Scala 代码如下:

```scala
//第 3 章/WindowOperatorDemo2.scala

import org.apache.flink.streaming.api.scala._
import org.apache.flink.streaming.api.windowing.assigners.SlidingProcessingTimeWindows
import org.apache.flink.streaming.api.windowing.time.Time

object WindowOperatorDemo2 {
  def main(args: Array[String]): Unit = {
    val host = "localhost"
    val port = 9999

    val env = StreamExecutionEnvironment.getExecutionEnvironment
    val text = env.socketTextStream(host, port)

    val counts = text
        .flatMap { _.toLowerCase.split("\\W+").filter{ _.nonEmpty } }
      .map { (_, 1) }
      .keyBy(_._1)                         //keyed stream
      .window(SlidingProcessingTimeWindows.of(Time.seconds(10),Time.seconds(5)))
                                           //滑动窗口
      .sum(1)                              //对元组的第 2 个字段求和

    counts.print()

    env.execute("Window Stream WordCount")
  }
}
```

Java 代码如下：

```java
//第3章/WindowOperatorDemo2.java

import org.apache.flink.api.common.functions.FlatMapFunction;
import org.apache.flink.api.java.tuple.Tuple2;
import org.apache.flink.streaming.api.datastream.DataStream;
import org.apache.flink.streaming.api.environment.StreamExecutionEnvironment;
import org.apache.flink.streaming.api.windowing.assigners.SlidingProcessingTimeWindows;
import org.apache.flink.streaming.api.windowing.time.Time;
import org.apache.flink.util.Collector;

public class WindowOperatorDemo2 {

    public static void main(String[] args) throws Exception {
        //设置流执行环境
        final StreamExecutionEnvironment env =
                StreamExecutionEnvironment.getExecutionEnvironment();

        //读取 Socket 数据源
        String host = "localhost";
        int port = 9999;

        DataStream<Tuple2<String, Integer>> dataStream = env
            .socketTextStream(host, port)
            .map(String::toLowerCase)
            .flatMap(new Splitter())
            .keyBy(t -> t.f0)              //Keyed Stream
            .window(SlidingProcessingTimeWindows.of(Time.seconds(10),Time.seconds(5)))
                                           //滑动窗口
            .sum(1);                       //对元组的第2个字段求和

        dataStream.print();

        env.execute("Window Stream WordCount");
    }

    //自定义 FlatMapFunction
    public static class Splitter
            implements FlatMapFunction<String,Tuple2<String, Integer>> {
        @Override
        public void flatMap(String sentence, Collector<Tuple2<String, Integer>> out) throws Exception {
            for (String word: sentence.split("\\W+")) {
                out.collect(new Tuple2<>(word, 1));
            }
        }
    }
}
```

【示例3-18】 使用3.5.3节创建的自定义数据源,统计每个用户每小时信用卡交易金额。
Scala代码如下:

```scala
//第3章/WindowOperatorDemo3.scala

import java.time.Duration
import com.xueai8.scala.ch03.entity.Transaction
import com.xueai8.scala.ch03.source.MyTransactionSource
import org.apache.flink.api.common.eventtime.{SerializableTimestampAssigner, WatermarkStrategy}
import org.apache.flink.streaming.api.scala._
import org.apache.flink.streaming.api.windowing.assigners.TumblingEventTimeWindows
import org.apache.flink.streaming.api.windowing.time.Time

/**
 * 使用滚动窗口,计算每个用户每小时信用卡交易金额
 */
object WindowOperatorDemo3 {

  def main(args: Array[String]) {
    //设置流执行环境
    val env = StreamExecutionEnvironment.getExecutionEnvironment

    //设置自定义数据源.参数false用于指定创建的是流数据源
    val stream = env
      .addSource(new MyTransactionSource(false))
      .name("transactions")

    //因为模拟数据没有时间戳,所以可用此方法添加事件时间戳和水印
    /* 使用匿名内部类 */
    stream.assignTimestampsAndWatermarks(WatermarkStrategy
      .forBoundedOutOfOrderness[Transaction](Duration.ofMillis(2000))
      .withTimestampAssigner(
        new SerializableTimestampAssigner[Transaction]() {
          //抽取交易事件中的timestamp字段作为事件时间戳
          override def extractTimestamp(t:Transaction, ts:Long):Long = {
            t.timestamp
          }
        })
      )
      .map(t => (t.accountId,t.timestamp,t.amount))
      .keyBy(_._1)                             //Keyed Stream
      .window(TumblingEventTimeWindows.of(Time.seconds(60 * 60)))  //滚动窗口
      .sum(1)                                  //对amount字段求和
      .print

    //执行
    env.execute("Flink Agg Demo")
  }
}
```

Java 代码如下：

```java
//第3章/WindowOperatorDemo3.java

import com.xueai8.java.ch03.entity.Transaction;
import com.xueai8.java.ch03.source.MyTransactionSource;
import org.apache.flink.api.common.eventtime.WatermarkStrategy;
import org.apache.flink.streaming.api.datastream.DataStream;
import org.apache.flink.streaming.api.environment.StreamExecutionEnvironment;
import org.apache.flink.streaming.api.windowing.assigners.TumblingEventTimeWindows;
import org.apache.flink.streaming.api.windowing.time.Time;

import Java.time.Duration;

/**
 * 使用滚动窗口,计算每个用户每小时信用卡交易金额
 */
public class WindowOperatorDemo3 {

    public static void main(String[] args) throws Exception {
        //设置流执行环境
        final StreamExecutionEnvironment env =
            StreamExecutionEnvironment.getExecutionEnvironment();

        //读取 Socket 数据源
        //设置自定义数据源.参数 false 用于指定创建的是流数据源
        DataStream<Transaction> stream = env
            .addSource(new MyTransactionSource(false))
            .name("transactions");

        //先应用水印策略,分配事件时间戳并指定 5s 水印
        Stream
            .assignTimestampsAndWatermarks(WatermarkStrategy
                .<Transaction>forBoundedOutOfOrderness(Duration.ofSeconds(5))
                .withTimestampAssigner((transaction, ts) -> transaction.timestamp))
            .keyBy(t -> t.accountId)          //Keyed Stream
            .window(TumblingEventTimeWindows.of(Time.seconds(60 * 60)))//滚动窗口
            .sum("amount")                    //对 amount 字段求和
            .print();

        env.execute("Window Agg Demo");
    }
}
```

通过前面的讲述和示例程序可知,使用 Flink 计算窗口分析依赖于以下两个主要的抽象:

(1) 将事件分配给窗口分配程序(Window Assigners，根据需要创建新的窗口对象)。
(2) 应用于分配给窗口的事件的窗口函数(Window Functions)。

在上面的例子中，基于窗口执行的聚合函数来自于 KeyedStream，详细介绍可参考 3.3.6 节"聚合转换"。

3.7.3 窗口函数

定义了窗口分配器之后，还需要指定要在每个窗口上执行的计算，也就是指定窗口函数。一旦系统确定某个窗口已准备好进行处理，该窗口函数将被用于处理窗口中的每个元素。

Flink 提供了如下几类窗口函数：
(1) ReduceFunction。
(2) AggregateFunction。
(3) ProcessWindowFunction。

前两个窗口函数的执行效率更高，因为 Flink 可以在每个元素到达窗口时增量地聚合元素，而 ProcessWindowFunction 的执行效率没有前两个窗口函数的执行效率高，因为 Flink 在调用函数之前必须在内部缓冲窗口的所有元素，然后以 Iterable 的形式传给它。

如何处理窗口内容，用户有以下 3 个基本的选择：
(1) 作为一个批处理，使用 ProcessWindowFunction，它将被传递一个带有窗口内容的 Iterable。
(2) 增量处理，在每个事件分配给窗口时调用 ReduceFunction 或 AggregateFunction。
(3) 或者两者结合使用，其中 ReduceFunction 或 AggregateFunction 的预聚合结果在窗口被触发时提供给 ProcessWindowFunction。

接下来，通过对温度传感器数据的分析，来理解和常用 Flink 所提供的这几类窗口函数的应用。首先从准备数据源开始。

【示例 3-19】 实现一个自定义的温度传感器数据源。

建议按以下步骤操作：
(1) 定义事件数据类型。

首先创建一个 Java POJO 类，用来封装从传感器采集到的事件数据，代码如下：

```
//第 3 章/SensorReading.java

public class SensorReading {

    public String id;                      //传感器 id
    public long timestamp;                 //读取时的时间戳
    public double temperature;             //读取的温度值

    public SensorReading() { }
```

```java
    public SensorReading(String id, long timestamp, double temperature) {
        this.id = id;
        this.timestamp = timestamp;
        this.temperature = temperature;
    }

    public String toString() {
        return "(" + this.id + ", " + this.timestamp + ", " + this.temperature + ")";
    }
}
```

(2) 自定义数据源,然后自定义传感器数据源,继承自 RichSourceFunction 类,代码如下:

```java
//第3章/SensorSource.java

import com.xueai8.java.ch03.entity.SensorReading;
import org.apache.flink.streaming.api.functions.source.RichSourceFunction;

import Java.util.Calendar;
import Java.util.Random;

/**
 * 自定义数据源
 */
public class SensorSource extends RichSourceFunction<SensorReading> {

    //标志变量,指示源是否仍在运行
    private boolean running = true;

    /** run()持续发射 SensorReadings,通过 SourceContext */
    @Override
    public void run(SourceContext<SensorReading> srcCtx) throws Exception {

        //初始化随机数发生器
        Random rand = new Random();

        //查找此并行任务(task)的索引
        int taskIdx = this.getRuntimeContext().getIndexOfThisSubtask();

        //初始化传感器 id 和温度
        String[] sensorIds = new String[10];
        double[] curFTemp = new double[10];
        for (int i = 0; i < 10; i++) {
            sensorIds[i] = "sensor_" + (taskIdx * 10 + i);
```

```java
            curFTemp[i] = 65 + (rand.nextGaussian() * 20);
        }

        while (running) {
            //获取当前时间
            long curTime = Calendar.getInstance().getTimeInMillis();

            //放射 SensorReadings
            for (int i = 0; i < 10; i++) {
                //修改当前温度
                curFTemp[i] += rand.nextGaussian() * 0.5;
                //发射读数
                srcCtx.collect(new SensorReading(sensorIds[i], curTime, curFTemp[i]));
            }

            //等待 100 ms
            Thread.sleep(100);
        }
    }

    /** 取消这个 SourceFunction. */
    @Override
    public void cancel() {
        this.running = false;
    }
}
```

注意：在上面的代码中，使用了 nextGaussian()方法,用于获得下一个伪随机、高斯分布（正态分布）的 double 值,均值为 0.0,标准偏差为 1.0。

(3) 测试数据源,接下来编写流测试程序,输出每个传感器的事件数据,代码如下：

```java
//第 3 章/SensorSourceTest.java
import com.xueai8.java.ch03.entity.SensorReading;
import com.xueai8.java.ch03.source.SensorSource;
import org.apache.flink.streaming.api.datastream.DataStream;
import org.apache.flink.streaming.api.environment.StreamExecutionEnvironment;

/**
 * 传感器流测试程序
 */
public class SensorSourceTest {

    public static void main(String[] args) throws Exception {
```

```
        //设置流执行环境
        final StreamExecutionEnvironment env =
              StreamExecutionEnvironment.getExecutionEnvironment();

        //读取流数据源
        DataStream<SensorReading> input = env.addSource(new SensorSource());

        input.print();

        //触发流程序执行
        env.execute("Flink Streaming Java API Skeleton");
    }
}
```

执行上面的代码,输出的传感器事件如下:

```
7> (sensor_3, 1615348339214, 34.79321478683425)
2> (sensor_6, 1615348339214, 48.92444755311628)
8> (sensor_4, 1615348339214, 96.07384090788655)
5> (sensor_1, 1615348339214, 28.681991445026327)
4> (sensor_0, 1615348339214, 72.66948362432143)
6> (sensor_2, 1615348339214, 43.846775241880266)
1> (sensor_5, 1615348339214, 71.07351152250708)
3> (sensor_7, 1615348339214, 71.6346953171324)
4> (sensor_8, 1615348339214, 81.4481576590041)
5> (sensor_9, 1615348339214, 41.89889773458169)
4> (sensor_6, 1615348339329, 48.850093468561276)
7> (sensor_1, 1615348339329, 28.694854674250678)
1> (sensor_3, 1615348339329, 34.39378107798509)
2> (sensor_4, 1615348339329, 96.46584713976698)
6> (sensor_0, 1615348339329, 72.29067790666107)
3> (sensor_5, 1615348339329, 71.89650023573446)
6> (sensor_8, 1615348339329, 81.58017900080814)
5> (sensor_7, 1615348339329, 71.47770075217043)
7> (sensor_9, 1615348339329, 42.49838055575979)
8> (sensor_2, 1615348339329, 43.78154120804511)
......
```

1. 使用 ReduceFunction

ReduceFunction 用于指定如何组合输入中的两个元素来生成相同类型的输出元素。Flink 使用 ReduceFunction 递增地聚合窗口的元素。

【示例 3-20】 找到每个传感器在每分钟的事件时间窗口内的温度峰值。

编写流处理程序(注意在窗口操作之前,要先分配事件时间戳)。

Scala 代码如下:

```scala
//第3章/SensorReduceFunDemo.scala
import com.xueai8.java.ch03.entity.SensorReading
import com.xueai8.java.ch03.source.SensorSource
import org.apache.flink.api.common.eventtime.{SerializableTimestampAssigner, WatermarkStrategy}
import org.apache.flink.api.common.functions.ReduceFunction
import org.apache.flink.streaming.api.scala._
import org.apache.flink.streaming.api.windowing.assigners.TumblingEventTimeWindows
import org.apache.flink.streaming.api.windowing.time.Time

object SensorReduceFunDemo {

  def main(args: Array[String]) {

    //设置流执行环境
    val env = StreamExecutionEnvironment.getExecutionEnvironment

    //设置并行度
    env.setParallelism(1)

    //读取流数据源
    val input = env.addSource(new SensorSource)

    //处理管道
    input
      //为事件分配时间戳和水印
      .assignTimestampsAndWatermarks(WatermarkStrategy
        .forMonotonousTimestamps[SensorReading]()
        .withTimestampAssigner(
          new SerializableTimestampAssigner[SensorReading] {
            override def extractTimestamp ( event: SensorReading, ts: Long): Long = event.timestamp
          })
      )
      .keyBy(sensorReading => sensorReading.id)
      .window(TumblingEventTimeWindows.of(Time.minutes(1)))
      .reduce(new ReduceMaxTemp())
      .print()

    //执行
    env.execute("Flink Streaming Job")
  }

  //窗口处理函数
  class ReduceMaxTemp extends ReduceFunction[SensorReading] {
```

```scala
    @throws[Exception]
    override def reduce(sr1: SensorReading, sr2: SensorReading): SensorReading = {
      if (sr1.temperature > sr2.temperature) sr1
      else sr2
    }
  }
}
```

Java 代码如下：

```java
//第3章/SensorReduceFunDemo.java

import com.xueai8.java.ch03.entity.SensorReading;
import com.xueai8.java.ch03.source.SensorSource;
import org.apache.flink.api.common.eventtime.WatermarkStrategy;
import org.apache.flink.api.common.functions.ReduceFunction;
import org.apache.flink.streaming.api.datastream.DataStream;
import org.apache.flink.streaming.api.environment.StreamExecutionEnvironment;
import org.apache.flink.streaming.api.windowing.assigners.TumblingEventTimeWindows;
import org.apache.flink.streaming.api.windowing.time.Time;

class SensorReduceFunDemo {

    public static void main(String[] args) throws Exception {
        //设置流执行环境
        final StreamExecutionEnvironment env =
            StreamExecutionEnvironment.getExecutionEnvironment();

        //读取流数据源
        DataStream<SensorReading> input = env.addSource(new SensorSource());

        input
            //分配时间戳和水印
            .assignTimestampsAndWatermarks(WatermarkStrategy
                .<SensorReading>forMonotonousTimestamps()
                .withTimestampAssigner((event,timestamp) -> event.timestamp)
            )
            //按传感器 id 分区
            //.keyBy("id")           //已弃用
            .keyBy(sensorReading -> sensorReading.id)
            //划分 1min 窗口
            .window(TumblingEventTimeWindows.of(Time.minutes(1)))
            //聚合
            .reduce(new ReduceMaxTemp())
            //输出
```

```java
            .print();

        //执行
        env.execute("flink streaming job");
    }

    //窗口处理函数
    public static class ReduceMaxTemp
                implements ReduceFunction<SensorReading>{

        @Override
        public SensorReading reduce(SensorReading sr1, SensorReading sr2) throws Exception {
            return sr1.temperature > sr2.temperature ? sr1 : sr2;
        }
    }
}
```

执行以上代码,输出结果如下:

```
8> (sensor_9, 1615348669229, 73.46634615051136)
3> (sensor_8, 1615348679677, 75.67541937317398)
1> (sensor_0, 1615348662489, 71.44518522080278)
7> (sensor_4, 1615348671540, 17.85601621623948)
2> (sensor_2, 1615348673449, 60.55270939353984)
7> (sensor_7, 1615348674656, 69.83861401937644)
5> (sensor_1, 1615348669531, 80.92630035619608)
6> (sensor_5, 1615348662690, 87.89684014528792)
1> (sensor_3, 1615348673650, 83.68706216364068)
6> (sensor_6, 1615348669732, 85.60202765682551)
......
```

2. 使用 AggregateFunction

AggregateFunction 是 ReduceFunction 的广义版本,它有 3 种参数类型:输入类型(IN)、累加器类型(ACC)和输出类型(OUT)。输入类型是输入流中元素的类型,AggregateFunction 有一种方法可以将一个输入元素添加到累加器中。该接口还具有创建初始累加器、将两个累加器合并到一个累加器及从累加器中提取输出(类型为 OUT)的方法。与 ReduceFunction 相同,Flink 将在窗口的输入元素到达时增量地聚合它们。

【示例 3-21】 找到每个传感器在每分钟的事件时间窗口内的平均温度,并生成包含(传感器 id、平均温度)的元组流。

编写流处理程序(注意在窗口操作之前,要先分配事件时间戳)。

Scala 代码如下:

```scala
//第3章/SensorAggregateFunDemo.scala

import com.xueai8.java.ch03.entity.SensorReading
import com.xueai8.java.ch03.source.SensorSource
import org.apache.flink.api.common.eventtime.{SerializableTimestampAssigner, WatermarkStrategy}
import org.apache.flink.api.common.functions.AggregateFunction
import org.apache.flink.streaming.api.scala._
import org.apache.flink.streaming.api.windowing.assigners.TumblingEventTimeWindows
import org.apache.flink.streaming.api.windowing.time.Time

object SensorAggregateFunDemo {

  def main(args: Array[String]) {

    //设置流执行环境
    val env = StreamExecutionEnvironment.getExecutionEnvironment

    //设置并行度
    env.setParallelism(1)

    //读取流数据源
    val input = env.addSource(new SensorSource)

    //处理管道
    input
      //为事件分配时间戳和水印
      .assignTimestampsAndWatermarks(WatermarkStrategy
        .forMonotonousTimestamps[SensorReading]()
        .withTimestampAssigner(
          new SerializableTimestampAssigner[SensorReading] {
            override def extractTimestamp(event: SensorReading, ts: Long): Long = event.timestamp
          })
      )
      .keyBy(sensorReading => sensorReading.id)
      .window(TumblingEventTimeWindows.of(Time.minutes(1)))
      .aggregate(new AggAvgTemp)
      .print()

    //执行
    env.execute("Flink Streaming Job")
  }

  //窗口聚合函数
```

```scala
class AggAvgTemp extends AggregateFunction[SensorReading, (String, Double, Long), (String, Double)] {
    //创建初始ACC
    override def createAccumulator = ("", 0.0, 0L)

    //累加每个传感器(每个分区)的事件
    override def add(sr: SensorReading, t3: (String, Double, Long)) = {
      (sr.id, t3._2 + sr.temperature, t3._3 + 1)
    }

    //分区合并
    override def merge(acc1: (String, Double, Long), acc2: (String, Double, Long)) = {
      (acc1._1, acc1._2 + acc2._2, acc1._3 + acc2._3)
    }

    //返回每个传感器的平均温度
    override def getResult(t3: (String, Double, Long)) = {
      (t3._1, t3._2 / t3._3)
    }
  }
}
```

Java代码如下:

```java
//第3章/SensorAggregateFunDemo.java

import com.xueai8.java.ch03.entity.SensorReading;
import com.xueai8.java.ch03.source.SensorSource;
import org.apache.flink.api.common.eventtime.WatermarkStrategy;
import org.apache.flink.api.common.functions.AggregateFunction;
import org.apache.flink.api.java.tuple.Tuple2;
import org.apache.flink.api.java.tuple.Tuple3;
import org.apache.flink.streaming.api.datastream.DataStream;
import org.apache.flink.streaming.api.environment.StreamExecutionEnvironment;
import org.apache.flink.streaming.api.windowing.assigners.TumblingEventTimeWindows;
import org.apache.flink.streaming.api.windowing.time.Time;

public class SensorAggregateFunDemo {

    public static void main(String[] args) throws Exception {
        //设置流执行环境
        final StreamExecutionEnvironment env =
                StreamExecutionEnvironment.getExecutionEnvironment();

        //读取流数据源
```

```java
        DataStream<SensorReading> input = env.addSource(new SensorSource());

    //管道
    input
        //分配时间戳和水印
        .assignTimestampsAndWatermarks(WatermarkStrategy
            .<SensorReading>forMonotonousTimestamps()
            .withTimestampAssigner((event,timestamp) -> event.timestamp)
        )
        //按传感器 id 分区
        //.keyBy("id")                              //已弃用
        .keyBy(sensorReading -> sensorReading.id)
        //划分 1min 窗口
        .window(TumblingEventTimeWindows.of(Time.minutes(1)))
        //聚合
        .aggregate(new AggAvgTemp())
        //输出
        .print();

        //执行
        env.execute("flink streaming job");
}

//窗口聚合函数
public static class AggAvgTemp
        implements AggregateFunction<
            SensorReading,                       //input
            Tuple3<String,Double,Long>,          //acc,<id,sum,count>
            Tuple2<String,Double>> {             //output,<id,avg>

    //创建初始 ACC
    @Override
    public Tuple3<String, Double, Long> createAccumulator() {
        return new Tuple3<>("",0.0,0L);
    }

    //累加每个传感器(每个分区)的事件
    @Override
    public Tuple3<String, Double, Long> add(SensorReading sr, Tuple3<String, Double, Long> t3) {
        return new Tuple3<>(sr.id, t3.f1 + sr.temperature,t3.f2 + 1);
    }

    //分区合并
    @Override
    public Tuple3<String, Double, Long> merge(
            Tuple3<String, Double, Long> acc1,
```

```java
                    Tuple3<String,Double,Long> acc2) {
            return new Tuple3<>(acc1.f0,acc1.f1 + acc2.f1,acc1.f2 + acc2.f2);
        }

        //返回每个传感器的平均温度
        @Override
        public Tuple2<String,Double> getResult(
                            Tuple3<String,Double,Long> t3){
            return new Tuple2<>(t3.f0, t3.f1/t3.f2);
        }
    }
}
```

执行以上代码,输出结果如下:

```
3> (sensor_8,42.08060591879755)
2> (sensor_2,71.6633424330926)
1> (sensor_0,77.84947694547725)
8> (sensor_9,82.01030900335661)
1> (sensor_3,59.51393327888449)
6> (sensor_5,79.5270624749896)
5> (sensor_1,49.93551500440228)
7> (sensor_4,93.84213670264313)
6> (sensor_6,81.73235148899215)
7> (sensor_7,107.1370199018038)
……
```

3. 使用ProcessWindowFunction

ProcessWindowFunction用于获取一个包含窗口所有元素的迭代器,以及一个具有时间和状态信息访问权的上下文对象(Context),这使它能够比其他窗口函数提供更大的灵活性,不过这是以性能和资源消耗为代价的,因为元素不能增量地聚合,而是需要在内部进行缓冲,直到认为窗口已经准备好进行处理。

在使用ProcessWindowFunction之前,首先了解一个该抽象类的定义,源码如下。
Scala代码如下:

```scala
//第3章/ProcessWindowFunction.scala

abstract class ProcessWindowFunction[IN, OUT, KEY, W <: Window] extends Function {

    def process(
        key: KEY,
        context: Context,
        elements: Iterable[IN],
```

```scala
      out: Collector[OUT])
  abstract class Context {
    def window: W

    def currentProcessingTime: Long

    def currentWatermark: Long

    def Windowstate: KeyedStateStore

    def globalState: KeyedStateStore
  }
}
```

Java 代码如下：

```java
//第 3 章/ProcessWindowFunction.java

public abstract class ProcessWindowFunction < IN, OUT, KEY, W extends Window > implements
Function {

    public abstract void process(
            KEY key,
            Context context,
            Iterable < IN > elements,
            Collector < OUT > out) throws Exception;

    public abstract class Context implements Java.io.Serializable {
        public abstract W window();

        public abstract long currentProcessingTime();

        public abstract long currentWatermark();

        public abstract KeyedStateStore Windowstate();

        public abstract KeyedStateStore globalState();
    }

}
```

注意：KEY 参数是通过 keyBy() 调用指定的 KeySelector 提取的 key。对于 tuple-index keys 或 string-field references，此 key 类型总是 Tuple，必须手动将其转换为正确大小的元组，以提取该 key 字段。

ProcessWindowFunction 被传递了一个 Context 上下文对象，该对象包含了关于窗口

的信息,用户可以从中获得关于窗口的信息。Context 接口的定义如下:

```java
public abstract class Context implements Java.io.Serializable {
    public abstract W window();

    public abstract long currentProcessingTime();
    public abstract long currentWatermark();

    public abstract KeyedStateStore Windowstate();
    public abstract KeyedStateStore globalState();
}
```

可以将 per-key、per-window 或 global per-key 的信息存储到 Windowstate 和 globalState 中。如果想记录关于当前窗口的信息,并在处理后续窗口时使用它,这会非常有用。

【示例 3-22】 找到在 1min 的事件时间窗口内每个传感器的温度峰值,并生成包含(传感器 id,窗口结束时间戳,温度峰值)的元组流。

Scala 代码如下:

```scala
//第 3 章/SensorProcessWindowFunDemo.scala

import com.xueai8.java.ch03.entity.SensorReading
import com.xueai8.java.ch03.source.SensorSource
import org.apache.flink.api.common.eventtime._
import org.apache.flink.api.java.tuple.Tuple3
import org.apache.flink.streaming.api.scala._
import org.apache.flink.streaming.api.scala.function.ProcessWindowFunction
import org.apache.flink.streaming.api.windowing.assigners.TumblingEventTimeWindows
import org.apache.flink.streaming.api.windowing.time.Time
import org.apache.flink.streaming.api.windowing.Windows.TimeWindow
import org.apache.flink.util.Collector

object SensorProcessWindowFunDemo {

  def main(args: Array[String]) {

    //设置流执行环境
    val env = StreamExecutionEnvironment.getExecutionEnvironment

    //设置并行度
    env.setParallelism(1)

    //读取流数据源
    val input = env.addSource(new SensorSource)
```

```
    //处理管道
    input
        //为事件分配时间戳和水印
        .assignTimestampsAndWatermarks(WatermarkStrategy
          .forMonotonousTimestamps[SensorReading]()
          .withTimestampAssigner(
            new SerializableTimestampAssigner[SensorReading] {
              override def extractTimestamp ( event: SensorReading, ts: Long ): Long = event.timestamp
            })
        )
        .keyBy(_.id)
        .window(TumblingEventTimeWindows.of(Time.minutes(1)))
        .process(new MyWastefulMax)
        .print()

    //执行
    env.execute("Flink Streaming Job")
  }

  //窗口处理函数
  /*
  注意,一定要导入
  org.apache.flink.streaming.api.scala.function.ProcessWindowFunction
  不要错误地导入
  org.apache.flink.streaming.api.functions.windowing.ProcessWindowFunction
  */
  class MyWastefulMax extends ProcessWindowFunction[SensorReading, (String, Long, Double), String, TimeWindow] {
    override def process(id: String,
                         context: Context,
                         events: Iterable[SensorReading],
                         out: Collector[(String, Long, Double)]): Unit = {
      var max = 0.0
      for (event <- events) {
        Math.max(event.temperature, max)
      }
      out.collect((id, context.window.getEnd, max))
    }
  }
}
```

Java 代码如下:

```
//第 3 章/SensorProcessWindowFunDemo.java

import com.xueai8.java.ch03.entity.SensorReading;
import com.xueai8.java.ch03.source.SensorSource;
```

```java
import org.apache.flink.api.common.eventtime.WatermarkStrategy;
import org.apache.flink.api.java.tuple.Tuple3;
import org.apache.flink.streaming.api.datastream.DataStream;
import org.apache.flink.streaming.api.environment.StreamExecutionEnvironment;
import org.apache.flink.streaming.api.functions.windowing.ProcessWindowFunction;
import org.apache.flink.streaming.api.windowing.assigners.TumblingEventTimeWindows;
import org.apache.flink.streaming.api.windowing.time.Time;
import org.apache.flink.streaming.api.windowing.windows.TimeWindow;
import org.apache.flink.util.Collector;

public class SensorProcessWindowFunDemo {

    public static void main(String[] args) throws Exception {
        //设置流执行环境
        final StreamExecutionEnvironment env =
                StreamExecutionEnvironment.getExecutionEnvironment();

        //读取流数据源
        DataStream<SensorReading> input = env.addSource(new SensorSource());

        input
            //分配事件时间戳
            .assignTimestampsAndWatermarks(WatermarkStrategy
              .<SensorReading> forMonotonousTimestamps()
              .withTimestampAssigner((event,timestamp) -> event.timestamp)
            )
            //.keyBy("id")                              //已弃用
            .keyBy(sensorReading -> sensorReading.id)
            .window(TumblingEventTimeWindows.of(Time.minutes(1)))
            .process(new MyWastefulMax())
            .print();

        //执行
        env.execute("flink streaming job");
    }

    //窗口处理函数
    public static class MyWastefulMax extends ProcessWindowFunction<
            SensorReading,                              //input type
            Tuple3<String, Long, Double>,               //output type
            String,                                     //key type
            TimeWindow> {                               //window type

        @Override
        public void process(String id,
```

```
                    Context context,
                    Iterable<SensorReading> events,
                    Collector<Tuple3<String, Long, Double>> out) {
        double max = 0;
        for (SensorReading event : events) {
            if (event.temperature > max) max = event.temperature;
        }
        out.collect(new Tuple3<>(id, context.window().getEnd(), max));
    }
  }
}
```

执行上面的程序,输出结果如下：

```
7> (sensor_25,1615343280000,87.72586269722625)
8> (sensor_35,1615343280000,43.142193856943834)
3> (sensor_21,1615343280000,67.3876390193542)
6> (sensor_37,1615343280000,69.2088403952636)
8> (sensor_9,1615343280000,51.63106770197567)
2> (sensor_24,1615343280000,58.968195282098684)
...
5> (sensor_64,1615343340000,57.68208910452411)
8> (sensor_19,1615343340000,29.283813257290152)
2> (sensor_43,1615343340000,85.09369114076935)
7> (sensor_25,1615343340000,110.71975425459033)
4> (sensor_72,1615343340000,79.27807830250289)
2> (sensor_51,1615343340000,81.77182221437575)
...
7> (sensor_52,1615343400000,40.71965827483373)
5> (sensor_57,1615343400000,61.80558008285541)
1> (sensor_48,1615343400000,110.79634444720006)
4> (sensor_46,1615343400000,67.34009796907614)
7> (sensor_33,1615343400000,57.52010144672712)
3> (sensor_53,1615343400000,102.57835200189706)
...
```

4. 使用增量 ReduceFunction ＋ ProcessWindowFunction

将 ProcessWindowFunction 用于简单的聚合（例如 count）是非常低效的。ProcessWindowFunction 可以与 ReduceFunction 组合使用,以便在元素到达窗口时对于其进行增量聚合。当窗口关闭时,ProcessWindowFunction 将提供聚合的结果。这样可以在访问 ProcessWindowFunction 的附加窗口元信息的同时递增地计算窗口。

【示例 3-23】 将增量 ReduceFunction 与 ProcessWindowFunction 组合起来,以返回窗口中最小的事件及窗口的开始时间。

Scala 代码如下：

```scala
//第3章/SensorReduceProcessFunDemo.scala
import com.xueai8.java.ch03.entity.SensorReading
import com.xueai8.java.ch03.source.SensorSource
import org.apache.flink.api.common.eventtime._
import org.apache.flink.api.common.functions.ReduceFunction
import org.apache.flink.streaming.api.scala._
import org.apache.flink.streaming.api.scala.function.ProcessWindowFunction
import org.apache.flink.streaming.api.windowing.assigners.TumblingEventTimeWindows
import org.apache.flink.streaming.api.windowing.time.Time
import org.apache.flink.streaming.api.windowing.windows.TimeWindow
import org.apache.flink.util.Collector

object SensorReduceProcessFunDemo {

  def main(args: Array[String]) {

    //设置流执行环境
    val env = StreamExecutionEnvironment.getExecutionEnvironment

    //设置并行度
    env.setParallelism(1)

    //读取流数据源
    val input = env.addSource(new SensorSource)

    //处理管道
    input
      //为事件分配时间戳和水印
      .assignTimestampsAndWatermarks(WatermarkStrategy
        .forMonotonousTimestamps[SensorReading]()
        .withTimestampAssigner(
          new SerializableTimestampAssigner[SensorReading] {
            override def extractTimestamp(event: SensorReading, ts: Long):
Long = event.timestamp
          })
      )
      .keyBy(_.id)
      .window(TumblingEventTimeWindows.of(Time.minutes(1)))
      .reduce(new ReduceMaxTemp, new ProcessMaxTemp)
      .print()

    //执行
    env.execute("Flink Streaming Job")
```

```scala
    }

    //窗口增量 reduce 函数
    class ReduceMaxTemp extends ReduceFunction[SensorReading] {
      override def reduce(sr1: SensorReading, sr2: SensorReading): SensorReading = {
        if (sr1.temperature > sr2.temperature) sr1 else sr2
      }
    }

    //窗口处理函数
    /*
      注意,一定要导入
      org.apache.flink.streaming.api.scala.function.ProcessWindowFunction
      不要错误地导入
      org.apache.flink.streaming.api.functions.windowing.ProcessWindowFunction
    */
    class ProcessMaxTemp extends ProcessWindowFunction[SensorReading, (String, Long, Double),
    String, TimeWindow] {
      override def process(id: String,
                           context: Context,
                           events: Iterable[SensorReading],
                           out: Collector[(String, Long, Double)]): Unit = {
        //注意,Iterable<SensorReading>将只包含一个读数
        //即 MyReduceFunction 计算出的预先聚合的最大值
        val max = events.iterator.next.temperature
        out.collect((id, context.window.getEnd, max))
      }
    }
}
```

Java 代码如下:

```java
//第 3 章/SensorReduceProcessFunDemo.java

import com.xueai8.java.ch03.entity.SensorReading;
import com.xueai8.java.ch03.source.SensorSource;
import org.apache.flink.api.common.eventtime.WatermarkStrategy;
import org.apache.flink.api.common.functions.ReduceFunction;
import org.apache.flink.api.java.tuple.Tuple3;
import org.apache.flink.streaming.api.datastream.DataStream;
import org.apache.flink.streaming.api.environment.StreamExecutionEnvironment;
import org.apache.flink.streaming.api.functions.windowing.ProcessWindowFunction;
import org.apache.flink.streaming.api.windowing.assigners.TumblingEventTimeWindows;
import org.apache.flink.streaming.api.windowing.time.Time;
import org.apache.flink.streaming.api.windowing.Windows.TimeWindow;
```

```java
import org.apache.flink.util.Collector;
public class SensorReduceProcessFunDemo {

    public static void main(String[] args) throws Exception {
        //设置流执行环境
        final StreamExecutionEnvironment env =
                StreamExecutionEnvironment.getExecutionEnvironment();

        //读取流数据源
        DataStream<SensorReading> input = env.addSource(new SensorSource());

        input
            //分配时间戳和水印
            .assignTimestampsAndWatermarks(WatermarkStrategy
                .<SensorReading>forMonotonousTimestamps()
                .withTimestampAssigner((event,timestamp) -> event.timestamp)
            )
            //按传感器 id 分区
            //.keyBy("id")                    //已弃用
            .keyBy(sensorReading -> sensorReading.id)
            //划分 1min 窗口
            .window(TumblingEventTimeWindows.of(Time.minutes(1)))
            //聚合
            .reduce(new ReduceMaxTemp(), new ProcessMaxTemp())
            //输出
            .print();

        //执行
        env.execute("flink streaming job");
    }

    //窗口增量 reduce 函数
    public static class ReduceMaxTemp
                    implements ReduceFunction<SensorReading> {
        @Override
        public SensorReading reduce(SensorReading sr1, SensorReading sr2) throws Exception {
            return sr1.temperature > sr2.temperature ? sr1 : sr2;
        }
    }

    //窗口处理函数
    public static class ProcessMaxTemp extends ProcessWindowFunction<
            SensorReading,                       //input type
            Tuple3<String, Long, Double>,        //output type
            String,                              //key type
            TimeWindow> {                        //window type
```

```java
        @Override
        public void process(String id,
                            Context context,
                            Iterable<SensorReading> events,
                            Collector<Tuple3<String, Long, Double>> out) {
            //注意,Iterable<SensorReading>将只包含一个读数
            //即 MyReduceFunction 计算出的预先聚合的最大值
            double max = events.iterator().next().temperature;
            out.collect(new Tuple3<>(id,context.window().getEnd(),max));
        }
    }
}
```

注意,Iterable<SensorReading>将只包含一个读数,即 MyReduceFunction 计算出的预先聚合的最大值。

执行上面的代码,输出结果如下:

```
8> (sensor_9,1615355520000,32.37375619031118)
3> (sensor_8,1615355520000,96.42499637334932)
5> (sensor_1,1615355520000,55.060786284631355)
6> (sensor_5,1615355520000,94.91489904737487)
2> (sensor_2,1615355520000,70.12297314504767)
1> (sensor_0,1615355520000,71.17846770042992)
7> (sensor_4,1615355520000,55.549531169074356)
6> (sensor_6,1615355520000,54.348664791183026)
1> (sensor_3,1615355520000,54.64384452116478)
7> (sensor_7,1615355520000,36.501163398755025)
...
```

5. 使用增量 AggregateFunction ＋ ProcessWindowFunction

ProcessWindowFunction 可以与 AggregateFunction 组合使用,以便在元素到达窗口时对其进行增量聚合。这样可以在访问 ProcessWindowFunction 的附加窗口元信息的同时增量地聚合窗口。

【示例 3-24】 将增量 AggregateFunction 与 ProcessWindowFunction 组合起来,计算在 1min 的事件时间窗口内每个传感器的温度平均值,并生成包含(sensor_id, end-of-window-timestamp, max_value)的元组流。

Scala 代码如下:

```scala
//第 3 章/SensorAggProcessFunDemo.scala

import com.xueai8.java.ch03.entity.SensorReading
import com.xueai8.java.ch03.source.SensorSource
```

```scala
import org.apache.flink.api.common.eventtime._
import org.apache.flink.api.common.functions.AggregateFunction
import org.apache.flink.streaming.api.scala._
import org.apache.flink.streaming.api.scala.function.ProcessWindowFunction
import org.apache.flink.streaming.api.windowing.assigners.TumblingEventTimeWindows
import org.apache.flink.streaming.api.windowing.time.Time
import org.apache.flink.streaming.api.windowing.Windows.TimeWindow
import org.apache.flink.util.Collector

object SensorAggProcessFunDemo {

  def main(args: Array[String]) {

    //设置流执行环境
    val env = StreamExecutionEnvironment.getExecutionEnvironment

    //设置并行度
    env.setParallelism(1)

    //读取流数据源
    val input = env.addSource(new SensorSource)

    //处理管道
    input
      //为事件分配时间戳和水印
      .assignTimestampsAndWatermarks(WatermarkStrategy
        .forMonotonousTimestamps[SensorReading]()
        .withTimestampAssigner(
          new SerializableTimestampAssigner[SensorReading] {
            override def extractTimestamp(event: SensorReading, ts: Long): Long = event.timestamp
          })
      )
      .keyBy(_.id)
      .window(TumblingEventTimeWindows.of(Time.minutes(1)))
      .aggregate(new AggAvgTemp, new ProcessAvgTemp)
      .print()

    //执行
    env.execute("Flink Streaming Job")
  }

  //窗口增量聚合函数
  class AggAvgTemp extends AggregateFunction[SensorReading, (Double, Long), Double] {
    //创建初始 ACC
    override def createAccumulator = (0.0, 0L)
```

```scala
//累加每个传感器(每个分区)的事件
override def add(sr: SensorReading, t2: (Double, Long)) = {
  (t2._1 + sr.temperature, t2._2 + 1)
}

//分区合并
override def merge(acc1: (Double, Long), acc2: (Double, Long)) = {
  (acc1._1 + acc2._1, acc1._2 + acc2._2)
}

//返回每个传感器的平均温度
override def getResult(t2: (Double, Long)): Double = t2._1 / t2._2
}

//窗口处理函数
/*
注意,一定要导入
    org.apache.flink.streaming.api.scala.function.ProcessWindowFunction
不要错误地导入
    org.apache.flink.streaming.api.functions.windowing.ProcessWindowFunction
*/
class ProcessAvgTemp extends ProcessWindowFunction[Double, (String, Long, Double), String, TimeWindow] {
  override def process(id: String,
                       context: Context,
                       events: Iterable[Double],
                       out: Collector[(String, Long, Double)]): Unit = {
    //注意,Iterable<SensorReading>将只包含一个读数
    //即 MyReduceFunction 计算出的预先聚合的平均值
    val average = events.iterator.next
    out.collect((id, context.window.getEnd, average))
  }
}
}
```

Java 代码如下:

```java
//第3章/SensorAggProcessFunDemo.java

import com.xueai8.java.ch03.entity.SensorReading;
import com.xueai8.java.ch03.source.SensorSource;
import org.apache.flink.api.common.eventtime.WatermarkStrategy;
import org.apache.flink.api.common.functions.AggregateFunction;
import org.apache.flink.api.java.tuple.Tuple2;
import org.apache.flink.api.java.tuple.Tuple3;
import org.apache.flink.streaming.api.datastream.DataStream;
```

```java
import org.apache.flink.streaming.api.environment.StreamExecutionEnvironment;
import org.apache.flink.streaming.api.functions.windowing.ProcessWindowFunction;
import org.apache.flink.streaming.api.windowing.assigners.TumblingEventTimeWindows;
import org.apache.flink.streaming.api.windowing.time.Time;
import org.apache.flink.streaming.api.windowing.windows.TimeWindow;
import org.apache.flink.util.Collector;

public class SensorAggProcessFunDemo {

    public static void main(String[] args) throws Exception {
        //设置流执行环境
        final StreamExecutionEnvironment env =
                StreamExecutionEnvironment.getExecutionEnvironment();

        //读取流数据源
        DataStream<SensorReading> input = env.addSource(new SensorSource());

        input
            //分配时间戳和水印
            .assignTimestampsAndWatermarks(WatermarkStrategy
              .<SensorReading>forMonotonousTimestamps()
              .withTimestampAssigner((event,timestamp) -> event.timestamp)
            )
            //按传感器 id 分区
            //.keyBy("id")            //已弃用
            .keyBy(sensorReading -> sensorReading.id)
            //划分 1min 窗口
            .window(TumblingEventTimeWindows.of(Time.minutes(1)))
            //聚合
            .aggregate(new AggAvgTemp(),new ProcessMaxTemp())
            //输出
            .print();

        //执行
        env.execute("flink streaming job");
    }

    //窗口增量聚合函数
    public static class AggAvgTemp implements AggregateFunction<
            SensorReading,                    //input
            Tuple2<Double,Long>,              //acc, <id, sum, count>
            Double> {                         //output

        //创建初始 ACC
        @Override
        public Tuple2<Double,Long> createAccumulator() {
```

```java
        return new Tuple2<>(0.0,0L);
    }

    //累加每个传感器(每个分区)的事件
    @Override
    public Tuple2<Double,Long> add(SensorReading sr, Tuple2<Double,Long> t2) {
        return new Tuple2<>(t2.f0 + sr.temperature,t2.f1 + 1);
    }

    //分区合并
    @Override
    public Tuple2<Double,Long> merge(
                Tuple2<Double,Long> acc1,
                Tuple2<Double,Long> acc2) {
        return new Tuple2<>(acc1.f0 + acc2.f0,acc1.f1 + acc2.f1);
    }

    //返回每个传感器的平均温度
    @Override
    public Double getResult(Tuple2<Double,Long> t2) {
        return t2.f0/t2.f1;
    }
}

//窗口处理函数
public static class ProcessMaxTemp extends ProcessWindowFunction<
                Double,                                  //input type
                Tuple3<String, Long, Double>,            //output type
                String,                                  //key type
                TimeWindow> {                            //window type
    @Override
    public void process(String id,
                Context context,
                Iterable<Double> events,
                Collector<Tuple3<String, Long, Double>> out) {
        double average = events.iterator().next();
        out.collect(new Tuple3<>(id,context.window().getEnd(), average));
    }
}
}
```

执行以上代码,输出结果如下:

```
5> (sensor_1,1615356240000,79.0445303156929)
1> (sensor_0,1615356240000,80.47726227716672)
```

```
7> (sensor_4,1615356240000,42.29616683871286)
6> (sensor_5,1615356240000,38.832404558254375)
2> (sensor_2,1615356240000,80.67415078224867)
7> (sensor_7,1615356240000,81.94299042105301)
6> (sensor_6,1615356240000,50.872792137481181)
8> (sensor_9,1615356240000,92.91812504642569)
3> (sensor_8,1615356240000,62.813793188352854)
1> (sensor_3,1615356240000,55.72952046921107)
......
```

6. 使用 WindowFunction（遗留的旧接口）

在一些可以使用 ProcessWindowFunction 的地方，还可以使用 WindowFunction。这是一个较老版本的 ProcessWindowFunction，它提供的上下文信息较少，并且缺少一些高级特性，例如 per-window keyed state（需要注意，这个接口将来某个时候可能会被弃用）。

WindowFunction 接口的签名如下。

Java 代码如下：

```java
public interface WindowFunction < IN, OUT, KEY, W extends Window > extends Function,
Serializable {

    /**
     * Evaluates the window and outputs none or several elements.
     *
     * @param key The key for which this window is evaluated.
     * @param window The window that is being evaluated.
     * @param input The elements in the window being evaluated.
     * @param out A collector for emitting elements.
     *
     * @throws Exception The function may throw exceptions to fail the program and trigger
     * recovery.
     */
    void apply(KEY key, W window, Iterable< IN > input, Collector< OUT > out) throws Exception;
}
```

Scala 代码如下：

```scala
trait WindowFunction[IN, OUT, KEY, W <: Window] extends Function with Serializable {

    /**
     * Evaluates the window and outputs none or several elements.
     *
     * @param key The key for which this window is evaluated.
     * @param window The window that is being evaluated.
```

```
 * @param input The elements in the window being evaluated.
 * @param out A collector for emitting elements.
 * @throws Exception The function may throw exceptions to fail the program and trigger
 * recovery.
 */
  def apply(key: KEY, window: W, input: Iterable[IN], out: Collector[OUT])
}
```

可以像下面这样来使用 WindowFunction 函数。

Scala 代码如下：

```
val input: DataStream[(String, Long)] = ...

input
  .keyBy(<key selector>)
  .window(<window assigner>)
  .apply(new MyWindowFunction())
```

Java 代码如下：

```
DataStream<Tuple2<String, Long>> input = ...;

input
  .keyBy(<key selector>)
  .window(<window assigner>)
  .apply(new MyWindowFunction());
```

【示例 3-25】 找到在 1min 的事件时间窗口内每个传感器的温度峰值，并生成包含（传感器 id，窗口结束时间戳，温度峰值）的元组流。

Scala 代码如下：

```scala
//第 3 章/SensorWindowFunctionDemo.scala

import com.xueai8.java.ch03.entity.SensorReading
import com.xueai8.java.ch03.source.SensorSource
import org.apache.flink.api.common.eventtime._
import org.apache.flink.streaming.api.scala.function.WindowFunction
import org.apache.flink.streaming.api.scala._
import org.apache.flink.streaming.api.windowing.assigners.TumblingEventTimeWindows
import org.apache.flink.streaming.api.windowing.time.Time
import org.apache.flink.streaming.api.windowing.windows.TimeWindow
import org.apache.flink.util.Collector

object SensorWindowFunctionDemo {
```

```scala
def main(args: Array[String]) {

    //设置流执行环境
    val env = StreamExecutionEnvironment.getExecutionEnvironment

    //设置并行度
    env.setParallelism(1)

    //读取流数据源
    val input = env.addSource(new SensorSource)

    //处理管道
    input
      //为事件分配时间戳和水印
      .assignTimestampsAndWatermarks(WatermarkStrategy
        .forMonotonousTimestamps[SensorReading]()
        .withTimestampAssigner(
          new SerializableTimestampAssigner[SensorReading] {
            override def extractTimestamp(event: SensorReading, ts: Long): Long = event.timestamp
          })
      )
      .keyBy(sensorReading => sensorReading.id)
      .window(TumblingEventTimeWindows.of(Time.minutes(1)))
      .apply(new MyWindowFunction)
      .print()

    //执行
    env.execute("Flink Streaming Job")
  }

  //窗口处理函数
  class MyWindowFunction extends WindowFunction[SensorReading, (String, Long, Double), String, TimeWindow] {
    override def apply(id: String,
                       timeWindow: TimeWindow,
                       events: Iterable[SensorReading],
                       out: Collector[(String, Long, Double)]): Unit = {
      var max = 0.0
      for (event <- events) {
        Math.max(event.temperature, max)
      }
      out.collect((id, timeWindow.getEnd, max))
    }
  }
}
```

Java 代码如下：

```java
//第3章/SensorWindowFunctionDemo.java

import com.xueai8.java.ch03.entity.SensorReading;
import com.xueai8.java.ch03.source.SensorSource;
import org.apache.flink.api.common.eventtime.WatermarkStrategy;
import org.apache.flink.api.java.tuple.Tuple3;
import org.apache.flink.streaming.api.datastream.DataStream;
import org.apache.flink.streaming.api.environment.StreamExecutionEnvironment;
import org.apache.flink.streaming.api.functions.windowing.WindowFunction;
import org.apache.flink.streaming.api.windowing.assigners.TumblingEventTimeWindows;
import org.apache.flink.streaming.api.windowing.time.Time;
import org.apache.flink.streaming.api.windowing.windows.TimeWindow;
import org.apache.flink.util.Collector;

public class SensorWindowFunctionDemo {

    public static void main(String[] args) throws Exception {
        //设置流执行环境
        final StreamExecutionEnvironment env =
                StreamExecutionEnvironment.getExecutionEnvironment();

        //读取流数据源
        DataStream<SensorReading> input = env.addSource(new SensorSource());

        //处理管道
        input
            //分配时间戳和水印
            .assignTimestampsAndWatermarks(WatermarkStrategy
                .<SensorReading>forMonotonousTimestamps()
                .withTimestampAssigner((event,timestamp) -> event.timestamp)
            )
            //按传感器 id 分区
            //.keyBy("id")                    //已弃用
            .keyBy(sensorReading -> sensorReading.id)
            //划分 1min 窗口
            .window(TumblingEventTimeWindows.of(Time.minutes(1)))
            //应用自定义窗口函数
            .apply(new MyWindowFunction())
            //输出
            .print();

        //触发流程序执行
        env.execute("Flink Sensor Stream");
    }
}
```

```java
//窗口处理函数
public static class MyWindowFunction implements WindowFunction<
        SensorReading,                          //input type
        Tuple3<String,Long,Double>,             //output type
        String,                                 //key type
        TimeWindow> {                           //window type
    @Override
    public void apply(String id,
                    TimeWindow timeWindow,
                    Iterable<SensorReading> events,
                    Collector<Tuple3<String,Long,Double>> out
        ) throws Exception {
        double max = 0;
        for (SensorReading event : events) {
            if (event.temperature > max) max = event.temperature;
        }
        out.collect(new Tuple3<>(id,timeWindow.getEnd(),max));
    }
}
```

执行以上代码,输出结果如下:

```
1> (sensor_0,1615357080000,78.73390698185324)
2> (sensor_2,1615357080000,96.7693075853522)
3> (sensor_8,1615357080000,74.20232753185753)
6> (sensor_5,1615357080000,33.620243996559225)
7> (sensor_4,1615357080000,87.31423973245688)
5> (sensor_1,1615357080000,80.09551023652499)
8> (sensor_9,1615357080000,76.00900462331602)
6> (sensor_6,1615357080000,92.86722930073302)
1> (sensor_3,1615357080000,123.3866054120469)
7> (sensor_7,1615357080000,76.97441380518192)
……
```

3.7.4 触发器

触发器(Trigger)用于确定一个窗口何时调用窗口函数。每个 WindowAssigner 都带有一个默认触发器。如果默认触发器不适合,用户则可以使用 trigger() 指定自定义触发器。

触发器接口有 5 种方法,允许触发器对不同的事件做出反应。

(1) onElement()方法:为添加到窗口的每个元素调用该方法。

(2) onEventTime()方法:当已注册的事件时间定时器触发时,将调用该方法。

(3) onProcessingTime()方法:当已注册的处理时间定时器触发时,将调用该方法。

（4）onMerge()方法：该方法与有状态触发器相关，并在两个触发器的相应窗口合并时合并它们的状态，例如在使用会话窗口时。

（5）clear()方法：执行删除相应窗口时所需的任何操作。

以上方法有以下两点需要注意：

（1）前 3 种方法通过返回 TriggerResult 来决定如何处理它们的调用事件。可以采取以下行动之一。

① CONTINUE：什么也不做。

② FIRE：触发计算。

③ PURGE：清除窗口中的元素。

④ FIRE_AND_PURGE：触发计算，然后清除窗口中的元素。

（2）这些方法中的任何一个都可以用来为将来的操作注册处理时间计时器或事件时间计时器。

1. 触发和清除

一旦触发器确定某个窗口已准备好进行处理，它就会触发，也就是说它返回 FIRE 或 FIRE_AND_PURGE。这是窗口运算符发送当前窗口结果的信号。给定一个带有 ProcessWindowFunction 的窗口，所有元素都被传递给 ProcessWindowFunction（可能在将它们传递给一个 Evictor 回收器之后）。带有 ReduceFunction、AggregateFunction 或 FoldFunction 的窗口简单地发送它们 eagerly 聚合的结果。

当触发器触发时，它可以是 FIRE 或 FIRE_AND_PURGE。当为 FIRE 时保留窗口的内容，而当为 FIRE_AND_PURGE 时则删除窗口的内容。默认情况下，预实现的触发器简单地 FIRE 而不清除窗口状态（清除将简单地删除窗口的内容，并保留关于窗口和任何触发器状态的任何潜在元信息）。

2. WindowAssigners 的默认触发器

WindowAssigner 的默认触发器适用于许多场景。例如，所有事件时间窗口分配器都有一个 EventTimeTrigger 作为默认触发器。一旦水印经过窗口的末端，这个触发器就会触发。

全局窗口（GlobalWindow）的默认触发器是永不触发的 NeverTrigger，因此，在使用全局窗口时，总是必须定义一个自定义触发器。

注意：通过 trigger() 指定触发器，会覆盖 WindowAssigner 的默认触发器。例如，如果为 TumblingEventTimeWindows 指定了一个 CountTrigger，则将不再根据时间进度获得窗口触发，而是只根据 count 计数。在这种情况下，如果想根据时间和计数都做出反应，就必须编写自己的自定义触发器。

---

3. 内置和自定义触发器

Flink 提供了以下 9 个内置的触发器。

(1) CountTrigger：一旦窗口中的元素数量超过给定的限制，CountTrigger 就会触发。
(2) EventTimeTrigger：基于由水印度量的事件时间进程触发。
(3) ProcessingTimeTrigger：根据处理时间触发。
(4) ContinuousEventTimeTrigger。
(5) ContinuousProcessingTimeTrigger。
(6) PurgingTrigger：将另一个触发器作为参数并将其转换为一个清除触发器。
(7) ProcessingTimeoutTrigger。
(8) DeltaTrigger。
(9) NeverTrigger。

注意：如果需要实现自定义触发器，则应该检查抽象 Trigger 类。需要注意，该 API 仍在发展中，并且在 Flink 的未来版本中可能会发生变化。

3.7.5 清除器

除 WindowAssigner 和 Trigger 之外，Flink 的窗口模型还允许指定可选的清除器 (Evictor)。这可以使用 evictor()方法实现。清除器能够在触发器触发之后及应用窗口函数之前和/或之后从窗口中删除元素。为此，Evictor 接口有两种方法，代码如下：

```
/**
 * Optionally evicts elements. Called before windowing function.
 *
 * @param elements The elements currently in the pane.
 * @param size The current number of elements in the pane.
 * @param window The {@link Window}
 * @param evictorContext The context for the Evictor
 */
void evictBefore ( Iterable < TimestampedValue < T > > elements, int size, W window,
EvictorContext evictorContext);

/**
 * Optionally evicts elements. Called after windowing function.
 *
 * @param elements The elements currently in the pane.
 * @param size The current number of elements in the pane.
 * @param window The {@link Window}
 * @param evictorContext The context for the Evictor
 */
void evictAfter(Iterable<TimestampedValue<T>> elements, int size, W window, EvictorContext
evictorContext);
```

其中 evictBefore() 包含要在窗口函数之前应用的清除逻辑，而 evictAfter() 包含要在窗口函数之后应用的清除逻辑。在应用窗口函数之前被移除的元素将不会被该窗口函数处理。

Flink 附带 3 个预实现的清除器，分别如下。

（1）CountEvictor：从窗口保留用户指定数量的元素，并从窗口缓冲区开始处丢弃剩余的元素。

（2）DeltaEvictor：接受一个 DeltaFunction 和一个阈值，计算窗口缓冲区中的最后一个元素与其余每个元素之间的增量，并删除增量大于或等于阈值的元素。

（3）TimeEvictor：将一个以毫秒为单位的 interval 作为参数，对于给定的窗口，它在其元素中找到最大时间戳 max_ts，并删除所有时间戳小于 max_ts－interval 的元素。

默认情况下，所有预实现的清除器都在窗口函数之前应用它们的逻辑。

注意：指定 evictor 可以防止任何预聚合，因为在应用计算之前，窗口的所有元素都必须传递给 evictor。这意味着带有 evictor 的窗口将创建更多的状态；Flink 不能保证窗口中元素的顺序。这意味着，尽管 evictor 可能从窗口的开始处删除元素，但这些元素不一定是最先到达或最后到达的元素。

3.7.6　处理迟到数据

当使用事件时间窗口时，可能会出现元素延迟到达的情况，即 Flink 用来跟踪事件时间进程的水印已经超过了元素所属窗口的结束时间戳。

注意：时延元素是在系统事件时间时钟（由水印表示）已经通过了时延元素的时间戳之后到达的元素。

默认情况下，当水印经过窗口末尾时，迟到的数据将被删除，然而，Flink 允许为窗口操作符指定允许的最大时延。允许时延指定元素在被删除之前可以延迟多长时间，其默认值为 0。在水印经过窗口末尾之后，但是在水印经过（窗口末尾＋允许的时延）之前到达的元素，仍然被添加到窗口中。根据使用的触发器的不同，延迟但未删除的元素可能会导致窗口再次触发。EventTimeTrigger 就是这种情况。

Flink 将会保持窗口的状态，直到它们允许的时延过期为止，然后，Flink 会删除窗口并删除它的状态。

默认情况下，允许的时延被设置为 0。也就是说，在水印后面到达的元素将被删除。可以像下面这样指定允许的时延。设置允许时延的代码如下。

Scala 代码如下：

```
val input: DataStream[T] = ...

input
    .keyBy(<key selector>)
    .window(<window assigner>)
    .allowedLateness(<time>)
    .<windowed transformation>(<window function>)
```

Java 代码如下:

```
DataStream<T> input = ...;

input
    .keyBy(<key selector>)
    .window(<window assigner>)
    .allowedLateness(<time>)
    .<windowed transformation>(<window function>);
```

注意：在使用 GlobalWindows 窗口分配器时，不会考虑数据时延，因为全局窗口的结束时间戳是 Long.MAX_VALUE。

当指定允许的时延大于 0 时，窗口及其内容将在水印通过窗口末端后保留。在这些情况下，当一个时延的未删除的元素到达时，可能会触发窗口的另一次触发。这些触发被称为时延触发，因为它们是由时延事件触发的，与主触发不同，主触发是窗口的第 1 次触发。

会话窗口基于可以合并的窗口抽象。每个元素最初都被分配给一个新窗口，在此之后，只要它们之间的差距足够小，就会合并窗口。通过这种方式，一个时延事件可以将两个之前分离的会话连接起来，从而产生一个时延合并。

使用 Flink 的侧输出特性，可以得到一个时延丢弃的数据流。

首先需要指定希望在窗口化的流上使用 sideOutputLateData(OutputTag)获取迟到数据，然后，可以在窗口操作的结果上得到侧输出流，实现的模板代码如下。

Scala 代码如下:

```
val lateOutputTag = OutputTag[T]("late-data")

val input: DataStream[T] = ...

val result = input
    .keyBy(<key selector>)
    .window(<window assigner>)
    .allowedLateness(<time>)
```

```
    .sideOutputLateData(lateOutputTag)
    .<windowed transformation>(<window function>)

val lateStream = result.getSideOutput(lateOutputTag)
```

Java 代码如下：

```
final OutputTag<T> lateOutputTag = new OutputTag<T>("late-data"){};

DataStream<T> input = ...;

SingleOutputStreamOperator<T> result = input
    .keyBy(<key selector>)
    .window(<window assigner>)
    .allowedLateness(<time>)
    .sideOutputLateData(lateOutputTag)
    .<windowed transformation>(<window function>);

DataStream<T> lateStream = result.getSideOutput(lateOutputTag);
```

3.7.7 处理窗口结果

窗口操作的结果还是 DataStream，在结果元素中没有保留窗口操作的信息，所以，如果想保持窗口的元信息，就必须在 ProcessWindowFunction 的结果元素中手动编码该信息。在结果元素上设置的唯一相关信息是元素时间戳。这被设置为已处理窗口的最大允许时间戳，即结束时间戳-1，因为窗口结束时间戳是不包括在内的。注意，这对于事件时间窗口和处理时间窗口都是一样的。例如，在窗口操作元素之后总有一个时间戳，但它可以是一个事件时间时间戳，也可以是一个处理时间时间戳。对于处理时间窗口，这没有特殊的含义，但是对于事件时间窗口，加上水印与窗口的交互方式，可以使用相同的窗口大小实现连续的窗口操作。

当水印到达窗口操作符时，这将触发以下两件事：

（1）当最大时间戳（结束时间戳-1）小于新水印时，水印触发所有窗口的计算。

（2）水印按原样转发给下游操作。

如前所述，计算窗口化结果的时间戳的方法及水印与窗口的交互方式允许将连续的窗口化操作串在一起。当用户想要执行两个连续的窗口操作，其中想要使用不同的 key，但仍然希望来自相同上游窗口的元素最终出现在相同的下游窗口中时，这将非常有用。可参考看下面的示例。

Scala 代码如下：

```scala
val input: DataStream[Int] = ...

val resultsPerKey = input
    .keyBy(<key selector>)
    .window(TumblingEventTimeWindows.of(Time.seconds(5)))
    .reduce(new Summer())

val globalResults = resultsPerKey
    .windowAll(TumblingEventTimeWindows.of(Time.seconds(5)))
    .process(new TopKWindowFunction())
```

Java 代码如下：

```java
DataStream<Integer> input = ...;

DataStream<Integer> resultsPerKey = input
    .keyBy(<key selector>)
    .window(TumblingEventTimeWindows.of(Time.seconds(5)))
    .reduce(new Summer());

DataStream<Integer> globalResults = resultsPerKey
    .windowAll(TumblingEventTimeWindows.of(Time.seconds(5)))
    .process(new TopKWindowFunction());
```

在本例中，第 1 个操作的时间窗[0,5]的结果也将在随后的窗口操作的时间窗[0,5]中结束。这允许计算每个 key 的和，然后在第 2 个操作中计算同一窗口中的 top-k 元素。

3.7.8 窗口连接

使用窗口连接(Window Join)将两个流的元素连接起来，这两个流共享一个公共 key，并且位于同一个窗口中。可以使用窗口分配程序定义这些窗口，并对来自这两个流的元素进行计算，然后将来自两边的元素传递给用户定义的 JoinFunction 或 FlatJoinFunction，用户可以在其中发出满足连接条件的结果。

一般用法可概括如下：

```
stream.join(otherStream)
    .where(<KeySelector>)
    .equalTo(<KeySelector>)
    .window(<WindowAssigner>)
    .apply(<JoinFunction>)
```

关于语义的一些注意事项：
(1) 两个流的元素成对组合的创建行为类似于内连接，这意味着如果一个流中的元素没有来自另一个流的相应元素要连接,则它们将不会被发出。

（2）那些确实被连接的元素将以仍然在各自窗口中的最大时间戳作为它们的时间戳。例如,以[5,10)为边界的窗口将导致合并的元素使用9作为时间戳。

下面将使用一些示例场景来概述不同类型的窗口连接的行为。

1. 滚动窗口连接

当执行一个滚动窗口连接时,所有具有公共key和公共滚动窗口的元素都以成对组合的方式连接,并传递给一个JoinFunction或FlatJoinFunction。因为这就像一个内连接,一个流的元素如果在其滚动窗口中没有来自另一个流的元素,就不会被发出。

例如,定义了一个大小为2ms的滚动窗口,它产生的窗口形式为[0,1],[2,3],…,如图3-39所示。

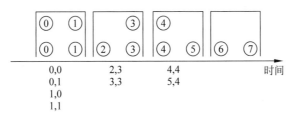

图3-39　滚动窗口连接

图3-39中显示了每个窗口中所有元素的成对组合,这些元素将被传递给JoinFunction。注意,在滚动窗口[6,7]中没有发出任何东西,因为第1排的流中不存在与第2排的元素⑥和⑦相连接的元素。

两个滚动窗口连接的模板代码如下。

Scala代码如下：

```
import org.apache.flink.streaming.api.windowing.assigners.TumblingEventTimeWindows;
import org.apache.flink.streaming.api.windowing.time.Time;

...
val orangeStream: DataStream[Integer] = ...
val greenStream: DataStream[Integer] = ...

orangeStream.join(greenStream)
    .where(elem => /* select key */)
    .equalTo(elem => /* select key */)
    .window(TumblingEventTimeWindows.of(Time.milliseconds(2)))
    .apply { (e1, e2) => e1 + "," + e2 }
```

Java代码如下：

```
import org.apache.flink.api.java.functions.KeySelector;
import org.apache.flink.streaming.api.windowing.assigners.TumblingEventTimeWindows;
```

```
import org.apache.flink.streaming.api.windowing.time.Time;

...
DataStream< Integer > orangeStream = ...
DataStream< Integer > greenStream = ...

orangeStream.join(greenStream)
   .where(< KeySelector >)
   .equalTo(< KeySelector >)
   .window(TumblingEventTimeWindows.of(Time.milliseconds(2)))
   .apply (new JoinFunction< Integer, Integer, String > (){
       @Override
       public String join(Integer first, Integer second) {
          return first + "," + second;
       }
   });
```

2. 滑动窗口连接

当执行滑动窗口连接时,所有具有公共key和公共滑动窗口的元素都作为成对组合连接,并传递给JoinFunction或FlatJoinFunction。在当前滑动窗口中,一个流的元素中没有来自另一个流的元素的元素不会被触发。注意,有些元素可能在一个滑动窗口中被连接,而在另一个窗口中却没有被连接。

例如,使用大小为2ms的滑动窗口,并将其滑动1ms,结果是滑动窗口[−1,0],[0,1],[1,2],[2,3],…,如图3-40所示。

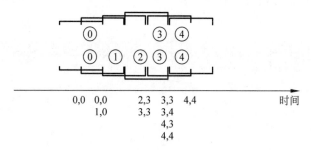

图 3-40 滑动窗口连接

在图3-40中,时间轴下面连接的元素是为每个滑动窗口传递给JoinFunction的元素。在这里还可以看到,例如,在窗口[2,3]中,第2排的②与第1排的③进行了join连接,但与窗口[1,2]中的任何内容都没有进行连接。

两个滑动窗口连接的模板代码如下。

Scala代码如下:

```
import org.apache.flink.streaming.api.windowing.assigners.SlidingEventTimeWindows;
import org.apache.flink.streaming.api.windowing.time.Time;

...
val orangeStream: DataStream[Integer] = ...
val greenStream: DataStream[Integer] = ...

orangeStream.join(greenStream)
    .where(elem => /* select key */)
    .equalTo(elem => /* select key */)
    .window(SlidingEventTimeWindows.of(Time.milliseconds(2) /* size */, Time.milliseconds(1) /* slide */))
    .apply { (e1, e2) => e1 + "," + e2 }
```

Java 代码如下：

```
import org.apache.flink.api.java.functions.KeySelector;
import org.apache.flink.streaming.api.windowing.assigners.SlidingEventTimeWindows;
import org.apache.flink.streaming.api.windowing.time.Time;

...
DataStream<Integer> orangeStream = ...
DataStream<Integer> greenStream = ...

orangeStream.join(greenStream)
    .where(<KeySelector>)
    .equalTo(<KeySelector>)
    .window(SlidingEventTimeWindows.of(Time.milliseconds(2) /* size */, Time.milliseconds(1) /* slide */))
    .apply (new JoinFunction<Integer, Integer, String> (){
        @Override
        public String join(Integer first, Integer second) {
            return first + "," + second;
        }
    });
```

3. 会话窗口连接

当执行会话窗口连接时，如果"组合"满足会话条件，则所有具有相同键的元素都将以成对的组合连接，并传递给 JoinFunction 或 FlatJoinFunction。同样，这将执行一个内连接，因此，如果会话窗口只包含来自一个流的元素，则不会发出任何输出。

例如，定义了一个会话窗口连接，其中每个会话至少间隔 1ms，如图 3-41 所示。

在图 3-41 中，有 3 个会话，在前两个会话中，来自两个流的连接元素被传递给 JoinFunction。在第 3 次会话中，第 1 排流中没有元素，所以⑧和⑨没有连接。

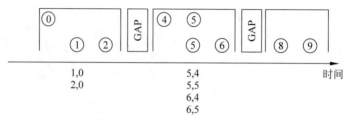

图 3-41 会话窗口连接

两个会话窗口连接的模板代码如下。
Scala 代码如下：

```scala
import org.apache.flink.streaming.api.windowing.assigners.EventTimeSessionWindows;
import org.apache.flink.streaming.api.windowing.time.Time;

...
val orangeStream: DataStream[Integer] = ...
val greenStream: DataStream[Integer] = ...

orangeStream.join(greenStream)
    .where(elem => /* select key */)
    .equalTo(elem => /* select key */)
    .window(EventTimeSessionWindows.withGap(Time.milliseconds(1)))
    .apply { (e1, e2) => e1 + "," + e2 }
```

Java 代码如下：

```java
import org.apache.flink.api.java.functions.KeySelector;
import org.apache.flink.streaming.api.windowing.assigners.EventTimeSessionWindows;
import org.apache.flink.streaming.api.windowing.time.Time;

...
DataStream<Integer> orangeStream = ...
DataStream<Integer> greenStream = ...

orangeStream.join(greenStream)
    .where(<KeySelector>)
    .equalTo(<KeySelector>)
    .window(EventTimeSessionWindows.withGap(Time.milliseconds(1)))
    .apply (new JoinFunction<Integer, Integer, String> (){
        @Override
        public String join(Integer first, Integer second) {
            return first + "," + second;
        }
    });
```

4. 间隔连接

间隔连接(Interval Join)使用一个公共 key 将两个流(称它们为 A 和 B)的元素连接起来,其中流 B 的元素的时间戳与流 A 中的元素的时间戳相对间隔。

这个也可以更正式地表示:

(1) b.timestamp ∈ [a.timestamp + lowerBound; a.timestamp + upperBound]。

(2) a.timestamp + lowerBound <= b.timestamp <= a.timestamp + upperBound。

其中 a 和 b 是 A 和 B 的元素,它们共用一个 key。只要下界总是小于或等于上界,下界和上界都可以是负的或正的。间隔连接目前仅执行内连接,并且仅支持事件时间。

当将一对元素传递给 ProcessJoinFunction 时,它们将被分配两个元素的较大时间戳(可以通过 ProcessJoinFunction.Context 访问该时间戳),如图 3-42 所示。

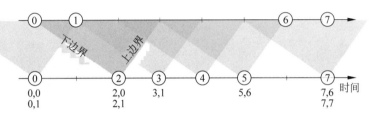

图 3-42 间隔连接

在图 3-42 中,用下界为 −2ms 和上界为 +1ms 连接第 2 排流和第 1 排流。默认情况下,这些边界是包含的,不过可以使用 .lowerBoundExclusive() 和 .upperBoundExclusive() 方法来改变默认行为。可以看到,流 A 的每个元素都会和流 B 的一定时间范围的元素进行连接,其中,上界和下界可以是负数,也可以是整数。Interval Join 目前只支持 INNER JOIN。将连接后的元素传递给 ProcessJoinFunction 时,时间戳变为两个元素中最大的那个时间戳。

用公式表示,表示图中三角形区域:

```
orangeElem.ts + lowerBound <= greenElem.ts <= orangeElem.ts + upperBound
```

两个间隔连接的模板代码如下。
Scala 代码如下:

```
import org.apache.flink.streaming.api.functions.co.ProcessJoinFunction;
import org.apache.flink.streaming.api.windowing.time.Time;

...
val orangeStream: DataStream[Integer] = ...
val greenStream: DataStream[Integer] = ...

orangeStream
    .keyBy(elem => /* select key */)
```

```
    .intervalJoin(greenStream.keyBy(elem => /* select key */))
    .between(Time.milliseconds(-2), Time.milliseconds(1))
    .process(new ProcessJoinFunction[Integer, Integer, String] {
        override def processElement(
            left: Integer,
            right: Integer,
            ctx: ProcessJoinFunction[Integer, Integer, String]#Context,
            out: Collector[String]): Unit = {
         out.collect(left + "," + right);
        }
    });
});
```

Java代码如下:

```
import org.apache.flink.api.java.functions.KeySelector;
import org.apache.flink.streaming.api.functions.co.ProcessJoinFunction;
import org.apache.flink.streaming.api.windowing.time.Time;

...
DataStream<Integer> orangeStream = ...
DataStream<Integer> greenStream = ...

orangeStream
    .keyBy(<KeySelector>)
    .intervalJoin(greenStream.keyBy(<KeySelector>))
    .between(Time.milliseconds(-2), Time.milliseconds(1))
    .process (new ProcessJoinFunction<Integer, Integer, String(){

        @Override
        public void processElement(Integer left, Integer right, Context ctx, Collector<String> out) {
            out.collect(first + "," + second);
        }
    });
```

3.8 低级操作

从之前的内容知道,Flink的转换操作是无法访问事件的时间戳信息和水印信息的。例如常用的MapFunction转换操作就无法访问时间戳或者当前事件的事件时间,而在一些应用场景下,访问事件的时间戳信息和水印信息极为重要,因此,Flink DataStream API提供了一系列的低级(Low-Level)转换操作,可以访问时间戳、水印及注册定时事件。还可以输出特定的一些事件,例如超时事件等。这一类的低级 API 被称为 ProcessFunction。

ProcessFunction 将事件处理与计时器和状态相结合,使其成为流处理应用程序的强大组件。这是使用 Flink 创建事件驱动应用程序的基础。

3.8.1 ProcessFunction

ProcessFunction 是一个低级的流处理操作,允许访问所有(非循环)流应用程序的基本组件,包括:

(1) events:数据流中的元素。

(2) state:状态,用于容错和一致性,仅用于 Keyed Stream。

(3) timer:定时器,支持事件时间和处理时间,仅用于 Keyed Stream。

ProcessFunction 可以被认为是一个访问 Keyed State 和定时器的 FlatMapFunction。通过为输入流中接收的每个事件调用它来处理事件。

ProcessFunction 用来构建事件驱动的应用及实现自定义的业务逻辑(使用之前的窗口函数和转换算子无法实现)。例如,Flink SQL 就是使用 ProcessFunction 实现的。

Flink 提供了 8 个 Process Function,分别如下。

(1) ProcessFunction:用于 DataStream。

(2) KeyedProcessFunction:用于 KeyedStream,keyBy 之后的流处理。

(3) CoProcessFunction:用于 connect 连接的流。

(4) ProcessJoinFunction:用于 join 流操作。

(5) BroadcastProcessFunction:用于广播。

(6) KeyedBroadcastProcessFunction:keyBy 之后的广播。

(7) ProcessWindowFunction:窗口增量聚合。

(8) ProcessAllWindowFunction:全窗口聚合。

ProcessFunction 有两个回调函数需要实现,分别是 processElement 和 onTimer。其源码定义如下:

```java
//第3章/ProcessFunction.java

/**
 * 一个处理流元素的函数
 *
 */
@PublicEvolving
public abstract class ProcessFunction < I, O > extends AbstractRichFunction {

    private static final long serialVersionUID = 1L;

    public abstract void processElement(I value, Context ctx, Collector < O > out) throws
Exception;
```

```java
    public void onTimer(long timestamp, OnTimerContext ctx, Collector<O> out) throws Exception {}

    public abstract class Context {

        public abstract Long timestamp();

        public abstract TimerService timerService();

        public abstract <X> void output(OutputTag<X> outputTag, X value);
    }

    public abstract class OnTimerContext extends Context {
        public abstract TimeDomain timeDomain();
    }
}
```

对每个传入的事件调用 processElement()方法；当计时器触发时调用 onTimer()方法。这些计时器可以是事件时间定时器，也可以是处理时间定时器。定时器可让应用程序对在处理时间和事件时间中的变化进行响应。processElement()和 onTimer()都提供了一个上下文对象，可用于与 TimerService（及其他内容）进行交互。这两个回调也都传入一个 Collector，该 Collector 可用于向下游发出结果。

要对两个输入流进行低级操作，应用程序可以使用 CoProcessFunction 或 KeyedCoProcessFunction。此函数可被绑定到两个不同的输入，并且对于来自这两个不同输入的记录，获取对 processElement1()和 processElement2()的单独调用。

实现低级别的 join 连接，通常遵循如下的模式：

(1) 为每个输入(或所有输入)创建一种状态对象。

(2) 从某个输入中接收元素时更新其状态。

(3) 接收到来自其他输入的元素后，探测状态并生成连接的结果。

例如，用户需要将客户数据连接到金融交易数据，并同时保留客户数据的状态。如果希望在面对无序事件时有完整和确定性的连接，则可以使用定时器在客户数据流的水印已经过了该交易的时间时，评估并发出该交易的连接。

3.8.2 KeyedProcessFunction 示例

KeyedProcessFunction 是 RichFunction 的一种。其源码定义如下：

```java
@PublicEvolving
public abstract class KeyedProcessFunction<K, I, O> extends AbstractRichFunction {

    private static final long serialVersionUID = 1L;
```

```java
    public abstract void processElement(I value, Context ctx, Collector<O> out) throws
Exception;

    public void onTimer(long timestamp, OnTimerContext ctx, Collector<O> out) throws
Exception {}

    public abstract class Context {

        public abstract Long timestamp();

        public abstract <X> void output(OutputTag<X> outputTag, X value);

        public abstract K getCurrentKey();
    }

    public abstract class OnTimerContext extends Context {
        public abstract TimeDomain timeDomain();

        @Override
        public abstract K getCurrentKey();
    }
}
```

每次调用 processElement() 函数都可以获得一个 Context 对象，通过该对象可以访问元素的事件时间时间戳及 TimerService。可以使用 TimerService 为将来的事件时间/处理时间实例注册回调。对于事件时间计时器，当前水印前进到或超过定时器的时间戳时，将调用 onTimer() 方法，而对于处理时间计时器，当挂钟时间达到指定时间时，将调用 onTimer() 方法。在调用期间，所有状态的范围再次限定为创建定时器所用的 key，从而允许定时器操作 Keyed State。

作为一个 RichFunction，它可以访问使用托管 Keyed State 所需的 open() 方法和 getRuntimeContext() 方法。对于容错状态，KeyedProcessFunction 可以通过 RuntimeContext 访问 Flink 的 Keyed State，这与其他有状态函数访问 Keyed State 的方式类似。

作为 ProcessFunction 的扩展，KeyedProcessFunction 在其 onTimer() 方法中通过 OnTimerContext 提供了对计时器 key 的访问。

Java 代码如下：

```java
@Override
public void onTimer(long timestamp,
                    OnTimerContext ctx,
                    Collector<OUT> out) throws Exception {
    K key = ctx.getCurrentKey();
    ...
}
```

Scala 代码如下：

```scala
override def onTimer(timestamp: Long,
                    ctx: OnTimerContext,
                    out: Collector[OUT]): Unit = {
  var key = ctx.getCurrentKey
  ...
}
```

在下面的示例中，KeyedProcessFunction 维护每个 key 的计数，并在每过 10s（以事件时间）而未更新该 key 时，发出一个 key/count 对，过程如下：

(1) 监听 Socket 数据源，获取输入字符串。

(2) 把 key（单词）、计数和最后修改时间戳存储在一个 ValueState 状态中，ValueState 的作用域是通过 key 隐式确定的。

(3) 对于每个记录，KeyedProcessFunction 递增计数器并设置最后修改时间戳。

(4) 该函数还安排了一个 10s 后的回调（以事件时间）。

(5) 在每次回调时，它根据存储的计数的最后修改时间检查回调的事件时间时间戳，并在它们匹配时发出 key/count（如果在 10s 内这个单词没有再次出现，就把这个单词和它出现的总次数发送到下游算子）。

【示例 3-26】 维护数据流中每个 key 的计数，并在每过 10s（以事件时间）而未更新该 key 时，发出一个 key/count 对。

Scala 代码如下：

```scala
//第 3 章/KeyedProcessFunDemo.scala

import Java.text.SimpleDateFormat
import Java.time.Duration
import Java.util.Date
import org.apache.flink.api.common.eventtime.{SerializableTimestampAssigner, WatermarkStrategy}
import org.apache.flink.api.common.state.{ValueState, ValueStateDescriptor}
import org.apache.flink.streaming.api.functions.KeyedProcessFunction
import org.apache.flink.streaming.api.scala._
import org.apache.flink.util.Collector

object KeyedProcessFunDemo {

  //存储在状态中的数据类型
  case class CountWithTimestamp(key: String, count: Long, lastModified: Long)

  def main(args: Array[String]) {
    //设置流执行环境
```

```scala
val env = StreamExecutionEnvironment.getExecutionEnvironment
//并行度为1
env.setParallelism(1)

//源数据流
val stream = env
  .socketTextStream("localhost", 9999)
  .flatMap(_.split("\\W+"))
  .map((_,1))

//为事件分配时间戳和水印
stream
  .assignTimestampsAndWatermarks(WatermarkStrategy
    .forMonotonousTimestamps[(String, Int)]()
    .withTimestampAssigner(new SerializableTimestampAssigner[(String, Int)] {
        override def extractTimestamp(t: (String, Int), ts: Long): Long = System.currentTimeMillis()
    })
    .withIdleness(Duration.ofSeconds(5))
  )
  .keyBy(_._1)
  .process(new CountWithTimeoutFunction())
  .print()

//执行流程序
env.execute("Process Function")
}

/**
 * KeyedProcessFunction 的子类,维护计数和超时.
 * 它的作用是将每个单词的最新出现时间记录到 backend,并创建定时器,
 * 定时器触发时,检查这个单词距离上次出现时是否已经达到10s,如果是,就发射给下游算子
 */
class CountWithTimeoutFunction extends KeyedProcessFunction[String, (String, Int), (String, Long)] {

  /**
   * 首先获得由这个处理函数(Process Function)维护的状态
   * 通过 RuntimeContext 访问 Flink 的 Keyed State
   */
  lazy val state: ValueState[CountWithTimestamp] = getRuntimeContext
    .getState(new ValueStateDescriptor[CountWithTimestamp](
        "myState", classOf[CountWithTimestamp])
    )
```

```scala
/**
 * 对于在输入流中接收的每个事件,此函数就会被调用以处理该事件
 * @param value        输入元素
 * @param ctx          上下文件环境
 * @param out
 * @throws Exception
 */
override def processElement(
        value: (String, Int),
        ctx: KeyedProcessFunction[String, (String, Int), (String, Long)]#Context,
        out: Collector[(String, Long)]): Unit = {

    //初始化或检索/更新状态
    val current = state.value match {
      case null =>
        //如果是第1个事件,则将初始计数设为1
        CountWithTimestamp(value._1, 1, ctx.timestamp)
      case CountWithTimestamp(key, count, lastModified) =>
        //先撤销之前的定时器
        ctx.timerService.deleteProcessingTimeTimer(lastModified + 10000)
        //如果不是第1个事件,则累加
        CountWithTimestamp(key, count + 1, ctx.timestamp)
    }

    //将修改过后的状态写回
    state.update(current)

    //从当前事件时间开始安排下一个定时器10s
    //为当前单词创建定时器,10s 后触发
    val timer = current.lastModified + 10000
    ctx.timerService.registerProcessingTimeTimer(timer)

    //打印所有信息,用于核对数据的正确性
    println(s"process, ${current.key}, ${current.count}, " +
      s"lastModified : ${current.lastModified}, (${time(current.lastModified)}), " +
      s"timer : ${timer} (${time(timer)})\n")
}

/** 定时器触发后执行的方法
 * 如果 1min 内没有新来的相同的单词,则发出 key/count 对
 *
 * @param timestamp 这段时间戳代表的是该定时器的触发时间
 * @param ctx
 * @param out
 * @throws Exception
 */
```

```scala
  override def onTimer(timestamp: Long,
                    ctx: KeyedProcessFunction[String, (String, Int), (String, Long)]
#OnTimerContext,
                    out: Collector[(String, Long)]): Unit = {
    //取得当前单词
    val currentKey = ctx.getCurrentKey

    //获取调度此计时器的 key 的状态(此单词的状态)
    val result = state.value

    //当前元素是否已经连续 10s 未出现的标志
    var isTimeout = false

    //timestamp 是定时器触发时间
    //如果等于最后一次更新时间 + 10s,就表示这 10s 内已经收到过该单词了
    //这种连续 10s 没有出现的元素,将被发送到下游算子
    result match {
      case CountWithTimestamp(key, count, lastModified) if timestamp >= lastModified +
10000 =>
        out.collect((key, count))          //发出 key/count 对
        isTimeout = true
      case _ =>
    }

    //打印数据,用于核对是否符合预期
    println(s"ontimer, ${currentKey}, ${result.count}, " +
      s"lastModified : ${result.lastModified}, (${time(result.lastModified)}), " +
      s"stamp : ${timestamp} (${time(timestamp)}),isTimeout: ${isTimeout}\n")
  }
}

  private def time(timeStamp: Long) = new SimpleDateFormat("yyyy-MM-dd HH:mm:ss").format
(new Date(timeStamp))
}
```

Java 代码如下:

```java
//第 3 章/KeyedProcessFunDemo.java

import org.apache.flink.api.common.eventtime.WatermarkStrategy;
import org.apache.flink.api.common.functions.FlatMapFunction;
import org.apache.flink.api.common.state.ValueState;
import org.apache.flink.api.common.state.ValueStateDescriptor;
import org.apache.flink.api.java.tuple.Tuple2;
import org.apache.flink.configuration.Configuration;
```

```java
import org.apache.flink.streaming.api.datastream.DataStream;
import org.apache.flink.streaming.api.environment.StreamExecutionEnvironment;
import org.apache.flink.streaming.api.functions.KeyedProcessFunction;
import org.apache.flink.util.Collector;
import org.apache.flink.util.StringUtils;

import java.text.SimpleDateFormat;
import java.time.Duration;
import java.util.Date;

public class KeyedProcessFunDemo {

    //存储在状态中的数据类型,POJO 类
    public static class CountWithTimestamp {
        public String key;
        public long count;
        public long lastModified;
    }

    public static void main(String[] args) throws Exception {
        //设置流执行环境
        final StreamExecutionEnvironment env =
            StreamExecutionEnvironment.getExecutionEnvironment();

        //并行度为 1
        env.setParallelism(1);

        //源数据流
        DataStream<Tuple2<String, Integer>> stream = env
            .socketTextStream("localhost",9999)
            .flatMap(new FlatMapFunction<String, Tuple2<String,Integer>>() {
                @Override
                public void flatMap(String s, Collector<Tuple2<String, Integer>> collector)
throws Exception {
                    if(StringUtils.isNullOrWhitespaceOnly(s)) {
                        System.out.println("invalid line");
                        return;
                    }

                    for(String word : s.split("\\W+")){
                        collector.collect(new Tuple2<>(word,1));
                    }
                }
            });

        stream
```

```java
            .assignTimestampsAndWatermarks(WatermarkStrategy
                    .<Tuple2<String, Integer>>forMonotonousTimestamps()
                    .withTimestampAssigner((event,ts) -> System.currentTimeMillis())
                    .withIdleness(Duration.ofSeconds(5))
            )
            .keyBy(t -> t.f0)
            .process(new CountWithTimeoutFunction())
            .print();

    //执行
    env.execute("flink streaming job");
}

/**
 * KeyedProcessFunction 的子类,维护计数和超时.
 * 它的作用是将每个单词的最新出现时间记录到 backend,并创建定时器.
 * 定时器触发时,检查这个单词距离上次出现时是否已经达到 10s,如果是,就发射给下游算子
 */
public static class CountWithTimeoutFunction
        extends KeyedProcessFunction<String,             //key
                Tuple2<String, Integer>,                 //input
                Tuple2<String, Long>> {                  //output

    //由此函数所维护的存储状态
    private ValueState<CountWithTimestamp> state;

    //首先获得由这个处理函数维护的状态
    //通过 RuntimeContext 访问 Flink 的 Keyed State
    @Override
    public void open(Configuration parameters) throws Exception {
        //初始化状态,状态名称是 myState
        state = getRuntimeContext().getState(
                new ValueStateDescriptor<>("myState", CountWithTimestamp.class));
    }

    /**
     * 对于在输入流中接收的每个事件,此函数会被调用以处理该事件
     * @param value                          输入元素
     * @param context                        上下文环境
     * @param out
     * @throws Exception
     */
    @Override
    public void processElement(
            Tuple2<String, Integer> value,
            Context context,
```

```java
                    Collector<Tuple2<String, Long>> out) throws Exception {

    //取得当前是哪个单词
    //String currentKey = context.getCurrentKey();

    //从 backend 取得当前单词的 myState 状态
    CountWithTimestamp current = state.value();

    //如果 myState 还从未赋值过,就在此初始化
    if (current == null) {
                    current = new CountWithTimestamp();
                    current.key = value.f0;
    }

    //先撤销之前的定时器
    //context.timerService()
    //.deleteEventTimeTimer(current.lastModified + 10000);
    context
        .timerService()
        .deleteProcessingTimeTimer(current.lastModified + 10000);

    //更新状态计数值(单词数量加 1)
    current.count++;

    //将状态的时间戳设置为记录分配的事件时间戳
    //取当前元素的时间戳,作为该单词最后一次出现的时间
    current.lastModified = context.timestamp();

    //重新保存到 backend,包括该单词出现的次数,以及最后一次出现的时间
    state.update(current);

    //为当前单词创建定时器,10s 后触发
    long timer = current.lastModified + 10000;

    context.timerService().registerEventTimeTimer(timer);

    //打印所有信息,用于核对数据的正确性
    System.out.println(String.format("process, %s, %d, lastModified : %d (%s), timer : %d (%s)\n", current.key, current.count, current.lastModified, time(current.lastModified), timer, time(timer)));
}

/** 定时器触发后执行的方法
 * 如果 1min 内没有新来的相同的单词,则发出 key/count 对
 *
 * @param timestamp        这段时间戳代表的是该定时器的触发时间
```

```java
 * @param ctx
 * @param out
 * @throws Exception
 */
@Override
public void onTimer(long timestamp,
                    OnTimerContext ctx,
                    Collector<Tuple2<String, Long>> out) throws Exception {

    //取得当前单词
    String currentKey = ctx.getCurrentKey();

    //获取调度此计时器的 key 的状态(此单词的状态)
    CountWithTimestamp result = state.value();

    //当前元素是否已经连续 10s 未出现的标志
    boolean isTimeout = false;

    //timestamp 是定时器触发时间
    //如果超过 10s 该单词(key)仍没有再出现,就发送给下游
    if (timestamp >= result.lastModified + 10000) {
        out.collect(new Tuple2<>(result.key, result.count));
        isTimeout = true;
    }

    //打印数据,用于核对是否符合预期
    System.out.println(String.format("ontimer, %s, %d, lastModified : %d (%s), stamp : %d (%s), isTimeout : %s\n\n", currentKey, result.count, result.lastModified, time(result.lastModified), timestamp, time(timestamp), String.valueOf(isTimeout)));
    }
}

private static String time(long timeStamp) {
    return new SimpleDateFormat("yyyy-MM-dd HH:mm:ss").format(new Date(timeStamp));
}
}
}
```

在 processElement 方法中, state.value() 可以取得当前单词的状态, state.update(current)可以设置当前单词的状态执行以上代码。

建议按以下步骤执行上面的程序:

(1) 在控制台执行命令 nc -lk 9999,这样就可以从控制台向 9999 端口发送字符串了。

(2) 在 IDEA 中直接执行流程序代码,程序运行后便可开始监听本机的 9999 端口。

(3) 在 netcat 的控制台输入字符串 aaa,然后按 Enter 键,连续输入两行,中间间隔不要超过 10s,然后输入字符串 ttt,再按 Enter 键,如图 3-43 所示。

图 3-43 从控制台向 9999 端口发送字符串

(4) 观察 IDEA 控制台的输出结果,如图 3-44 所示。

图 3-44 观察输出结果

3.8.3 案例:服务器故障检测报警程序

对于 Hadoop 运维人员,需要监控生产环境大数据的各个组件的状态信息,由于服务器经常会宕机,因此需要监控服务器的状态信息,每次服务器上线和下线都会发送一种状态信息。如果服务器宕机后,服务自启动功能在 30s 以内没有将服务器拉起(这里只是为了测试,所以使用 30s),就进行持续告警,直到收到上线消息,告警取消。具体过程说明如下:

(1) 这里的消息模拟从 Socket 接收服务器告警消息,消息格式包含 3 个字段,分别是:主机名 hostname、告警时间 time 和状态 status(RUNNING 服务正常,DEAD 服务停止)。

(2) 接收消息之后对数据流按照主机名进行分组,对于状态为 DEAD 的消息,设置定时器 30s 以内如果状态不恢复为 RUNNING,则定时进行告警,如果 30s 内恢复 RUNNING 状态,则认为上一条消息是误报的,所以需要删除定时器,取消报警。

【示例 3-27】 编写 Flink 流处理程序,实时检测服务器故障信息并予以报警。

Scala 代码如下:

```
//第3章/KeyedProcessFunDemo2.scala

import org.apache.flink.api.common.state._
import org.apache.flink.streaming.api.functions.KeyedProcessFunction
import org.apache.flink.streaming.api.scala._
import org.apache.flink.util.Collector

object KeyedProcessFunDemo2 {

  //服务器信息类型:主机名,时间,状态(RUNNING 表示正常,DEAD 表示宕机)
```

```scala
case class MessageInfo(hostname: String, msgTime: String, status: String)

def main(args: Array[String]) {
  //设置流执行环境
  val env = StreamExecutionEnvironment.getExecutionEnvironment

  //并行度为1
  env.setParallelism(1)

  //源数据流
  val stream = env
    .socketTextStream("localhost", 9999)
    .filter(line => line match{
        case null | "" => false
        case row       => row.split(",").length == 3
    })
    .map(line => {
      val lines = line.split(",")
      MessageInfo(lines(0), lines(1), lines(2))
    })
    .keyBy(msg => msg.hostname)
    .process(new AlertDownFunction)
    .print()

  //执行流程序
  env.execute("Process Function")
}

/**
 * KeyedProcessFunction 的子类
 */
class AlertDownFunction extends KeyedProcessFunction[String, MessageInfo, String] {

  /**
   * 首先获得由这个处理函数(process function)维护的状态
   * 通过 RuntimeContext 访问 Flink 的 Keyed State
   */
  lazy val lastStatus: ValueState[String] = getRuntimeContext.getState(
    new ValueStateDescriptor[String]("lastStatus", classOf[String])
  )
  lazy val warningTimer: ValueState[Long] = getRuntimeContext.getState(
    new ValueStateDescriptor[Long]("warning-timer", classOf[Long])
  )

  /**
   * 对于在输入流中接收的每个事件,此函数会被调用以处理该事件
```

```scala
 * @param value                      输入元素
 * @param ctx                        上下文件环境
 * @param out
 * @throws Exception
 */
override def processElement(value: MessageInfo,
        ctx: KeyedProcessFunction[String, MessageInfo, String]#Context,
        out: Collector[String]): Unit = {

  //获取当前的状态和上次的定时器时间
  val currentStatus = value.status                    //当前消息中的状态值
  val currentTimer = warningTimer.value               //上次的定时器时间

  println("\ncurrentStatus:" + currentStatus)         //当前事件状态
  println("lastStatus:" + lastStatus.value)           //上次事件状态

  //连续两次状态都是 DEAD,说明是宕机状态,新建定时器,30s 后进行告警
  if ("DEAD" == currentStatus && "DEAD" == lastStatus.value) {
    val timeTs = ctx.timerService.currentProcessingTime + 30000L
    ctx.timerService.registerProcessingTimeTimer(timeTs)
    warningTimer.update(timeTs)                       //将状态更新为定时器时间
  }
  else { //如果不是连续告警,则认为是误报警,删除定时器
    if ("RUNNING" == currentStatus && "DEAD" == lastStatus.value) {
      if (null != currentTimer)
         ctx.timerService.deleteProcessingTimeTimer(currentTimer)
      warningTimer.clear()
    }
  }

  //状态更新:将当前状态设为 last status
  lastStatus.update(currentStatus)
}

/** 定时器触发后执行的方法
 *
 * @param timestamp 这段时间戳代表的是该定时器的触发时间
 * @param ctx
 * @param out
 * @throws Exception
 */
override def onTimer(timestamp: Long,
             ctx: KeyedProcessFunction[String, MessageInfo, String]#OnTimerContext,
             out: Collector[String]): Unit = {
  //输出报警信息,Regionserver 两次状态监测为 DEAD(宕机)
  val hostname = ctx.getCurrentKey
  out.collect("主机 IP:" + hostname + ",服务器状态监测连续宕机,请排查!")
}
}
```

Java 代码如下：

```java
//第 3 章/KeyedProcessFunDemo2.java

import org.apache.flink.api.common.functions.FilterFunction;
import org.apache.flink.api.common.functions.MapFunction;
import org.apache.flink.api.common.state.ValueState;
import org.apache.flink.api.common.state.ValueStateDescriptor;
import org.apache.flink.api.java.functions.KeySelector;
import org.apache.flink.configuration.Configuration;
import org.apache.flink.streaming.api.datastream.DataStream;
import org.apache.flink.streaming.api.environment.StreamExecutionEnvironment;
import org.apache.flink.streaming.api.functions.KeyedProcessFunction;
import org.apache.flink.util.Collector;

public class KeyedProcessFunDemo2 {

    //POJO 类,服务器告警信息类型
    public static class MessageInfo {
        public String hostname;             //主机名
        public String msgTime;              //时间
        public String status;               //RUNNING 表示正常,DEAD 表示宕机

        public MessageInfo(){}

        public MessageInfo(String hostname,String msgTime,String status){
            this.hostname = hostname;
            this.msgTime = msgTime;
            this.status = status;
        }
    }

    public static void main(String[] args) throws Exception {
        //设置流执行环境
        final StreamExecutionEnvironment env =
            StreamExecutionEnvironment.getExecutionEnvironment();

        //这里为了便于理解,将并行度设置为 1
        env.setParallelism(1);

        //指定数据源从 Socket 的 9999 端口接收数据,先进行了不合法数据的过滤
        DataStream<String> sourceDS = env
            .socketTextStream("localhost", 9999)
            .filter(new FilterFunction<String>() {
                @Override
                public boolean filter(String line) throws Exception {
```

```java
            if (null == line || "".equals(line)) {
                return false;
            }
            String[] lines = line.split(",", -1);
            return lines.length == 3;
        }
    });

    //做了一个简单的map转换,将数据转换成MessageInfo格式
    DataStream<String> warningDS = sourceDS
        .map(new MapFunction<String, MessageInfo>() {
            @Override
            public MessageInfo map(String line) throws Exception {
                String[] lines = line.split(",");
                return new MessageInfo(lines[0], lines[1], lines[2]);
            }
        })
        .keyBy(msg -> msg.hostname)
        .process(new AlertDownFunction());

    //打印报警信息
    warningDS.print();

    //执行流程序
    env.execute("Process Alert Down");
}

/**
 * ProcessFunction 实现,处理告警信息
 * KeyedProcessFunction<K,I,O>
 */
public static class AlertDownFunction
        extends KeyedProcessFunction<String, MessageInfo, String> {

    //声明两种状态
    private ValueState<String> lastStatus;
    private ValueState<Long> warningTimer;

    //初始函数:首先获得之前(上次)已经存储的状态信息
    @Override
    public void open(Configuration parameters) throws Exception {
        super.open(parameters);
        lastStatus = getRuntimeContext().getState(
                new ValueStateDescriptor<>("lastStatus", String.class));
        warningTimer = getRuntimeContext().getState(
                new ValueStateDescriptor<>("warning-timer", Long.class));
```

```java
        }
        //对于在输入流中接收的每个事件,此函数会被调用以处理该事件
        //获得当前事件的状态,与之前保存的状态进行比较
        @Override
        public void processElement(MessageInfo value, Context ctx, Collector<String> out)
throws Exception {

            //获取当前的状态和上次的定时器时间
            String currentStatus = value.status;                    //当前消息中的状态值
            Long currentTimer = warningTimer.value();               //上次的定时器时间

            System.out.println("\ncurrentStatus:" + currentStatus);  //当前事件状态
            System.out.println("lastStatus:" + lastStatus.value()); //上次事件状态

            //连续两次状态都是 DEAD,说明是宕机状态,新建定时器,30s 后进行告警
            if("DEAD".equals(currentStatus) && "DEAD".equals(lastStatus.value())){
                long timeTs = ctx.timerService().currentProcessingTime() + 30000L;
                ctx.timerService().registerProcessingTimeTimer(timeTs);
                warningTimer.update(timeTs);                        //将状态更新为定时器时间
            }
            //如果不是连续告警,则认为是误报警,删除定时器
            else if(("RUNNING".equals(currentStatus) && "DEAD".equals(lastStatus.value()))){
                if(null != currentTimer){
                    ctx.timerService().deleteProcessingTimeTimer(currentTimer);
                }
                warningTimer.clear();
            }

            //状态更新:将当前状态设为 last status
            lastStatus.update(currentStatus);
        }

        //定时器
        @Override
        public void onTimer(long timestamp,
                            OnTimerContext ctx,
                            Collector<String> out) throws Exception {
            //输出报警信息,Regionserver 两次状态监测为 DEAD(宕机)
            out.collect("主机 IP:" + ctx.getCurrentKey() + ",服务器状态监测连续宕机,请排查!");
        }
    }
}
```

建议按以下步骤执行程序。

(1) 启动 netcat 服务器,执行命令如下:

```
$ nc -lk 9999
```

(2) 在命令行依次输入以下信息,模拟服务器监控事件:

```
192.168.1.101,2019-04-07 21:00,RUNNING
192.168.1.101,2019-04-07 21:02,DEAD
192.168.1.101,2019-04-07 21:03,DEAD
192.168.1.101,2019-04-07 21:04,RUNNING
```

(3) 最终的执行结果如下:

```
currentStatus:RUNNING
lastStatus:null

currentStatus:DEAD
lastStatus:RUNNING

currentStatus:DEAD
lastStatus:DEAD
```

主机 IP:192.168.1.101,两次 Regionserver 状态监测宕机,请监测!

3.9 状态和容错

状态函数和运算符在各个元素/事件的处理过程中存储数据,这使状态成为任何类型的复杂操作的关键组成部分。
(1) 当应用程序搜索某些事件模式时,状态将存储到目前为止遇到的事件序列。
(2) 当以每分钟/小时/天的速度聚合事件时,状态将保留挂起的聚合。
(3) 当在一个数据点流上训练一个机器学习模型时,状态保持模型参数的当前版本。
(4) 当需要管理历史数据时,该状态允许对过去发生的事件进行有效访问。

3.9.1 状态运算

Flink 的操作可以是无状态的,也可以是有状态的,如图 3-45 所示。

如果处理事件的结果只与事件本身的内容相关,则称为无状态操作。无状态程序查看每个单独的事件,并基于最后一个事件创建一些输出。例如,流程序可能从传感器接收温度读数,并在温度超过 90℃ 时发出警报。

如果结果与先前处理的事件相关,则称为有状态操作。有状态操作意味着一个事件的处理方式可能取决于它之前发生的所有事件的累积效果。状态可以用于简单的事情，例如计算仪表板上显示的每分钟事件数，也可以用于更复杂的事情，例如计算欺诈检测模型的功能。有状态程序基于多个事件一起创建输出。任何重要的数据处理（如基本聚合）都是有状态的处理。

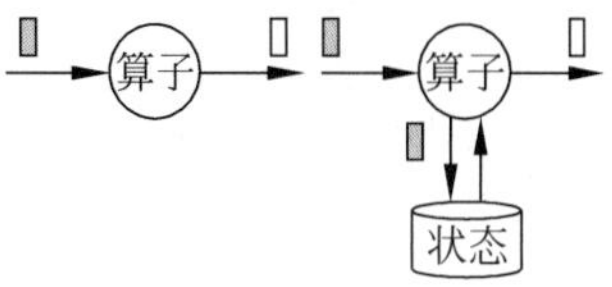

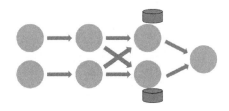

图 3-45　无状态流和有状态流

有状态程序的场景包括

（1）所有类型的窗口。例如，获取传感器在过去一小时内报告的平均温度是一个有状态的计算。当按分钟/小时/天聚合事件时，状态保存着挂起的聚合。

（2）用于复杂事件处理（CEP）的各种状态机。例如，在接收到两个在 1min 内相差超过 20℃ 的温度读数后创建一个警报是一个有状态的计算。

（3）当应用程序搜索某些事件模式时，状态将存储到目前为止遇到的事件序列。

（4）当通过数据点流训练机器学习模型时，状态保持模型参数的当前版本。

（5）流之间及流与静态或缓慢变化的表之间的各种连接。

（6）当需要管理历史数据时，状态允许对过去发生的事件进行有效访问。

Flink 应用程序在分布式集群上并行运行。给定操作符的不同并行实例将在不同的线程中独立执行，并且通常运行在不同的机器上。

状态操作符的并行实例集实际上是一个分片键值存储。每个并行实例负责处理特定键组的事件，这些键的状态保存在本地。

例如，有一个包含无状态操作符和有状态操作符的作业，如图 3-46 所示。

图 3-46　作业中包含无状态操作符和有状态操作符

图 3-46 中显示了在作业图的前 3 个操作符中并行度为 2 的作业运行，在并行度为 1 的接收器中终止。第 3 个操作符是有状态的，可以看到在第 2 个和第 3 个操作符之间发生了全连接的网络 shuffle。这样做是为了通过某个键对流进行分区，以便所有需要一起处理的事件都被一起处理。

状态总是在本地访问，Flink 针对本地状态访问进行了优化，这有助于 Flink 应用程序实现高吞吐量和低时延。可以选择将状态保存在 JVM 堆中，或者如果它太大，则在访问高效的磁盘数据结构中维护，如图 3-47 所示。

3.9.2　状态的类型

Flink 提供了两种基本类型的状态：Keyed State 和 Operator State。

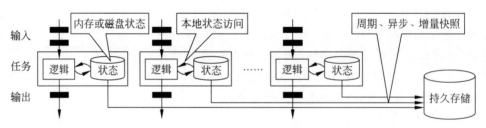

图 3-47 本地状态可保存在持久存储中

(1) Keyed State 始终是相对于 key 的,并且只能在 Keyed Stream 上的函数和算子中使用。

(2) 使用 Operator State(或 non-keyed state),每个算子状态都绑定到一个并行算子实例。

1. Keyed State

Keyed State 是专门针对 Keyed Stream 能够使用的状态。换句话说,就是按照 key 进行分区之后的操作才能使用状态。Keyed State 被维护在可以被认为是嵌入式键/值存储的地方。状态被严格地与状态操作符读取的流一起划分和分布,因此,访问 key-value 状态只可能在 Keyed Stream 上,即在执行 keyBy 分区数据交换之后,并且仅限于与当前事件 key 相关的值,如图 3-48 所示。

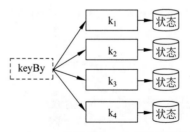

图 3-48 Keyed State 是按照 key 分区后能够使用的状态

Keyed State 可以确保所有的状态更新都是本地操作,从而保证了状态数据的一致性而不会产生事务开销。因为 Keyed State 是按照 key 进行维护和访问的,所以 Flink 会为每个 key 都维护一种状态实例,该状态实例总是位于处理该 key 记录的算子任务上,因此同一个 key 的记录可以访问一样的状态。这种对流和状态的 key 进行对齐的方式还允许 Flink 重新分配状态并透明地调整流分区,如图 3-49 所示。

Flink 为 Keyed State 提供了不同的存储结构类型,这些存储结构官方称为"状态原语"。可用的状态原语如图 3-50 所示。

这些状态原语的具体说明如下。

(1) ValueState<T>:这将保留一个可以更新和检索的值(如前所述,作用域为输入元素的键,因此操作看到的每个键可能都有一个值)。可以使用 update(T) 设置该值,并使用

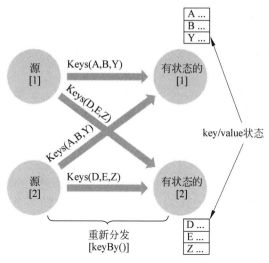

图 3-49　Flink 会为每个 key 都维护一种状态实例

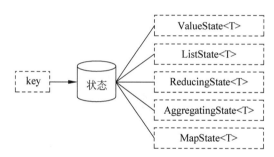

图 3-50　Flink 为 Keyed State 提供了不同的存储结构类型

T value()检索该值。

（2）ListState<T>：它保存了一个元素列表。可以在当前存储的所有元素上追加元素并检索一个 Iterable。使用 add(T)或 addAll(List<T>)添加元素，可使用 Iterable<T> get()检索 Iterable。还可以使用 update(List<T>)覆盖现有列表。

（3）ReducingState<T>：这将保留一个值，该值表示添加到状态的所有值的聚合。该接口类似于 ListState，但使用 add(T)添加的元素可使用指定的 ReduceFunction 缩减为聚合。

（4）AggregatingState<IN，OUT>：这将保持单个值，该值表示添加到状态的所有值的聚合，聚合类型可能与添加到该状态的元素的类型不同。这个接口与 ListState 类似，但使用 add(IN)添加的元素是使用指定的 AggregateFunction 聚合的。

（5）MapState<UK，UV>：这保存了一个映射列表。可以将 key-value 对放入状态，并在所有当前存储的映射上检索一个 Iterable。使用 put(UK，UV)或 putAll(Map<UK，UV>)添加映射。可以使用 get(UK)检索与用户 key 关联的值。可以使用 entries()、keys()、 values()分别检索映射、keys 和 values 的可迭代视图。还可以使用 isEmpty()来检查该 map 是否包含任何 key-value 映射。

所有类型的状态都有一种方法 clear(),该方法为当前活动的 key(输入元素的 key)清除状态。

务必记住,这些状态对象仅用于与状态交互。状态不一定存储在内部,但可能驻留在磁盘或其他地方。另外,从状态获取的值取决于输入元素的 key,因此,如果涉及的 key 不同,则在一次用户函数调用中得到的值可能与在另一次调用中得到的值不同。

2. Operator State

Operator State(算子状态)是一种 non-keyed state,与并行的操作算子实例相关联。Operator State 的作用域是某个算子任务,这意味着所有在同一个并行任务之内的记录都能访问相同的状态。算子状态不能通过其他任务访问,如图 3-51 所示。

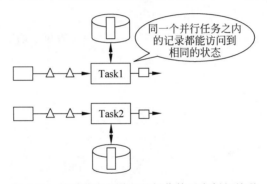

图 3-51 算子状态与并行的操作算子实例相关联

而当并行度发生调整时,需要在操作符(算子)的并行度上重新分配状态。在大多数的流处理程序开发中,用户很少用到算子状态,因为用户定义的函数通常不需要 non-keyed state。这个特性最常用于实现 source 和 sink,而 source、sink 都是没有 key 的,这就需要用到算子状态了,如图 3-52 所示。

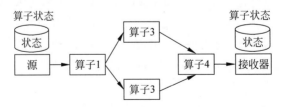

图 3-52 Operator State 最常用于实现 source 和 sink

3. Keyed State VS Operator State

Keyed State 和 Operator State 的对比见表 3-12。

表 3-12　Keyed State 和 Operator State 的对比（1）

区别	Keyed State	Operator State
使用范围	只能在 KeyedStream 算子中使用，每个 key 对应一个 state	可以用于所有算子，常用在 source 和 sink。一个 Operator 实例对应一个 state
扩缩容模式	Flink 把所有键值分为不同的 key 组。当并发改变时，Flink 会重新分配 Keyed State，以 key 组为单位，将键值分配给不同的任务	当并发改变时，有多种方式进行重分配，例如 ListState 使用均匀分配模式，BroadcastState 会把状态复制到全部新任务上
访问方式	实现 RichFunction，通过 getRuntimeContext() 返回的 RuntimeContext 获取	实现 CheckpointedFunction 或者 ListCheckpoint 的接口
数据结构	JP2ValueState、ListState、ReducingState、AggregatingState、MapState	ListState、BroadcastState 等

4. 原生状态和托管状态

根据状态数据是否受 Flink 管理，Flink 又将状态（Keyed State 和 Operator State）分为两种形式：托管状态和原生状态。

托管状态在由 Flink 运行时控制的数据结构中表示，如内部哈希表或 RocksDB。例如 ValueState、ListState 等。Flink 的运行时对状态进行序列化编码并将它们写入检查点。Flink 管理的状态存储在状态后端（State Backend）。

原生状态是运算符在自己的数据结构中保持的状态。当检查点时，它们只向检查点写入一字节序列。Flink 对状态的数据结构一无所知，只看到原生字节。

不同状态类型下托管状态和原生状态所使用的数据结构见表 3-13。

表 3-13　Keyed State 和 Operator State 的对比（2）

状态类型	托管状态	原生状态
Keyed State	ValueState\<T\> ListState\<T\> MapState\<UK, UV\> ReducingState AggregatingState	字节数组：Byte[]
Operator State	ListState BroadcastState	字节数组：Byte[]

所有 DataStream 函数都可以使用托管状态，但是原生状态接口只能在实现自定义运算符时使用。建议使用托管状态而不是原生状态，因为通过托管状态，当并行性发生变化时，Flink 能够自动重新分配状态，而且还可以进行更好的内存管理。

托管状态与原生状态的区别见表 3-14。

表 3-14 Keyed State 和 Operator State 的对比(3)

状态类型	托管状态	原生状态
状态管理方式	Flink Runtime 托管，自动存储、自动恢复，内存管理上有优化	用户自己管理，需要自己进行序列化
数据结构	Flink 提供了常用数据结构，如 ValueState、ListState、MapState 等	字节数组：Byte[]
使用场景	所有 DataStream 函数都可以使用	自定义 Operator 时

Flink 为它管理的状态提供了如下一些特性。

(1) 本地化：Flink 状态保持在处理它的机器的本地，并且可以以内存速度访问。

(2) 持久化：Flink 状态是容错的，也就是说，它会定期自动地检查点，并在失败时还原。

(3) 垂直扩展：Flink 状态可以存储在嵌入的 RocksDB 实例中，这些实例通过添加更多的本地磁盘进行扩展。

(4) 水平扩展：Flink 状态会随着集群的增长和收缩而重新分布。

(5) 可查询的：可以通过 Queryable State API 从外部查询 Flink 状态。

Flink 提供了简单易用的 API 来存储和获取状态。

3.9.3 使用托管的 Keyed State

托管的 Keyed State 接口提供了对不同类型状态的访问，这些状态的范围都是当前输入元素的 key。这意味着这种状态只能在 Keyed Stream 上使用，而 Keyed Stream 可以通过 stream.keyBy() 创建。

要获得状态句柄，必须创建一个 StateDescriptor，它代表状态描述符。StateDescriptor 持有状态的名称(名称必须唯一)、状态的值的类型，另外可能还有一个指定的函数，例如 ReduceFunction。每种状态都对应各自的描述符，根据希望检索的状态类型，可以创建 ValueStateDescriptor、ListStateDescriptor、ReducingStateDescriptor、MapStateDescriptor 或 AggregatingStateDescriptor 等。

通过描述符使用 RuntimeContext 访问相应的状态，而 RuntimeContext 只有在 RichFunction 中才能获取。在 RichFunction 中可以使用 RuntimeContext，它具有以下访问状态的方法：

(1) ValueState<T> getState(ValueStateDescriptor<T>)。

(2) ReducingState<T> getReducingState(ReducingStateDescriptor<T>)。

(3) ListState<T> getListState(ListStateDescriptor<T>)。

(4) AggregatingState<IN, OUT> getAggregatingState(AggregatingStateDescriptor<IN, ACC, OUT>)。

(5) MapState<UK, UV> getMapState(MapStateDescriptor<UK, UV>)。

所以要想使用 Keyed State,用户编写的类必须继承 RichFunction 或者其子类。

当计算依赖于累积给定 key 的数据时,需要将所有这些数据合并在一起。这可以通过两种方式完成:

(1) 将数据存储在 ListState 容器中,等待会话结束,并在会话结束时将所有数据合并在一起。

(2) 使用 ReducingState 在每个新事件到达时,将其与之前的事件合并。

使用哪种状态取决于在 WindowedStream 上运行的功能:使用 ProcessWindowFunction 的 process()方法调用将使用 ListState,而使用 ReduceFunction 的 reduce()方法调用则将使用 ReducingState。

ReducingState 的优点非常明显:不存储窗口处理之前的所有数据,而是在单个记录中不断地对其进行聚合。这通常会导致状态更小,取决于在 reduce 操作期间会丢弃多少数据,但是,在特定的情况下,它在存储方面几乎没有改善,例如,与历史会话存储的 7 天数据相比,该状态的大小可以忽略不计。这时,通过 ListState 反而可以提高性能。原因是,每次新事件到来时,连续的 reduce 操作都需要对数据进行反序列化和序列化。这可以在 RocksDBReducingState 的 add()函数中看到,该函数会调用 getInternal()方法,从而导致数据反序列化。

下面通过几个示例来理解和掌握托管的 Keyed State 应用。

【示例 3-28】 数据去重。假设有一个想要去重的事件流,要求只保留每个 key 的第 1 个事件。在下面这个应用程序中,通过名为 Deduplicator 的 RichFlatMapFunction 实现去重操作。

分析:这需要使用 Flink 的 Keyed State 接口来记住每个 key 是否已经有过一个事件。像这样使用 Keyed Stream 时,Flink 将为所管理的每种状态项维护一个 key-value 存储(托管状态)。Flink 支持几种不同类型的 Keyed State,本例使用最简单的一种,即 ValueState。这意味着对于每个键,Flink 将存储单个对象,如在本例中是一个 Boolean 类型的对象。

实现了 RichFlatMapFunction 接口的 Deduplicator 类有两种方法:open() 和 flatMap(),其中 open()方法通过定义 ValueStateDescriptor < Boolean >来建立托管状态的使用。构造函数的参数为这个 Keyed State 项指定了一个名称,即 keyHasBeenSeen,并提供可用于序列化这些对象的信息(在本例中为 Types.BOOLEAN)。

Scala 代码如下:

```
//第 3 章/DeduplicatorDemo.scala

import com.xueai8.java.ch03.DeduplicatorDemo.Deduplicator
import org.apache.flink.api.common.functions.RichFlatMapFunction
import org.apache.flink.api.common.state.{ValueState, ValueStateDescriptor}
import org.apache.flink.api.common.typeinfo.Types
import org.apache.flink.configuration.Configuration
import org.apache.flink.streaming.api.scala._
```

```scala
import org.apache.flink.util.Collector

object DeduplicatorDemo {

  //POJO 类,事件类型
  case class Event(key:String, timestamp:Long)

  def main(args: Array[String]) {
    //设置流执行环境
    val env = StreamExecutionEnvironment.getExecutionEnvironment

    //一批模拟流数据
    val eventList = List(
      Event("a", 123L),
      Event("b", 223L),
      Event("a", 323L),
      Event("b", 423L),
      Event("c", 523L)
    )

    //流管道
    env.fromCollection(eventList)
      .keyBy(_.key)
      .flatMap(new Deduplicator)
      .print

    //触发流程序执行
    env.execute("Flink Streaming Scala API Skeleton")
  }

  //去重处理
  class Deduplicator extends RichFlatMapFunction[Event, Event] {
    //状态存储对象(注意要声明数据类型)
    var keyHasBeenSeen: ValueState[Boolean] = _

    override def open(conf: Configuration): Unit = {
      val desc = new ValueStateDescriptor[Boolean]("keyHasBeenSeen", classOf[Boolean])
      keyHasBeenSeen = getRuntimeContext.getState(desc)
    }

    @throws[Exception]
    override def flatMap(event: Event, out: Collector[Event]): Unit = {
      //每个 key 只发送一次
      if (keyHasBeenSeen.value == null) {
        out.collect(event)
        keyHasBeenSeen.update(true)
      }
    }
  }
}
```

Java 代码如下：

```java
//第 3 章/DeduplicatorDemo.java
import org.apache.flink.api.common.functions.RichFlatMapFunction;
import org.apache.flink.api.common.state.ValueState;
import org.apache.flink.api.common.state.ValueStateDescriptor;
import org.apache.flink.api.common.typeinfo.Types;
import org.apache.flink.configuration.Configuration;
import org.apache.flink.streaming.api.environment.StreamExecutionEnvironment;
import org.apache.flink.util.Collector;

import Java.util.ArrayList;
import Java.util.List;

public class DeduplicatorDemo {

    //POJO 类,事件类型
    public static class Event {
        public String key;
        public long timestamp;

        public Event(){}

        public Event(String key, long timestamp) {
            this.key = key;
            this.timestamp = timestamp;
        }

        @Override
        public String toString() {
            return "Event{" +
                    "key = '" + key + '\'' +
                    ", timestamp = " + timestamp +
                    '}';
        }
    }

    public static void main(String[] args) throws Exception {
        //设置流执行环境
        StreamExecutionEnvironment env =
            StreamExecutionEnvironment.getExecutionEnvironment();

        //一批模拟流数据
        List< Event > eventList = new ArrayList<>();
        eventList.add(new Event("a", 123L));
```

```java
        eventList.add(new Event("b", 223L));
        eventList.add(new Event("a", 323L));
        eventList.add(new Event("b", 423L));
        eventList.add(new Event("c", 523L));

        //流管道
        env.fromCollection(eventList)
            .keyBy(e -> e.key)
            .flatMap(new Deduplicator())
            .print();

        //执行
        env.execute();
    }

    //去重处理
    public static class Deduplicator
                        extends RichFlatMapFunction<Event,Event> {

        //状态存储对象
        private transient ValueState<Boolean> keyHasBeenSeen;

        @Override
        public void open(Configuration conf) {
            ValueStateDescriptor<Boolean> desc =
                new ValueStateDescriptor<>("keyHasBeenSeen", Types.BOOLEAN);
            keyHasBeenSeen = getRuntimeContext().getState(desc);
        }

        @Override
        public void flatMap(Event event, Collector<Event> out)
                        throws Exception {
            //每个 key 只发送一次
            if (keyHasBeenSeen.value() == null) {
                out.collect(event);
                keyHasBeenSeen.update(true);
            }
        }
    }
}
```

执行以上代码,输出结果如下:

```
6> Event{key = 'a', timestamp = 123}
2> Event{key = 'b', timestamp = 223}
4> Event{key = 'c', timestamp = 523}
```

当 flatMap()方法调用 keyHasBeenSeen.value()时,Flink 的运行时会在上下文中查找

键的这段状态的值，只有当它为 null 时，才会继续将事件收集到输出中。在本例中，它还将 keyHasBeenSeen 的值更新为 true。

当部署到分布式集群时，将有许多这个 Deduplicator 的实例，每个实例负责整个键空间的一个不相交子集，因此，当用户看到 ValueState 的单个项时，例如

```
ValueState < Boolean > keyHasBeenSeen;
```

这不仅表示一个 Boolean 值，而且表示一个分布式、分片的 key-value 存储。

【示例 3-29】 实现一个简单计数窗口，统计每两个整数元素的平均值。

Scala 代码如下：

```scala
//第 3 章/CountWindowAverageDemo.scala

import org.apache.flink.api.common.functions.RichFlatMapFunction
import org.apache.flink.api.common.state.{ValueState, ValueStateDescriptor}
import org.apache.flink.configuration.Configuration
import org.apache.flink.streaming.api.scala._
import org.apache.flink.util.Collector

object CountWindowAverageDemo {

  def main(args: Array[String]) {
    //设置流执行环境
    val env = StreamExecutionEnvironment.getExecutionEnvironment

    env.fromCollection(List(
      (1L, 3L),
      (1L, 5L),
      (1L, 7L),
      (1L, 4L),
      (1L, 2L)
    )).keyBy(_._1)
      .flatMap(new CountWindowAverage())
      .print()
    //输出将会是(1,4) 和 (1,5)

    //执行
    env.execute("ExampleManagedState")
  }

  class CountWindowAverage
              extends RichFlatMapFunction[(Long, Long), (Long, Long)] {

    private var sum: ValueState[(Long, Long)] = _
```

```scala
    override def flatMap(input: (Long, Long), out: Collector[(Long, Long)]): Unit = {

      //访问状态值
      val tmpCurrentSum = sum.value

      //如果它以前没有被使用过,则它将是null
      val currentSum = if (tmpCurrentSum != null) {
        tmpCurrentSum
      } else {
        (0L, 0L)
      }

      //更新计数
      val newSum = (currentSum._1 + 1, currentSum._2 + input._2)

      //更新状态
      sum.update(newSum)

      //如果计数达到2,则发出平均值并清除状态
      if (newSum._1 >= 2) {
        out.collect((input._1, newSum._2 / newSum._1))
        sum.clear()
      }
    }

    override def open(parameters: Configuration): Unit = {
      sum = getRuntimeContext.getState(
        new ValueStateDescriptor[(Long, Long)]("average", createTypeInformation[(Long, Long)])
      )
    }
  }
}
```

Java 代码如下:

```java
//第3章/CountWindowAverageDemo.java

import org.apache.flink.api.common.functions.RichFlatMapFunction;
import org.apache.flink.api.common.state.ValueState;
import org.apache.flink.api.common.state.ValueStateDescriptor;
import org.apache.flink.api.common.typeinfo.TypeHint;
import org.apache.flink.api.common.typeinfo.TypeInformation;
import org.apache.flink.api.java.tuple.Tuple2;
import org.apache.flink.configuration.Configuration;
```

```java
import org.apache.flink.streaming.api.environment.StreamExecutionEnvironment;
import org.apache.flink.util.Collector;

/**
 * 实现一个简单计数窗口,统计每两个整数元素的平均值
 *
 * keyed state 示例
 */
public class CountWindowAverageDemo {

    public static void main(String[] args) throws Exception {
        //设置流执行环境
        final StreamExecutionEnvironment env =
            StreamExecutionEnvironment.getExecutionEnvironment();

        //流数据管道
        env
            .fromElements(Tuple2.of(1L, 3L),
                          Tuple2.of(1L, 5L),
                          Tuple2.of(1L, 7L),
                          Tuple2.of(1L, 4L),
                          Tuple2.of(1L, 2L))
            .keyBy(t -> t.f0)
            .flatMap(new CountWindowAverage())
            .print();                    //输出将是(1,4) 和(1,5)

        //触发流程序执行
        env.execute("Flink File Source");
    }

    public static class CountWindowAverage
         extends RichFlatMapFunction<Tuple2<Long,Long>,Tuple2<Long,Long>> {

        //存储状态的 ValueState
        private transient ValueState<Tuple2<Long, Long>> sum;

        @Override
        public void open(Configuration config) {
            ValueStateDescriptor<Tuple2<Long, Long>> descriptor =
              new ValueStateDescriptor<>("average", //状态名称
                TypeInformation.of(new TypeHint<Tuple2<Long,Long>>(){}) //类型信息
            );
            //状态初始化
            sum = getRuntimeContext().getState(descriptor);
        }
```

```java
@Override
public void flatMap(Tuple2<Long, Long> input,
        Collector<Tuple2<Long, Long>> out) throws Exception {

    //访问状态值
    Tuple2<Long, Long> currentSum = sum.value();

    //如果还没有状态值,则初始化一种状态值
    if(currentSum == null){
        currentSum = Tuple2.of(0L, 0L);
    }

    //更新计数
    currentSum.f0 += 1;

    //累加输入值的第2个字段
    currentSum.f1 += input.f1;

    //更新状态
    sum.update(currentSum);

    //如果计数达到2,则发出平均值并清除状态
    if (currentSum.f0 >= 2) {
        out.collect(new Tuple2<>(input.f0,currentSum.f1/currentSum.f0));
        sum.clear();
    }
}
```

这个示例实现了一个简单计数窗口。通过第1个字段输入元组(在本例中,所有的 key 值都是1)。一旦计数达到2,它就会发出平均值并清除状态,这样就可以从0开始。注意,如果元组在第1个字段中有不同的值,则将为每个不同的输入 key 保留不同的状态值。

1. 状态生命周期

对于任何类型 Keyed State 都可以设定状态的生命周期(Time-To-Live, TTL),即状态的存活时间,以此来配置 Keyed State 的有效期,以确保能够在规定时间内及时地清理状态数据。如果配置了状态的 TTL,则当某个 key 的状态数据过期时,存储的状态会被清除。如果状态是集合类型,则 TTL 是单独针对每个元素设置的,也就是说每个 List 元素或者 Map 的 entry 都有独立的 TTL。对于 TTL 的理解如图3-53所示。

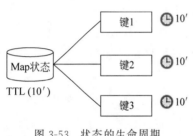

图3-53 状态的生命周期

为了使用状态 TTL，必须首先构建 StateTtlConfig 配置对象，然后，可以在任何状态描述符中通过传递该配置对象来启用 TTL 功能，这是通过将 StateTtlConfig 配置传入 StateDescriptor 中的 enableTimeToLive() 方法实现的。下面是启用和配置 TTL 的模板代码。

Scala 代码如下：

```scala
import org.apache.flink.api.common.state.StateTtlConfig
import org.apache.flink.api.common.state.ValueStateDescriptor
import org.apache.flink.api.common.time.Time

val ttlConfig = StateTtlConfig
  .newBuilder(Time.seconds(1))  //Time-To-Live Value,必需的
  .setUpdateType(StateTtlConfig.UpdateType.OnCreateAndWrite)    //或者 OnReadAndWrite
  .setStateVisibility(StateTtlConfig.StateVisibility.NeverReturnExpired) //未清理的过期
                                                                          //值也不返回
  .build

val stateDescriptor =
        new ValueStateDescriptor[String]("text state", classOf[String])
stateDescriptor.enableTimeToLive(ttlConfig)
```

Java 代码如下：

```java
import org.apache.flink.api.common.state.StateTtlConfig;
import org.apache.flink.api.common.state.ValueStateDescriptor;
import org.apache.flink.api.common.time.Time;

StateTtlConfig ttlConfig = StateTtlConfig
  .newBuilder(Time.seconds(1))
  .setUpdateType(StateTtlConfig.UpdateType.OnCreateAndWrite)
  .setStateVisibility(StateTtlConfig.StateVisibility.NeverReturnExpired)
  .build();

ValueStateDescriptor<String> stateDescriptor =
        new ValueStateDescriptor<>("text state", String.class);
stateDescriptor.enableTimeToLive(ttlConfig);
```

简单来讲就是在创建状态描述符时，添加 StateTtlConfig 配置。在创建 TTL 配置时，newBuilder() 方法的第 1 个参数是强制性的，它代表的是生存时间值。

可以进行以下配置，指定当状态 TTL 刷新时的更新类型。

（1）StateTtlConfig.UpdateType.OnCreateAndWrite：只在 key 的状态创建和写入时更新 TTL（默认）。

(2) StateTtlConfig.UpdateType.OnReadAndWrite：读取和写入状态时可以更新TTL。

如果要指定状态可见性配置是否在读取时返回过期值（如果它还没有被清除），则可指定如下返回类型。

(1) StateTtlConfig.StateVisibility.NeverReturnExpired：过期的值永远不会返回。

(2) StateTtlConfig.StateVisibility.ReturnExpiredIfNotCleanedUp：如果还可用，则返回。

默认为NeverReturnExpired。在NeverReturnExpired的情况下，过期状态的行为就像它不再存在一样，即使它仍然必须被删除。该选项对于数据在TTL之后无法进行严格的读访问的情况非常有用，例如使用隐私敏感数据的应用程序。另一个选项ReturnExpiredIfNotCleanedUp允许在清理之前返回过期状态。

注意：目前只支持与处理时间相关的TTL。

2. 清除过期状态

默认情况下，在读取时显式地删除过期值（如 ValueState#value）；如果配置的状态后端支持，则在后台定期回收过期值。后台清理可以在 StateTtlConfig 中禁用，代码模板如下。

Scala 代码如下：

```
import org.apache.flink.api.common.state.StateTtlConfig

val ttlConfig = StateTtlConfig
  .newBuilder(Time.seconds(1))
  .disableCleanupInBackground
  .build
```

Java 代码如下：

```
import org.apache.flink.api.common.state.StateTtlConfig;

StateTtlConfig ttlConfig = StateTtlConfig
  .newBuilder(Time.seconds(1))
  .disableCleanupInBackground()
  .build();
```

此外，可以在获取完整状态快照时激活清理，这将减小快照的大小。在当前实现下，本地状态不会被清除，但当从上一个快照恢复时，它将不包括已删除的过期状态。可以在StateTtlConfig中配置。模板代码如下。

Scala 代码如下：

```
import org.apache.flink.api.common.state.StateTtlConfig
import org.apache.flink.api.common.time.Time

val ttlConfig = StateTtlConfig
  .newBuilder(Time.seconds(1))
  .cleanupFullSnapshot
  .build
```

Java 代码如下:

```
import org.apache.flink.api.common.state.StateTtlConfig;
import org.apache.flink.api.common.time.Time;

StateTtlConfig ttlConfig = StateTtlConfig
  .newBuilder(Time.seconds(1))
  .cleanupFullSnapshot()
  .build();
```

注意:这个选项不适用于 RocksDB 状态后端的增量检查点。

3. 在 Scala DataStream API 中的状态

除了上面描述的接口,对于只有一个 ValueState 的 Keyed Stream 流,Scala API 还提供了有状态 map()或 flatMap()函数的简捷方式。用户函数可获取 Option 中 ValueState 的当前值,并且必须返回一个更新后的值,该值将用于更新状态。模板代码如下:

```
val stream: DataStream[(String, Int)] = ...

val counts: DataStream[(String, Int)] = stream
  .keyBy(_._1)
  .mapWithState((in: (String, Int), count: Option[Int]) =>
    count match {
      case Some(c) => ( (in._1, c), Some(c + in._2) )
      case None => ( (in._1, 0), Some(in._2) )
    })
```

3.9.4 使用托管 Operator State

要使用托管运算符状态,状态函数可以实现更一般的 CheckpointedFunction 接口,也可以实现 ListCheckpointed＜T extends Serializable＞接口。

1. CheckpointedFunction

CheckpointedFunction 接口通过不同的重分布方案提供对 non-keyed state 的访问。它

需要实现两种方法,这两种方法的声明如下:

```
void snapshotState(FunctionSnapshotContext context) throws Exception;

void initializeState(FunctionInitializationContext context) throws Exception;
```

无论何时执行检查点,都会调用 snapshotState()方法。对应的 initializeState()方法在每次初始化用户定义的函数时都会被调用,无论是在函数第 1 次初始化时,还是在函数从之前的检查点实际恢复时,因此,initializeState()方法不仅是初始化不同类型状态的地方,也是包含状态恢复逻辑的地方。

目前,支持列表样式的托管操作符状态。状态应该是一个可序列化对象的 List 列表,它们彼此独立,因此可以在重新调节(大小)时进行重新分布。换句话说,这些对象是可以重新分布 non-keyed state 的最细粒度。根据状态访问方法,定义了以下重分布方式。

(1) Even-split Redistribution:每个操作符返回一种状态元素列表。整种状态在逻辑上是所有列表的连接。在恢复/再分配中,只要有并行操作符,列表就被均匀地划分为尽可能多的子列表。每个操作符都有一个子列表,它可以是空的,也可以包含一个或多个元素。例如,如果在 parallelism 1 中,操作符的 checkpoint 状态包含元素 element1 和 element2,当将并行度增加到 2 时,element1 可能会出现在操作符实例 0 中,而 element2 将出现在操作符实例 1 中。

(2) Union Redistribution:每个操作符返回一种状态元素列表。整种状态在逻辑上是所有列表的连接。在恢复/重新分配时,每个操作符都获得状态元素的完整列表。

下面是一个有状态 SinkFunction 的例子,它在将事件元素发送到外部系统之前使用 CheckpointedFunction 来缓冲元素。它演示了基本的 Even-split Redistribution 列表状态。

Scala 代码如下:

```scala
//第 3 章/BufferingSink.scala

class BufferingSink(threshold: Int = 0) extends SinkFunction[(String, Int)]
    with CheckpointedFunction {

  @transient
  private var checkpointedState: ListState[(String, Int)] = _

  private val bufferedElements = ListBuffer[(String, Int)]()

  override def invoke(value: (String, Int), context: Context): Unit = {
    bufferedElements += value
    if (bufferedElements.size == threshold) {
      for (element <- bufferedElements) {
        //将元素发送到 sink
      }
```

```scala
      bufferedElements.clear()
    }
  }

  override def snapshotState(context: FunctionSnapshotContext): Unit = {
    checkpointedState.clear()
    for (element <- bufferedElements) {
      checkpointedState.add(element)
    }
  }

  override def initializeState(context: FunctionInitializationContext): Unit = {
    val descriptor = new ListStateDescriptor[(String, Int)](
      "buffered-elements",
      TypeInformation.of(new TypeHint[(String, Int)]() {})
    )

checkpointedState = context
    .getOperatorStateStore
    .getListState(descriptor)

    if(context.isRestored) {
      for(element <- checkpointedState.get()) {
        bufferedElements += element
      }
    }
  }

}
```

Java 代码如下:

```java
//第 3 章/BufferingSink.java

public class BufferingSink implements
          SinkFunction<Tuple2<String, Integer>>, CheckpointedFunction {

  private final int threshold;

  private transient ListState<Tuple2<String, Integer>> checkpointedState;

  private List<Tuple2<String, Integer>> bufferedElements;

  public BufferingSink(int threshold) {
      this.threshold = threshold;
```

```java
        this.bufferedElements = new ArrayList<>();
    }

    @Override
    public void invoke(Tuple2<String, Integer> value, Context contex) throws Exception {
        bufferedElements.add(value);
        if (bufferedElements.size() == threshold) {
            for (Tuple2<String, Integer> element: bufferedElements) {
                //将元素发送到sink
            }
            bufferedElements.clear();
        }
    }

    @Override
    public void snapshotState(FunctionSnapshotContext context) throws Exception {
        checkpointedState.clear();
        for (Tuple2<String, Integer> element : bufferedElements) {
            checkpointedState.add(element);
        }
    }

    @Override
    public void initializeState(FunctionInitializationContext context) throws Exception {
        ListStateDescriptor<Tuple2<String, Integer>> descriptor =
            new ListStateDescriptor<>(
                "buffered-elements",
                TypeInformation.of(new TypeHint<Tuple2<String, Integer>>() {}));

        checkpointedState = context
            .getOperatorStateStore()
            .getListState(descriptor);

        if (context.isRestored()) {
            for (Tuple2<String, Integer> element : checkpointedState.get()) {
                bufferedElements.add(element);
            }
        }
    }
}
```

在上面的代码中，initializeState()方法以FunctionInitializationContext作为参数。这用于初始化non-keyed state容器。这些是ListState类型的容器，non-keyed state对象将在检查点时存储在其中。

需要注意状态是如何被初始化的，类似于Keyed State，使用一个StateDescriptor状态描述符，其中包含状态名称和关于状态所持有的值的类型的信息。

Scala 代码如下：

```scala
val descriptor = new ListStateDescriptor[(String, Long)](
    "buffered-elements",
    TypeInformation.of(new TypeHint[(String, Long)]() {})
)

checkpointedState = context.getOperatorStateStore.getListState(descriptor)
```

Java 代码如下：

```java
ListStateDescriptor<Tuple2<String, Integer>> descriptor =
    new ListStateDescriptor<>(
        "buffered-elements",
        TypeInformation.of(new TypeHint<Tuple2<Long, Long>>() {}));

checkpointedState = context
    .getOperatorStateStore()
    .getListState(descriptor);
```

在初始化容器之后，使用上下文的 isRestore() 方法来检查在发生故障后是否正在恢复。如果这是真的，即正在恢复，则应用恢复逻辑。

如修改后的 BufferingSink 代码所示，在状态初始化期间恢复的这个 ListState 保存在一个类变量中，供 snapshotState() 将来使用。在那里，ListState 被清除了前一个检查点包含的所有对象，然后填充希望检查点的新对象。

作为补充说明，Keyed State 也可以在 initializeState() 方法中初始化。可以使用提供的 FunctionInitializationContext 来完成。

2．ListCheckpointed 接口

ListCheckpointed 接口是 CheckpointedFunction 的一个更受限的变体，它只支持恢复时带有 Even-split Redistribution 方案的列表样式的状态。它还需要实现以下两种方法：

```java
List<T> snapshotState(long checkpointId, long timestamp) throws Exception;

void restoreState(List<T> state) throws Exception;
```

在 snapshotState() 方法中，操作符应该向检查点返回一个对象列表，而 restoreState() 方法必须在恢复时处理这样的列表。如果状态不可重分区，则始终可以在 snapshotState() 方法中返回一个 Collections.singletonList(MY_STATE)。

3．有状态的源函数

与其他运算符相比，有状态的源需要更多的关注。为了对状态和输出集合进行原子性

更新（对于故障/恢复的精确一次性语义来讲,这是必需的),用户需要从源的上下文中获取一个锁。使用有状态源的代码模板如下所示。

Scala 代码如下：

```scala
//第3章/CounterSource.scala

class CounterSource
    extends RichParallelSourceFunction[Long]
    with ListCheckpointed[Long] {

  @volatile
  private var isRunning = true

  private var offset = 0L

  override def run(ctx: SourceFunction.SourceContext[Long]): Unit = {
    val lock = ctx.getCheckpointLock

    while (isRunning) {
      //输出及状态更新都是原子的
      lock.synchronized({
        ctx.collect(offset)

        offset += 1
      })
    }
  }

  override def cancel(): Unit = isRunning = false

  override def restoreState(state: util.List[Long]): Unit =
    for (s <- state) {
      offset = s
    }

  override def snapshotState(checkpointId: Long, timestamp: Long): util.List[Long] =
    Collections.singletonList(offset)

}
```

Java 代码如下：

```java
//第3章/CounterSource.java

public static class CounterSource
```

```java
        extends RichParallelSourceFunction<Long>
        implements ListCheckpointed<Long> {

    /** 用于精确一次性语义的当前偏移量 */
    private Long offset = 0L;

    /** 用于作业撤销的标志变量 */
    private volatile boolean isRunning = true;

    @Override
    public void run(SourceContext<Long> ctx) {
        final Object lock = ctx.getCheckpointLock();

        while (isRunning) {
            //输出及状态更新都是原子的
            synchronized (lock) {
                ctx.collect(offset);
                offset += 1;
            }
        }
    }

    @Override
    public void cancel() {
        isRunning = false;
    }

    @Override
    public List<Long> snapshotState(long checkpointId, long checkpointTimestamp) {
        return Collections.singletonList(offset);
    }

    @Override
    public void restoreState(List<Long> state) {
        for (Long s : state)
            offset = s;
    }
}
```

当一个检查点被 Flink 完全确认时，一些操作符可能需要这些信息来与外部世界进行通信。在这种情况下，可查看 org.apache.flink.runtime.state.CheckpointListener 接口。

3.9.5 广播状态

广播状态（Broadcast State）是 Operator State 的一种特殊类型。广播状态被引入以支持这样的应用场景：一个流的记录需要广播给所有下游任务，它们用于在所有子任务之间保持相同的状态，然后可以在处理第2个流的记录时访问此状态。举个例子，广播状态可以自然地出现，可以想象一个低吞吐量的流包含一组规则，用户想要对来自另一个流的所有元素进行评估。考虑到上述类型的应用场景，广播状态与其他操作符状态的区别在于：

（1）它有一个map格式。

（2）它只对以broadcasted流和non-broadcasted流作为输入的特定操作符可用。

（3）这样的操作符可以有多个名称不同的广播状态。

假设有一个具有不同颜色和形状的对象流，并且用户希望找到遵循特定模式（例如，一个矩形后面跟着一个三角形）的相同颜色的对象对。假设一组有趣的模式会随着时间的推移而发展变化。

在本例中，需要用到两个流，其中一个流将包含Item类型（带有Color和Shape属性）的元素，而另一个流将包含规则（模式）。

对于第1个Items流，需要将Color定为key，这将确保相同颜色的元素最终出现在相同的物理机上。实现代码如下：

```
//将Color设为key
KeyedStream<Item, Color> colorPartitionedStream = itemStream.keyBy(new KeySelector<Item, Color>(){...});
```

另一个包含规则的流应该被广播到所有下游任务，这些下游任务应该将它们存储在本地，以便可以根据所有传入的Items对它们进行评估。下面的代码片段将广播规则流，并使用提供的MapStateDescriptor创建存储规则的广播状态，代码如下：

```
//用于存储规则(字符串)名称和规则本身的映射描述符
MapStateDescriptor<String, Rule> ruleStateDescriptor = new
MapStateDescriptor<>(
                "RulesBroadcastState",
                BasicTypeInfo.STRING_TYPE_INFO,
                TypeInformation.of(new TypeHint<Rule>() {}));

//广播规则并创建广播状态
BroadcastStream<Rule> ruleBroadcastStream = ruleStream.broadcast(ruleStateDescriptor);
```

最后，为了针对来自Items流的输入元素评估规则，用户需要连接两个流，并指定匹配检测逻辑。这可以通过调用非广播流上的connect()将流（keyed或non-keyed）与BroadcastStream连接起来，该BroadcastStream作为connect()方法的参数。这将返回一个BroadcastConnectedStream，用户可以在其上调用process()方法，使用特殊类型的CoProcessFunction。

CoProcessFunction将包含用户的匹配逻辑，它的确切类型取决于非广播流的类型：

（1）如果它是keyed的，则这个函数就是一个KeyedBroadcastProcessFunction。

(2) 如果它是 non-keyed 的, 则该函数是 BroadcastProcessFunction。

鉴于这里的非广播流是 keyed 流, 以下片段包括上述调用:

```
DataStream < String > output = colorPartitionedStream
    .connect(ruleBroadcastStream)
    .process(
        //KeyedBroadcastProcessFunction 中的类型参数表示
        //1. Keyed Stream 的 key
        //2. 非广播端元素的类型
        //3. 广播端元素的类型
        //4. 结果的类型, 这里是一个字符串
        new KeyedBroadcastProcessFunction < Color, Item, Rule, String >() {
                //匹配逻辑代码
        }
    );
```

注意: 连接应该在非广播流上调用, 以 BroadcastStream 作为参数。

对于 BroadcastProcessFunction 和 KeyedBroadcastProcessFunction, 它们有两个要实现的处理方法, 说明如下。

(1) processBroadcastElement(): 负责处理广播流中的传入元素。

(2) processElement(): 用于非广播流。

BroadcastProcessFunction 和 KeyedBroadcastProcessFunction 的定义如下:

```
public abstract class BroadcastProcessFunction < IN1, IN2, OUT >
        extends BaseBroadcastProcessFunction {

    public abstract void processElement(IN1 value, ReadOnlyContext ctx, Collector < OUT > out)
throws Exception;

    public abstract void processBroadcastElement(IN2 value, Context ctx, Collector < OUT > out)
throws Exception;
}

public abstract class KeyedBroadcastProcessFunction < KS, IN1, IN2, OUT > {

    public abstract void processElement(IN1 value, ReadOnlyContext ctx, Collector < OUT > out)
throws Exception;

    public abstract void processBroadcastElement(IN2 value, Context ctx, Collector < OUT > out)
throws Exception;
```

```
    public void onTimer(long timestamp, OnTimerContext ctx, Collector<OUT> out) throws
Exception;
}
```

这两种方法的不同在于它们所提供的上下文环境。非广播方法包含一个 ReadOnlyContext 对象,而广播方法包含一个 Context 对象。

这两个上下文对象(以下列举中的 ctx)具备以下功能。

(1) 给出对广播状态的访问:例如,调用 ctx.getBroadcastState(MapStateDescriptor<K, V> stateDescriptor)。

(2) 允许查询元素的时间戳:例如,调用 ctx.timestamp()。

(3) 获取当前水印:例如,调用 ctx.currentWatermark()。

(4) 获取当前的处理时间:例如,调用 ctx.currentProcessingTime()。

(5) 将元素发送到边输出:例如,调用 ctx.output(OutputTag<X> outputTag, X value)。

在上下文对象的 getBroadcastState() 方法中的参数 stateDescriptor(表示状态描述符)应该与 broadcast(ruleStateDescriptor) 中的状态描述符是同一个。不同之处在于对广播状态的访问类型不同。广播方法对其具有读写访问权,而非广播方法具有只读访问权(名称)。

最后,由于 KeyedBroadcastProcessFunction 是在 Keyed Stream 流上操作的,所以它公开了 BroadcastProcessFunction 不可用的一些功能,包括:

(1) 在 processElement() 方法中的 ReadOnlyContext 参数允许访问 Flink 的底层计时器服务,该服务允许注册事件时间和/或处理时间计时器。当计时器触发时,onTimer() 方法被调用,带有 OnTimerContext 作为参数。OnTimerContext 公开与 ReadOnlyContext plus 相同的功能,以及能够询问触发的计时器是事件时间还是处理时间和查询与计时器关联的 key。

(2) 在 KeyedBroadcastProcessFunction 的 processBroadcastElement() 方法中的 Context 包含方法 applyToKeyedState(StateDescriptor<S, VS> stateDescriptor, KeyedStateFunction<KS, S> function)。这允许注册一个 KeyedStateFunction,应用到与所提供的状态描述符相关联的所有 key 的所有状态。

注意:只有在 KeyedBroadcastProcessFunction 的 processElement() 处才能注册计时器,而且只能在那里注册。这在 processBroadcastElement() 方法中是不可能的,因为没有与广播元素相关联的 key。

回到最初的例子,KeyedBroadcastProcessFunction 看起来可能像下面这样:

```
new KeyedBroadcastProcessFunction<Color, Item, Rule, String>() {

    //存储部分匹配,即第1个元素等待第2个元素
```

```java
//保留一个列表,因为可能有许多第 1 个元素在等待
private final MapStateDescriptor<String, List<Item>> mapStateDesc =
    new MapStateDescriptor<>(
        "items",
        BasicTypeInfo.STRING_TYPE_INFO,
        new ListTypeInfo<>(Item.class));

//与上面的 ruleStateDescriptor 相同
private final MapStateDescriptor<String, Rule> ruleStateDescriptor =
    new MapStateDescriptor<>(
        "RulesBroadcastState",
        BasicTypeInfo.STRING_TYPE_INFO,
        TypeInformation.of(new TypeHint<Rule>() {}));

@Override
public void processBroadcastElement(Rule value,
                                    Context ctx,
                                    Collector<String> out) throws Exception {
    ctx.getBroadcastState(ruleStateDescriptor).put(value.name, value);
}

@Override
public void processElement(Item value,
                           ReadOnlyContext ctx,
                           Collector<String> out) throws Exception {

    final MapState<String, List<Item>> state =
                    getRuntimeContext().getMapState(mapStateDesc);
    final Shape shape = value.getShape();

    for (Map.Entry<String, Rule> entry :
        ctx
          .getBroadcastState(ruleStateDescriptor)
          .immutableEntries()) {
            final String ruleName = entry.getKey();
            final Rule rule = entry.getValue();

            List<Item> stored = state.get(ruleName);
            if (stored == null) {
                stored = new ArrayList<>();
            }

            if (shape == rule.second && !stored.isEmpty()) {
                for (Item i : stored) {
                    out.collect("MATCH: " + i + " - " + value);
                }
```

```
            stored.clear();
        }
        
        //如果rule.first == rule.second,则没有else{}
        if (shape.equals(rule.first)) {
            stored.add(value);
        }
        
        if (stored.isEmpty()) {
            state.remove(ruleName);
        } else {
            state.put(ruleName, stored);
        }
    }
}
```

3.9.6 状态后端

在使用状态时,需要了解 Flink 的状态后端。

Flink 提供了不同的状态后端,用于指定状态存储的方式和位置。状态可以位于 Java 的堆上,也可以位于堆外。根据用户的状态后端,Flink 也可以管理应用程序的状态,这意味着 Flink 处理内存管理(如果需要,则可能会溢出到磁盘),以允许应用程序保存非常大的状态。默认情况下,配置文件 flink-conf.yaml 会确定所有 Flink 作业的状态后端。

1. 可用的状态后端

状态后端有两种实现,一种基于 RocksDB,它是一个嵌入式键/值存储,将其工作状态保存在磁盘上;另一种基于堆的状态后端,将其工作状态保存在 Java 堆的内存中。状态在内部是如何表示的,以及它在检查点上是如何及在哪里持久化的都取决于所选择的状态后端。

Apache Flink 打包了如下这些状态后端:

(1) HashMapStateBackend。

(2) EmbeddedRocksDBStateBackend。

如果没有其他配置,系统将使用 HashMapStateBackend 状态后端。这两种状态后端的特点见表 3-15。

表 3-15 不同状态后端的特点

名 称	工作状态	快照	特 点
EmbeddedRocksDBStateBackend	本地磁盘 (tmp 目录)	全量/增量	支持大于可用内存的状态 经验法则比基于堆的状态后端慢 10 倍
HashMapStateBackend	JVM 堆	全量	速度快,要求大内存 支持 GC

HashMapStateBackend 将数据作为 Java 堆上的对象在内部保存。键/值状态和窗口操作符维护存储值、触发器等的哈希表。当使用基于堆的状态后端保存的状态时，访问和更新涉及读和写堆上的对象。

HashMapStateBackend 适用于以下场景：

(1) 具有大状态、长窗口、大键/值状态的作业。

(2) 所有高可用性模式。

还建议将托管内存设置为 0。这将确保为 JVM 上的用户代码分配最大的内存量。

HashMapStateBackend 总执行异步快照，这意味着它们可以在不妨碍正在进行的流处理的情况下拍摄快照。

EmbeddedRocksDBStateBackend 在 RocksDB 数据库中保存动态数据，RocksDB 数据库默认存储在 TaskManager 本地数据目录中。与在 HashMapStateBackend 中存储 Java 对象不同，数据被存储为序列化的字节数组，主要由类型序列化器定义，导致键比较按字节进行，而不是使用 Java 的 hashCode() 和 equals() 方法。

EmbeddedRocksDBStateBackend 总执行异步快照，这意味着它们可以在不妨碍正在进行的流处理的情况下拍摄快照。

EmbeddedRocksDBStateBackend 的限制：由于 RocksDB 的 JNI 桥接 API 基于 Byte[]，每个键和每个值支持的最大大小为 2^31 字节（2GB）。在 RocksDB 中使用 merge 操作的状态（例如 ListState）可以默默地积累值大小大于 2^31 字节，然后在下一次检索时失败。这是 RocksDB JNI 目前的局限性。

EmbeddedRocksDBStateBackend 被鼓励用于以下场景：

(1) 具有大状态、长窗口、大键/值状态的作业。

(2) 所有高可用性模式。

注意，EmbeddedRocksDBStateBackend 可以保留的状态量仅受可用磁盘空间的限制。与将状态保存在内存中的 HashMapStateBacken 相比，这允许保持非常大的状态，但是对于在 EmbeddedRocksDBStateBackend 中保存的对象，访问和更新涉及序列化和反序列化，因此开销要大得多。还需要注意的是，只有 EmbeddedRocksDBStateBackend 能够进行增量快照，这对于具有大量缓慢变化状态的应用程序来讲是一个显著的好处。

2. 选择正确的状态后端

在 HashMapStateBackend 和 RocksDB 之间进行选择时，需要在性能和可扩展性之间做出选择。

HashMapStateBackend 非常快，因为每种状态访问和更新操作在 Java 堆上的对象上，但是，状态大小受集群内可用内存的限制。

RocksDB 可以根据可用磁盘空间进行扩展，并且是唯一支持增量快照的状态后端，然而，每种状态访问和更新都需要（反）序列化和可能从磁盘读取，这导致平均性能比内存状态后端慢一个数量级。

在 Flink 1.13 中，统一了 Flink 保存点（savepoint）的二进制格式。这意味着可以获取一个保存点,然后使用不同的状态后端从它恢复。所有的状态后端仅从版本1.13开始产生一种通用格式。

3. 配置状态后端

默认状态后端可以在 flink-conf.yaml 文件中配置,由键 state.backend 指定。值可以是 hashmap（HashMapStateBackend）、rocksdb（EmbeddedRocksDBStateBackend），或者实现状态后端工厂类 StateBackendFactory 的全限定类名,例如,用于 EmbeddedRocksDBStateBackend 的 org.apache.flink.contrib.streaming.state.EmbeddedRocksDBStateBackendFactory。默认的状态后端可以被代码中的配置覆盖。

配置项 state.checkpoints.dir 定义了所有后端写入检查点数据和元数据文件的目录。配置文件中的示例部分如下：

```
#用于存储操作符状态检查点的后端
state.backend: hashmap

#存储检查点的目录
state.checkpoints.dir: hdfs://localhost:8020/flink/checkpoints
```

可以通过 StreamExecutionEnvironment 为每个 job 设置状态后端。
Scala 代码如下：

```scala
val env = StreamExecutionEnvironment.getExecutionEnvironment()
env.setStateBackend(new HashMapStateBackend())
```

Java 代码如下：

```java
StreamExecutionEnvironment env = 
            StreamExecutionEnvironment.getExecutionEnvironment();
env.setStateBackend(new HashMapStateBackend());
```

如果想在集成开发工具 IDE 中使用 EmbeddedRocksDBStateBackend 或者在 Flink 工作中以编程方式配置它,则需要将以下依赖配置到 Flink 项目中：

```xml
<dependency>
    <groupId>org.apache.flink</groupId>
    <artifactId>flink-statebackend-rocksdb_2.12</artifactId>
    <version>1.13.2</version>
    <scope>provided</scope>
</dependency>
```

由于 RocksDB 是默认的 Flink 发行版的一部分,如果在作业中没有使用任何 RocksDB 代码,并在 flink-conf.yaml 文件中通过 state.backend 配置了状态后端及进一步的

checkpointing 和 RocksDB 特定的参数，就不需要这个依赖。

对于 RocksDB 状态后端，虽然官方鼓励在大状态下使用增量检查点，但需要手动启用这个功能，可通过在 flink-conf.yaml 文件中配置以下内容启用此功能：

```
state.backend.incremental: true
```

也可以直接在代码中配置它（覆盖 config 默认值），代码如下：

```
EmbeddedRocksDBStateBackend backend =
            new EmbeddedRocksDBStateBackend(true);
```

注意：一旦启用了增量检查点，在 Web UI 中显示的 Checkpointed Data Size 仅代表该检查点的增量检查点数据大小，而不是完整状态大小。

当选择 RocksDB State Backend 时，计时器也默认存储在 RocksDB 中，然而，在 RocksDB 中维护计时器可能会有一定的成本，因此 Flink 提供了将计时器存储在 JVM 堆上的选项，即使 RocksDB 用于存储其他状态。当计时器数量较少时，基于堆的计时器有更好的性能。

将配置项 state.backend.rocksdb.timer-service.factory 设为 heap（而不是默认的 rocksdb），以便将计时器存储在堆上。

注意：RocksDB 状态后端与基于堆的计时器的组合目前不支持计时器状态的异步快照。其他状态（如键态）仍然支持异步快照。

3.9.7 检查点机制

Flink 中的每个函数和操作符都可以是有状态的。有状态函数在处理单个元素/事件时存储数据，使状态成为任何类型更精细操作的关键构建块。

Flink 是一个分布式的流处理引擎，而流处理的其中一个特点就是 7×24 小时持续运行。那么，如何保障 Flink 作业的持续运行呢？Flink 的内部会将应用状态存储到本地内存或者嵌入式的 kv 数据库（RocksDB）中。由于采用的是分布式架构，为了使状态容错，Flink 需要对本地生成的状态进行持久化存储，以避免因应用或者节点机器故障等原因导致数据的丢失，Flink 是通过 checkpoint（检查点）的方式将状态定期和异步地写入远程的持久化存储，从而就可以实现不同语义的结果保障，如图 3-54 所示。

检查点是 Flink 实现容错的核心功能。Flink 定期在每个操作符中获取所有状态的持久快照，并将这些快照复制到更持久的地方，例如分布式文件系统。在出现故障时，Flink 可以重新启动应用程序，并从最新的检查点加载其状态，从而恢复应用程序的完整状态，并

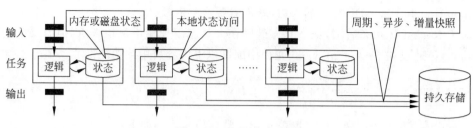

图 3-54 Flink 状态存储

继续处理,就像没有发生任何错误一样,从而确保了在发生故障时的精确一次状态一致性(exactly-once)。

1. 检查点存储位置

Flink 的检查点机制在计时器和有状态操作符(包括连接器、窗口和任何用户定义的状态)中存储所有状态的一致快照。存储这些快照的位置是通过作业检查点存储(Checkpoint Storage)定义的。检查点存储有两种实现方式,一种是将状态快照持久化到分布式文件系统,另一种使用 JobManager 的堆内存,这两种实现方式见表 3-16。

表 3-16 Flink 检查点存储实现

名 称	状 态 备 份	特 点
FileSystemCheckpointStorage	分布式文件系统	支持非常大的状态 高持久性 推荐用于生产部署
HashMapStateBackend	JobManagerJVM 堆	适合于小状态(本地)的测试和实验

默认情况下,检查点存储在 JobManager 的堆内存中。为了正确地持久化大状态,Flink 支持在其他位置检查点状态的各种方法。

可以通过 StreamExecutionEnvironment.getCheckpointConfig().setCheckpointStorage()配置选择检查点存储。强烈建议将检查点存储在用于生产部署的高可用文件系统中。

Flink 目前仅为没有迭代的作业提供处理保证。在迭代作业上启用检查点会导致异常。为了在迭代程序上强制检查点,用户需要在启用检查点时设置一个特殊的标志,代码如下:

```
env.enableCheckpointing(interval, CheckpointingMode.EXACTLY_ONCE, force = true);
```

2. 检查点的生成

以电商用户实时购买行为数据流为例,如图 3-55 所示。

图 3-55 中输入流是用户购买行为数据,包括购买(buy)和加入购物车(cart)两种,每种行为数据都有一个偏移量,用于统计每种行为的个数。在这个数据流中,触发检查点的步骤

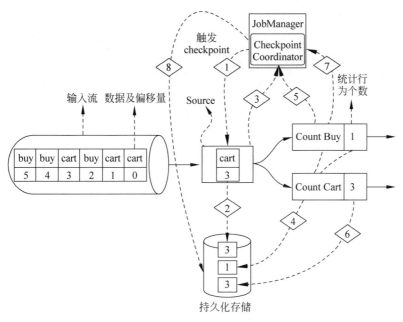

图 3-55　电商用户实时购买行为数据流

如下。

第 1 步：JobManager checkpoint coordinator 触发 checkpoint。

第 2 步：假设当消费到[cart，3]这条数据时，触发了 checkpoint。此时数据源会把消费的偏移量 3 写入持久化存储。

第 3 步：当写入结束后，source 会将 State Handle（状态存储路径）反馈给 JobManager 的 Checkpoint Coordinator。

第 4 步：接着算子 Count Buy 与 Count Cart 也会执行同样的操作步骤。

第 5 步：等所有的算子都完成了上述步骤之后，即当 Checkpoint Coordinator 收集齐所有任务的 State Handle，就认为这一次的 checkpoint 全局完成了，向持久化存储中再备份一个 Checkpoint Meta 元文件，那么整个 checkpoint 也就完成了，如果中间有一个不成功，则本次 checkpoint 就宣告失败。

3. 检查点的恢复

如果 Flink 程序出现故障，就可以从上一次 checkpoint 中进行状态恢复，从而提供容错保障。下面来看一下如何从检查点恢复。

当任务失败时，如图 3-56 所示。

这时 Flink 会重启作业，如图 3-57 所示。

并从最新的检查点加载其状态，从而恢复应用程序的完整状态，如图 3-58 所示。

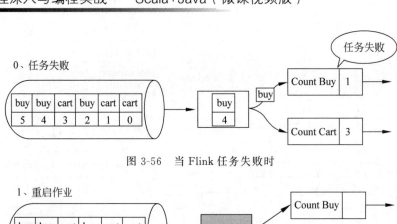

图 3-56　当 Flink 任务失败时

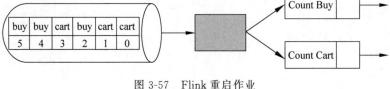

图 3-57　Flink 重启作业

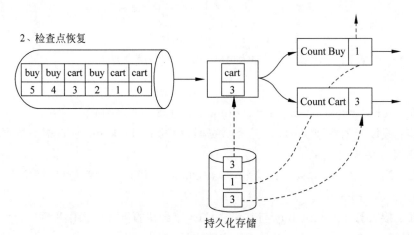

图 3-58　从检查点加载状态并恢复是

然后继续处理,就像没有发生任何错误一样,从而确保了在发生故障时的精确一次状态一致性,如图 3-59 所示。

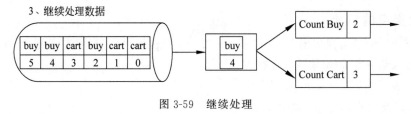

图 3-59　继续处理

上述过程可概括为

第 1 步:重启作业。

第 2 步：从上一次检查点恢复状态数据。

第 3 步：继续处理新的数据。

4．启用和配置检查点

默认情况下，检查点是禁用的。要启用检查点，在 StreamExecutionEnvironment 上调用 enableCheckpointing(n) 方法，其中 n 是检查点间隔时间（以毫秒为单位）。

检查点的其他参数包括：

（1）检查点存储：可以设置使检查点快照持久的位置。默认情况下，Flink 将使用 JobManager 的堆。对于生产部署，建议使用持久的文件系统。

（2）精确一次性和至少一次性：可以选择将模式传递给 enableCheckpointing(n) 方法，以在两个保证级别之间进行选择。对大多数应用程序来讲，精确一次性是更好的选择。至少一次可能与某些超低时延（持续几毫秒）应用程序相关。

（3）检查点超时：进程中的检查点中止的时间，在此之后，如果某个正在进行的检查点尚未完成，则中止该检查点。

（4）检查点之间的最短时间：为了确保流应用程序在检查点之间的进度，可以定义检查点之间需要多长时间。例如，如果将该值设置为 5000，则下一个检查点将在前一个检查点完成后不超过 5s 启动，而不考虑检查点持续时间和检查点间隔。注意，这意味着检查点间隔永远不会小于这个参数。并且这个值还意味着并发检查点的数量为 1。

（5）并发检查点数量：默认情况下，当一个检查点还在进行时，系统不会触发另一个检查点。这可以确保拓扑不会在检查点上花费太多时间，也不会在处理流方面取得进展。可以允许多个重叠的检查点，这对有一定处理时延（例如，因为函数调用外部服务，需要一些时间来回应）的管道很有意义，但是仍然需要频繁地检查点（100ms）来在失败时重新处理很少的数据。当定义了检检查点之间的最短时间时，不能使用此选项。

（6）外部化的检查点：可以将定期检查点配置为在外部持久化。外部化检查点将其元数据写到持久存储中，并且在作业失败时不会自动清除。这样，如果作业失败了，就会有一个检查点来恢复工作。

（7）检查点错误导致任务失败/继续：这将确定在执行任务的检查点的过程中发生错误时任务是否会失败。这是默认的行为。或者，当禁用此选项时，任务将简单地将检查点拒绝给检查点协调器，并继续运行。

（8）选择检查点进行恢复：这决定了一个作业是否会退回最新的检查点，以便使有更多的最近的保存点可用来潜在地减少恢复时间。

下面是启用和配置检查点的代码示例。

Scala 代码如下：

```
val env = StreamExecutionEnvironment.getExecutionEnvironment()

//每1000ms启动一个检查点
env.enableCheckpointing(1000)
```

```
//高级选项

//设置模式 exactly-once(这是默认的)
env.getCheckpointConfig.setCheckpointingMode(CheckpointingMode.EXACTLY_ONCE)

//确保在检查点之间有 500ms 的进度
env.getCheckpointConfig.setMinPauseBetweenCheckpoints(500)

//检查点必须在 1min 内完成,否则将被丢弃
env.getCheckpointConfig.setCheckpointTimeout(60000)

//防止任务失败,如果检查点发生错误,则检查点将被拒绝
env.getCheckpointConfig.setFailTasksOnCheckpointingErrors(false)

//只允许一个检查点在同一时间进行
env.getCheckpointConfig.setMaxConcurrentCheckpoints(1)

//启用实验性的未对齐检查点
env.getCheckpointConfig().enableUnalignedCheckpoints();

//设置将写入检查点快照的检查点存储
env.getCheckpointConfig().setCheckpointStorage("hdfs://checkpoint/dir")
```

Java 代码如下:

```
StreamExecutionEnvironment env =
        StreamExecutionEnvironment.getExecutionEnvironment();

//每 1000ms 启动一个检查点
//checkpoint 的时间间隔,如果状态比较大,则可以适当地调大该值
env.enableCheckpointing(1000);

//高级选项

//将模式设为 exactly-once(这是默认的模式)
env.getCheckpointConfig().setCheckpointingMode(CheckpointingMode.EXACTLY_ONCE);

//两个 checkpoint 之间的最小时间间隔,防止因 checkpoint 时间过长导致 checkpoint 积压
env.getCheckpointConfig().setMinPauseBetweenCheckpoints(500);    //确保在检查点之间有 500ms
                                                                  //的进度

//checkpoint 执行的上限时间,如果超过该阈值,则会中断 checkpoint
env.getCheckpointConfig().setCheckpointTimeout(60000);    //检查点必须在一分钟内完成,否则将
                                                          //被丢弃

//最大并行执行的检查点数量,默认为 1,可以指定多个,从而同时触发多个 checkpoint,提升效率
```

```
env.getCheckpointConfig().setMaxConcurrentCheckpoints(1); //同时只允许一个检查点在运行

//启用周期性外部检查点,将状态数据持久化到外部系统中.外部检查点在作业取消后会被保留
env.getCheckpointConfig().enableExternalizedCheckpoints(
ExternalizedCheckpointCleanup.RETAIN_ON_CANCELLATION);

//启用实验性的未对齐检查点
env.getCheckpointConfig().enableUnalignedCheckpoints();

//设置将写入检查点快照的检查点存储
env.getCheckpointConfig().setCheckpointStorage("hdfs://checkpoint/dir")
```

可以通过 conf/flink-confl.yaml 设置更多的参数和/或默认值,可参考官网文档。

3.9.8 状态快照

Flink 需要知道状态,以便使用检查点和保存点容错。

Flink 能够通过状态快照和流重放的组合提供容错的、精确一次的语义。这些快照用于捕获分布式管道的整个状态,将偏移量记录到输入队列中,以及在此之前由于摄入数据而导致的整个作业图的状态。当故障发生时,源被重绕(rewind),状态被还原(restore),处理被恢复(resume)。如上所述,这些状态快照是异步捕获的,不会妨碍正在进行的处理。

1. 状态快照工作原理

Flink 容错机制的核心部分是绘制分布式数据流和操作符状态的一致快照。这些快照充当一致的检查点,系统在出现故障时可以回退到这些检查点。

Flink 分布式快照的一个核心元素是 Stream Barriers。这些 barriers 被注入数据流中,并将记录作为数据流的一部分进行流处理。barriers 永远不会超过记录,它们严格地在一条线上流动。一个 barrier 将数据流中的记录分离为进入当前快照的记录集和进入下一个快照的记录集。每个 barriers 携带着被它把记录推到前面的快照的 ID。barriers 不会中断流的流动,因此非常轻量。来自不同快照的多个 barriers 可以同时在流中,这意味着各种快照可能同时发生。

Flink 使用了 Chandy-Lamport 算法的一种变体,称为 Asynchronous Barrier Snapshotting。

当检查点协调器(Checkpoint Coordinator,JobManager 的一部分)指示一个 TaskManager 开始 checkpoint 时,这个 TaskManager 会让所有源记录它们的偏移量,并将编号的 checkpoint barriers 插入它们的流中。这些 barriers 贯穿 Job Graph,指明每个 checkpoint 之前和之后的流部分。这个过程如图 3-60 所示。

Checkpoint n 将包含每个操作符(由于在 Checkpoint Barrier n 之前消费了所有事件,而在 Checkpoint Barrier n 之后没有消费任何事件)的状态。

当 Job Graph 中的每个操作符(算子)接收到其中一个 barrier 时,它记录其状态(到状

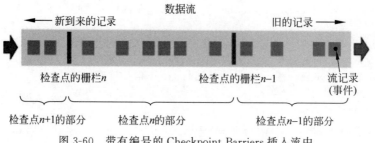

图 3-60 带有编号的 Checkpoint Barriers 插入流中

态后端)。具有两个输入流(如 CoProcessFunction)的操作符执行 barrier 对齐,以便快照反映从两个输入流(但不超过两个 barrier)消费事件所产生的状态,如图 3-61 所示。

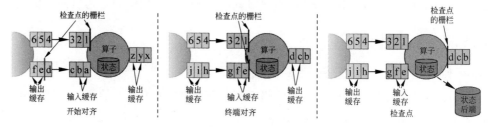

图 3-61 具有两个输入流的操作符执行 barrier 对齐

Flink 的状态后端使用 copy-on-write 机制,允许在异步快照旧版本状态时不受阻碍地继续流处理。只有当快照被持久地持久化时,这些旧版本的状态才会被作为垃圾回收,如图 3-62 所示。

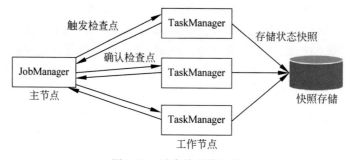

图 3-62 异常快照持久化

2. Flink 内部精确一次实现

Flink 提供了精确一次的处理语义。精确一次的处理语义可以理解为数据可能会重复计算,但是结果状态只有一个。

Flink 通过 Checkpoint 机制实现了精确一次的处理语义。Flink 在触发 checkpoint 时会向 Source 端插入 Checkpoint Barrier,Checkpoint Barriers 是从 Source 端插入的,并且会向下游算子进行传递。Checkpoint Barriers 携带一个 Checkpoint ID,用于标识属于哪一个

checkpoint,Checkpoint Barriers 将流逻辑分为两部分。对于双流的情况,通过 barrier 对齐的方式实现精确一次的处理语义。

关于什么是 Checkpoint Barrier,可以看一下 CheckpointBarrier 类的源码描述,源码如下:

```
/**
 * Checkpoint Barriers 用来在数据流中实现 checkpoint 对齐
 * Checkpoint Barrier 由 JobManager 的 Checkpoint Coordinator 插入 Source 中,
 * Source 会把 barrier 广播发送到下游算子,当一个算子接收到了其中一个输入流的 Checkpoint
 * Barrier 时,
 * 它就会知道已经处理完了本次 checkpoint 与上次 checkpoint 之间的数据。
 *
 * 一旦某个算子接收到了所有输入流的 Checkpoint Barrier,
 * 意味着该算子的已经处理完了截止当前 checkpoint 的数据,
 * 可以触发 checkpoint,并将 barrier 向下游传递
 *
 * 根据用户选择的处理语义,在 checkpoint 完成之前会缓存后一次 checkpoint 的数据,
 * 直到本次 checkpoint 完成(Exactly Once)
 *
 * Checkpoint Barrier 的 ID 是严格单调递增的
 *
 */
public class CheckpointBarrier extends RuntimeEvent {...}
```

可以看出,Checkpoint Barrier 的主要功能是实现 checkpoint 对齐,从而可以实现精确一次处理语义。

3. 完全一次保证

当流处理应用程序出现问题时,可能会丢失或重复结果。检查点模式定义了在出现故障时系统所保证的一致性。当检查点被激活时,数据流将被重放,从而重复处理过程中丢失的部分。Flink 支持以下两种检查点模式:

(1) 不丢失数据,但可能会有重复的结果(CheckpointingMode. AT_LEAST_ONCE,至少一次)。

(2) 既不会丢失数据,也不会有重复的结果(CheckpointingMode. EXACTLY_ONCE,完全一次性)。

注意,这里所讲的精确一次性,并不是说每个事件恰好处理一次,而是意味着每个事件将只影响一次 Flink 所管理的状态。

只有提供完全一次保证才需要 barrier 对齐。如果不需要完全一次保证,则可以通过配置 Flink 使用 CheckpointingMode. AT_LEAST_ONCE 来获得一些性能,这时会禁用 barrier 对齐,代码如下:

```
//设置模式
env.getCheckpointConfig.setCheckpointingMode(CheckpointingMode.AT_LEAST_ONCE)
```

4. 完全一次端到端

为了实现端到端的精确一次，以便来自源的每个事件都精确地影响sink一次，以下条件必须成立：

(1) 源必须是可重放的(replayable)。

(2) sink必须是事务性的(必须是事务性的或幂等的)。

3.10 侧输出流

除了DataStream操作产生的主流之外，还可以生成任意数量的附加侧输出(Side Output)结果流。

3.10.1 什么是侧输出流

很多场景下需要从一个Flink操作符输出一个以上的输出流，例如：

(1) 异常。

(2) 格式不正确的事件。

(3) 时延事件。

(4) 操作警报，例如到外部服务的超时连接。

除了错误报告之外，侧输出也是实现流的N路拆分的好方法。例如，在某个应用场景中，用户需要将从Kafka过来的告警和恢复数据进行分类拆分，然后将每种数据再分为告警数据和恢复数据，如图3-63所示。

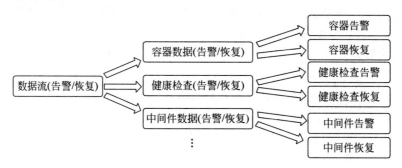

图3-63 数据流的多路拆分

每个侧输出通道都与一个OutputTag<T>关联，用于标记具有与侧输出的DataStream类型对应的泛型类型，并且具有名称。具有相同名称的两个OutputTag应该具有相同的类型，并将引用相同的侧输出。结果流中的数据类型不必与主流中的数据类型匹配，不同端输出的类型也可以不同。当希望分割一个数据流时，这个操作非常有用。

当使用侧输出时，首先需要定义一个 OutputTag，用于标识侧输出流。

Java 代码如下：

```
//这需要是一个匿名的内部类，以便可以分析类型
OutputTag<String> outputTag = new OutputTag<String>("side-output") {};
```

Scala 代码如下：

```
val outputTag = OutputTag[String]("side-output")
```

需要注意 OutputTag 是如何根据侧输出流所包含的元素类型进行类型化的。

可以通过以下函数将数据发送到侧输出：

(1) ProcessFunction。
(2) KeyedProcessFunction。
(3) CoProcessFunction。
(4) ProcessWindowFunction。
(5) ProcessAllWindowFunction。

可以使用上面函数中向用户公开的 Context 参数，将数据发送到 OutputTag 标识的侧输出。下面是一个从 ProcessFunction 发出侧输出数据的例子。

Scala 代码如下：

```
val input: DataStream[Int] = ...
val outputTag = OutputTag[String]("side-output")

val mainDataStream = input
  .process(new ProcessFunction[Int, Int] {
    override def processElement(
        value: Int,
        ctx: ProcessFunction[Int, Int]#Context,
        out: Collector[Int]): Unit = {
      //将数据发送到常规输出
      out.collect(value)

      //将数据发送到侧输出
      ctx.output(outputTag, "sideout-" + String.valueOf(value))
    }
  })
```

Java 代码如下：

```
DataStream<Integer> input = ...;

final OutputTag<String> outputTag = new OutputTag<String>("side-output"){};
```

```java
SingleOutputStreamOperator<Integer> mainDataStream = input
    .process(new ProcessFunction<Integer, Integer>() {

        @Override
        public void processElement(
            Integer value,
            Context ctx,
            Collector<Integer> out) throws Exception {
            //将数据发送到常规输出
            out.collect(value);

            //将数据发送到侧输出
            ctx.output(outputTag, "sideout-" + String.valueOf(value));
        }
    });
```

要获得侧输出流，可以在 DataStream 操作的结果流上调用 getSideOutput(OutputTag)方法，这将得到一个 DataStream，它是输入侧输出流的结果。参看下面的模板代码。

Scala 代码如下：

```scala
val outputTag = OutputTag[String]("side-output")

val mainDataStream = ...

val sideOutputStream: DataStream[String] = mainDataStream.getSideOutput(outputTag)
```

Java 代码如下：

```java
final OutputTag<String> outputTag = new OutputTag<String>("side-output"){};

SingleOutputStreamOperator<Integer> mainDataStream = ...;

DataStream<String> sideOutputStream = mainDataStream.getSideOutput(outputTag);
```

3.10.2 侧输出流应用示例

下面通过两个示例来理解侧输出流的应用方法。

【示例 3-30】 在这个示例中，扩展了经典的单词计数程序，使其只计算至少 5 个字母长的单词，并将较短的单词发送到一个侧输出。

Scala 代码如下：

```scala
//第 3 章/SideOutputDemo1.scala

import org.apache.flink.util.Collector
import org.apache.flink.streaming.api.functions.ProcessFunction
import org.apache.flink.streaming.api.scala._

/**
 * 侧输出示例:计算至少 5 个字母长的单词,并将较短的单词发送到一个侧输出
 */
object SideOutputDemo1 {

  def main(args: Array[String]): Unit = {
    //设置流执行环境
    val env = StreamExecutionEnvironment.getExecutionEnvironment

    //定义一个侧输出的标签
    val shortWordsTag:OutputTag[String] = new OutputTag[String]("short") {}

    val tokenized = env
      .fromElements("Born as the bright summer flowers",
        "Died as the quiet beauty of autumn leaves")
      .process(new ProcessFunction[String, (String, Int)] {
        override def processElement(value: String,
                ctx: ProcessFunction[String, (String, Int)]#Context,
                out: Collector[(String,Int)]): Unit = {
          //规范化并分隔每行
          val tokens = value.toLowerCase.split("\\W+")

          //分流输出
          for (token <- tokens) {
            if (token.length < 5) {        //将短的词发送到一个侧输出
              ctx.output(shortWordsTag, token)
            }
            else if (token.length > 0) {    //发出 key - value 对
              out.collect((token, 1))
            }
          }
        }
      })

    //输出侧输出流(长度小于 5 的词)
    val shortWords = tokenized.getSideOutput[String](shortWordsTag)
    shortWords.print("short word")
```

```scala
        //对主输出流执行单词计数运算,并输出结果
        val wordCounts = tokenized.keyBy(t => t._1).sum(1)
        wordCounts.print("long word")

        //执行
        env.execute("flink Side Output")
    }
}
```

Java 代码如下:

```java
//第3章/SideOutputDemo1.java
import org.apache.flink.api.java.tuple.Tuple2;
import org.apache.flink.streaming.api.datastream.DataStream;
import org.apache.flink.streaming.api.datastream.SingleOutputStreamOperator;
import org.apache.flink.streaming.api.environment.StreamExecutionEnvironment;
import org.apache.flink.streaming.api.functions.ProcessFunction;
import org.apache.flink.util.Collector;
import org.apache.flink.util.OutputTag;

/**
 * 侧输出示例:只计算至少5个字母长的单词,并将较短的单词发送到一个侧输出
 */
public class SideOutputDemo1 {
    public static void main(String[] args) throws Exception {
        //设置流执行环境
        final StreamExecutionEnvironment env =
                StreamExecutionEnvironment.getExecutionEnvironment();

        //定义一个侧输出的标签
        final OutputTag<String> shortWordsTag =
                new OutputTag<String>("short") {};

        //分割数据流,返回主输出流
        SingleOutputStreamOperator<Tuple2<String, Integer>> tokenized = env
            .fromElements("Born as the bright summer flowers",
                    "Died as the quiet beauty of autumn leaves")
            .process(new ProcessFunction<String, Tuple2<String, Integer>>() {
                @Override
                public void processElement(String value,
                    Context ctx,
                    Collector<Tuple2<String, Integer>> out) throws Exception {
                    //规范化并分隔每行
                    String[] tokens = value.toLowerCase().split("\\W+");
```

```
            for (String token : tokens) {
                if (token.length() < 5) {
                    //将短的词发送到一个侧输出
                    ctx.output(shortWordsTag, token);
                } else if (token.length() > 0) {
                    //发出 key-value 对
                    out.collect(new Tuple2<>(token, 1));
                }
            }
        }
    });

    //输出侧输出流(长度小于 5 的词)
    DataStream<String> shortWords =
                    tokenized.getSideOutput(shortWordsTag);
    shortWords.print("short word");

    //对主输出流执行单词计数运算,并输出结果
    DataStream<Tuple2<String, Integer>> wordCounts = tokenized
        .keyBy(t -> t.f0)
        .sum(1);
    wordCounts.print("long word");

    //执行
    env.execute("flink Side Output");
    }
}
```

在上面的代码中,使用传递给 processElement()方法的 Context 来写入侧输出,使用 OutputTag 标签指定要写入哪个侧输出。在本例中,短单词被发送到侧输出进行收集,而剩余的单词作为元组(以典型的单词计数样式)随主收集器一起发送。

执行上面的代码,输出结果如下:

```
short word:8 > born
short word:1 > died
short word:1 > as
short word:1 > the
short word:1 > of
short word:8 > as
short word:8 > the
long word:1 > (beauty,1)
long word:8 > (autumn,1)
long word:6 > (quiet,1)
long word:3 > (leaves,1)
```

```
long word:5 > (summer,1)
long word:7 > (bright,1)
long word:5 > (flowers,1)
```

【示例3-31】 下面是一个从ProcessFunction发出侧输出数据的例子,将数据集中的负数挑选出来,输出到侧输出中。

Scala代码如下:

```scala
//第3章/SideOutputDemo2.scala

import org.apache.flink.streaming.api.functions.ProcessFunction
import org.apache.flink.streaming.api.scala._
import org.apache.flink.util.Collector

/**
 * 侧输出示例:将数据集中的负数挑选出来,输出到侧输出中
 */
object SideOutputDemo2 {

  def main(args: Array[String]): Unit = {
    //设置流执行环境
    val env = StreamExecutionEnvironment.getExecutionEnvironment

    //侧输出
    val numbers = List(1, 2, -3, 4, 5, -6, 7, 8, -9, 10)
    val ds = env.fromElements(numbers:_*)

    //定义用来标记侧输出的标签,注意泛型参数是侧输出数据的类型
    val outputTag = new OutputTag[Int]("side-output"){}

    //分流处理
    val mainDataStream = ds.process(new ProcessFunction[Int, Int] {
      override def processElement(value: Int,
            ctx: ProcessFunction[Int, Int]#Context,
            out: Collector[Int]): Unit = {
        if (value > 0) out.collect(value)         //将数据发送到常规输出
        else ctx.output(outputTag, value)          //向侧输出发送负数
      }
    })

    //获取侧输出结果
    val sideOutputStream = mainDataStream.getSideOutput[Int](outputTag)

    //打印主数据流
```

```
        mainDataStream.print("主数据流")

        //打印侧输出流
        sideOutputStream.print("侧输出流")

        //执行
        env.execute("flink transformatiion")
    }
}
```

Java 代码如下:

```
//第 3 章/SideOutputDemo2.java

import org.apache.flink.streaming.api.datastream.DataStream;
import org.apache.flink.streaming.api.datastream.SingleOutputStreamOperator;
import org.apache.flink.streaming.api.environment.StreamExecutionEnvironment;
import org.apache.flink.streaming.api.functions.ProcessFunction;
import org.apache.flink.util.Collector;
import org.apache.flink.util.OutputTag;

/**
 * 侧输出示例:将数据集中的负数挑选出来,输出到侧输出中
 */
public class SideOutputDemo2 {
    public static void main(String[] args) throws Exception {
        //设置流执行环境
        final StreamExecutionEnvironment env =
                StreamExecutionEnvironment.getExecutionEnvironment();

        //侧输出
        DataStream< Integer > ds = env.fromElements(1,2, -3,4,5, -6,7,8, -9,10);

        //定义用来标记侧输出的标签,注意泛型参数是侧输出数据的类型
        final OutputTag< Integer > outputTag =
                new OutputTag< Integer >("side - output"){};

        //分流处理
        SingleOutputStreamOperator< Integer > mainDataStream = ds
            .process(new ProcessFunction< Integer, Integer >() {
                @Override
                public void processElement(Integer value, Context ctx, Collector< Integer > out)
throws Exception {
                    if(value > 0){
                        out.collect(value);              //将数据发送到常规输出
```

```
                    }else{
                        ctx.output(outputTag, value);        //向侧输出发送负数
                    }
                }
            });

//获取侧输出结果
DataStream < Integer > sideOutputStream =
    mainDataStream.getSideOutput(outputTag);

//打印主输出流
//mainDataStream.print("主数据流");

//打印侧输出流
sideOutputStream.print("侧输出流");

//执行
env.execute("flink transformatiion");
    }
}
```

执行以上程序,输出结果如下:

```
侧输出流:8 > -3
侧输出流:3 > -6
侧输出流:6 > -9
```

3.11　Flink 流连接器

从之前学习的内容中,我们了解到 Flink 内置了一些基本的数据源和接收器。预定义的数据源包括从文件、目录和 Socket 套接字中读取数据,以及从集合和迭代器中获取数据。预定义的数据接收器支持向文件、stdout 和 stderr 及 Socket 套接字写入数据。

除了上述内置 API 外,Flink 中还支持(捆绑)了系列的 Connectors 连接器。连接器提供了用于与各种第三方系统交互的代码,用于将计算结果输入常用的存储系统或者消息中间件中。Flink 目前捆绑了如下连接器:

(1) Apache Kafka (source/sink)。

(2) Apache Cassandra (sink)。

(3) Amazon Kinesis Streams (source/sink)。

(4) Elasticsearch (sink)。

(5) FileSystem (Hadoop included)-Streaming only (sink)。

(6) FileSystem (Hadoop included)-Streaming and Batch (sink)。
(7) RabbitMQ (source/sink)。
(8) Apache NiFi (source/sink)。
(9) Twitter Streaming API (source)。
(10) Google PubSub (source/sink)。
(11) JDBC (sink)。

注意,虽然这里列出的流连接器是Flink项目的一部分,并且包含在源版本中,但是它们没有包含在二进制发行版中,因此需要单独下载,或者在Maven项目中添加相应的依赖。

Flink提供了addSink()方法,用来调用自定义的Sink或者第三方的连接器。

3.11.1 Kafka连接器

Kafka是一个发布-订阅、分布式、消息队列系统,允许用户将消息发布到某个主题,然后将其分发给主题的订阅者。

Flink为读写Kafka主题提供了专门的Kafka连接器,支持将Kafka作为Flink流应用程序的数据源和Data Sink。Flink Kafka消费者集成了Flink的检查点机制来提供精确一次的处理语义。为了实现这一点,Flink并不完全依赖Kafka的消费者组的偏移跟踪,而是在内部跟踪和检查这些偏移。

Flink Kafka连接器的工作原理如图3-64所示。

图3-64 Flink Kafka连接器的工作原理

因为Flink Kafka流连接器目前不是二进制发行版的一部分,所以需要使用一个特定的JAR文件。通常在项目中使用以下Maven依赖项来加载和管理连接器JAR包。在Maven项目中导入指定版本的Kafka连接器,需要在pom.xml文件中添加以下依赖项:

```
<dependency>
    <groupId>org.apache.flink</groupId>
    <artifactId>flink-connector-kafka_2.12</artifactId>
    <version>1.13.2</version>
    <scope>compile</scope>
</dependency>
```

注意：用模板生成 Maven 项目，由于已经自带了 Kafka 连接器的配置，只是被注释掉了，所以取消注释就可以了。

如果使用的是 Kafka 源，则 flink-connector-base 也需要作为依赖项，内容如下：

```xml
<dependency>
    <groupId>org.apache.flink</groupId>
    <artifactId>flink-connector-base</artifactId>
    <version>1.13.2</version>
</dependency>
```

1. Kafka 源

Kafka 源提供了一个构建类来构造 KafkaSource 的实例。下面的代码片段展示了如何构建一个 KafkaSource 来消费主题 input-topic 的最早偏移量的消息，带有消费组 my-group，并且只将 message 的值反序列化为字符串，代码如下：

```java
KafkaSource<String> source = KafkaSource.<String>builder()
    .setBootstrapServers(brokers)
    .setTopics("input-topic")
    .setGroupId("my-group")
    .setStartingOffsets(OffsetsInitializer.earliest())
    .setValueOnlyDeserializer(new SimpleStringSchema())
    .build();

env.fromSource(source, WatermarkStrategy.noWatermarks(), "Kafka Source");
```

以下属性是构建 KafkaSource 所必需的：

(1) Bootstrap 服务器，由 setBootstrapServers(String) 配置。
(2) 用户组 ID，由 setGroupId(String) 配置。
(3) 要订阅的主题/分区。
(4) 解析 Kafka 消息的反序列化器。

Kafka 源提供了 3 种主题分区订阅方式，分别为：
(1) 主题列表，从主题列表中的所有分区订阅消息，代码如下：

```java
KafkaSource.builder().setTopics("topic-a", "topic-b")
```

(2) 主题模式，从名称与提供的正则表达式匹配的所有主题订阅消息，代码如下：

```java
KafkaSource.builder().setTopicPattern("topic.*")
```

(3) 分区集,订阅提供的分区集中的分区,代码如下:

```
final HashSet<TopicPartition> partitionSet = new HashSet<>(Arrays.asList(
    new TopicPartition("topic-a", 0),          //"topic-a"的分区 0
    new TopicPartition("topic-b", 5)));        //"topic-b"的分区 5
KafkaSource.builder().setPartitions(partitionSet)
```

解析:Kafka 消息需要提供一个反序列化器(反序列化模式)。反序列化器可以通过 setDeserializer()方法进行配置,该方法接收一个参数 KafkaRecordDeserializationSchema, 它定义了如何反序列化一个 Kafka 的 ConsumerRecord。如果只需 Kafka ConsumerRecord 的值,则可以在构建器中使用 setValueOnlyDeserializer(DeserializationSchema)方法,其中 DeserializationSchema 定义了如何反序列化 Kafka 消息值的二进制文件。

也可以使用 Kafka Deserializer 来反序列化 Kafka 消息值,例如,使用 StringDeserializer 将 Kafka 消息值反序列化为 String 字符串,代码如下:

```
import org.apache.kafka.common.serialization.StringDeserializer;

KafkaSource.<String>builder()
    .setDeserializer(KafkaRecordDeserializationSchema.valueOnly(StringSerializer.class));
```

Kafka 源可以通过指定 OffsetsInitializer 来消费从不同偏移量开始的消息。内置的 OffsetsInitializer 设定项包括多种偏移策略,使用代码如下:

```
KafkaSource.builder()
    //从消费组的提交偏移量开始,没有重置策略
    .setStartingOffsets(OffsetsInitializer.committedOffsets())
    //从提交的偏移量开始,如果提交的偏移量不存在,则可使用 EARLIEST 作为重置策略
    .setStartingOffsets(OffsetsInitializer.committedOffsets(OffsetResetStrategy.EARLIEST))
    //从第 1 个时间戳大于或等于指定时间戳的记录开始
    .setStartingOffsets(OffsetsInitializer.timestamp(1592323200L))
    //从最早偏移点开始
    .setStartingOffsets(OffsetsInitializer.earliest())
    //从最近的偏移量开始
    .setStartingOffsets(OffsetsInitializer.latest())
```

如果未指定偏移初始化器,则默认使用 OffsetsInitializer.earliest()。如果上面的内置初始化器不能满足需求,则可以实现自定义偏移量初始化器。

默认情况下,记录将使用 Kafka ConsumerRecord 中嵌入的时间戳作为事件时间。可以定义自己的水印策略来从记录本身提取事件时间,并向下游发送水印,代码如下:

```
env.fromSource(kafkaSource, new CustomWatermarkStrategy(), "带有自定义水印策略的 Kafka 源")
```

下面通过一个示例演示如何编写一个 Flink 流程序作为 Kafka 的消费者程序,将 Kafka 作为数据源。

【示例 3-32】 编写一个 Flink 流应用程序,消费 Kafka 中指定主题 words 的数据。使用 Kafka 自带的生产者脚本向 words 主题写数据。

建议按以下步骤操作:

(1) 在 IntelliJ IDEA 中创建一个 Flink 项目,使用 flink-quickstart-scala 或 flink-quickstart-Java 项目模板(Flink 项目的创建过程可参考 2.2 节)。

(2) 设置依赖。在 pom.xml 文件中添加如下依赖:

```xml
<dependency>
    <groupId>org.apache.flink</groupId>
    <artifactId>flink-scala_2.12</artifactId>
    <version>1.13.2</version>
    <scope>provided</scope>
</dependency>
<dependency>
    <groupId>org.apache.flink</groupId>
    <artifactId>flink-streaming-scala_2.12</artifactId>
    <version>1.13.2</version>
    <scope>provided</scope>
</dependency>
<dependency>
    <groupId>org.apache.flink</groupId>
    <artifactId>flink-Java</artifactId>
    <version>1.13.2</version>
    <scope>provided</scope>
</dependency>
<dependency>
    <groupId>org.apache.flink</groupId>
    <artifactId>flink-streaming-Java_2.12</artifactId>
    <version>1.13.2</version>
    <scope>provided</scope>
</dependency>
<dependency>
    <groupId>org.apache.flink</groupId>
    <artifactId>flink-connector-kafka_2.12</artifactId>
    <version>1.13.2</version>
    <scope>compile</scope>
</dependency>
```

(3) 创建流应用程序主类。

Scala 代码如下:

```scala
//第3章/KafkaSourceDemo.scala
import org.apache.flink.api.common.eventtime.WatermarkStrategy
import org.apache.flink.api.common.functions.FlatMapFunction
import org.apache.flink.api.common.serialization.SimpleStringSchema
import org.apache.flink.connector.kafka.source.KafkaSource
import org.apache.flink.connector.kafka.source.enumerator.initializer.OffsetsInitializer
import org.apache.flink.streaming.api.scala._
import org.apache.flink.util.Collector
import org.apache.flink.api.scala._

object KafkaSourceDemo {

  def main(args: Array[String]) {
    //设置流执行环境
    val env = StreamExecutionEnvironment.getExecutionEnvironment

    val source = KafkaSource.builder[String]
      .setBootstrapServers("localhost:9092")
      .setTopics("words")
      .setGroupId("group-test")
      .setStartingOffsets(OffsetsInitializer.earliest)
      .setValueOnlyDeserializer(new SimpleStringSchema)
      .build

    env
      //指定 Kafka 数据源
      .fromSource(source, WatermarkStrategy.noWatermarks[String], "Kafka Source")
      .flatMap(new FlatMapFunction[String, String]() {
        @throws[Exception]
        override def flatMap(s:String,collector:Collector[String]): Unit = {
          for (word <- s.split("\\W+")) {
            collector.collect(word)
          }
        }
      })
      //输出从 Kafka Words Topic 拉取的数据
      .print

    //触发流程序执行
    env.execute("Flink Kafka Source")
  }
}
```

Java 代码如下：

```java
//第3章/KafkaSourceDemo.java

import org.apache.flink.api.common.eventtime.WatermarkStrategy;
import org.apache.flink.api.common.functions.FlatMapFunction;
import org.apache.flink.api.common.serialization.SimpleStringSchema;
import org.apache.flink.connector.kafka.source.KafkaSource;
import org.apache.flink.connector.kafka.source.enumerator.initializer.OffsetsInitializer;
import org.apache.flink.streaming.api.environment.StreamExecutionEnvironment;
import org.apache.flink.util.Collector;

public class KafkaSourceDemo {

    public static void main(String[] args) throws Exception {
        //设置流执行环境
        final StreamExecutionEnvironment env =
                StreamExecutionEnvironment.getExecutionEnvironment();

        KafkaSource<String> source = KafkaSource.<String>builder()
            .setBootstrapServers("localhost:9092")
            .setTopics("words")
            .setGroupId("group-test")
            .setStartingOffsets(OffsetsInitializer.earliest())
            .setValueOnlyDeserializer(new SimpleStringSchema())
            .build();

        env
            //指定Kafka数据源
            .fromSource(source, WatermarkStrategy.noWatermarks(), "Kafka Source")
            //进行flatMap转换
            .flatMap(new FlatMapFunction<String, String>() {
                @Override
                public void flatMap(String s, Collector<String> collector) throws Exception {
                    for(String word : s.split("\\W+")){
                        collector.collect(word);
                    }
                }
            })
            //输出从Kafka Words Topic拉取的数据
            .print();

        //触发流程序执行
        env.execute("Flink Kafka Source");
    }
}
```

(4) 执行程序。建议按以下步骤执行：

① 启动 ZooKeeper 服务和 Kafka 服务。打开一个终端窗口，启动 ZooKeeper（不要关闭），命令如下：

```
$ ./bin/zookeeper-server-start.sh config/zookeeper.properties
```

② 打开另一个终端窗口，启动 Kafka 服务（不要关闭），命令如下：

```
$ ./bin/kafka-server-start.sh config/server.properties
```

③ 打开第 3 个终端窗口，在 Kafka 中创建一个名为 words 的主题（topic），命令如下：

```
$ ./bin/kafka-topics.sh --create --Bootstrap-server localhost:9092 --replication-factor 1 --partitions 1 --topic words
```

④ 查看已经创建的 Topic，命令如下：

```
$ ./bin/kafka-topics.sh --list --Bootstrap-server localhost:9092
$ ./bin/kafka-topics.sh --describe --Bootstrap-server localhost:9092 --topic words
$ ./bin/kafka-topics.sh --delete --Bootstrap-server localhost:9092 --topic words
# 删除 topic
```

⑤ 运行上面编写的流执行程序（相当于 Kafka 的消费者程序）。
⑥ 在第 3 个终端窗口，执行 Kafka 自带的生产者脚本，命令如下：

```
$ ./bin/kafka-console-producer.sh --broker-list localhost:9092 --topic words
```

⑦ 随意输入一些句子，单词之间以空格分隔开。例如，输入以下内容：

```
good good study
day day up
```

⑧ 可以得到输出结果如下：

```
8> good
8> good
8> study
8> day
8> day
8> up
```

2. Kafka 生产者

Flink 的 Kafka 生产者被称为 FlinkKafkaProducer，通过它可以将记录流写入一个或多个 Kafka 主题。FlinkKafkaProducer 的构造函数接受以下参数：

(1) 一个默认的输出主题，事件应该在其中写入。

(2) SerializationSchema/KafkaSerializationSchema 用于将数据序列化到 Kafka 中。

(3) Kafka 客户端的属性。以下属性是必需的：Bootstrap.servers（用逗号分隔的 Kafka Broker 列表）。

(4) 一个容错语义。

创建 FlinkKafkaProducer 的模板。

Scala 代码如下：

```scala
val stream: DataStream[String] = ...

val properties = new Properties
properties.setProperty("Bootstrap.servers", "localhost:9092")

val myProducer = new FlinkKafkaProducer[String](
    "my-topic",                                    //目标 topic
    new SimpleStringSchema(),                      //序列化模式
    properties,                                    //producer 配置
    FlinkKafkaProducer.Semantic.EXACTLY_ONCE)      //容错性

stream.addSink(myProducer)
```

Java 代码如下：

```java
DataStream<String> stream = ...

Properties properties = new Properties();
properties.setProperty("Bootstrap.servers", "localhost:9092");

FlinkKafkaProducer<String> myProducer = new FlinkKafkaProducer<>(
    "my-topic",                                    //目标 topic
    new SimpleStringSchema(),                      //序列化 schema
    properties,                                    //producer 配置
    FlinkKafkaProducer.Semantic.EXACTLY_ONCE);     //容错性

stream.addSink(myProducer);
```

【示例 3-33】 编写一个 Flink 流应用程序，将数据写入 Kafka 中的指定主题 words。使用 Kafka 自带的消费者脚本来查看写入的内容。

建议按以下步骤操作：

(1) 在 IntelliJ IDEA 中创建一个 Flink 项目，使用 flink-quickstart-scala 或 flink-quickstart-Java 项目模板（Flink 项目的创建过程可参考 2.2 节）。

(2) 设置依赖。在 pom.xml 文件中添加如下依赖：

```xml
<dependency>
    <groupId>org.apache.flink</groupId>
    <artifactId>flink-scala_2.12</artifactId>
    <version>1.13.2</version>
    <scope>provided</scope>
</dependency>
<dependency>
    <groupId>org.apache.flink</groupId>
    <artifactId>flink-streaming-scala_2.12</artifactId>
    <version>1.13.2</version>
    <scope>provided</scope>
</dependency>
<dependency>
    <groupId>org.apache.flink</groupId>
    <artifactId>flink-Java</artifactId>
    <version>1.13.2</version>
    <scope>provided</scope>
</dependency>
<dependency>
    <groupId>org.apache.flink</groupId>
    <artifactId>flink-streaming-Java_2.12</artifactId>
    <version>1.13.2</version>
    <scope>provided</scope>
</dependency>
<dependency>
    <groupId>org.apache.flink</groupId>
    <artifactId>flink-connector-kafka_2.12</artifactId>
    <version>1.13.2</version>
    <scope>compile</scope>
</dependency>
```

(3) 创建流应用程序主类。

Scala 代码如下：

```scala
//第3章/KafkaProducerDemo.scala

import Java.util.Properties
import org.apache.flink.api.common.functions.FlatMapFunction
import org.apache.flink.api.common.serialization.SimpleStringSchema
import org.apache.flink.streaming.api.scala._
import org.apache.flink.streaming.connectors.kafka.FlinkKafkaProducer
import org.apache.flink.util.Collector
```

```scala
object KafkaProducerDemo {
  def main(args: Array[String]) {
    //设置流执行环境
    val env = StreamExecutionEnvironment.getExecutionEnvironment

    //启用检查点
    env.enableCheckpointing(50000)

    val properties = new Properties()
    properties.setProperty("Bootstrap.servers", "localhost:9092")

    //构造 Flink Kafka Sink,默认使用 FlinkKafkaProducer.Semantic.AT_LEAST_ONCE 语义
    val myProducer = new FlinkKafkaProducer[String](
        "words",                     //目标 topic
        new SimpleStringSchema,      //序列化 schema
        properties                   //producer 配置
        )

    //将数据写入 Kafka 的 words 主题
    env
      .fromElements("good good study", "day day up")
      .flatMap(new FlatMapFunction[String, String]() {
        @throws[Exception]
        override def flatMap(s: String, out: Collector[String]): Unit = {
          for (word <- s.split("\\W+")) {
            out.collect(word)
          }
        }
      })
      .addSink(myProducer)

    //触发流程序执行
    env.execute("Flink Streaming Scala API Skeleton")
  }
}
```

Java 代码如下:

```java
//第 3 章/KafkaProducerDemo.java

import org.apache.flink.api.common.functions.FlatMapFunction;
import org.apache.flink.api.common.serialization.SimpleStringSchema;
import org.apache.flink.streaming.api.datastream.DataStream;
import org.apache.flink.streaming.api.environment.StreamExecutionEnvironment;
import org.apache.flink.streaming.connectors.kafka.FlinkKafkaProducer;
```

```java
import org.apache.flink.streaming.connectors.kafka.KafkaSerializationSchema;
import org.apache.flink.util.Collector;
import org.apache.kafka.clients.producer.ProducerRecord;
import Java.nio.charset.StandardCharsets;
import Java.util.Properties;

public class KafkaProcuderDemo {

    public static void main(String[] args) throws Exception {
        //设置流执行环境
        final StreamExecutionEnvironment env =
            StreamExecutionEnvironment.getExecutionEnvironment();

        //启用检查点
        env.enableCheckpointing(50000);

        Properties properties = new Properties();
        properties.setProperty("Bootstrap.servers", "localhost:9092");

        //构造 Flink Kafka Sink
        //默认使用 FlinkKafkaProducer.Semantic.AT_LEAST_ONCE 语义
        FlinkKafkaProducer<String> myProducer =
            new FlinkKafkaProducer<String>(
                "words",                        //目标 topic
                new SimpleStringSchema(),       //序列化 schema
                properties                      //producer 配置
            );

        //将数据写入 Kafka 的 words 主题
        env
            .fromElements("good good study","day day up")
            .flatMap(new FlatMapFunction<String, String>() {
                @Override
                public void flatMap(String s, Collector<String> out) throws Exception {
                    for(String word : s.split("\\W+")){
                        out.collect(word);
                    }
                }
            })
            .addSink(myProducer);

        //触发流程序执行
        env.execute("Kafka Flink Producer Demo");
    }
}
```

默认 FlinkKafkaProducer 使用的是 AT_LEAST_ONCE 语义,这要求启用检查点。随着 Flink 的检查点启用,Flink Kafka 消费者将消费一个主题的记录,并定期检查点所有的 Kafka 偏移,以及其他操作的状态。如果检查点被禁用,则 Kafka 消费者会定期向 ZooKeeper 提交偏移量。

（4）执行程序。建议按以下步骤执行:

① 启动 ZooKeeper 服务和 Kafka 服务。打开一个终端窗口,启动 ZooKeeper(不要关闭),命令如下:

```
$ ./bin/zookeeper-server-start.sh config/zookeeper.properties
```

② 打开另一个终端窗口,启动 Kafka 服务(不要关闭),命令如下:

```
$ ./bin/kafka-server-start.sh config/server.properties
```

③ 打开第 3 个终端窗口,在 Kafka 中创建一个名为 words 的主题(topic),命令如下:

```
$ ./bin/kafka-topics.sh --create --Bootstrap-server localhost:9092 --replication-factor 1 --partitions 1 --topic words
```

④ 查看已经创建的 Topic,命令如下:

```
$ ./bin/kafka-topics.sh --list --Bootstrap-server localhost:9092
```

⑤ 在第 3 个终端窗口,执行 Kafka 自带的消费者脚本,命令如下:

```
$ ./bin/kafka-console-consumer.sh --Bootstrap-server localhost:9092 --topic words
```

⑥ 运行上面编写的流执行程序(相当于 Kafka 的生产者程序)。
⑦ 在第 3 个终端窗口(运行 Kafka 消费者脚本的窗口),可以看到以下的输出内容:

```
day
day
up
good
good
study
```

在上面的代码中,使用默认的 Semantic.AT_LEAST_ONCE 语义,这时可以简单地将序列化模式指定为 SimpleStringSchema,但是,如果要指定 FlinkKafkaProducer 使用 EXACTLY_ONCE 语义,就需要自定义序列化模式。自定义序列化模式需要实现 KafkaSerializationSchema < T > 接口。

Flink Kafka Producer 需要知道如何将 Java/Scala 对象转换成二进制数据。KafkaSerializationSchema 允许用户指定这样的模式。对于流中的每条记录会调用 ProducerRecord<Byte[], Byte[]> serialize(T element, @Nullable Long timestamp)方法,生成一个写入 Kafka 的 ProducerRecord。

这让用户可以细粒度地控制数据如何写入 Kafka。通过生产者记录,可以:

(1) 设置 header 值。

(2) 为每个记录定义 key。

(3) 指定自定义数据分区。

现在重构上面的代码,在构造 FlinkKafkaProducer 时使用 Semantic.EXACTLY_ONCE 语义,因此,需要使用自定义的字符串序列化器 MyStringSerializationSchema。重构后的代码如下。

Scala 代码如下:

```scala
//第3章/KafkaProducerDemo2.scala

import Java.nio.charset.StandardCharsets
import Java.util.Properties

import org.apache.flink.api.common.functions.FlatMapFunction
import org.apache.flink.streaming.api.scala._
import org.apache.flink.streaming.connectors.kafka.{FlinkKafkaProducer, KafkaSerializationSchema}
import org.apache.flink.util.Collector
import org.apache.kafka.clients.producer.ProducerRecord

object KafkaProducerDemo2 {
  def main(args: Array[String]) {
    //设置流执行环境
    val env = StreamExecutionEnvironment.getExecutionEnvironment
    env.setParallelism(1)

    //启用检查点
    env.enableCheckpointing(50000)

    val properties = new Properties()
    properties.setProperty("Bootstrap.servers", "localhost:9092")
    //Kafka Brokers 默认的最大事务超时(transaction.max.timeout.ms)为
    //15 minutes,当使用 Semantic.EXACTLY_ONCE 语义时,下面这个属性值不能
    //超过 15min(默认为 1 hour)
    properties.setProperty("transaction.timeout.ms", String.valueOf(5 * 60 * 1000))

    //构造 Flink Kafka Sink,默认使用
```

```scala
//FlinkKafkaProducer.Semantic.AT_LEAST_ONCE 语义
val myProducer = new FlinkKafkaProducer[String](
    "words",                                            //目标 topic
    new MyStringSerializationSchema("words"),           //序列化 schema
    properties,                                         //producer 配置
    FlinkKafkaProducer.Semantic.EXACTLY_ONCE            //容错性
)

//将数据写入 Kafka 的 words 主题
env
  .fromElements("good good study", "day day up")
  .flatMap(new FlatMapFunction[String, String]() {
    @throws[Exception]
    override def flatMap(s: String, out: Collector[String]): Unit = {
      for (word <- s.split("\\W+")) {
        out.collect(word)
      }
    }
  })
  .addSink(myProducer)

//触发流程序执行
env.execute("Kafka Flink Producer Demo")
}

//自定义字符串序列化器
class MyStringSerializationSchema(topic: String)
                    extends KafkaSerializationSchema[String] {
  override def serialize(element: String,
                  timestamp: Java.lang.Long
              ): ProducerRecord[Array[Byte], Array[Byte]] = {
    new ProducerRecord[Array[Byte], Array[Byte]](topic, element.getBytes
(StandardCharsets.UTF_8))
  }
}
}
```

Java 代码如下：

```java
//第3章/KafkaProducerDemo2.java

import org.apache.flink.api.common.functions.FlatMapFunction;
import org.apache.flink.streaming.api.environment.StreamExecutionEnvironment;
import org.apache.flink.streaming.connectors.kafka.FlinkKafkaProducer;
import org.apache.flink.streaming.connectors.kafka.KafkaSerializationSchema;
```

```java
import org.apache.flink.util.Collector;
import org.apache.kafka.clients.producer.ProducerRecord;

import Java.nio.charset.StandardCharsets;
import Java.util.Properties;

public class KafkaProcuderDemo2 {

    public static void main(String[] args) throws Exception {
        //设置流执行环境
        final StreamExecutionEnvironment env =
            StreamExecutionEnvironment.getExecutionEnvironment();

        //启用检查点
        env.enableCheckpointing(50000);

        Properties properties = new Properties();
        properties.setProperty("Bootstrap.servers", "localhost:9092");
        properties.setProperty("transaction.timeout.ms", String.valueOf(5 * 60 * 1000));

        //构造 Flink Kafka Sink,默认使用
        //FlinkKafkaProducer.Semantic.AT_LEAST_ONCE 语义
        FlinkKafkaProducer<String> myProducer = new FlinkKafkaProducer<String>(
            "words",                                      //目标 topic
            new MyStringSerializationSchema("words"),     //序列化 schema
            properties,                                   //producer 配置
            FlinkKafkaProducer.Semantic.EXACTLY_ONCE      //容错性
        );

        //将数据写入 Kafka 的 words 主题
        env
            .fromElements("good good study","day day up")
            .flatMap(new FlatMapFunction<String, String>() {
                @Override
                public void flatMap(String s, Collector<String> out) throws Exception {
                    for(String word : s.split("\\W+")){
                        out.collect(word);
                    }
                }
            })
            .addSink(myProducer);

        //触发流程序执行
        env.execute("Kafka Flink Producer Demo");
    }
}
```

```java
//自定义字符串序列化器
public static class MyStringSerializationSchema
        implements KafkaSerializationSchema<String> {

    private String topic;

    public MyStringSerializationSchema(String topic) {
        super();
        this.topic = topic;
    }

    @Override
    public ProducerRecord<Byte[], Byte[]> serialize(String element, Long timestamp) {
        return new ProducerRecord<>(topic, element.getBytes(StandardCharsets.UTF_8));
    }
}
```

需要注意在上面的代码中的粗体字部分。Kafka Brokers 默认的最大事务超时 (transaction.max.timeout.ms)为 15min，生产者在设置事务超时时不允许大于这个值，但是在默认情况下，FlinkKafkaProducer 将事务超时属性 (transaction.timeout.ms)设置为 1h，超过了 Kafka Brokers 默认的最大事务超时 15min，因此，当使用 Semantic.EXACTLY_ONCE 语义时，会遇到包含下面类似信息的异常：

```
……
org.apache.kafka.common.KafkaException: org.apache.kafka.common.KafkaException: Unexpected error in InitProducerIdResponse; The transaction timeout is larger than the maximum value allowed by the broker (as configured by transaction.max.timeout.ms).
……
```

因此，需要修改 FlinkKafkaProducer 的事务超时属性 transaction.timeout.ms 来解决这个问题。例如，将 transaction.timeout.ms 从 1h 降为 5min，代码如下：

```
properties.setProperty(
    "transaction.timeout.ms", String.valueOf(5 * 60 * 1000));
```

如果写入 Kafka 的不是字符串而是 Java 对象，则需要自定义一个 SerializatonSchema 序列化器，该序列化需要实现接口 KafkaSerializationSchema。例如，在下面的代码中，定义了一个将 Book 对象序列化到 Kafka 中的自定义序列化对象。

【示例 3-34】 编写一个 Flink 流应用程序，将流数据（Book 对象类型）写入 Kafka 中的指定主题 books。使用 Kafka 自带的消费者脚本来查看写入的内容。

Scala 代码如下：

```scala
//第 3 章/KafkaProducerDemo3.scala

import Java.util.Properties

import org.apache.flink.shaded.jackson2.com.fasterxml.jackson.core.jsonProcessingException
import org.apache.flink.shaded.jackson2.com.fasterxml.jackson.databind.ObjectMapper
import org.apache.flink.streaming.api.scala._
import org.apache.flink.streaming.connectors.kafka.{FlinkKafkaProducer, KafkaSerializationSchema}
import org.apache.kafka.clients.producer.ProducerRecord

object KafkaProducerDemo3 {

  //事件元素类型(注意:这里不能使用 case class,应该是 Flink API 自身的问题)
  class Book(id: Long, title: String, authors: String, year: Integer) {
    def getId:Long = id
    def getTitle:String = title
    def getAuthors:String = authors
    def getYear:Integer =  year
  }

  def main(args: Array[String]) {
    //设置流执行环境
    val env = StreamExecutionEnvironment.getExecutionEnvironment
    env.setParallelism(1)

    //启用检查点
    env.enableCheckpointing(50000)

    val properties = new Properties()
    properties.setProperty("Bootstrap.servers", "localhost:9092")
    properties.setProperty("transaction.timeout.ms", String.valueOf(5 * 60 * 1000))

    val myProducer = new FlinkKafkaProducer[Book](
        "books",                                   //目标 topic
        new ObjSerializationSchema("books"),       //序列化 schema
        properties,                                //producer 配置
        FlinkKafkaProducer.Semantic.EXACTLY_ONCE   //容错性
    )

    //将数据写入 Kafka 的 rds 主题
    env
      .fromElements(
        new Book(101L, "Java 从初学到精通", "xinliwei", 2019),
```

```scala
        new Book(102L, "流处理系统", "someone", 2018),
        new Book(103L, "Hadoop 大数据处理技术", "zhangsan", 2017),
        new Book(104L, "Spark 大数据处理技术", "xinliwei", 2021)
      )
      .addSink(myProducer)

    //触发流程序执行
    env.execute("Kafka Flink Producer Demo")
  }

  //自定义的序列化模式
  class ObjSerializationSchema(topic: String)
                    extends KafkaSerializationSchema[Book] {

    private var mapper = new ObjectMapper()

    override def serialize(obj: Book,
                    timestamp: Java.lang.Long
                  ): ProducerRecord[Array[Byte], Array[Byte]] = {
      var b:Array[Byte] = null
      try{
        b = mapper.writeValueAsBytes(obj)
      }catch {
        case e: JsonProcessingException => //TODO
      }
      new ProducerRecord[Array[Byte], Array[Byte]](topic, b)
    }
  }
}
```

Java 代码如下:

```java
//第 3 章/KafkaProducerDemo3.java

import org.apache.flink.shaded.jackson2.com.fasterxml.jackson.core.JsonProcessingException;
import org.apache.flink.shaded.jackson2.com.fasterxml.jackson.databind.ObjectMapper;
import org.apache.flink.streaming.api.environment.StreamExecutionEnvironment;
import org.apache.flink.streaming.connectors.kafka.FlinkKafkaProducer;
import org.apache.flink.streaming.connectors.kafka.KafkaSerializationSchema;
import org.apache.kafka.clients.producer.ProducerRecord;
import Java.util.Properties;

public class KafkaProcuderDemo3 {
```

```java
//POJO类,事件类型
public static class Book {
    public Long id;
    public String title;
    public String authors;
    public Integer year;

    public Book(){}

    public Book(Long id, String title, String authors, Integer year) {
        this.id = id;
        this.title = title;
        this.authors = authors;
        this.year = year;
    }

    @Override
    public String toString() {
        return "Book{" +
                "id = " + id +
                ", title = '" + title + '\'' +
                ", authors = '" + authors + '\'' +
                ", year = " + year +
                '}';
    }
}

public static void main(String[] args) throws Exception {
    //设置流执行环境
    final StreamExecutionEnvironment env =
            StreamExecutionEnvironment.getExecutionEnvironment();

    //启用检查点
    env.enableCheckpointing(50000);

    Properties properties = new Properties();
    properties.setProperty("Bootstrap.servers", "localhost:9092");
    properties.setProperty("transaction.timeout.ms", String.valueOf(5 * 60 * 1000));

    //构造 Flink Kafka Sink
    FlinkKafkaProducer < Book > myProducer = new FlinkKafkaProducer < Book >(
        "books",                                    //目标 topic
        new ObjSerializationSchema("books"),        //序列化 schema
        properties,                                 //producer 配置
        FlinkKafkaProducer.Semantic.EXACTLY_ONCE    //容错性
    );
```

```java
//将数据写入 Kafka 的 words 主题
env
    .fromElements(
        new Book(101L, "Java 从初学到精通", "xinliwei", 2019),
        new Book(102L, "流处理系统", "someone", 2018),
        new Book(103L, "Hadoop 大数据处理技术", "zhangsan", 2017),
        new Book(104L, "Spark 大数据处理技术", "xinliwei", 2021)
    )
    .addSink(myProducer);

//触发流程序执行
env.execute("Kafka Flink Producer Demo");
}

//自定义的序列化模式
public static class ObjSerializationSchema
                implements KafkaSerializationSchema<Book>{

    private String topic;
    private ObjectMapper mapper;

    public ObjSerializationSchema(String topic) {
        super();
        this.topic = topic;
    }

    @Override
    public ProducerRecord<Byte[], Byte[]> serialize(Book obj, Long timestamp) {
        Byte[] b = null;
        if (mapper == null) {
            mapper = new ObjectMapper();
        }
        try {
            b = mapper.writeValueAsBytes(obj);
        } catch (JsonProcessingException e) {
            //TODO
        }
        return new ProducerRecord<>(topic, b);
    }
}
}
```

执行程序。建议按以下步骤执行:

(1) 启动 ZooKeeper 服务和 Kafka 服务。打开一个终端窗口,启动 ZooKeeper(不要关闭),命令如下:

```
$ ./bin/zookeeper-server-start.sh config/zookeeper.properties
```

(2) 打开另一个终端窗口,启动 Kafka 服务(不要关闭),命令如下:

```
$ ./bin/kafka-server-start.sh config/server.properties
```

(3) 打开第 3 个终端窗口,在 Kafka 中创建一个名为 books 的主题(topic),命令如下:

```
$ ./bin/kafka-topics.sh --create --Bootstrap-server localhost:9092 --replication-factor 1 --partitions 1 --topic books
```

(4) 查看已经创建的 Topic,命令如下:

```
$ ./bin/kafka-topics.sh --list --Bootstrap-server localhost:9092
```

(5) 在第 3 个终端窗口,执行 Kafka 自带的消费者脚本,命令如下:

```
$ ./bin/kafka-console-consumer.sh --Bootstrap-server localhost:9092 --topic books
```

(6) 运行上面编写的流执行程序(相当于 Kafka 的生产者程序)。
(7) 在第 3 个终端窗口(运行 Kafka 消费者脚本的窗口),可以看到以下的输出内容:

```
{"id":103,"title":"Hadoop 大数据处理技术","authors":"zhangsan","year":2017}
{"id":101,"title":"Java 从初学到精通","authors":"xinliwei","year":2019}
{"id":102,"title":"流处理系统","authors":"someone","year":2018}
{"id":104,"title":"Spark 大数据处理技术","authors":"xinliwei","year":2021}
```

启用 Flink 的检查点后,FlinkKafkaProducer 可以提供精确一次的交付保证。

除了启用 Flink 的检查点,还可以通过传递适当的 semantic 语义参数给 FlinkKafkaProducer 来选择以下 3 种不同的操作模式。

(1) Semantic.NONE:Flink 将不会做任何保证。生成的记录可能会丢失,也可能会重复。

(2) Semantic.AT_LEAST_ONCE:默认设置。这保证了不会丢失任何记录(尽管它们可以被重复)。

(3) Semantic.EXACTLY_ONCE:使用 Kafka 事务提供精确一次的语义。当使用事务写入 Kafka 时,不要忘记为任何消费 Kafka 记录的应用程序设置期望的 isolation.level 值(read_committed 或 read_uncommitted,后者是默认值)。

3.11.2 JDBC 连接器

JDBC Connector 提供了一个接收器,用于将数据写入 JDBC 数据库。要使用它,应将以下依赖项添加到项目中：

```xml
<dependency>
    <groupId>org.apache.flink</groupId>
    <artifactId>flink-connector-JDBC_2.12</artifactId>
    <version>1.13.2</version>
</dependency>
<dependency>
    <groupId>mysql</groupId>
    <artifactId>mysql-connector-Java</artifactId>
    <version>5.1.48</version>
</dependency>
```

注意：MySQL的版本可以根据自己的实际安装进行修改。

创建的JDBC接收器提供了"至少一次"的保证。有效使用upsert语句或幂等更新可以实现精确一次保证。其用法如下：

```
JDBCSink.sink(
        sqlDmlStatement,            //必需的
        JDBCStatementBuilder,       //必需的
        JDBCExecutionOptions,       //可选的
        JDBCConnectionOptions       //必需的
);
```

SQL DML语句是用户提供的SQL语句，通常带有占位符。

JDBC语句构建器从用户提供的SQL字符串构建一个JDBC预编译语句，并重复调用用户指定的函数来使用流中的每个元素值更新这个预编译语句。

SQL DML语句是批量执行的，可以选择使用以下实例配置这些语句：

```
JDBCExecutionOptions.builder()
    .withBatchIntervalMs(200)       //可选的: default = 0,这意味着没有基于时间的执行
    .withBathSize(1000)             //可选的: default = 5000
    .withMaxRetries(5)              //可选的: default = 3
.build()
```

只要满足以下条件之一，就会执行JDBC批处理：
（1）已经过配置的批处理间隔时间。
（2）已达到最大批大小。
（3）Flink检查点已经启动。

数据库连接是用JDBCConnectionOptions实例配置的。

> **注意**：从 Flink 1.13 开始，Flink JDBC 接收器支持精确一次模式。该实现依赖于 XA 标准的 JDBC 驱动程序支持；在 Flink 1.13 中，对于 MySQL 或其他不支持每个连接多个 XA 事务的数据库，Flink JDBC 接收器不支持精确一次模式。

【示例 3-35】 使用 JDBCSink 将流数据写入 MySQL 数据库。
建议按以下步骤执行：
(1) 首先，在 MySQL 中执行如下脚本，创建用来接收写入数据的 books 数据表。

```
#创建数据库 xueai8
mysql> create database xueai8;

#切换到数据库 xueai8
mysql> use xueai8;

#创建数据表 books
create table books(
    id bigint,
    title varchar(100),
    authors varchar(50),
    year int
);
```

(2) 然后，编写 Flink 流处理程序，将数据流中的图书信息写入上面创建的 books 数据表中。

Scala 代码如下：

```
//第 3 章/JDBCSinkDemo.scala

import Java.sql.PreparedStatement

import org.apache.flink.connector.JDBC._
import org.apache.flink.connector.JDBC.JDBCConnectionOptions
import org.apache.flink.streaming.api.scala._

object JDBCSinkDemo {

    //事件元素类型
    case class Book(id: Long, title: String, authors: String,year: Integer)

    def main(args: Array[String]) {
        //设置流执行环境
        val env = StreamExecutionEnvironment.getExecutionEnvironment
```

```scala
    env
      .fromElements(
        Book(101L, "Java 从初学到精通", "xinliwei", 2019),
        Book(102L, "流处理系统", "someone", 2018),
        Book(103L, "Hadoop 大数据处理技术", "zhangsan", 2017),
        Book(104L, "Spark 大数据处理技术", "xinliwei", 2021)
      )
      .addSink(
        JDBCSink.sink(
          //SQL 语句,带有占位符
          "insert into books (id, title, authors, year) values (?, ?, ?, ?)",
          //预编译语句
          new JDBCStatementBuilder[Book]{
            //重复调用这个函数,用流的每个值更新这个预编译语句
            override def accept(ps:PreparedStatement, book:Book): Unit = {
              ps.setLong(1, book.id)
              ps.setString(2, book.title)
              ps.setString(3, book.authors)
              ps.setInt(4, book.year)
            }
          },
          //可选,JDBC 批处理参数
          JDBCExecutionOptions.builder
            .withBatchSize(1000)
            .withBatchIntervalMs(200)
            .withMaxRetries(5)
            .build,
          //JDBC 连接配置信息
          new JDBCConnectionOptions.JDBCConnectionOptionsBuilder()
            .withUrl("JDBC: mysql://localhost: 3306/xueai8? characterEncoding = UTF - 8&useSSL = false")
            .withDriverName("com.mysql.JDBC.Driver")
            .withUsername("root")
            .withPassword("admin")
            .build()
        )
      )

    //触发流程序执行
    env.execute("Flink Streaming Scala API Skeleton")
  }
}
```

Java 代码如下:

```java
//第 3 章/JDBCSinkDemo.java
import org.apache.flink.connector.JDBC.JDBCConnectionOptions;
import org.apache.flink.connector.JDBC.JDBCExecutionOptions;
import org.apache.flink.connector.JDBC.JDBCSink;
import org.apache.flink.streaming.api.environment.StreamExecutionEnvironment;

public class JDBCSinkDemo1 {

    //POJO 类,事件类型
    public static class Book {
        public Long id;
        public String title;
        public String authors;
        public Integer year;

        public Book(){}

        public Book(Long id, String title, String authors, Integer year) {
            this.id = id;
            this.title = title;
            this.authors = authors;
            this.year = year;
        }
    }

    public static void main(String[] args) throws Exception {
        //设置流执行环境
        final StreamExecutionEnvironment env =
                StreamExecutionEnvironment.getExecutionEnvironment();

        env
            .fromElements(
                new Book(101L, "Java 从初学到精通", "xinliwei", 2019),
                new Book(102L, "Streaming Systems", "Tyler Akidau, Slava Chernyak, Reuven Lax", 2018),
                new Book(103L, "Designing Data - Intensive Applications", "Martin Kleppmann", 2017),
                new Book(104L, "Spark 大数据处理技术", "xinliwei", 2021)
            )
            .addSink(JDBCSink.sink(
                //SQL 语句,带有占位符
                "insert into books (id, title, authors, year) values (?, ?, ?, ?)",
                //重复调用这个函数,用流的每个值更新这个预编译语句
                (statement, book) -> {
                    statement.setLong(1, book.id);
```

```
                statement.setString(2, book.title);
                statement.setString(3, book.authors);
                statement.setInt(4, book.year);
            },
            //可选,JDBC 批处理参数
            JDBCExecutionOptions
                .builder()
                .withBatchSize(1000)
                .withBatchIntervalMs(200)
                .withMaxRetries(5)
                .build(),
            //JDBC 连接配置信息
            new JDBCConnectionOptions.JDBCConnectionOptionsBuilder()
                .withUrl("JDBC: mysql://localhost: 3306/xueai8? characterEncoding = UTF -
8&useSSL = false")
                .withDriverName("com.mysql.JDBC.Driver")
                .withUsername("root")
                .withPassword("admin")
                .build()
        ));

        //触发流程序执行
        env.execute("JDBC Sink");
    }
}
```

(3) 执行以上程序。

(4) 在 MySQL 中查询 books 表,可以看到数据已经被正确地写入了,如图 3-65 所示。

图 3-65　写入 MySQL 中的数据

3.12　其他

下面是一些不便归于特定主题的内容,包括如何在应用程序中使用日志、如何加载外部配置参数及如何尽量避免使用大型的元组类型。

3.12.1　在应用程序中使用日志

在任何 Flink 应用程序中使用 SLF4J 时,需要导入依赖包和类,并用类名初始化日志程序，代码如下：

```
import org.slf4j.LoggerFactory
import org.slf4j.Logger

Logger LOG = LoggerFactory.getLogger(MyClass.class)
```

使用占位符机制而不是使用字符串格式化程序进行日志记录也是一种最佳实践。占位符机制有助于避免不必要的字符串格式化，而它只执行字符串连接。使用占位符的代码如下：

```
LOG.info("Value of a = {}, value of b = {}", myobject.a, myobject.b);
```

也可以在异常处理中使用占位符日志，代码如下：

```
catch(Exception e){
    LOG.error("Error occurred {}", e);
}
```

3.12.2　使用 ParameterTool

从 Flink 0.9 开始，在 Flink 中就有了一个内置的 ParameterTool，它可以帮助从外部源（如命令行参数、系统属性或属性文件）获取参数。在内部，它是字符串的 map 映射，保留 key 作为参数名，value 作为参数值。

例如，可以考虑在 DataStream API 示例中使用 ParameterTool，用户需要在其中设置 Kafka 属性，代码如下：

```
String kafkaproperties = "/path/to/kafka.properties";
ParameterTool parameter = ParameterTool.fromPropertiesFile(propertiesFile);
```

（1）从系统属性加载，即可以读取系统变量中定义的属性。在初始化系统属性文件之前，需要通过设置 Dinput=hdfs://myfile 来传递系统属性文件，然后，用户可以在 ParameterTool 中读取所有相关属性，代码如下：

```
ParameterTool parameters = ParameterTool.fromSystemProperties();
```

（2）从命令行参数加载，即还可以从命令行参数中读取参数。必须在调用应用程序之前设置--elements。从命令行参数中读取参数，代码如下：

```
ParameterTool parameters = ParameterTool.fromArgs(args);
```

(3) 从.properties文件加载,即还可以从.properties文件中读取参数,代码如下:

```
String propertiesFile = "/my.properties";
ParameterTool parameters = ParameterTool.fromPropertiesFile(propertiesFile);
```

加载配置参数之后,可以在Flink程序中读取它们,代码如下:

```
parameter.getRequired("key");
parameter.get("paramterName", "myDefaultValue");
parameter.getLong("expectedCount", -1L);
parameter.getNumberOfParameters()
```

3.12.3 命名大型TupleX类型

众所周知,元组是一种用于表示复杂数据结构的复杂数据类型。它是各种基本数据类型的组合。一般情况下,建议不要使用大元组;相反,建议使用Java POJO。如果想使用元组,则建议使用一些自定义POJO类型来命名它。

为大型元组创建自定义类型非常容易。例如,如果想使用Tuple8,则可以这样定义它:

```
//初始化Record Tuple
RecordTuple rc = new RecordTuple(value0, value1, value2, value3, value4, value5, value6, value7);

//定义RecordTuple而不是使用Tuple8
public static class RecordTuple extends Tuple8 < String, String, Integer, String, Integer, Integer, Integer, Integer > {
    public RecordTuple() {
        super();
    }

    public RecordTuple(String value0, String value1, Integervalue2, String value3, Integer value4, Integer value5, Integer value6, Integer value7) {
        super(value0, value1, value2, value3, value4, value5, value6, value7);
    }
}
```

3.13 Flink流处理案例

到目前为止,已经了解了DataStream API的各方面。接下来尝试使用这些概念来解决实际的问题。

3.13.1 处理 IoT 事件流

假设一台机器上安装了传感器，用户希望从这些传感器收集数据，并每 5min 计算每个传感器的平均温度，其架构如图 3-66 所示。

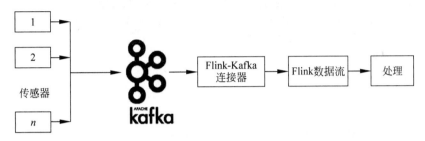

图 3-66　IoT 事件处理架构

在这个场景中，假设传感器将信息发送给 Kafka 的主题 temp，数据格式为（传感器 id、时间戳、温度）。这里假设以字符串的形式接收 Kafka 主题中的事件，部分数据如下：

```
//第 3 章/sensortemp.csv

sensor_1,1629943899014,51.087254019871054
sensor_9,1629943899014,70.44743245583899
sensor_7,1629943899014,65.53215956486392
sensor_0,1629943899014,53.210570822216546
sensor_8,1629943899014,93.12876931817556
sensor_3,1629943899014,57.55153052162809
sensor_2,1629943899014,107.61249366604993
sensor_5,1629943899014,92.02083744773739
sensor_4,1629943899014,95.7688424087137
sensor_6,1629943899014,95.04398353316257
……
```

现在需要编写 Flink 流处理代码从 Kafka 的 temp 主题读取这些数据，并使用 Flink 转换处理数据，因此 Kafka 作为数据源，Flink 流处理程序作为 Kafka 的消费者。

这里要考虑的是，既然有来自传感器的时间戳值，那么可以使用事件时间计算时间因素。这意味着可以处理乱序的传感器数据。

Scala 代码如下：

```scala
//第 3 章/KafkaIotDemo.scala

import java.time.Duration
import org.apache.flink.api.common.eventtime.{SerializableTimestampAssigner, WatermarkStrategy}
```

```scala
import org.apache.flink.api.common.functions.AggregateFunction
import org.apache.flink.api.common.serialization.SimpleStringSchema
import org.apache.flink.connector.kafka.source.KafkaSource
import org.apache.flink.connector.kafka.source.enumerator.initializer.OffsetsInitializer
import org.apache.flink.streaming.api.scala._
import org.apache.flink.streaming.api.scala.function.ProcessWindowFunction
import org.apache.flink.streaming.api.windowing.assigners.TumblingEventTimeWindows
import org.apache.flink.streaming.api.windowing.time.Time
import org.apache.flink.streaming.api.windowing.Windows.TimeWindow
import org.apache.flink.util.Collector

object KafkaIotDemo {

  //case class,流数据类型
  case class SensorReading(id:String, timestamp:Long, temperature:Double)

  def main(args: Array[String]) {
    //设置流执行环境
    val env = StreamExecutionEnvironment.getExecutionEnvironment

    val source = KafkaSource.builder[String]
      .setBootstrapServers("localhost:9092")
      .setTopics("temp")
      .setGroupId("group-test")
      .setStartingOffsets(OffsetsInitializer.earliest)
      .setValueOnlyDeserializer(new SimpleStringSchema)
      .build

    //水印策略
    val watermarkStrategy = WatermarkStrategy
      .forBoundedOutOfOrderness[String](Duration.ofSeconds(1))
      .withTimestampAssigner(new SerializableTimestampAssigner[String]() {
        override def extractTimestamp(s: String, l: Long): Long =
          s.split(",")(1).toLong
      })

    env
      //读取 Kafka 数据源
      .fromSource(source, watermarkStrategy, "Sensor temperature Source")
      //转换流数据类型
      .map(s => {
        val fields: Array[String] = s.split(",")
        SensorReading(fields(0), fields(1).toLong, fields(2).toDouble)
      })
      //按 key 分区
      .keyBy(sr => sr.id)
```

```scala
    //开大小为 5min 的滚动窗口(这里为了测试,设置为 5s)
    .window(TumblingEventTimeWindows.of(Time.seconds(5)))
    //执行增量聚合
    .aggregate(new AggAvgTemp(),new ProcessAvgTemp())
    //输出结果
    .print()

  //触发流程序执行
  env.execute("Flink Sensor Temperature Demo")
}

//增量处理函数
class AggAvgTemp extends AggregateFunction[SensorReading, (Double, Long), Double] {
  //创建初始 ACC
  override def createAccumulator() = (0.0, 0L)

  //累加每个传感器(每个分区)的事件
  override def add(sr: SensorReading, acc: (Double, Long)) =
    (sr.temperature + acc._1, acc._2 + 1L)

  //分区合并
  override def merge(acc1: (Double, Long), acc2: (Double, Long)) =
    (acc1._1 + acc2._1, acc1._2 + acc2._2)

  //返回每个传感器的平均温度
  override def getResult(acc: (Double, Long)): Double = acc._1 / acc._2
}

//窗口处理函数(注意这里引入的 ProcessWindowFunction 不要错引了 Java 的)
class ProcessAvgTemp extends ProcessWindowFunction[Double, (String, Long, Double), String, TimeWindow] {
  override def process(key: String,
                       context: Context,
                       elements: Iterable[Double],
                       out: Collector[(String, Long, Double)]): Unit = {
    //计算平均温度
    val average = Math.round(elements.iterator.next * 100) / 100.0
    //发送到下游算子
    out.collect((key, context.window.getEnd, average))
  }
}
```

Java 代码如下:

```java
//第3章/KafkaIotDemo.java

import org.apache.flink.api.common.eventtime.SerializableTimestampAssigner;
import org.apache.flink.api.common.eventtime.WatermarkStrategy;
import org.apache.flink.api.common.functions.AggregateFunction;
import org.apache.flink.api.common.functions.MapFunction;
import org.apache.flink.api.common.serialization.SimpleStringSchema;
import org.apache.flink.api.java.functions.KeySelector;
import org.apache.flink.api.java.tuple.Tuple2;
import org.apache.flink.api.java.tuple.Tuple3;
import org.apache.flink.connector.kafka.source.KafkaSource;
import org.apache.flink.connector.kafka.source.enumerator.initializer.OffsetsInitializer;
import org.apache.flink.streaming.api.environment.StreamExecutionEnvironment;
import org.apache.flink.streaming.api.functions.windowing.ProcessWindowFunction;
import org.apache.flink.streaming.api.windowing.assigners.TumblingEventTimeWindows;
import org.apache.flink.streaming.api.windowing.time.Time;
import org.apache.flink.streaming.api.windowing.Windows.TimeWindow;
import org.apache.flink.util.Collector;
import Java.time.Duration;

public class KafkaIotDemo {

    //POJO类,温度数据类型
    public static class SensorReading {

        public String id;                    //传感器 id
        public long timestamp;               //读取时的时间戳
        public double temperature;           //读取的温度值

        public SensorReading() { }

        public SensorReading(String id, long timestamp, double temperature) {
            this.id = id;
            this.timestamp = timestamp;
            this.temperature = temperature;
        }

        public String toString() {
            return "(" + this.id + ", "
                    + this.timestamp + ", "
                    + this.temperature + ")";
        }
    }

    public static void main(String[] args) throws Exception {
        //设置流执行环境
```

```java
final StreamExecutionEnvironment env =
    StreamExecutionEnvironment.getExecutionEnvironment();

//数据源
KafkaSource<String> source = KafkaSource.<String>builder()
    .setBootstrapServers("localhost:9092")
    .setTopics("temp")
    .setGroupId("group-test")
    .setStartingOffsets(OffsetsInitializer.earliest())
    .setValueOnlyDeserializer(new SimpleStringSchema())
    .build();

//水印策略
WatermarkStrategy<String> watermarkStrategy = WatermarkStrategy
    .<String>forBoundedOutOfOrderness(Duration.ofSeconds(2))
    .withTimestampAssigner(new SerializableTimestampAssigner<String>(){
        @Override
        public long extractTimestamp(String s, long l) {
            return Long.parseLong(s.split(",")[1]);
        }
    });

env
    //指定 Kafka 数据源
    .fromSource(source, watermarkStrategy, "Sensor temperature Source")
    //转换为 DataStream<SensorReading>
    .map(new MapFunction<String, SensorReading>() {
        @Override
        public SensorReading map(String s) throws Exception {
            String[] fields = s.split(",");
            return new SensorReading(fields[0], Long.parseLong(fields[1]),
Double.parseDouble(fields[2]));
        }
    })
    //转换为 KeyedStream
    .keyBy(new KeySelector<SensorReading, String>() {
        @Override
        public String getKey(SensorReading sensorReading) throws Exception {
            return sensorReading.id;
        }
    })
    //开大小为 5min 的滚动窗口(这里为了测试,设置为 5s)
    .window(TumblingEventTimeWindows.of(Time.seconds(5)))
    //执行增量聚合
    .aggregate(new AggAvgTemp(),new ProcessAvgTemp())
    .print();
```

```java
    //触发流程序执行
    env.execute("Flink Sensor Temperature Demo");
}

//增量处理函数
public static class AggAvgTemp implements AggregateFunction<
        SensorReading,                          //input
        Tuple2<Double,Long>,                    //acc, <sum, count>
        Double> {                               //output, avg

    //创建初始ACC
    @Override
    public Tuple2<Double,Long> createAccumulator() {
        return new Tuple2<>(0.0,0L);
    }

    //累加每个传感器(每个分区)的事件
    @Override
    public Tuple2<Double,Long> add(SensorReading sr, Tuple2<Double,Long> acc) {
        return new Tuple2<>(sr.temperature + acc.f0,acc.f1 + 1);
    }

    //分区合并
    @Override
    public Tuple2<Double,Long> merge(
                Tuple2<Double,Long> acc1,
                Tuple2<Double,Long> acc2) {
        return new Tuple2<>(acc1.f0 + acc2.f0,acc1.f1 + acc2.f1);
    }

    //返回每个传感器的平均温度
    @Override
    public Double getResult(Tuple2<Double,Long> t2) {
        return t2.f0/t2.f1;
    }
}

//窗口处理函数
public static class ProcessAvgTemp extends ProcessWindowFunction<
        Double,                                 //input type
        Tuple3<String, Long, Double>,           //output type
        String,                                 //key type
        TimeWindow> {                           //window type

    @Override
    public void process(String id,
```

```
                        Context context,
                        Iterable<Double> events,
                        Collector<Tuple3<String, Long, Double>> out) {
            double average = Math.round(events.iterator().next() * 100) / 100.0;
            out.collect(new Tuple3<>(id,context.window().getEnd(),average));
        }
    }
}
```

对以上程序打 JAR 包。在命令行下,执行的命令如下:

```
$ mvn clean package
```

要执行作业,首先需要启动 ZooKeeper 集群和 Kafka 服务,并创建主题 temp。建议按以下步骤操作:

(1) 启动 ZooKeeper 服务,启动 Kafka 服务。

打开一个终端窗口,启动 ZooKeeper(不要关闭),命令如下:

```
$ ./bin/zookeeper-server-start.sh config/zookeeper.properties
```

打开另一个终端窗口,启动 Kafka 服务(不要关闭),命令如下:

```
$ ./bin/kafka-server-start.sh config/server.properties
```

(2) 在 Kafka 中创建一个名为 temp 的主题(topic),命令如下:

```
$ ./bin/kafka-topics.sh --create --Bootstrap-server localhost:9092 --replication-factor 1 --partitions 1 --topic temp
```

查看已经创建的 Topic,命令如下:

```
$ ./bin/kafka-topics.sh --list --Bootstrap-server localhost:9092
```

要将作业提交到 Flink 集群上运行,建议按以下步骤操作:

(1) 编写 Shell 脚本 streamiot.sh,读取每行 IoT 数据,发送给 Kafka 的 temp 主题,代码如下:

```
#!/bin/bash
BROKER=$1
if [ -z "$1" ]; then
    BROKER="localhost:9092"
fi
```

```
cat sensortemp.csv | while read line; do
    echo "$line"
    sleep 0.1
done | ~/bigdata/kafka_2.12-2.4.1/bin/kafka-console-producer.sh --broker-list
$BROKER --topic temp
```

注意：streamiot.sh 脚本应该具有可执行权限。如果没有，则可使用以下命令添加执行权限：$ chmod a+x streamiot.sh

(2) 执行脚本 streamiot.sh，命令如下：

```
$ ./streamiot.sh localhost:9092
```

(3) 将程序 JAR 包提交到集群上运行，抓取 Kafka 的 temp 主题中的消息，并输出在控制台，命令如下：

```
$ cd ~/bigdata/flink-1.13.2/
$ ./bin/flink run --class com.xueai8.java.ch03.StreamingJob ~/flinkdemos/FlinkJavaDemo-1.0-SNAPSHOT.jar
```

当将作业提交到 Flink 集群上运行时，标准输出其实是到了 flink-hduser-taskexecutor-0-localhost.out 文件中，因此要查看此结果，需要查看该文件。使用 Kafka 消费者脚本测试 Kafka 中的主题内容，命令如下：

```
$ ./bin/kafka-console-consumer.sh --Bootstrap-server localhost:9092 --topic temp --consumer-property group.id=test
```

可观察到输出的结果如下：

```
2> (sensor_2,1629943900000,107.4)
7> (sensor_7,1629943900000,66.11)
1> (sensor_0,1629943900000,52.86)
8> (sensor_9,1629943900000,71.53)
3> (sensor_8,1629943900000,93.09)
1> (sensor_3,1629943900000,57.93)
7> (sensor_4,1629943900000,96.48)
5> (sensor_1,1629943900000,51.64)
6> (sensor_5,1629943900000,92.0)
1> (sensor_0,1629943905000,50.59)
3> (sensor_8,1629943905000,91.96)
8> (sensor_9,1629943905000,71.15)
2> (sensor_2,1629943905000,107.48)
```

```
3> (sensor_8,1629943910000,91.13)
1> (sensor_3,1629943905000,60.1)
6> (sensor_6,1629943900000,96.08)
7> (sensor_7,1629943905000,64.93)
5> (sensor_1,1629943905000,53.84)
2> (sensor_2,1629943910000,109.88)
3> (sensor_8,1629943915000,94.77)
1> (sensor_3,1629943910000,62.18)
5> (sensor_1,1629943910000,59.87)
7> (sensor_4,1629943905000,97.99)
5> (sensor_1,1629943915000,63.32)
6> (sensor_6,1629943905000,95.86)
8> (sensor_9,1629943910000,71.55)
7> (sensor_7,1629943910000,67.59)
1> (sensor_0,1629943910000,50.96)
2> (sensor_2,1629943915000,105.35)
7> (sensor_4,1629943910000,100.87)
8> (sensor_9,1629943915000,75.48)
6> (sensor_5,1629943905000,91.47)
7> (sensor_7,1629943915000,69.73)
1> (sensor_0,1629943915000,51.91)
...
```

3.13.2 运输公司车辆超速实时监测

设计一个物流公司车队管理解决方案，其中车队中的车辆启用了无线网络功能。每辆车定期报告其地理位置和操作参数，如燃油水平、速度、加速度、轴承、发动机温度等。物流公司希望利用这一遥测数据流实现一系列应用程序，以帮助他们管理业务的运营和财务方面。例如，监测运输车辆是否超速。

接下来创建一个简单的实时流应用程序来计算运输车辆每几秒的平均速度。在本方案中，使用 Kafka 作为流数据源，将从 Kafka 的 cars 主题来读取这些事件。同时也将 Kafka 作为流的 Data Sink，检测出的超速事件将写入 Kafka 的 fastcars 主题。车辆超速实时监测流程序的架构如图 3-67 所示。

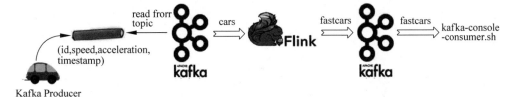

图 3-67　车辆超速实时监测流程序架构

为了模拟车辆发送传感器数据,首先将创建一个 Kafka Producer,它将 id、speed、acceleration 和 timestamp 写入 Kafka 的 cars 主题。

Scala 代码如下:

```scala
//第3章/RandomCarsKafkaProducer.scala

import Java.util.Properties
import org.apache.kafka.clients.producer.{KafkaProducer, ProducerRecord}
import scala.annotation.tailrec
import scala.util.{Random => r}

object RandomCarsKafkaProducer {

  def main(args: Array[String]): Unit = {
    val props = new Properties()
    props.put("Bootstrap.servers", "localhost:9092")
    props.put("key.serializer",
        "org.apache.kafka.common.serialization.StringSerializer")
    props.put("value.serializer",
        "org.apache.kafka.common.serialization.StringSerializer")

    val producer = new KafkaProducer[String, String](props)
    val interval = 1000
    val topic = "cars"
    val numRecsToProduce: Option[Int] = None          //None = infinite
    //val numRecsToProduce: Option[Int] = 1000         //连续产生1000条数据

    @tailrec
    def produceRecord(numRecToProduce: Option[Int]): Unit = {
      //每次调用下面这种方法,将一条车辆行驶数据发送给 Kafka "cars" topic
      def generateCarRecord(topic: String): ProducerRecord[String,String] = {
        val carName = s"car${r.nextInt(10)}"
        val speed = r.nextInt(150)
        val acc = r.nextFloat * 100

        val value = s"$carName,$speed,$acc,${System.currentTimeMillis()}"
        print(s"Writing $value\n")
        val d = r.nextFloat() * 100
        if (d < 2) {
          //产生随机时延
          println("抱歉! 有一些网络时延!")
          Thread.sleep((d * 100).toLong)
        }
        new ProducerRecord[String, String](topic,"key", value)
      }
```

```scala
    numRecToProduce match {
      case Some(x) if x > 0 =>
          //生成一条数据,发送一条数据
        producer.send(generateCarRecord(topic))
        Thread.sleep(interval)
        produceRecord(Some(x - 1))

      case None =>
        producer.send(generateCarRecord(topic))
        Thread.sleep(interval)
        produceRecord(None)

      case _ =>
    }
  }

  produceRecord(numRecsToProduce)
 }

}
```

Java 代码如下:

```java
//第 3 章/RandomCarsKafkaProducer.java

import org.apache.kafka.clients.producer.KafkaProducer;
import org.apache.kafka.clients.producer.ProducerRecord;
import Java.util.Properties;
import Java.util.Random;

public class RandomCarsKafkaProducer {

    private static final String TOPIC_CARS = "cars";      //topic

    public static void main(String[] args) {
        Properties props = new Properties();
        props.put("Bootstrap.servers", "localhost:9092");
        props.put("key.serializer",
            "org.apache.kafka.common.serialization.StringSerializer");
        props.put("value.serializer",
            "org.apache.kafka.common.serialization.StringSerializer");

        KafkaProducer<String, String> producer = new KafkaProducer<>(props);
        //int numRecsToProduce = -1;                      //-1 = infinite
        int numRecsToProduce = 1000;                      //连续产生 1000 条数据
```

```java
            produceRecord(producer,numRecsToProduce);
    }

    private static void produceRecord(KafkaProducer<String, String> producer, int recordNum){
        int interval = 1000;

        //生成有限数据记录
        if(recordNum > 0){
            //生成一条数据,发送一条数据
            producer.send(generateCarRecord(TOPIC_CARS));
            try {
                Thread.sleep(interval);
            } catch (InterruptedException e) {
                e.printStackTrace();
            }
            produceRecord(producer, recordNum - 1);
        }
        //生成无限数据记录
        else if(recordNum < 0){
            producer.send(generateCarRecord(TOPIC_CARS));
            try {
                Thread.sleep(interval);
            } catch (InterruptedException e) {
                e.printStackTrace();
            }
            produceRecord(producer, -1);
        }
    }

    //生成一条车辆监测信息的方法
    private static ProducerRecord<String, String> generateCarRecord(String topic){
        Random r = new Random();
        String carName = "car" + r.nextInt(10);
        int speed = r.nextInt(150);
        float acc = r.nextFloat() * 100;
        long ts = System.currentTimeMillis();

        String value = carName + "," + speed + "," + acc + "," + ts;
        System.out.println("== Writing == :" + value);
        float d = r.nextFloat() * 100;
        if (d < 2) {
            //产生随机时延
            System.out.println("抱歉! 有一些网络时延!");
            try {
                Thread.sleep(Float.valueOf(d * 100).longValue());
```

```
            } catch (InterruptedException e) {
                e.printStackTrace();
            }
        }
        return new ProducerRecord<>(topic,"key", value);
    }
}
```

注意,这里的时间戳是在源处生成事件(消息)的时间。

下面是流处理管道的实现。

Scala 代码如下:

```
//第3章/FastCarsDetect.scala

import Java.sql.Timestamp
import Java.time.Duration
import Java.util.Properties
import org.apache.flink.api.common.eventtime.{SerializableTimestampAssigner, WatermarkStrategy}
import org.apache.flink.api.common.functions.AggregateFunction
import org.apache.flink.api.common.serialization.SimpleStringSchema
import org.apache.flink.connector.kafka.source.KafkaSource
import org.apache.flink.connector.kafka.source.enumerator.initializer.OffsetsInitializer
import org.apache.flink.shaded.jackson2.com.fasterxml.jackson.core.JsonProcessingException
import org.apache.flink.shaded.jackson2.com.fasterxml.jackson.databind.ObjectMapper
import org.apache.flink.streaming.api.scala._
import org.apache.flink.streaming.api.scala.function.ProcessWindowFunction
import org.apache.flink.streaming.api.windowing.assigners.SlidingEventTimeWindows
import org.apache.flink.streaming.api.windowing.time.Time
import org.apache.flink.streaming.api.windowing.Windows.TimeWindow
import org.apache.flink.streaming.connectors.kafka.{FlinkKafkaProducer, KafkaSerializationSchema}
import org.apache.flink.util.Collector
import org.apache.kafka.clients.producer.ProducerRecord

/**
 * 物流公司车辆超速检测程序
 */
object FastCarsDetect {

  //定义流数据类型,这里用了伴生类和伴生对象
  case class CarEvent(carId: String, speed: Int, acceleration: Double, timestamp: Long)

  object CarEvent {
    def apply(rawStr: String): CarEvent = {
```

```scala
    val parts = rawStr.split(",")
    CarEvent(parts(0), parts(1).toInt, parts(2).toDouble, parts(3).toLong)
  }
}

//平均速度的流数据类型
case class CarAvgEvent(carId: String, avgSpeed: Double, start: String, end: String){

  def getCarId:String = carId
  def getAvgSpeed:Double = avgSpeed
  def getStart:String = start
  def getEnd:String = end
}

def main(args: Array[String]) {
  //设置流执行环境
  val env = StreamExecutionEnvironment.getExecutionEnvironment

  //启用检查点
  env.enableCheckpointing(50000)

  //Kafka Source
  val source = KafkaSource.builder[String]
    .setBootstrapServers("localhost:9092")
    .setTopics("cars")
    .setGroupId("group-test")
    .setStartingOffsets(OffsetsInitializer.earliest)
    .setValueOnlyDeserializer(new SimpleStringSchema)
    .build

  //Kafka Sink
  val properties = new Properties()
  properties.setProperty("Bootstrap.servers", "localhost:9092")
  //Kafka Brokers 默认的最大事务超时(transaction.max.timeout.ms)为15min
  //当使用 Semantic.EXACTLY_ONCE 时,下面这个属性值不能超过 15min(默认为1h)
  properties.setProperty("transaction.timeout.ms", String.valueOf(5 * 60 * 1000))
  val myProducer = new FlinkKafkaProducer[CarAvgEvent](
    "fastcars",                                       //目标 topic
    new ObjSerializationSchema("fastcars"),           //序列化 schema
    properties,                                       //producer 配置
    FlinkKafkaProducer.Semantic.EXACTLY_ONCE          //容错性
  )

  myProducer.setWriteTimestampToKafka(true)

  //水印策略
```

```scala
    val watermarkStrategy = WatermarkStrategy
      .forBoundedOutOfOrderness[String](Duration.ofSeconds(1))
      .withTimestampAssigner(new SerializableTimestampAssigner[String]() {
        override def extractTimestamp(s: String, l: Long): Long = s.split(",")(3).toLong
      })

    env
      //指定Kafka数据源
      .fromSource(source, watermarkStrategy, "from cars topic")
      .map(r => CarEvent(r.toString))
      .keyBy(carEvent => carEvent.carId)
      //大小为5s,滑动为2s的滑动窗口
      .window(SlidingEventTimeWindows.of(Time.seconds(5),Time.seconds(2)))
      .aggregate(new AvgSpeedAggFun(), new AvgSpeedProcessFun())
      .filter(carAvgEvent => carAvgEvent.avgSpeed > 120.0)
      //输出
      //.print
      .addSink(myProducer)

    //触发流程序执行
    env.execute("Flink Kafka Source")
  }

  //增量处理函数
  class AvgSpeedAggFun extends AggregateFunction[CarEvent, (Double, Long), Double] {
    //创建初始ACC
    override def createAccumulator() = (0.0, 0L)

    //累加每个传感器(每个分区)的事件
    override def add(carEvent: CarEvent, acc: (Double, Long)) =
      (carEvent.speed + acc._1, acc._2 + 1L)

    //分区合并
    override def merge(acc1: (Double, Long), acc2: (Double, Long)) =
      (acc1._1 + acc2._1, acc1._2 + acc2._2)

    //返回每辆车的平均速度
    override def getResult(acc: (Double, Long)): Double = acc._1 / acc._2
  }

  //窗口处理函数(注意这里引入的ProcessWindowFunction不要错引了Java的)
  //ProcessWindowFunction[IN, OUT, KEY, W]
    class AvgSpeedProcessFun extends ProcessWindowFunction[Double, CarAvgEvent, String, TimeWindow] {
      override def process(key: String,
                           context: Context,
```

```scala
                    elements: Iterable[Double],
                    out: Collector[CarAvgEvent]): Unit = {
    //计算平均速度
    val average = Math.round(elements.iterator.next * 100) / 100.0
    //发送到下游算子
    out.collect(CarAvgEvent(key, average,
      new Timestamp(context.window.getStart).toString,
      new Timestamp(context.window.getEnd).toString))
  }
}

//自定义的序列化模式
//将 CarAvgEvent 作为 json 保存到 Kafka
class ObjSerializationSchema(topic: String)
        extends KafkaSerializationSchema[CarAvgEvent] {
  private var mapper = new ObjectMapper()

  override def serialize(obj: CarAvgEvent,
                    timestamp: Java.lang.Long
                  ): ProducerRecord[Array[Byte], Array[Byte]] = {
    var b:Array[Byte] = null
    try{
      b = mapper.writeValueAsBytes(obj)
    }catch {
      case e: JsonProcessingException => //TODO
    }
    new ProducerRecord[Array[Byte], Array[Byte]](topic, b)
  }
}
```

Java 代码如下：

```java
//第 3 章/FastCarsDetect.java

import org.apache.flink.api.common.eventtime.SerializableTimestampAssigner;
import org.apache.flink.api.common.eventtime.WatermarkStrategy;
import org.apache.flink.api.common.functions.AggregateFunction;
import org.apache.flink.api.common.functions.FilterFunction;
import org.apache.flink.api.common.functions.MapFunction;
import org.apache.flink.api.common.serialization.SimpleStringSchema;
import org.apache.flink.api.java.functions.KeySelector;
import org.apache.flink.api.java.tuple.Tuple2;
import org.apache.flink.connector.kafka.source.KafkaSource;
import org.apache.flink.connector.kafka.source.enumerator.initializer.OffsetsInitializer;
import org.apache.flink.shaded.jackson2.com.fasterxml.jackson.core.jsonProcessingException;
```

```java
import org.apache.flink.shaded.jackson2.com.fasterxml.jackson.databind.ObjectMapper;
import org.apache.flink.streaming.api.environment.StreamExecutionEnvironment;
import org.apache.flink.streaming.api.functions.windowing.ProcessWindowFunction;
import org.apache.flink.streaming.api.windowing.assigners.SlidingEventTimeWindows;
import org.apache.flink.streaming.api.windowing.time.Time;
import org.apache.flink.streaming.api.windowing.windows.TimeWindow;
import org.apache.flink.streaming.connectors.kafka.FlinkKafkaProducer;
import org.apache.flink.streaming.connectors.kafka.KafkaSerializationSchema;
import org.apache.flink.util.Collector;
import org.apache.kafka.clients.producer.ProducerRecord;

import Java.sql.Timestamp;
import Java.time.Duration;
import Java.util.Properties;

/**
 * 物流公司车辆超速检测程序
 */
public class FastCarsDetect {

    //POJO 类,检测到的车速数据类型
    public static class CarEvent {

        public String carId;                //车辆 id
        public int speed;                   //速度
        public double acceleration;         //加速度
        public long timestamp;              //时间戳

        public CarEvent() { }

        public CarEvent(String carId, int speed, double acceleration, long timestamp) {
            this.carId = carId;
            this.speed = speed;
            this.acceleration = acceleration;
            this.timestamp = timestamp;
        }

        @Override
        public String toString() {
            return "CarEvent{" +
                    "carId = '" + carId + '\'' +
                    ", speed = " + speed +
                    ", acceleration = " + acceleration +
                    ", timestamp = " + timestamp +
                    '}';
        }
```

```java
    }

    //POJO类,检测到的平均车速数据类型
    public static class CarAvgEvent {

        public String carId;              //车辆 id
        public double avgSpeed;           //平均速度
        public String start;              //计算平均值的时间范围下限
        public String end;                //计算平均值的时间范围上限

        public CarAvgEvent() { }

        public CarAvgEvent(String carId, double avgSpeed, String start, String end) {
            this.carId = carId;
            this.avgSpeed = avgSpeed;
            this.start = start;
            this.end = end;
        }

        @Override
        public String toString() {
            return "CarEvent{" +
                    "carId = '" + carId + '\'' +
                    ", avgSpeed = " + avgSpeed +
                    ", start = '" + start + '\'' +
                    ", end = '" + end + '\'' +
                    '}';
        }
    }

    public static void main(String[] args) throws Exception {
        //设置流执行环境
        final StreamExecutionEnvironment env =
                StreamExecutionEnvironment.getExecutionEnvironment();

        //Kafka Source
        KafkaSource<String> source = KafkaSource.<String>builder()
            .setBootstrapServers("localhost:9092")
            .setTopics("cars")
            .setGroupId("group-test")
            .setStartingOffsets(OffsetsInitializer.earliest())
            .setValueOnlyDeserializer(new SimpleStringSchema())
            .build();

        //Kafka Sink
        Properties properties = new Properties();
```

```java
properties.setProperty("Bootstrap.servers", "localhost:9092");
//Kafka Brokers 默认的最大事务超时(transaction.max.timeout.ms)为 15min
//当使用 Semantic.EXACTLY_ONCE 语义时,下面这个属性值不能超过 15min
properties.setProperty("transaction.timeout.ms", String.valueOf(5 * 60 * 1000));
FlinkKafkaProducer myProducer = new FlinkKafkaProducer<CarAvgEvent>(
        "fastcars",                                 //目标 topic
        new ObjSerializationSchema("fastcars"),     //序列化 schema
        properties,                                 //producer 配置
        FlinkKafkaProducer.Semantic.EXACTLY_ONCE    //容错性
    );

myProducer.setWriteTimestampToKafka(true);

//水印策略
WatermarkStrategy<String> watermarkStrategy = WatermarkStrategy
 .<String>forBoundedOutOfOrderness(Duration.ofSeconds(2))
 .withTimestampAssigner(new SerializableTimestampAssigner<String>() {
    @Override
    public long extractTimestamp(String s, long l) {
        return Long.parseLong(s.split(",")[3]);
    }
 });

env
    //指定 Kafka 数据源
    .fromSource(source, watermarkStrategy, "from cars topic")
    //转换为 DataStream<SensorReading>
    .map(new MapFunction<String, CarEvent>() {
        @Override
        public CarEvent map(String s) throws Exception {
            String[] fields = s.split(",");
            return new CarEvent(fields[0],
                    Integer.parseInt(fields[1]),
                    Double.parseDouble(fields[2]),
                    Long.parseLong(fields[3]));
        }
    })
    //转换为 KeyedStream
    .keyBy(new KeySelector<CarEvent, String>() {
        @Override
        public String getKey(CarEvent carEvent) throws Exception {
            return carEvent.carId;
        }
    })
    //大小为 5s,滑动为 2s 的滑动窗口
    .window(SlidingEventTimeWindows.of(Time.seconds(5),Time.seconds(2)))
```

```java
        //执行增量聚合
        .aggregate(new AvgSpeedAggFun(),new AvgSpeedProcessFun())
        .filter(new FilterFunction<CarAvgEvent>() {
            @Override
            public boolean filter(CarAvgEvent carAvgEvent) throws Exception {
                return carAvgEvent.avgSpeed > 120.0;
            }
        })
        //.print()
        .addSink(myProducer);

    //触发流程序执行
    env.execute("Flink Sensor Temperature Demo");
}

//增量处理函数
public static class AvgSpeedAggFun implements AggregateFunction<
        CarEvent,                           //input
        Tuple2<Double,Long>,                //acc, <sum, count>
        Double> {                           //output, avg

    //创建初始ACC
    @Override
    public Tuple2<Double,Long> createAccumulator() {
        return new Tuple2<>(0.0,0L);
    }

    //累加每个传感器(每个分区)的事件
    @Override
    public Tuple2<Double,Long> add(CarEvent carEvent, Tuple2<Double,Long> acc) {
        return new Tuple2<>(carEvent.speed + acc.f0, acc.f1 + 1);
    }

    //分区合并
    @Override
    public Tuple2<Double,Long> merge(
                Tuple2<Double,Long> acc1,
                Tuple2<Double,Long> acc2) {
        return new Tuple2<>(acc1.f0 + acc2.f0, acc1.f1 + acc2.f1);
    }

    //返回每个车辆的平均温度
    @Override
    public Double getResult(Tuple2<Double,Long> t2) {
        return t2.f0/t2.f1;
    }
}
```

```java
}

//窗口处理函数
public static class AvgSpeedProcessFun extends ProcessWindowFunction<
        Double,                         //input type
        CarAvgEvent,                    //output type
        String,                         //key type
        TimeWindow> {

    @Override
    public void process(String id,
                        Context context,
                        Iterable<Double> events,
                        Collector<CarAvgEvent> out) {
        double average = Math.round(events.iterator().next() * 100) / 100.0;
        out.collect(new CarAvgEvent(id, average,
                new Timestamp(context.window().getStart()).toString(),
                new Timestamp(context.window().getEnd()).toString()));
    }
}

//自定义的序列化模式
public static class ObjSerializationSchema
            implements KafkaSerializationSchema<CarAvgEvent> {

    private String topic;
    private ObjectMapper mapper;

    public ObjSerializationSchema(String topic) {
        super();
        this.topic = topic;
    }

    @Override
    public ProducerRecord<Byte[], Byte[]> serialize(CarAvgEvent obj, Long timestamp) {
        Byte[] b = null;
        if (mapper == null) {
            mapper = new ObjectMapper();
        }
        try {
            b = mapper.writeValueAsBytes(obj);
        } catch (JsonProcessingException e) {
            //TODO
        }
        return new ProducerRecord<>(topic, b);
    }
}
}
```

要运行以上程序,建议按以下步骤执行:
(1) 启动 ZooKeeper,命令如下:

```
$ ./bin/zookeeper-server-start.sh config/zookeeper.properties
```

(2) 启动 Kafka,命令如下:

```
$ ./bin/kafka-server-start.sh config/server.properties
```

(3) 创建两个 topic,命令如下:

```
$ ./bin/kafka-topics.sh --list --zookeeper localhost:2181
$ ./bin/kafka-topics.sh --zookeeper localhost:2181 --replication-factor 1 --partitions 1 --create --topic cars
$ ./bin/kafka-topics.sh --zookeeper localhost:2181 --replication-factor 1 --partitions 1 --create --topic fastcars
```

(4) 先在一个新的终端窗口中执行消费者脚本,以此来拉取 fastcars topic 数据,命令如下:

```
$ ./bin/kafka-console-consumer.sh --Bootstrap-server localhost:9092 --topic fastcars
```

建议保持窗口运行。
(5) 执行前面编写的流计算程序。
(6) 执行前面编写的数据源程序,命令如下:

```
$ cd cars
$ Java RandomCarsKafkaProducer
```

(7) 回到消息者脚本执行窗口(第 4 步的窗口),查看超速数据。如果一切正常,则应该看到在 fastcars 主题收到的超速数据如下(部分):

```
……
{"carId":"car2","avgSpeed":144.0,"start":"2021-08-27 12:20:48.0","end":"2021-08-27 12:20:53.0"}
{"carId":"car7","avgSpeed":130.0,"start":"2021-08-27 12:20:48.0","end":"2021-08-27 12:20:53.0"}
{"carId":"car7","avgSpeed":130.0,"start":"2021-08-27 12:20:50.0","end":"2021-08-27 12:20:55.0"}
{"carId":"car4","avgSpeed":148.0,"start":"2021-08-27 12:20:56.0","end":"2021-08-27 12:21:01.0"}
{"carId":"car4","avgSpeed":148.0,"start":"2021-08-27 12:20:58.0","end":"2021-08-27 12:21:03.0"}
```

```
{"carId":"car4","avgSpeed":148.0,"start":"2021 - 08 - 27 12:21:00.0","end":"2021 - 08 - 27 12:21:05.0"}
{"carId":"car1","avgSpeed":126.5,"start":"2021 - 08 - 27 12:21:00.0","end":"2021 - 08 - 27 12:21:05.0"}
{"carId":"car1","avgSpeed":134.0,"start":"2021 - 08 - 27 12:21:02.0","end":"2021 - 08 - 27 12:21:07.0"}
{"carId":"car1","avgSpeed":134.0,"start":"2021 - 08 - 27 12:21:04.0","end":"2021 - 08 - 27 12:21:09.0"}
{"carId":"car7","avgSpeed":149.0,"start":"2021 - 08 - 27 12:21:12.0","end":"2021 - 08 - 27 12:21:17.0"}
{"carId":"car7","avgSpeed":149.0,"start":"2021 - 08 - 27 12:21:14.0","end":"2021 - 08 - 27 12:21:19.0"}
{"carId":"car7","avgSpeed":149.0,"start":"2021 - 08 - 27 12:21:16.0","end":"2021 - 08 - 27 12:21:21.0"}
{"carId":"car6","avgSpeed":125.0,"start":"2021 - 08 - 27 12:21:16.0","end":"2021 - 08 - 27 12:21:21.0"}
{"carId":"car1","avgSpeed":139.0,"start":"2021 - 08 - 27 12:21:20.0","end":"2021 - 08 - 27 12:21:25.0"}
{"carId":"car1","avgSpeed":139.0,"start":"2021 - 08 - 27 12:21:22.0","end":"2021 - 08 - 27 12:21:27.0"}
{"carId":"car0","avgSpeed":144.0,"start":"2021 - 08 - 27 12:21:40.0","end":"2021 - 08 - 27 12:21:45.0"}
{"carId":"car0","avgSpeed":144.0,"start":"2021 - 08 - 27 12:21:42.0","end":"2021 - 08 - 27 12:21:47.0"}
{"carId":"car0","avgSpeed":144.0,"start":"2021 - 08 - 27 12:21:44.0","end":"2021 - 08 - 27 12:21:49.0"}
{"carId":"car4","avgSpeed":130.0,"start":"2021 - 08 - 27 12:21:46.0","end":"2021 - 08 - 27 12:21:51.0"}
{"carId":"car4","avgSpeed":130.0,"start":"2021 - 08 - 27 12:21:48.0","end":"2021 - 08 - 27 12:21:53.0"}
{"carId":"car4","avgSpeed":130.0,"start":"2021 - 08 - 27 12:21:50.0","end":"2021 - 08 - 27 12:21:55.0"}
...
```

第 4 章 开发 Flink 批数据处理程序
CHAPTER 4

对于 Flink，批处理是流处理的一种特殊情况。Flink 支持通过 DataSet API 对数据进行批处理。Flink 中的 DataSet 程序是实现数据集转换（如过滤、映射、连接、分组）的常规程序。

4.1 Flink 批处理程序编程模型

6min

在深入了解 Flink 批数据处理程序开发之前，先通过一个简单示例来了解使用 Flink 的 DataSet API 构建批处理应用程序的过程。Flink 程序可以在不同的上下文中运行，可以独立运行，也可以嵌入其他程序中。执行可以发生在本地 JVM 中，也可以发生在多台机器的集群中。

注意：从 Flink 1.12 开始，DataSet API 已经被软弃用。官方建议使用 Table API 和 SQL 在一个完全统一的 API 中运行高效的批处理管道。Table API 与普通批处理连接器和目录集成得很好。或者，还可以使用带有 BATCH 执行模式的 DataStream API。随着开发的进展，DataSet API 最终将被删除。有关这一决定的背景信息，可参阅 FLIP-131。

4.1.1 批应用程序实现

下面的程序是一个完整的 Flink 批处理示例。
【示例 4-1】 应用 Flink DataSet API，实现经典的单词计数程序。
Scala 代码如下：

```
//第 4 章/WordCount.scala

import org.apache.flink.api.scala._

object WordCount {
```

```scala
def main(args: Array[String]) {
  //设置批处理执行环境
  val env = ExecutionEnvironment.getExecutionEnvironment

  val text = env.fromElements(
    "good good study",
    "day day up")

  val wordCounts = text
    .flatMap { _.toLowerCase.split("\\W+") filter { _.nonEmpty } }
    .map { (_, 1) }
    .groupBy(0)
    .sum(1)

  wordCounts.print()
 }
}
```

Java 代码如下:

```java
//第 4 章/WordCount.java

import org.apache.flink.api.common.functions.FlatMapFunction;
import org.apache.flink.api.java.DataSet;
import org.apache.flink.api.java.ExecutionEnvironment;
import org.apache.flink.api.java.tuple.Tuple2;
import org.apache.flink.util.Collector;

public class WordCount {

    public static void main(String[] args) throws Exception {
        //设置批处理执行环境
        final ExecutionEnvironment env =
                ExecutionEnvironment.getExecutionEnvironment();

        DataSet<String> text = env.fromElements(
                    "good good study",
                    "day day up");

        DataSet<Tuple2<String, Integer>> wordCounts = text
            .flatMap(new LineSplitter())
            .groupBy(0)
            .sum(1);

        wordCounts.print();
    }
```

```java
    public static class LineSplitter
            implements FlatMapFunction<String, Tuple2<String, Integer>> {
        @Override
        public void flatMap(String line,Collector<Tuple2<String,Integer>> out){
            for (String word : line.split(" ")) {
                out.collect(new Tuple2<String, Integer>(word, 1));
            }
        }
    }
}
```

执行以上程序,输出结果如下:

```
(up,1)
(day,2)
(good,2)
(study,1)
```

4.1.2 批应用程序剖析

所有的 Flink 批处理程序都以特定的步骤来工作,如图 4-1 所示。

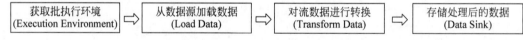

图 4-1 Flink 批处理程序执行步骤

也就是说,每个 Flink 批处理程序都由相同的部分组成:
(1) 获取一个执行环境。
(2) 加载/创建初始数据。
(3) 指定对该数据的转换。
(4) 指定将计算结果放在哪里。

1. 获取一个执行环境

为了开始编写 Flink 程序,用户首先需要获得一个现有的执行环境,如果没有,就需要先创建一个。根据目的的不同,Flink 支持以下几种方式:
(1) 获得一个已经存在的 Flink 环境。
(2) 创建本地环境。
(3) 创建远程环境。

Flink 批处理程序的入口点是 ExecutionEnvironment 类的一个实例,它定义了程序执行的上下文。例如,可以通过静态方法获得 ExecutionEnvironment 的一个实例,代码如下:

```
ExecutionEnvironment.getExecutionEnvironment()
ExecutionEnvironment.createLocalEnvironment()
ExecutionEnvironment.createRemoteEnvironment(String host, int port, String... jarFiles)
```

要获得执行环境,通常只需调用 getExecutionEnvironment()方法。这将根据用户的上下文选择正确的执行环境。如果正在 IDE 中的本地环境上执行,则它将启动一个本地执行环境。如果是从程序中创建了一个 JAR 文件,并通过命令行调用它,则 Flink 集群管理器将执行 main()方法,getExecutionEnvironment()将返回用于在集群上以分布式方式执行程序的执行环境。

例如,在 4.1.1 节的示例中,要获得批处理程序的执行环境。

Scala 代码如下:

```
//设置批处理执行环境
val env = ExecutionEnvironment.getExecutionEnvironment
```

Java 代码如下:

```
//设置批处理执行环境
final ExecutionEnvironment env =
        ExecutionEnvironment.getExecutionEnvironment();
```

2. 加载/创建初始数据

执行环境可以从多种数据源读取数据,包括文本文件、CSV 文件等,也可以使用自定义的数据输入格式。例如,要将文本文件读取为行序列。

Scala 代码如下:

```
val env = ExecutionEnvironment.getExecutionEnvironment
val text = env.readTextFile("file:///path/to/file")
```

Java 代码如下:

```
final ExecutionEnvironment env =
        ExecutionEnvironment.getExecutionEnvironment();
DataSet<String> text = env.readTextFile("file:///path/to/file");
```

数据被逐行读到内存后,Flink 会将它们组织到 DataSet 中,这是 Flink 中用来表示批数据的特殊类。

在 4.1.1 节的示例程序中,使用 fromElements()方法读取内存集合数据,并将读取的数据存储为 DataSet 类型。

Scala 代码如下:

```
val text = env.fromElements("good good study","day day up")
```

Java 代码如下：

```
DataSet<String> text = env.fromElements("good good study", "day day up");
```

3. 指定对该数据的转换

每个 Flink 程序都可对分布式数据集合执行转换操作。Flink 的 DataSet API 提供了多种数据转换功能，包括过滤、映射、连接、分组和聚合。在 4.1.1 节的示例程序中，连续使用了多个不同的转换，将原始数据集转换为新的 DataSet。

Scala 代码如下：

```scala
//对读取的每行文本,进行分割、过滤、转换等操作,然后进行汇总统计
val counts = text
    .flatMap { _.toLowerCase.split("\\W+") filter { _.nonEmpty } }
    .map { (_, 1) }
    .groupBy(0)
    .sum(1)
```

Java 代码如下：

```java
DataSet<Tuple2<String, Integer>> wordCounts = text
    .flatMap(new LineSplitter())
    .groupBy(0)
    .sum(1);
```

暂时不必了解每个转换的具体含义，后面会详细介绍每个转换操作。需要强调的是，Flink 中的转换是延迟执行(惰性)的，在调用 sink 操作之前不会真正执行。

4. 指定将计算结果放在哪里

一旦有了包含最终结果的 DataSet，就可以通过创建接收器(sink)将其写入外部系统。在 4.1.1 节的示例程序中，将计算结果打印输出到控制台上。

Scala 代码如下：

```scala
wordCounts.print()
```

Java 代码如下：

```java
wordCounts.print()
```

Flink 中的接收器(sink)操作可触发批计算的执行，以生成程序所需的结果，例如将结

果保存到文件系统或将其打印到标准输出。上面的代码使用 counts.print() 将结果打印到任务管理器日志中(在 IDE 中运行时,任务管理器日志将显示在 IDE 的控制台中)。这将对批的每个元素调用其 toString() 方法。

4.2 数据源

数据源是 DataSet API 希望从其中获取数据的地方。它可以以文件或 Java 集合的形式出现。DataSet API 支持许多内置的数据源函数。它还支持编写自定义数据源函数,因此不支持的任何数据格式都可以通过编程轻松处理。

DataSet API 支持从多个数据源将批数据集读取到 Flink 系统,并将它们转换为 DataSet 数据集。它主要包括 3 种类型:基于文件的数据源、基于集合的数据源和通用数据源。同时,可以在 DataSet API 中定制 InputFormat/RichInputFormat 接口,以访问不同数据格式的数据源,如 CsvInputFormat、TextInputFormat 等。从 ExecutionEnvironment 类提供的方法中,可以看到支持的数据源方法,如图 4-2 所示。

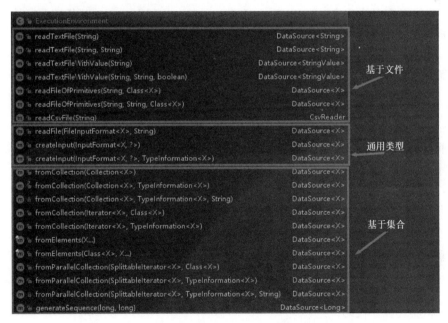

图 4-2 ExecutionEnvironment 类提供的数据源方法

接下来,通过示例程序来分别了解这些数据源函数。

4.2.1 基于文件的数据源

Flink 支持从文件中读取数据。它逐行读取数据并将其作为字符串返回。以下是可以用来读取数据的内置函数。

(1) readTextFile(String path)/TextInputFormat：从路径指定的文件中传输数据。默认情况下，它将读取 TextInputFormat 并逐行读取字符串。

(2) readTextFileWithValue(String path)/TextValueInputFormat：从路径指定的文件中传输数据。它返回可变字符串。

(3) readCsvFile(String path)/CsvInputFormat：从逗号分隔的文件中读取数据。它返回 Java POJO、tuples 或 case class 对象。

(4) readFileofPremitives(path, class)/readFileofPremitives(path, delimiter)/PrimitiveInputFormat：这将把新行解析为基本数据类型，如字符串或整数。

(5) readFileofPremitives(path, delimiter, class)/PrimitiveInputFormat：这将按指定的分割符，把新行解析为基本类型，如字符串或整数。

(6) readHadoopFile(FileInputFormat, Key, Value, path)：它使用给定的 FileInputFormat、Key 类和 Value 类从指定路径读取文件。它将解析后的值作为元组 Tuple2<Key,Value>返回。

(7) readSequenceFile(Key, Value, path)/SequenceFileInputFormat：创建一个 JobConf 并使用给定的 SequenceFileInputFormat、Key 类和 Value 类从指定路径读取文件。它将解析后的值作为元组 Tuple2<Key,Value>返回。

注意：对于文件数据源，Flink 支持递归遍历给定路径中指定的文件夹。为了使用这个功能，需要设置一个环境变量，并在读取数据时将其作为参数传递。要设置的变量是 recursive.file.enumeration。需要将这个变量设置为 true，以便启用递归遍历。

目前，集合数据源要求数据类型和迭代器实现 Serializable。此外，集合数据源不能并行执行(parallelism=1)。

【示例 4-2】 编写 Flink 批处理程序，读取文本文件数据源中的数据。

为了演示这些数据转换函数，需要先准备好示例数据。在项目的 input 目录下，新建一个名为 wc.txt 的文本文件，编辑以下内容，并保存：

```
Good good study
Day day up
```

接下来实现程序代码。
Scala 代码如下：

```scala
//第 4 章/FileSourceDemo.scala

import org.apache.flink.api.scala._

/**
 * 文件数据源
```

```scala
 */
object FileSourceDemo {
  def main(args: Array[String]): Unit = {
    //设置批处理执行环境
    val env = ExecutionEnvironment.getExecutionEnvironment

    //得到输入数据
    val textPath = "input/wc.txt"
    val text = env.readTextFile(textPath)

    //对数据进行转换
    //text.map( _.toLowerCase).print()
    text.flatMap { _.toLowerCase.split("\\W+") }
        .map { (_, 1) }
        .groupBy(0)
        .sum(1)
        .print()
  }
}
```

Java 代码如下：

```java
//第4章/FileSourceDemo.java

import org.apache.flink.api.common.functions.FlatMapFunction;
import org.apache.flink.api.java.DataSet;
import org.apache.flink.api.java.ExecutionEnvironment;
import org.apache.flink.api.java.tuple.Tuple2;
import org.apache.flink.util.Collector;

/**
 * 文件数据源
 */
public class FileSourceDemo {
    public static void main(String[] args) throws Exception {
        //设置批处理执行环境
        final ExecutionEnvironment env =
                ExecutionEnvironment.getExecutionEnvironment();

        //首先从环境中获取一些数据,例如
        String textPath = "input/wc.txt";
        //env.readTextFile(textPath).map(String::toLowerCase).print();
        env
            .readTextFile(textPath)
            .flatMap(new FlatMapFunction< String, Tuple2< String, Integer >>() {
```

```java
            @Override
            public void flatMap(String s, Collector<Tuple2<String, Integer>> out) throws Exception {
                String line = s.toLowerCase();
                for(String word : line.split(" ")){
                    out.collect(new Tuple2<>(word, 1));
                }
            }
        })
        .groupBy(0)
        .sum(1)
        .print();
    }
}
```

执行以上程序,输出结果如下:

```
(up,1)
(day,2)
(good,2)
(study,1)
```

【示例 4-3】 编写 Flink 批处理程序,读取 CSV 文件数据源中的数据。

首先准备数据源。在项目的 input 目录下创建一个 CSV 文件 peoples.csv,并编辑其内容如下:

```
张三,男,23
李四,女,18
王老五,男,35
```

接下来实现批处理程序。

Scala 代码如下:

```scala
//第 4 章/FileSourceDemo2.scala

import org.apache.flink.api.scala._

/**
 * 读取 CSV 文件数据源中的数据
 */
object FileSourceDemo2 {

  //case class
  case class Person(pName:String, pGender:String, pAge: String)
```

```scala
def main(args: Array[String]): Unit = {
  //设置批处理执行环境
  val env = ExecutionEnvironment.getExecutionEnvironment

  //读取 CSV 文件数据源
  val csvFilePath = "input/peoples.csv"
  val csvInput = env
    .readCsvFile[Person](filePath = csvFilePath, fieldDelimiter = ",")

  //输出结果
  csvInput.print
 }
}
```

Java 代码如下：

```java
//第 4 章/FileSourceDemo2.java

import org.apache.flink.api.java.DataSet;
import org.apache.flink.api.java.ExecutionEnvironment;

/**
 * 读取 CSV 文件数据源中的数据
 */
public class FileSourceDemo2 {
    //POJO 类
    public static class Person{
        public String pName;
        public String pGender;
        public String pAge;

        public Person(){}

        public Person(String pName, String pGender, String pAge) {
            this.pName = pName;
            this.pGender = pGender;
            this.pAge = pAge;
        }

        @Override
        public String toString() {
            return "Person(" + pName + "," + pGender + "," + pAge + ")";
        }
    }
```

```java
    public static void main(String[] args) throws Exception {
        //设置批处理执行环境
        final ExecutionEnvironment env =
                ExecutionEnvironment.getExecutionEnvironment();

        //读取 CSV 文件数据源
        String csvFilePath = "input/peoples.csv";
        DataSet<Person> csvInput = env
          .readCsvFile(csvFilePath)
          .fieldDelimiter(",")
          .pojoType(Person.class, "pName", "pGender", "pAge");

        //输出结果
        csvInput.print();
    }
}
```

执行程序,输出结果如下:

```
Person(李四,女,18)
Person(王老五,男,35)
Person(张三,男,23)
```

也可以不使用POJO,直接将读取的每行数据转换为一个元组。参看下面的示例。

【示例4-4】 编写 Flink 批处理程序,读取 CSV 文件数据源中的数据。

Scala 代码如下:

```scala
//第 4 章/FileSourceDemo3.scala

import org.apache.flink.api.scala._

/**
 * 读取 CSV 文件数据源中的数据
 *
 * 也可以不使用 POJO,直接将读取的每行数据转换为一个元组
 */
object FileSourceDemo3 {

  def main(args: Array[String]): Unit = {
    //设置批处理执行环境
    val env = ExecutionEnvironment.getExecutionEnvironment

    //读取 CSV 文件数据源
    val csvFilePath = "input/peoples.csv"
```

```scala
    val csvInput = env
      .readCsvFile[(String,String,String)](filePath = csvFilePath, fieldDelimiter = ",")

    //输出结果
    csvInput.print
  }
}
```

Java 代码如下：

```java
//第 4 章/FileSourceDemo3.java

import org.apache.flink.api.java.DataSet;
import org.apache.flink.api.java.ExecutionEnvironment;
import org.apache.flink.api.java.tuple.Tuple3;

/**
 * 读取 CSV 文件数据源中的数据
 *
 * 也可以不使用 POJO,直接将读取的每行数据转换为一个元组
 */
public class FileSourceDemo3 {

    public static void main(String[] args) throws Exception {
        //设置批处理执行环境
        final ExecutionEnvironment env =
                ExecutionEnvironment.getExecutionEnvironment();

        //读取 CSV 文件数据源
        String csvFilePath = "input/peoples.csv";
        DataSet<Tuple3<String,String,Integer>> csvInput = env
            .readCsvFile(csvFilePath)
            .fieldDelimiter(",")
            .ignoreInvalidLines()
            .types(String.class, String.class, Integer.class);

        //输出结果
        csvInput.print();
    }
}
```

执行以上代码,输出结果如下：

(张三,男,23)
(李四,女,18)
(王老五,男,35)

4.2.2 基于集合的数据源

使用 Flink DataSet API,用户还可以从基于 Java 的集合中读取数据。以下是可以用来读取数据的函数。

(1) fromCollection(Collection):从基于 Java 的集合创建一个数据集。

(2) fromCollection(Iterable):从一个迭代创建一个数据集。由 Iterable 返回的所有元素必须是相同类型的。

(3) fromCollection(Iterator,Class):从迭代器创建数据集。迭代器的元素属于 Class 类参数给定的类型。

(4) fromElements(elements:_*):从给定的对象序列创建一个数据集。所有对象必须具有相同的类型。

(5) fromParallelCollection(SplittableIterator,Class):并行地从迭代器创建数据集。Class 参数表示对象类型。

(6) generateSequence(from,to):生成给定区间之间的数字序列,以并行方式生成。

在下面的示例程序中,演示了如何使用基于集合的数据源方法。

【示例 4-5】 编写 Flink 批处理程序,使用基于集合的数据源示例。

Scala 代码如下:

```scala
//第4章/CollectSourceDemo.scala

import org.apache.flink.api.scala._
/**
 * 基于集合的数据源
 */
object CollectSourceDemo {

  def main(args: Array[String]): Unit = {
    //设置批处理执行环境
    val env = ExecutionEnvironment.getExecutionEnvironment

//env.fromElements("Good good study", "Day day up")
//.map(_.toLowerCase)
//.print

    //或者
```

```scala
    val list = List("Good good study","Day day up")
    env.fromCollection(list).map(_.toLowerCase).print
  }
}
```

Java 代码如下：

```java
//第 4 章/CollectSourceDemo.java

import org.apache.flink.api.java.ExecutionEnvironment;
import Java.util.ArrayList;
import Java.util.Arrays;
import Java.util.List;

/**
 * 集合数据源
 */
public class CollectSourceDemo {
  public static void main(String[] args) throws Exception {
        /设置批处理执行环境
        final ExecutionEnvironment env = ExecutionEnvironment.getExecutionEnvironment();

        //从内存数据构造数据源
//        env.fromElements("Good good study","Day day up")
//                            .map(String::toLowerCase)
//                            .print();

        //或者
        List<String> list = new ArrayList<>(Arrays.asList("Good good study","Day day up"));
        env.fromCollection(list)
            .map(String::toLowerCase)
            .print();
    }
}
```

执行以上代码，输出结果如下：

```
good good study
day day up
```

1. 文件数据源模板代码

更多读取文件数据的用法，可参考下面的模板代码。

Scala 代码如下：

```scala
val env = ExecutionEnvironment.getExecutionEnvironment

//从本地文件系统读取文本文件
val localLines = env.readTextFile("file:///path/to/my/textfile")

//从运行在 nnHost:nnPort 上的 HDFS 读取文本文件
val hdfsLines = env.readTextFile("hdfs://nnHost:nnPort/path/to/my/textfile")

//读取包含 3 个字段的 CSV 文件
val csvInput = env.readCsvFile[(Int, String, Double)]("hdfs://the/CSV/file")

//读取包含 5 个字段的 CSV 文件,只取其中的两个字段
val csvInput = env.readCsvFile[(String, Double)](
  "hdfs://the/CSV/file",
  includedFields = Array(0, 3))           //使用第 1 个和第 4 个字段

//CSV 输入还可以与 case class 一起使用
case class MyCaseClass(str: String, dbl: Double)
val csvInput = env.readCsvFile[MyCaseClass](
  "hdfs://the/CSV/file",
  includedFields = Array(0, 3))           //使用第 1 个和第 4 个字段

//将包含 3 个字段的 CSV 文件读入包含相应字段的 POJO (Person)
val csvInput = env.readCsvFile[Person](
  "hdfs://the/CSV/file",
  pojoFields = Array("name", "age", "zipcode"))

//从一些给定的元素创建一个集合
val values = env.fromElements("Foo", "bar", "foobar", "fubar")

//生成一个数字序列
val numbers = env.generateSequence(1, 10000000)

//从类型为 SequenceFileInputFormat 的指定路径读取文件
val tuples = env.createInput(HadoopInputs.readSequenceFile(classOf[IntWritable], classOf[Text],
"hdfs://nnHost:nnPort/path/to/file"))
```

Java 代码如下：

```java
ExecutionEnvironment env = ExecutionEnvironment.getExecutionEnvironment();

//从本地文件系统读取文本文件
DataSet<String> localLines = env.readTextFile("file:///path/to/my/textfile");

//从运行在 nnHost:nnPort 上的 HDFS 读取文本文件
DataSet<String> hdfsLines = env.readTextFile("hdfs://nnHost:nnPort/path/to/my/textfile");

//读取包含 3 个字段的 CSV 文件
DataSet<Tuple3<Integer, String, Double>> csvInput = env.readCsvFile("hdfs:///the/CSV/file").types(Integer.class, String.class, Double.class);

//读取包含 5 个字段的 CSV 文件,只取其中的两个字段
DataSet<Tuple2<String, Double>> csvInput =
    env.readCsvFile("hdfs:///the/CSV/file")
                    .includeFields("10010") //取第 1 个和第 4 个字段
                    .types(String.class, Double.class);

//将具有 3 个字段的 CSV 文件读入具有相应字段的 POJO (Person.class)
DataSet<Person>> csvInput = env
                    .readCsvFile("hdfs:///the/CSV/file")
                    .pojoType(Person.class, "name", "age", "zipcode");

//从 SequenceFileInputFormat 类型的指定路径读取文件
DataSet<Tuple2<IntWritable, Text>> tuples = env.createInput(
    HadoopInputs.readSequenceFile(IntWritable.class, Text.class, "hdfs://nnHost:nnPort/path/to/file"));

//从一些给定的元素创建一个集合
DataSet<String> value = env.fromElements("Foo", "bar", "foobar", "fubar");

//生成一个数字序列
DataSet<Long> numbers = env.generateSequence(1, 10000000);

//使用 JDBC 输入格式从关系数据库读取数据
DataSet<Tuple2<String, Integer> dbData = env.createInput(
    JDBCInputFormat.buildJDBCInputFormat()
                    .setDrivername("org.apache.derby.jdbc.EmbeddedDriver")
                    .setDBUrl("jdbc:derby:memory:persons")
                    .setQuery("select name, age from persons")
                    .setRowTypeInfo(new RowTypeInfo(BasicTypeInfo.STRING_TYPE_INFO, BasicTypeInfo.INT_TYPE_INFO))
                    .finish()
);
```

需要注意,Flink 的程序编译器需要推断由 InputFormat 返回的数据项的数据类型。如果不能自动推断此信息,则有必要手动提供以上示例所示的类型信息。

2. 递归遍历输入路径目录

对于基于文件的输入,当输入路径是目录时,默认情况下不会枚举嵌套文件。相反,只读取基本目录中的文件,而忽略嵌套文件。可以通过 recursive.file.enumeration 配置参数来启用嵌套文件的递归枚举。

Scala 代码如下:

```
//启用嵌套输入文件的递归枚举
val env = ExecutionEnvironment.getExecutionEnvironment

//创建一个配置对象
val parameters = new Configuration

//设置递归枚举参数
parameters.setBoolean("recursive.file.enumeration", true)

//将配置传递给数据源
env.readTextFile("file://path/with.nested/files")
  .withParameters(parameters)
```

Java 代码如下:

```
//启用嵌套输入文件的递归枚举
ExecutionEnvironment env = ExecutionEnvironment.getExecutionEnvironment();

//创建一个配置对象
Configuration parameters = new Configuration();

//设置递归枚举参数
parameters.setBoolean("recursive.file.enumeration", true);

//将配置传递给数据源
DataSet<String> logs = env
    .readTextFile("file://path/with.nested/files")
    .withParameters(parameters);
```

4.2.3 通用的数据源

DataSet API 支持两个读取数据的通用函数,分别介绍如下。

(1) readFile(inputFormat,path)/FileInputFormat:从给定路径创建 FileInputFormat 类型的数据集。

(2) createInput(inputFormat)/InputFormat:创建一个通用输入格式的数据集。

4.2.4 压缩文件

Flink 支持在读取带有适当扩展名的文件时对其进行透明解压缩。我们不需要进行任何其他配置来读取压缩文件，任何 FileInputFormat 都支持压缩，包括自定义输入格式。如果检测到扩展名正确的文件，则 Flink 会自动解压并将其发送给流进一步处理。

> **注意**：这里需要注意的是，压缩文件可能不会并行读取，从而影响作业的可伸缩性。

Flink 当前支持的压缩方法见表 4-1。

表 4-1 Flink 支持的文件压缩方法

压缩算法	文件扩展名	是否并行
GZip	.gz, .gzip	不并行
DEFLATE	.deflate	不并行
Bzip2	.bz2	不并行
XZ	.xz	不并行

4.3 数据转换

数据转换将数据集从一种形式转换为另一种形式。输入可以是一个或多个数据集，输出也可以是零个、一个或多个数据集。程序可以将多个数据集转换组合成复杂的程序集。

4.3.1 map 转换

map 转换在数据集的每个元素上应用用户定义的映射函数。它实现了一对一的映射，也就是说，函数必须返回一个元素。这是最简单的转换之一，其中输入的是一个数据集，输出的也是一个数据集。例如，应用 map 转换将读取的数据都转换为小写。

Scala 代码如下：

```scala
//第 4 章/TransformerDemo01.scala

import org.apache.flink.api.scala._

object TransformerDemo01 {
  def main(args: Array[String]): Unit = {
    //设置批处理执行环境
    val env = ExecutionEnvironment.getExecutionEnvironment
```

```
        env.fromElements("Good good study", "Day day up")
          .map(_.toLowerCase)
          .print
    }
}
```

Java 代码如下：

```java
//第4章/TransformerDemo01.java

import org.apache.flink.api.java.ExecutionEnvironment;

public class TransformerDemo01 {
    public static void main(String[] args) throws Exception {
        //设置批处理执行环境
        final ExecutionEnvironment env =
                ExecutionEnvironment.getExecutionEnvironment();

        //map 转换：
        env.fromElements("Good good study","Day day up")
           .map(String::toLowerCase)          //Java 的方法引用
           .print();
    }
}
```

对于简单的 map 转换，Flink 可以推测其类型。对于复杂的 map 转换，则需要指定 return type，或者构造一个 MapFunction，还可以 extends 自 Tuple2<Integer, Integer>。

执行以上代码，输出结果如下：

```
good good study
day day up
```

4.3.2 flatMap 转换

flatMap 转换在数据集的每个元素上应用用户定义的 flat-map 函数。flatMap 接收一条记录并输出零条、一条或多条记录。map 函数的这种变体可以为每个输入元素返回任意多个结果元素(包括 none)。下面是进行 flatMap 转换的示例代码。

Scala 代码如下：

```scala
//第4章/TransformerDemo02.scala

import org.apache.flink.api.scala._
```

```scala
object TransformerDemo02 {
  def main(args: Array[String]): Unit = {
    //设置批处理执行环境
    val env = ExecutionEnvironment.getExecutionEnvironment

    env.fromElements("Good good study", "Day day up")
        .map(_.toLowerCase)
        .flatMap(_.split("\\W+"))            //相当于先执行 map,再执行 flatMap
        .map((_,1))
        .print
  }
}
```

Java 代码如下：

```java
//第 4 章/TransformerDemo02.java
import org.apache.flink.api.common.functions.FlatMapFunction;
import org.apache.flink.api.common.typeinfo.Types;
import org.apache.flink.api.java.ExecutionEnvironment;
import org.apache.flink.api.java.tuple.Tuple2;
import org.apache.flink.util.Collector;

public class TransformerDemo02 {
    public static void main(String[] args) throws Exception {
        //设置批处理执行环境
        final ExecutionEnvironment env =
                ExecutionEnvironment.getExecutionEnvironment();

        //首先从环境中获取一些数据,再执行 map 和 flatMap 转换
        /* 使用匿名内部类
        env.fromElements("Good good study","Day day up")
            .map(String::toLowerCase)
            .flatMap(new FlatMapFunction<String, Tuple2<String,Integer>>() {
                @Override
                public void flatMap(String value, Collector<Tuple2<String, Integer>> out)
        throws Exception {
                    for(String word : value.split("\\W+")){
                        out.collect(new Tuple2<>(word,1));
                    }
                }
            })
            .print();
         */
```

```java
            //或者使用 Lambda 函数
            env.fromElements("Good good study","Day day up")
                .map(String::toLowerCase)
                .flatMap((FlatMapFunction<String, Tuple2<String,Integer>>) (value, out) -> {
                    for(String word: value.split("\\W+")){
                        out.collect(new Tuple2<>(word,1));
                    }
                }).returns(Types.TUPLE(Types.STRING, Types.INT))
                .print();
        }
    }
```

对于 flatMap 的支持是无法猜测出来类型的,必须通过 returns(Types.STRING) 指定具体的返回值类型。

执行以上代码,输出结果如下:

```
(good,1)
(good,1)
(study,1)
(day,1)
(day,1)
(up,1)
```

4.3.3　mapPartition 转换

mapPartition 在单个函数调用中转换并行分区。该函数以 Iterable 流的形式获取分区,并可以生成任意数量的结果值。每个分区中的元素数量取决于并行度和之前的操作。它类似于 map,但是一次 map 一整个并行分区。

例如,应用 mapPartition 转换将文本行数据集转换为每个分区的计数数据集。

Scala 代码如下:

```scala
//第 4 章/TransformerDemo03.scala

import org.apache.flink.api.scala._

object TransformerDemo03 {

  def main(args: Array[String]): Unit = {
    //设置批处理执行环境
    val env = ExecutionEnvironment.getExecutionEnvironment

    val wordList = env.fromElements(1, 2, 3, 4, 5, 6, 7, 8).setParallelism(2)
```

```scala
        val counts = wordList.mapPartition(in => Some(in.size))

        counts.print()
    }
}
```

Java 代码如下：

```java
//第4章/TransformerDemo03.java

import org.apache.flink.api.common.functions.MapPartitionFunction;
import org.apache.flink.api.java.DataSet;
import org.apache.flink.api.java.ExecutionEnvironment;
import org.apache.flink.util.Collector;

public class TransformerDemo03 {
    public static void main(String[] args) throws Exception {
        //设置批处理执行环境
        final ExecutionEnvironment env =
                ExecutionEnvironment.getExecutionEnvironment();

        DataSet<Integer> wordList = env
            .fromElements(1, 2, 3, 4, 5, 6, 7, 8)
            .setParallelism(2);

        DataSet<Long> counts = wordList.mapPartition(
            new MapPartitionFunction<Integer, Long>() {
                @Override
                public void mapPartition(Iterable<Integer> values, Collector<Long> out)
throws Exception {
                    long count = 0;
                    for (Integer v : values) {
                        count++;
                    }
                    out.collect(count);
                }
            });

        counts.print();
    }
}
```

执行以上代码，输出结果如下：

```
4
4
```

注意：mapPartition 比 map 更高效，在性能调优中，经常会被建议尽量用 mapPartition 操作去替代 map 操作。

4.3.4 filter 转换

filter 转换在数据集的每个元素上应用一个用户定义的 filter 函数，并仅保留该函数的返回值为 true 的那些元素。filter 函数对条件进行评估，如果结果为 true，则该条数据输出。filter 函数可以输出 0 个记录。下面是进行 filter 转换的示例代码。

Scala 代码如下：

```scala
//第 4 章/TransformerDemo04.scala

import org.apache.flink.api.scala._

object TransformerDemo04 {
  def main(args: Array[String]): Unit = {
    //设置批处理执行环境
    val env = ExecutionEnvironment.getExecutionEnvironment

    //得到输入数据，然后执行 filter 转换
    env.fromElements("Good good study", "Day day up")
      .map(_.toLowerCase)
      .filter(_.contains("study"))
      .print()
  }
}
```

Java 代码如下：

```java
//第 4 章/TransformerDemo04.java

import org.apache.flink.api.java.ExecutionEnvironment;

public class TransformerDemo04 {
  public static void main(String[] args) throws Exception {
    //设置批处理执行环境
    final ExecutionEnvironment env =
        xecutionEnvironment.getExecutionEnvironment();

    //得到输入数据，然后执行 filter 转换
    env.fromElements("Good good study","Day day up")
      .map(String::toLowerCase)
```

```
        .filter(s -> s.contains("study"))
        .print();
    }
}
```

执行以上代码,输出结果如下:

```
good good study
```

4.3.5 reduce 转换

reduce 转换通过缩减当前值和最后一个 reduced 过的值来滚动输出。下面的代码对 DataSet 进行求和。

Scala 代码如下:

```
//第 4 章/TransformerDemo05.scala

import org.apache.flink.api.scala._

object TransformerDemo05 {

  def main(args: Array[String]): Unit = {
    //设置批处理执行环境
    val env = ExecutionEnvironment.getExecutionEnvironment

    //得到输入数据,reduce 转换
    val collection1 = env.fromElements(1,5,7,8)
    collection1.reduce(_ + _).print() //21

    val collection2 = env.fromElements((1,1),(2,2),(3,3))
    collection2.reduce((a,b) => (a._1 + b._1, a._2 * b._2)).print() //(6,6)
  }
}
```

Java 代码如下:

```
//第 4 章/TransformerDemo05.java

import org.apache.flink.api.common.functions.ReduceFunction;
import org.apache.flink.api.java.ExecutionEnvironment;
import org.apache.flink.api.java.operators.DataSource;
import org.apache.flink.api.java.tuple.Tuple2;
```

```java
public class TransformerDemo05 {
    public static void main(String[] args) throws Exception {
        //设置批处理执行环境
        final ExecutionEnvironment env =
                ExecutionEnvironment.getExecutionEnvironment();

        //reduce 转换
        DataSource<Integer> collection1 = env.fromElements(1,5,7,8);
        collection1
            .reduce((ReduceFunction<Integer>) (t1, t2) -> t1 + t2)
            .print();                       //21

        DataSource<Tuple2<Integer,Integer>> collection2 = env
            .fromElements(new Tuple2<>(1,1),new Tuple2<>(2,2),new Tuple2<>(3,3));
        collection2
            .reduce((ReduceFunction<Tuple2<Integer, Integer>>) (t1, t2) ->
                new Tuple2<>(t1.f0 + t2.f0, t1.f1 * t2.f1))
            .print();                       //(6,6)
    }
}
```

执行以上代码,输出结果如下:

```
21
(6,6)
```

4.3.6 在分组数据集上的 reduce 转换

既可以将 reduce 操作用在分组的 DataSet 上,也可以用在不分组的 DataSet 上。应用于分组 DataSet 的 reduce 转换使用用户定义的 reduce()函数将每个组减少为单个元素,相当于 SQL 语言中的分组聚合。对于每组输入元素,reduce()函数连续地将元素对组合成一个元素,直到每个组只剩下一个元素。

注意,对于 ReduceFunction,返回对象的 key 字段应与输入值匹配。这是因为 reduce()是可隐式组合(combine)的,并且从 combine()运算符发出的对象在传递给 reduce()运算符时再次按 key 分组。

Scala 代码如下:

```scala
//第 4 章/TransformerDemo06.scala

import org.apache.flink.api.scala._

object TransformerDemo06 {
```

```scala
//case class 类
case class WC(word:String,count:Int)

def main(args: Array[String]): Unit = {
  //设置批处理执行环境
  val env = ExecutionEnvironment.getExecutionEnvironment

  //用法一
  val words = env.fromElements(("good",1),("good",1),("study",1))
  val wordcounts = words.groupBy(_._1).reduce((a,b) => (a._1,a._2 + b._2))
    //按元组索引
  //val wordcounts = words.groupBy(0).reduce((a,b) => (a._1,a._2 + b._2))
  //或者按 key 的位置索引
  wordcounts.print()

  //用法二
  val words2 = env.fromElements(WC("good",1),WC("good",1), WC("study",1))
  val wordCounts2 = words2
    .groupBy(0).reduce((wc1,wc2) => WC(wc1.word,wc1.count + wc2.count))

  wordCounts2.print()
  }
}
```

Java 代码如下：

```java
//第 4 章/TransformerDemo06.java

import org.apache.flink.api.common.functions.ReduceFunction;
import org.apache.flink.api.java.DataSet;
import org.apache.flink.api.java.ExecutionEnvironment;
import org.apache.flink.api.java.tuple.Tuple2;

public class TransformerDemo07 {
    public static void main(String[] args) throws Exception {
        //设置批处理执行环境
        final ExecutionEnvironment env = ExecutionEnvironment.getExecutionEnvironment();

        //在分组数据集上的 reduce 转换
        //用法一
        Tuple2<String, Integer> word01 = new Tuple2<>("good",1);
        Tuple2<String, Integer> word02 = new Tuple2<>("good",1);
        Tuple2<String, Integer> word03 = new Tuple2<>("study",1);

        DataSet<Tuple2<String, Integer>> words1 = env.fromElements(word01,word02,word03);
```

```java
        DataSet<Tuple2<String, Integer>> wordCounts1 = words1
                .groupBy(0)
                .reduce((ReduceFunction<Tuple2<String, Integer>>) (t1, t2) -> new Tuple2<>(t1.f0,t1.f1 + t2.f1));

        wordCounts1.print();

        //用法二
        DataSet<WC> words2 = env.fromElements(new WC("good",1),new WC("good",1),new WC("study",1));
        DataSet<WC> wordCounts2 = words2
                .groupBy(wc -> wc.word)
                .reduce((ReduceFunction<WC>) (wc1, wc2) -> new WC(wc1.word, wc1.count + wc2.count));

        wordCounts2.print();
    }

    public static class WC {
        String word;
        int count;

        WC(String word, int count) {
            this.word = word;
            this.count = count;
        }

        @Override
        public String toString() {
            return "WC(" + word + "," + count + ")";
        }
    }
}
```

执行以上代码,输出结果如下:

```
(good,2)
(study,1)


WC(good,2)
WC(study,1)
```

4.3.7 在分组数据集上的 GroupReduce 转换

将一组元素组合成一个或多个元素。GroupReduce 可以应用于完整的数据集,也可以应用于分组的数据集。在按字段位置 key 分组的数据集上执行 GroupReduce，仅适用于 Tuple DataSet。

下面的代码展示了如何从按整数分组的数据集中删除重复的字符串。

Scala 代码如下：

```scala
//第 4 章/TransformerDemo07.scala

import org.apache.flink.api.scala._
import org.apache.flink.util.Collector
import scala.collection.mutable

object TransformerDemo07 {
  def main(args: Array[String]): Unit = {
    //设置批处理执行环境
    val env = ExecutionEnvironment.getExecutionEnvironment

    //聚合操作
    val ds = env.fromElements(
      ("河南","郑州市"),
      ("河北","石家庄市"),
      ("河南","开封市"),
      ("河南","开封市"),
      ("河北","邯郸市"),
      ("河北","邯郸市")
    )

    val output = ds.groupBy(0)
      .reduceGroup( (in:Iterator[(String, String)], out:Collector[(String, String)]) => {
        val city: mutable.Set[String] = mutable.Set[String]()
        var key: String = null

        for(t <- in){
          key = t._1
          city.add(t._2)
        }

        //发送给下游
        out.collect((key, "[" + city.toList.mkString(",") + "]"))
      })

    output.print()
  }
}
```

Java 代码如下：

```java
//第4章/TransformerDemo07.java

import org.apache.flink.api.common.functions.GroupReduceFunction;
import org.apache.flink.api.java.DataSet;
import org.apache.flink.api.java.ExecutionEnvironment;
import org.apache.flink.api.java.tuple.Tuple2;
import org.apache.flink.util.Collector;

import Java.util.HashSet;
import Java.util.Set;

public class TransformerDemo07 {
    public static void main(String[] args) throws Exception {
        //设置批处理执行环境
        final ExecutionEnvironment env =
                ExecutionEnvironment.getExecutionEnvironment();

        //构造DataSet
        Tuple2<String,String> word01 = new Tuple2<>("河南","郑州市");
        Tuple2<String,String> word02 = new Tuple2<>("河北","石家庄市");
        Tuple2<String,String> word03 = new Tuple2<>("河南","开封市");
        Tuple2<String,String> word04 = new Tuple2<>("河南","开封市");
        Tuple2<String,String> word05 = new Tuple2<>("河北","邯郸市");
        Tuple2<String,String> word06 = new Tuple2<>("河北","邯郸市");

        DataSet<Tuple2<String,String>> ds = env
                .fromElements(word01,word02,word03,word04,word05,word06);

        //转换操作
        DataSet<Tuple2<String,String>> result = ds
            .groupBy(0)                    //按第1个字段来分组数据集
            .reduceGroup(new GroupReduceFunction<Tuple2<String,String>, Tuple2<String,String>>(){
                @Override
                public void reduce(Iterable<Tuple2<String,String>> in,
                        Collector<Tuple2<String,String>> out) throws Exception {
                    Set<String> city = new HashSet<>();
                    String key = null;

                    //将同一组的所有value添加到一个HashSet中
                    for (Tuple2<String,String> t : in) {
                        key = t.f0;
                        city.add(t.f1);
                    }

                    //发送给下游
```

```
                out.collect(new Tuple2<>(key, city.toString())));
            }
        });
        result.print();
    }
}
```

执行以上代码,输出结果如下:

```
(河南,[开封市, 郑州市])
(河北,[邯郸市, 石家庄市])
```

4.3.8 在分组数据集上的 GroupCombine 转换

GroupCombine 转换是可组合的 GroupReduceFunction 中 combine 步骤的泛化形式。它允许将输入类型 I 与任意输出类型 O 相结合,而 GroupReduce 中 combine 步骤只允许从输入类型 I 到输出类型 I 的组合。

在某些应用程序中,在执行其他转换(例如,减小数据大小)之前,最好将数据集组合成中间格式。这可以通过一个成本非常低的 GroupCombine 转换实现。例如,将 GroupCombine 转换用于另一种 WordCount 实现。

Scala 代码如下:

```
//第4章/TransformerDemo08.scala

import org.apache.flink.api.scala._
import org.apache.flink.util.Collector

import scala.collection.mutable

object TransformerDemo08 {

  def main(args: Array[String]): Unit = {
    //设置批处理执行环境
    val env = ExecutionEnvironment.getExecutionEnvironment

    val combinedWords = env
      .fromElements("Good good study", "Day day up")
      .map(_.toLowerCase)
      .flatMap(_.split("\\W+"))        //相当于先执行 map,再执行 flatMap
      .groupBy((s: String) => s)
      //combine
```

```scala
      .combineGroup((in:Iterator[String], out:Collector[(String,Int)]) => {
        var key:String = null
        var count:Int = 0

        for (word <- in) {
          key = word
          count += 1
        }
        //将带有 word 和 count 的元组发送给下游
        out.collect((key, count))
      })

  val output = combinedWords
    .groupBy(0)
    .reduceGroup( (in:Iterator[(String, Int)], out:Collector[(String, Int)]) => {
      var key:String = null
      var count:Int = 0

      for(t <- in){
        key = t._1
        count += t._2
      }

      //发送给下游
      out.collect((key, count))
    })

  output.print()
  }
}
```

Java 代码如下：

```java
//第 4 章/TransformerDemo08.java

import org.apache.flink.api.common.functions.FlatMapFunction;
import org.apache.flink.api.common.functions.GroupCombineFunction;
import org.apache.flink.api.common.functions.GroupReduceFunction;
import org.apache.flink.api.java.DataSet;
import org.apache.flink.api.java.ExecutionEnvironment;
import org.apache.flink.api.java.tuple.Tuple2;
import org.apache.flink.util.Collector;

public class TransformerDemo08 {
    public static void main(String[] args) throws Exception {
```

```java
//设置批处理执行环境
final ExecutionEnvironment env =
    ExecutionEnvironment.getExecutionEnvironment();

//map 转换
DataSet<String> input = env.fromElements("Good good study day","Day day up study");

DataSet<Tuple2<String, Integer>> combinedWords = input
    .map(String::toLowerCase)
    .flatMap(new FlatMapFunction<String, String>() {
        @Override
        public void flatMap(String value, Collector<String> out) throws Exception {
            for(String word : value.split("\\W+")){
                out.collect(word);
            }
        }
    })
    .groupBy(s -> s)            //分组相同的单词
    //combine
    .combineGroup(new GroupCombineFunction<String, Tuple2<String, Integer>>() {
        @Override
        public void combine(Iterable<String> words, Collector<Tuple2<String, Integer>> out) {
            String key = null;
            int count = 0;

            for (String word : words) {
                key = word;
                count++;
            }
            //将带有 word 和 count 的元组发送给下游
            out.collect(new Tuple2(key, count));
        }
    });

DataSet<Tuple2<String, Integer>> output = combinedWords
    .groupBy(0)                    //再次按词分组
    //具有全部数据交换的 GroupReduce
    .reduceGroup(new GroupReduceFunction<Tuple2<String, Integer>, Tuple2<String, Integer>>() {
        @Override
        public void reduce(Iterable<Tuple2<String, Integer>> iterable,
            Collector<Tuple2<String,Integer>> collector) throws Exception {
            String key = null;
            int count = 0;
```

```java
                    for (Tuple2<String, Integer> word : iterable) {
                        key = word.f0;
                        count += word.f1;
                    }
                    //将带有 word 和 count 的元组发送给下游
                    collector.collect(new Tuple2(key, count));
                }
            });

        output.print();
    }
}
```

执行以上代码,输出结果如下:

```
(up,1)
(day,3)
(good,2)
(study,2)
```

4.3.9 在分组元组数据集上执行聚合

聚合转换非常常见。聚合转换提供了如下的内置聚合函数:sum、min 和 max。聚合转换只能应用于元组数据集,并且只支持用于分组的字段位置键。例如,对按字段位置键分组的数据集应用聚合转换。

Scala 代码如下:

```scala
//第 4 章/TransformerDemo09.scala

import org.apache.flink.api.java.aggregation.Aggregations.{MIN, SUM}
import org.apache.flink.api.scala._

object TransformerDemo09 {
  def main(args: Array[String]): Unit = {
    //设置批处理执行环境
    val env = ExecutionEnvironment.getExecutionEnvironment

    //聚合操作
    val ds = env.fromElements(
      (1,"财务",12000.00),
      (1,"财务",22000.00),
      (1,"市场",18000.00)
    )
```

```scala
    val output = ds.groupBy(1)              //按第 2 个字段分组
      .aggregate(SUM, 0)                    //对第 1 个字段求和
      .and(MIN, 2)                          //计算第 3 个字段的最小值

    output.print()
  }
}
```

Java 代码如下：

```java
//第 4 章/TransformerDemo09.java

import org.apache.flink.api.java.DataSet;
import org.apache.flink.api.java.ExecutionEnvironment;
import org.apache.flink.api.java.tuple.Tuple3;

import static org.apache.flink.api.java.aggregation.Aggregations.MIN;
import static org.apache.flink.api.java.aggregation.Aggregations.SUM;

public class TransformerDemo09 {
    public static void main(String[] args) throws Exception {
        //设置批处理执行环境
        final ExecutionEnvironment env =
            ExecutionEnvironment.getExecutionEnvironment();

        //分组聚合操作
        Tuple3<Integer, String, Double> user01 = new Tuple3<>(1,"财务", 12000.00);
        Tuple3<Integer, String, Double> user02 = new Tuple3<>(1,"财务", 22000.00);
        Tuple3<Integer, String, Double> user03 = new Tuple3<>(1,"市场", 18000.00);

        DataSet<Tuple3<Integer, String, Double>> ds =
                    env.fromElements(user01,user02,user03);

        DataSet<Tuple3<Integer, String, Double>> output = ds
            .groupBy(1)              //按第 2 个字段分组
            .aggregate(SUM, 0)       //对第 1 个字段求和
            .and(MIN, 2);            //计算第 3 个字段的最小值

        output.print();
    }
}
```

执行以上代码，输出结果如下：

```
(1,市场,18000.0)
(2,财务,12000.0)
```

要在一个数据集上应用多个聚合，必须在第 1 次聚合之后使用 and() 函数，这意味着

aggregate(SUM, 0).and(MIN, 2)可生成原始数据集的字段 0 的和及字段 2 的最小值。与此相对照，aggregate(SUM, 0).aggregate(MIN, 2)将对聚合应用聚合。在给定的例子中，它将在计算按字段 1 分组字段 0 的和后产生字段 2 的最小值。

4.3.10　在分组元组数据集上执行 minBy 转换

minBy 转换为每组元组选择一个元组。所选的元组是一个或多个指定字段值最小的元组。用于比较的字段必须是有效的 key 字段，即具有可比性。如果多个元组具有最小字段值，则返回这些元组的任意元组。

例如，从数据集< Tuple3 < Integer，String，Double >>中为每组具有相同字符串值的元组选择 Integer 和 Double 字段的最小值的元组。

Scala 代码如下：

```
val input: DataSet[(Int, String, Double)] = //[...]
val output: DataSet[(Int, String, Double)] = input
    .groupBy(1)          //在第 2 个字段上聚合数据
    .minBy(0, 2)         //选择第 1 个字段和第 3 个字段具有最小值的元组
```

Java 代码如下：

```
DataSet< Tuple3< Integer, String, Double >> input = //[...]
DataSet< Tuple3< Integer, String, Double >> output  =  input
    .groupBy(1)          //在第 2 个字段上聚合数据
    .minBy(0, 2);        //选择第 1 个字段和第 3 个字段具有最小值的元组
```

参看下面的应用示例。
Scala 代码如下：

```
//第 4 章/TransformerDemo10.scala

import org.apache.flink.api.scala._

object TransformerDemo10 {

  def main(args: Array[String]): Unit = {
    //设置批处理执行环境
    val env = ExecutionEnvironment.getExecutionEnvironment

    //聚合操作
    val ds = env.fromElements(
        (2, "财务", 12000.00),
        (1, "财务", 22000.00),
        (1, "财务", 18000.00))
```

```scala
    val output = ds
      .groupBy(1)                    //按第 2 个字段分组
      .minBy(0, 2)                   //选择第 1 列和第 3 列最小值所在的元组

    output.print()
  }
}
```

Java 代码如下：

```java
//第 4 章/TransformerDemo10.java

import org.apache.flink.api.java.DataSet;
import org.apache.flink.api.java.ExecutionEnvironment;
import org.apache.flink.api.java.tuple.Tuple3;

public class TransformerDemo10 {
    public static void main(String[] args) throws Exception {
        //设置批处理执行环境
        final ExecutionEnvironment env =
            ExecutionEnvironment.getExecutionEnvironment();

        //minBy
        Tuple3<Integer, String, Double> user01 = new Tuple3<>(2,"财务", 12000.00);
        Tuple3<Integer, String, Double> user02 = new Tuple3<>(1,"财务", 22000.00);
        Tuple3<Integer, String, Double> user03 = new Tuple3<>(1,"财务", 18000.00);

        DataSet<Tuple3<Integer, String, Double>> ds =
                    env.fromElements(user01,user02,user03);
        DataSet<Tuple3<Integer, String, Double>> output = ds
                .groupBy(1)                     //按第 2 个字段分组
                .minBy(0, 2);                   //选择第 1 列和第 3 列最小值所在的元组

        output.print();
    }
}
```

执行以上代码，输出结果如下：

(1,财务,18000.0)

4.3.11　在分组元组数据集上执行 maxBy 转换

maxBy 转换为每组元组选择一个元组，所选的元组是一个或多个指定字段值最大值的

元组。用于比较的字段必须是有效的 key 字段，即具有可比性。如果多个元组具有最大字段值，则返回这些元组的任意元组。

Scala 代码如下：

```scala
//第 4 章/TransformerDemo11.scala

import org.apache.flink.api.scala._

object TransformerDemo11 {

  def main(args: Array[String]) {
    //设置批处理执行环境
    val env = ExecutionEnvironment.getExecutionEnvironment

    //聚合操作
    val ds = env.fromElements(
        (2, "财务", 12000.00),
        (2, "财务", 22000.00),
        (1, "财务", 18000.00))
    val output = ds.groupBy(1)            //按第 2 个字段分组
          .maxBy(0, 2)                    //选择第 1 列和第 3 列最大值所在的元组

    output.print()
  }
}
```

Java 代码如下：

```java
//第 4 章/TransformerDemo11.java

import org.apache.flink.api.java.DataSet;
import org.apache.flink.api.java.ExecutionEnvironment;
import org.apache.flink.api.java.tuple.Tuple3;

import static org.apache.flink.api.java.aggregation.Aggregations.MIN;
import static org.apache.flink.api.java.aggregation.Aggregations.SUM;

public class TransformerDemo11{

    public static void main(String[] args) throws Exception {
        //设置批处理执行环境
        final ExecutionEnvironment env =
            ExecutionEnvironment.getExecutionEnvironment();
```

```
        Tuple3 < Integer, String, Double > user01 = new Tuple3 <>(2,"财务", 12000.00);
        Tuple3 < Integer, String, Double > user02 = new Tuple3 <>(2,"财务", 22000.00);
        Tuple3 < Integer, String, Double > user03 = new Tuple3 <>(1,"财务", 18000.00);

        DataSet < Tuple3 < Integer, String, Double >> ds =
                    env.fromElements(user01,user02,user03);

        DataSet < Tuple3 < Integer, String, Double >> output = ds
                .groupBy(1)              //按第 2 个字段分组
                .maxBy(0, 2);            //选择第 1 列和第 3 列最大值所在的元组

        output.print();
    }
}
```

执行以上代码,输出结果如下:

```
(2,财务,22000.0)
```

4.3.12 在全部元组数据集上执行聚合操作

有一些常用的聚合操作,聚合转换提供以下内置聚合功能:max、min 和 sum。聚合转换只能应用于元组数据集。例如,在一个完整的数据集上应用聚合转换。

Scala 代码如下:

```
//第 4 章/TransformerDemo12.scala

import org.apache.flink.api.java.aggregation.Aggregations.{MIN, SUM}
import org.apache.flink.api.scala._

object TransformerDemo12 {

  def main(args: Array[String]) {
    //设置批处理执行环境
    val env = ExecutionEnvironment.getExecutionEnvironment

    //聚合操作
    val ds = env.fromElements((1,"财务",12000.00),(1,"财务",22000.00),(1,"市场",18000.00))
    val output = ds.aggregate(SUM, 0)           //对第 1 个字段求和
                   .and(MIN, 2)                 //计算第 3 个字段的最小值

    output.print()
  }
}
```

Java代码如下：

```java
//第4章/TransformerDemo12.java

import org.apache.flink.api.java.DataSet;
import org.apache.flink.api.java.ExecutionEnvironment;
import org.apache.flink.api.java.tuple.Tuple3;

import static org.apache.flink.api.java.aggregation.Aggregations.MIN;
import static org.apache.flink.api.java.aggregation.Aggregations.SUM;

public class TransformerDemo11 {

    public static void main(String[] args) throws Exception {
        //设置批处理执行环境
        final ExecutionEnvironment env = ExecutionEnvironment.getExecutionEnvironment();

        //在全部数据集上执行聚合操作
        Tuple3<Integer, String, Double> user01 = new Tuple3<>(1,"财务", 12000.00);
        Tuple3<Integer, String, Double> user02 = new Tuple3<>(1,"财务", 22000.00);
        Tuple3<Integer, String, Double> user03 = new Tuple3<>(1,"市场", 18000.00);

        DataSet<Tuple3<Integer, String, Double>> ds = env.fromElements(user01,user02,user03);
        DataSet<Tuple3<Integer, String, Double>> output = ds
                .aggregate(SUM, 0)          //对第1个字段求和
                .and(MIN, 2);               //计算第3个字段的最小值

        output.print();
    }
}
```

执行以上代码，输出结果如下：

```
(3,市场,12000.0)
```

4.3.13　distinct转换

distinct转换用来从数据集中删除所有重复的元素。
Scala代码如下：

```scala
//第4章/TransformerDemo13.scala

import org.apache.flink.api.scala._
```

```scala
object TransformerDemo13 {

  def main(args: Array[String]) {
    //设置批处理执行环境
    val env = ExecutionEnvironment.getExecutionEnvironment

    //去重操作
    val ds = env.fromElements(
            (1, "财务", 12000.00),
            (1, "财务", 12000.00),
            (1, "市场", 18000.00))
    val output = ds.distinct

    output.print()
  }
}
```

Java 代码如下：

```java
//第 4 章/TransformerDemo13.java
import org.apache.flink.api.java.DataSet;
import org.apache.flink.api.java.ExecutionEnvironment;
import org.apache.flink.api.java.tuple.Tuple3;

import static org.apache.flink.api.java.aggregation.Aggregations.MIN;
import static org.apache.flink.api.java.aggregation.Aggregations.SUM;

public class TransformerDemo13 {

    public static void main(String[] args) throws Exception {
        //设置批处理执行环境
        final ExecutionEnvironment env =
                ExecutionEnvironment.getExecutionEnvironment();

        //在全部数据集上执行聚合操作
        Tuple3<Integer, String, Double> user01 = new Tuple3<>(1,"财务", 12000.00);
        Tuple3<Integer, String, Double> user02 = new Tuple3<>(1,"财务", 12000.00);
        Tuple3<Integer, String, Double> user03 = new Tuple3<>(1,"市场", 18000.00);

        DataSet<Tuple3<Integer, String, Double>> ds =
                env.fromElements(user01,user02,user03);
        DataSet<Tuple3<Integer, String, Double>> output = ds.distinct();

        output.print();
    }
}
```

执行以上代码,输出结果如下:

```
(1,市场,18000.0)
(1,财务,12000.0)
```

4.3.14　join 连接转换

连接转换将两个数据集连接到一个数据集中。连接条件可以定义为每个数据集中的一个 key。两个数据集的元素在一个或多个 key 上进行连接,这些 key 可以这样指定:

(1) 一个 key 表达式。
(2) 一个 key-selector 函数。
(3) 一个或多个字段位置 key(仅限元组数据集)。
(4) Case Class 字段。

下面是进行 join 转换的示例代码。

Scala 代码如下:

```scala
//第4章/TransformerDemo14.scala

import org.apache.flink.api.scala._

object TransformerDemo14 {

  def main(args: Array[String]) {
    //设置批处理执行环境
    val env = ExecutionEnvironment.getExecutionEnvironment

    //join连接
    val employees = env.fromElements(("张三",1),("李四",2))
    val depts = env.fromElements(("市场部",1),("财务部",3))

    val result = employees.join(depts).where(1).equalTo(1)
    result.print()
  }
}
```

Java 代码如下:

```java
//第4章/TransformerDemo14.scala

import org.apache.flink.api.common.functions.MapFunction;
import org.apache.flink.api.java.DataSet;
import org.apache.flink.api.java.ExecutionEnvironment;
```

```java
import org.apache.flink.api.java.tuple.Tuple2;

public class TransformerDemo14 {

    public static class Employee { public String empName; public int deptId; }
    public static class Dept { public String deptName; public int id; }

    public static void main(String[] args) throws Exception {
        //设置批处理执行环境
        final ExecutionEnvironment env =
                ExecutionEnvironment.getExecutionEnvironment();

        //构造员工数据集
        Employee emp01 = new Employee();
        emp01.empName = "张三";
        emp01.deptId = 1;

        Employee emp02 = new Employee();
        emp02.empName = "李四";
        emp02.deptId = 2;

        DataSet<Employee> employees = env.fromElements(emp01,emp02);

        //构造部门数据集
        Dept dept01 = new Dept();
        dept01.deptName = "市场部";
        dept01.id = 1;

        Dept dept02 = new Dept();
        dept02.deptName = "财务部";
        dept02.id = 3;

        DataSet<Dept> depts = env.fromElements(dept01,dept02);

        //两个数据集执行join连接
        DataSet<Tuple2<Employee, Dept>> result = employees.join(depts)
                .where("deptId")
                .equalTo("id");

        result.map(new MapFunction<Tuple2<Employee,Dept>, String>() {
            @Override
            public String map(Tuple2<Employee, Dept> t) throws Exception {
                return t.f0.empName + "-" + t.f0.deptId + "-" + t.f1.id + "-" + t.f1.deptName;
            }
        }).print();
    }
}
```

执行以上代码,输出结果如下:

```
张三-1-1-市场部
```

4.3.15 union 转换

union 函数执行两个或多个数据集的联合,这将并行地组合数据集。如果将一个数据集与它自己组合起来,则它将输出每个记录两次,生成两个必须具有相同类型的数据集的并集。一个超过两个数据集的 union 可以通过调用多个 union 来实现。

Scala 代码如下:

```scala
val vals1: DataSet[(String, Int)] = //[...]
val vals2: DataSet[(String, Int)] = //[...]
val vals3: DataSet[(String, Int)] = //[...]

val unioned = vals1.union(vals2).union(vals3)
```

Java 代码如下:

```java
DataSet<Tuple2<String, Integer>> vals1 = //[...]
DataSet<Tuple2<String, Integer>> vals2 = //[...]
DataSet<Tuple2<String, Integer>> vals3 = //[...]
DataSet<Tuple2<String, Integer>> unioned = vals1.union(vals2).union(vals3);
```

例如,使用 union 转换合并两个数据集运算,代码如下:
Scala 代码如下:

```scala
//第4章/TransformerDemo15.scala

import org.apache.flink.api.scala._

object TransformerDemo15 {

  def main(args: Array[String]) {
    //设置批处理执行环境
    val env = ExecutionEnvironment.getExecutionEnvironment

    //union合并
    val ds1 = env.fromElements("good good study")
              .flatMap(_.toLowerCase.split("\\W+"))
              .map( (_, 1) )

    val ds2 = env.fromElements("day day up")
```

```
            .flatMap(_.toLowerCase.split("\\W+"))
            .map { (_, 1) }

        ds1.union(ds2).print()
    }
}
```

Java 代码如下：

```java
//第4章/TransformerDemo15.java

import org.apache.flink.api.common.functions.FlatMapFunction;
import org.apache.flink.api.java.DataSet;
import org.apache.flink.api.java.ExecutionEnvironment;
import org.apache.flink.api.java.tuple.Tuple2;
import org.apache.flink.util.Collector;

public class TransformerDemo15 {
    public static void main(String[] args) throws Exception {
        //设置批处理执行环境
        final ExecutionEnvironment env =
                ExecutionEnvironment.getExecutionEnvironment();

        //union 转换
        DataSet<Tuple2<String,Integer>> ds1 = env
            .fromElements("good good study")
            .flatMap(new FlatMapFunction<String, Tuple2<String,Integer>>() {
                @Override
                public void flatMap(String value, Collector<Tuple2<String,Integer>> out)
throws Exception {
                    for(String word: value.split("\\W+")){
                        out.collect(new Tuple2<>(word, 1));
                    }
                }
            });


        DataSet<Tuple2<String,Integer>> ds2 = env
            .fromElements("day day up")
            .flatMap(new FlatMapFunction<String, Tuple2<String, Integer>>() {
                @Override
                public void flatMap(String value, Collector<Tuple2<String,Integer>> out)
throws Exception {
                    for(String word: value.split("\\W+")){
                        out.collect(new Tuple2<>(word, 1));
```

```
                }
            }
        });

        //合并
        DataSet<Tuple2<String, Integer>> ds1_and_ds2 = ds1.union(ds2);

        //输出
        ds1_and_ds2.print();
    }
}
```

执行以上代码,输出结果如下:

```
(good,1)
(day,1)
(good,1)
(day,1)
(study,1)
(up,1)
```

4.3.16 project 转换

project 转换用于删除或移动元组数据集的元组字段。project(int)方法选择应该由其索引保留的元组字段,并在输出元组中定义它们的顺序。使用 project 转换将一个元组的元素移除或移动到另一个元组中。这可以用来对特定的元素进行选择性处理(相当于 SQL 语句中的投影概念)。

注意:在 Scala 中,不支持 project 转换。

例如,对数据集执行 project 转换,只保留想要的字段。
Java 代码如下:

```
//第 4 章/TransformerDemo16.java

import org.apache.flink.api.java.DataSet;
import org.apache.flink.api.java.ExecutionEnvironment;
import org.apache.flink.api.java.tuple.Tuple3;

public class TransformerDemo16 {
```

```java
    public static void main(String[] args) throws Exception {
        //设置批处理执行环境
        final ExecutionEnvironment env =
                ExecutionEnvironment.getExecutionEnvironment();

        //构造数据集
        Tuple3<Integer, String, Double> user01 = new Tuple3<>(1,"张三", 12000.00);
        Tuple3<Integer, String, Double> user02 = new Tuple3<>(2,"李四", 22000.00);
        Tuple3<Integer, String, Double> user03 = new Tuple3<>(3,"王老五", 18000.00);
        DataSet<Tuple3<Integer, String, Double>> ds =
                env.fromElements(user01,user02,user03);

        //project 转换
        DataSet<Tuple3<Integer, String, Double>> ds_select = ds.project(1,2);
        ds_select.print();
    }
}
```

执行以上代码,输出结果如下:

```
(张三,12000.0)
(李四,22000.0)
(王老五,18000.0)
```

4.3.17 first-n 转换

first-n 转换可任意地返回数据集的前 n 个元素。first-n 可应用于常规数据集、分组数据集或分组排序数据集。分组键可以被指定为键选择器函数或字段位置键。

例如,只想要返回一个数据集的前 3 个元素,使用 first-n 转换。

Scala 代码如下:

```scala
//第 4 章/TransformerDemo17.scala

import org.apache.flink.api.scala.ExecutionEnvironment

object TransformerDemo17 {

  def main(args: Array[String]): Unit = {
    //设置批处理执行环境
    val env = ExecutionEnvironment.getExecutionEnvironment

    //first-n 转换
    env.fromElements("All", "ours", "love", "Flink")
```

```
            .map(_.toLowerCase)          //转换为小写
            .first(3)                    //返回数据集中的前3个元素
            .print
    }
}
```

Java代码如下:

```
//第4章/TransformerDemo17.java

import org.apache.flink.api.java.ExecutionEnvironment;

public class TransformerDemo17 {

    public static void main(String[] args) throws Exception {
        //设置批处理执行环境
        final ExecutionEnvironment env =
                ExecutionEnvironment.getExecutionEnvironment();

        //first-n 转换
        env.fromElements("All","ours","love","Flink")
                .map(String::toLowerCase)    //Java的方法引用
                .first(3)                    //返回数据集中的前3个元素
                .print();
    }
}
```

执行以上代码,输出结果如下:

```
all
ours
love
```

4.4 数据接收器

在完成数据转换之后,用户需要将结果保存到某个地方。数据接收器(Data Sink)接收数据集并用于存储或返回它们。

4.4.1 将计算结果保存到文本文件

下面是Flink DataSet Java API提供的一些保存结果的选项。

(1) writeAsText()/TextOutputFormat:将元素按行写为字符串。这些字符串是通过调用每个元素的 toString()方法获得的。

（2）writeAsFormattedText()/TextOutputFormat：将元素按行写成字符串。通过为每个元素调用用户定义的 format()方法来获得字符串。

（3）writeAsCsv()/CsvOutputFormat：将元组写入以逗号分隔的值文件。行和字段分隔符是可配置的。每个字段的值来自对象的 toString()方法。

（4）print()/printToErr()/print(String msg)/printToErr(String msg)：在标准输出/标准错误流中打印每个元素的 toString()值。可选地，可以提供前缀（msg）作为输出的前缀。这有助于区分不同的打印调用。如果并行度大于 1，则输出也将以产生输出的任务的标识符作为前缀。

（5）write()/FileOutputFormat：自定义文件输出的方法和基类。支持自定义对象到字节的转换。

（6）output()/OutputFormat：大多数通用输出方法，用于非基于文件的数据接收器（如将结果存储在数据库中）。

【示例 4-6】 编写 Flink 批处理程序，执行单词计数任务，并将批处理结果保存到文本文件中。

Scala 代码如下：

```scala
//第 4 章/DataSinkDemo01.scala

import org.apache.flink.api.java.aggregation.Aggregations.SUM
import org.apache.flink.api.scala._

object DataSinkDemo01 {

  def main(args: Array[String]) {
    //设置批处理执行环境
    val env = ExecutionEnvironment.getExecutionEnvironment

    //获得数据，执行 map 和 flatMap 转换
    env
      .fromElements("Good good study", "Day day up")
      .map(_.toLowerCase)
      .flatMap(_.split("\\W+"))
      .map((_,1))
      .groupBy(0)
      .aggregate(SUM,1)
      .setParallelism(1)
      .writeAsText("output/word-output");

    //注意：这时一定要有 execute()方法，来触发写操作的执行
    env.execute("write to txt")
  }
}
```

Java 代码如下：

```java
//第4章/DataSinkDemo01.java
import org.apache.flink.api.common.functions.FlatMapFunction;
import org.apache.flink.api.java.ExecutionEnvironment;
import org.apache.flink.api.java.aggregation.Aggregations;
import org.apache.flink.api.java.tuple.Tuple2;
import org.apache.flink.util.Collector;

public class DataSinkDemo01 {

    public static void main(String[] args) throws Exception {
        //设置批处理执行环境
        final ExecutionEnvironment env =
                ExecutionEnvironment.getExecutionEnvironment();

        //获得数据,执行 map 和 flatMap 转换
        env.fromElements("Good good study","Day day up")
            .map(String::toLowerCase)
            .flatMap(new FlatMapFunction<String, Tuple2<String,Integer>>() {
                @Override
                public void flatMap(String s, Collector<Tuple2<String,Integer>> collector) throws Exception {
                    for(String word : s.split("\\W+")){
                        collector.collect(new Tuple2<>(word,1));
                    }
                }
            })
            .groupBy(0)
            .aggregate(Aggregations.SUM,1)
            .setParallelism(1)
            .writeAsText("output/word-output");

        //注意:这时一定要有 execute()方法,来触发写操作的执行
        env.execute("write to txt");
    }
}
```

注意,当要将批处理结果写到文件中时,一定要调用 env.execute()方法,来触发写操作的执行。上面的程序执行以后,可以在项目根目录下看到生成的结果文件 result.txt。打开它,可以看到以下的结果：

```
(up,1)
(day,2)
(good,2)
(study,1)
```

4.4.2 将计算结果保存到 JDBC

也可以使用自定义格式。例如，将单词计数统计结果保存到 MySQL 数据库中。要实现这样的要求，建议按以下步骤操作：

首先，在项目的 pom.xml 文件中添加如下依赖：

```xml
<dependency>
    <groupId>org.apache.flink</groupId>
    <artifactId>flink-JDBC_2.12</artifactId>
    <version>1.13.2</version>
</dependency>

<dependency>
    <groupId>mysql</groupId>
    <artifactId>mysql-connector-Java</artifactId>
    <version>5.1.46</version>
</dependency>
```

在 MySQL 数据库中，创建表 wc，用来接收 Flink 单词计数结果，SQL 语句如下：

```
mysql> create table wc(word varchar(20), cnt int);
```

然后使用 Flink 自带的 JDBCOutputFormat 来写入 MySQL。

无 Scala 代码实现。

Java 代码如下：

```java
//第4章/DataSinkDemo02.java

import org.apache.flink.api.common.functions.FlatMapFunction;
import org.apache.flink.api.common.functions.MapFunction;
import org.apache.flink.api.java.ExecutionEnvironment;
import org.apache.flink.api.java.aggregation.Aggregations;
import org.apache.flink.api.java.tuple.Tuple2;
import org.apache.flink.connector.JDBC.JDBCOutputFormat;
import org.apache.flink.types.Row;
import org.apache.flink.util.Collector;

import Java.sql.Types;

public class DataSinkDemo02 {

    private static String dbUrl = "JDBC:mysql://localhost:3306/xueai8?" +
            "characterEncoding=UTF-8&useSSL=false";
```

```java
    private static String driverClass = "com.mysql.JDBC.Driver";
    private static String userNmae = "root";
    private static String passWord = "admin";

    public static void main(String[] args) throws Exception {
        //设置批处理执行环境
        final ExecutionEnvironment env =
                ExecutionEnvironment.getExecutionEnvironment();

        //获得数据,执行map和flatMap转换
        env.fromElements("Good good study","Day day up")
          .map(String::toLowerCase)
          .flatMap(new FlatMapFunction<String, Tuple2<String, Integer>>() {
              @Override
              public void flatMap(String s, Collector<Tuple2<String, Integer>> collector)
                      throws Exception {
                  for(String word : s.split("\\W+")){
                      collector.collect(new Tuple2<>(word,1));
                  }
              }
          })
          .groupBy(0)
          .aggregate(Aggregations.SUM,1)
          //必须转换为DataSet<Row>
          .map(new MapFunction<Tuple2<String,Integer>, Row>() {
              @Override
              public Row map(Tuple2<String, Integer> t) throws Exception {
                  Row row = new Row(2);
                  row.setField(0, t.f0);
                  row.setField(1, t.f1);
                  return row;
              }
          })
          .setParallelism(1)
          //将数据集写入关系数据库
          .output(
              //构建和配置OutputFormat
              JDBCOutputFormat.buildJDBCOutputFormat()
                .setDBUrl(dbUrl)
                .setDrivername(driverClass)
                .setUsername(userNmae)
                .setPassword(passWord)
                .setQuery("insert into wc (word, cnt) values (?,?)")
                .setSqlTypes(new int[]{Types.NCHAR, Types.INTEGER})
                .finish()
          );
```

```
        //注意:这时一定要有execute()方法,来触发写操作的执行
        env.execute("write to JDBC");
    }
}
```

执行以上代码,然后在 MySQL 数据库中执行查询操作,查看写入的结果,如图 4-3 所示。

```
MariaDB [xueai8]> select * from wc;
+-------+-----+
| word  | cnt |
+-------+-----+
| good  |   2 |
| study |   1 |
| up    |   1 |
| day   |   2 |
+-------+-----+
4 rows in set (0.01 sec)
```

图 4-3 Flink 状态存储

4.4.3 标准 DataSink 方法

下面是标准 DataSink 方法的模板代码。
Scala 代码如下:

```scala
//文本数据
val textData: DataSet[String] = //[...]

//将 DataSet 写入本地文件系统的文件中
textData.writeAsText("file:///my/result/on/localFS")

//将 DataSet 写入 HDFS 的文件中
textData.writeAsText("hdfs://nnHost:nnPort/my/result/on/localFS")

//将 DataSet 写入文件,如果文件存在,则重写该文件
textData.writeAsText("file:///my/result/on/localFS", WriteMode.OVERWRITE)

//元组作为行,管道符号为分隔符 a|b|c
val values: DataSet[(String, Int, Double)] = //[...]
values.writeAsCsv("file:///path/to/the/result/file", "\n", "|")

//将元组以文本格式(a, b, c)写入,而不是作为 CSV 行
values.writeAsText("file:///path/to/the/result/file")

//使用用户定义的格式将值写入字符串
values map { tuple => tuple._1 + " - " + tuple._2 } .writeAsText("file:///path/to/the/result/file")
```

Java 代码如下：

```java
//文本数据
DataSet<String> textData = //[...]

//将 DataSet 写入本地文件系统的文件中
textData.writeAsText("file://my/result/on/localFS");

//将 DataSet 写入 HDFS 的文件中
textData.writeAsText("hdfs://nnHost:nnPort/my/result/on/localFS");

//将 DataSet 写入文件,如果文件存在,则重写该文件
textData.writeAsText("file://my/result/on/localFS", WriteMode.OVERWRITE);

//元组作为行,管道符号为分隔符 a|b|c
DataSet<Tuple3<String, Integer, Double>> values = //[...]
values.writeAsCsv("file://path/to/the/result/file", "\n", "|");

//将元组以文本格式(a, b, c)写入,而不是作为 CSV 行
values.writeAsText("file://path/to/the/result/file");

//使用用户定义的 TextFormatter 对象将值作为字符串写入
values.writeAsFormattedText("file://path/to/the/result/file",
    new TextFormatter<Tuple2<Integer, Integer>>() {
        public String format (Tuple2<Integer, Integer> value) {
            return value.f1 + " - " + value.f0;
        }
    });

//使用自定义输出格式
DataSet<Tuple3<String, Integer, Double>> myResult = [...]

//将元组数据集写入关系数据库
myResult.output(
    //构建和配置 OutputFormat
    JDBCOutputFormat.buildJDBCOutputFormat()
        .setDrivername("org.apache.derby.JDBC.EmbeddedDriver")
        .setDBUrl("JDBC:derby:memory:persons")
        .setQuery("insert into persons (name, age, height) values (?,?,?)")
        .finish()
);
```

4.4.4 本地排序输出

数据接收器的输出可以使用元组字段位置或字段表达式对指定顺序中的指定字段进行

本地排序，这适用于每种输出格式。下面的例子展示了如何使用这个功能。

Scala 代码如下：

```scala
val tData: DataSet[(Int, String, Double)] = //[...]
val pData: DataSet[(BookPojo, Double)] = //[...]
val sData: DataSet[String] = //[...]

//对 String 字段的输出按升序排序
tData.sortPartition(1, Order.ASCENDING).print()

//对 Double 字段按降序排序,对 Int 字段按升序排序
tData.sortPartition(2, Order.DESCENDING).sortPartition(0, Order.ASCENDING).print()

//对嵌套 BookPojo 的 author 字段的输出按降序排序
pData.sortPartition("_1.author", Order.DESCENDING).writeAsText(...)

//按升序对完整元组的输出进行排序
tData.sortPartition("_", Order.ASCENDING).writeAsCsv(...)

//对原子类型(String)输出按降序排序
sData.sortPartition("_", Order.DESCENDING).writeAsText(...)
```

Java 代码如下：

```java
DataSet<Tuple3<Integer, String, Double>> tData = //[...]
DataSet<Tuple2<BookPojo, Double>> pData = //[...]
DataSet<String> sData = //[...]

//对 String 字段的输出按升序排序
tData.sortPartition(1, Order.ASCENDING).print();

//对 Double 字段按降序排序,对 Int 字段按升序排序
tData.sortPartition(2, Order.DESCENDING).sortPartition(0, Order.ASCENDING).print();

//对嵌套 BookPojo 的 author 字段的输出按降序排序
pData.sortPartition("f0.author", Order.DESCENDING).writeAsText(...);

//按升序对完整元组的输出进行排序
tData.sortPartition("*", Order.ASCENDING).writeAsCsv(...);

//对原子类型(String)输出按降序排序
sData.sortPartition("*", Order.DESCENDING).writeAsText(...);
```

注意：目前还不支持全局排序的输出。

4.5 广播变量

广播变量是分布式计算框架中常用的一种数据共享方法。它的主要功能是通过网络传输小数据集,在每台机器上保持一个只读缓存变量,计算节点的实例可以直接读取本地内存中的广播数据集,以避免在数据计算过程中多次从其他节点读取小数据集,从而提高整个任务的计算效率。

广播变量可以理解为一个公共共享变量,它可以广播数据集,使不同的任务可以读取数据,每个节点上只保存一份广播数据。如果不使用广播变量,则会在任务中的每个节点上复制数据集的副本,从而导致内存浪费。

广播数据集通过 withBroadcastSet(DataSet,String)按名称被注册,并且可以通过目标操作符上的 getRuntimeContext().getBroadcastVariable(String)访问。

下面的代码片段展示了如何广播数据集并根据需要使用它。

Scala 代码如下:

```scala
//第4章/BroadcastDemo.scala

import org.apache.flink.api.common.functions.RichMapFunction
import org.apache.flink.api.scala._
import org.apache.flink.configuration.Configuration
import scala.collection.javaConverters._

object BroadcastDemo {

  def main(args: Array[String]) {
    //设置批处理执行环境
    val env = ExecutionEnvironment.getExecutionEnvironment

    //1. 获得一个要被广播的数据集
    val toBroadcast = env.fromElements(1, 2, 3)

    //基本内存集合的数据集
    val data = env.fromElements("India", "USA", "UK")

    val bcData = data.map(new RichMapFunction[String, String]() {
      var toBroadcast: Traversable[Int] = null

      override def open(config: Configuration): Unit = {
        //将广播数据集作为集合访问
        toBroadcast = getRuntimeContext()
          .getBroadcastVariable[Int]("country")
```

```scala
          .asScala
      }

      override def map(input: String): String = {
        var sum = 0
        for (a <- toBroadcast) {
          sum = a + sum
        }
        input.toUpperCase + sum
      }
    }).withBroadcastSet(toBroadcast, "country") //2.广播数据集

    bcData.print()
  }
}
```

Java 代码如下：

```java
//第4章/BroadcastDemo.java
import org.apache.flink.api.common.functions.RichMapFunction;
import org.apache.flink.api.java.DataSet;
import org.apache.flink.api.java.ExecutionEnvironment;
import org.apache.flink.configuration.Configuration;
import Java.util.List;

public class BroadCastDemo {

    public static void main(String[] args) throws Exception {
        //设置批处理执行环境
        final ExecutionEnvironment env =
                ExecutionEnvironment.getExecutionEnvironment();

        //1. 获得一个要被广播的数据集
        DataSet<Integer> toBroadcast = env.fromElements(1, 2, 3);

        //基本内存集合的数据集
        DataSet<String> data = env.fromElements("India", "USA", "UK");

        DataSet<String> bcData = data.map(new RichMapFunction<String,String>() {
            private List<Integer> toBroadcast;

            //必须使用open()方法从上下文获取广播集
            @Override
            public void open(Configuration parameters) throws Exception {
```

```
                //获取广播集,可作为集合使用
                this.toBroadcast =
                    getRuntimeContext().getBroadcastVariable("country");
            }

            @Override
            public String map(String input) throws Exception {
                int sum = 0;
                for (int a : toBroadcast) {
                    sum = a + sum;
                }
                return input.toUpperCase() + sum;
            }
        }).withBroadcastSet(toBroadcast, "country");   //2.广播该数据集,带有名称

        bcData.print();
    }
}
```

执行以上代码,输出结果如下:

```
INDIA6
USA6
UK6
```

当用户有要用于转换的查找条件且查找数据集相对较小时,广播变量非常有用。

注意:由于广播变量的内容保存在每个节点的内存中,所以它不应该太大。对于更简单的内容,例如标量值,可以简单地将参数作为函数闭包的一部分,或者使用withParameters()方法来传递配置。

4.6 分布式缓存

Flink 提供了一个分布式缓存,类似于 Apache Hadoop,使文件在本地可被用户函数的并行实例访问。此功能可用于共享包含静态外部数据(如字典或机器学习的回归模型)的文件。

缓存的工作方式如下:程序将本地或远程文件系统(如 HDFS 或 S3)的文件或目录注册到其执行环境中的特定名称下,作为缓存文件。当程序执行时,Flink 自动将文件或目录复制到所有 TaskManager 节点的本地文件系统。用户可以通过注册别名找到该文件或目录,然后在 TaskManager 节点的本地文件系统中访问该文件。

获取缓存文件的方法类似于广播变量。它还实现了 RichFunction 接口,通过 RichFunction 接口获取 RuntimeContext 对象,然后通过 RuntimeContext 提供的接口获取相应的本地缓存文件。

使用分布式缓存技术的模板代码如下。

Scala 代码如下:

```scala
//在 ExecutionEnvironment 中注册文件或目录
val env = ExecutionEnvironment.getExecutionEnvironment

//从 HDFS 注册一个文件
env.registerCachedFile("hdfs:///path/to/your/file", "hdfsFile")

//注册一个本地可执行文件(script, executable, ...)
env.registerCachedFile("file:///path/to/exec/file", "localExecFile", true)

//定义程序并执行
...
val input: DataSet[String] = ...
val result: DataSet[Integer] = input.map(new MyMapper())
...
env.execute()

//在用户函数(这里是 MapFunction)中访问缓存的文件
//该函数必须扩展 RichFunction 类,因为它需要访问 RuntimeContext
class MyMapper extends RichMapFunction[String, Int] {

  override def open(config: Configuration): Unit = {

    //通过 RuntimeContext 和 DistributedCache 访问缓存的文件
    val myFile: File =
            getRuntimeContext.getDistributedCache.getFile("hdfsFile")
    //读取文件(或导航目录)
    ...
  }

  override def map(value: String): Int = {
    //使用缓存文件的内容
    ...
  }
}
```

Java 代码如下:

```java
//在 ExecutionEnvironment 中注册文件或目录
ExecutionEnvironment env = ExecutionEnvironment.getExecutionEnvironment();
//从 HDFS 注册一个文件
```

```
env.registerCachedFile("hdfs://path/to/your/file", "hdfsFile")

//注册一个本地可执行文件(script, executable, ...)
env.registerCachedFile("file://path/to/exec/file", "localExecFile", true)

//定义程序并执行
...
DataSet<String> input = ...
DataSet<Integer> result = input.map(new MyMapper());
...
env.execute();

//在用户函数(这里是 MapFunction)中访问缓存的文件
//该函数必须扩展 RichFunction 类,因为它需要访问 RuntimeContext
public final class MyMapper extends RichMapFunction<String, Integer> {

    @Override
    public void open(Configuration config) {

        //通过 RuntimeContext 和 DistributedCache 访问缓存的文件
        File myFile =
            getRuntimeContext().getDistributedCache().getFile("hdfsFile");
        //读取文件(或导航目录)
        ...
    }

    @Override
    public Integer map(String value) throws Exception {
        //使用缓存文件的内容
        ...
    }
}
```

【示例 4-7】 Flink 批处理分布式缓存使用示例。
首先在本地创建一个文本文件,例如 E:/目录下,内容如下:

```
1,张三
2,李四
3,王老五
```

然后将其进行分布式缓存。
Scala 代码如下:

```
//第 4 章/DistributeCacheDemo.scala

import org.apache.flink.api.common.functions.RichMapFunction
import org.apache.flink.api.scala._
```

```scala
import org.apache.flink.configuration.Configuration

import scala.collection.mutable
import scala.io.Source

object DistributeCacheDemo {

  def main(args: Array[String]) {
    //设置批处理执行环境
    val env = ExecutionEnvironment.getExecutionEnvironment

    //注册一个本地文件
    val cacheFile = "/flinkbook/cache/userinfo.txt"
    env.registerCachedFile(cacheFile, "localFileUserInfo", executable = true)

    val rawUserAount = List((1, 1000.00), (2, 500.20), (3, 800.50))

    //处理数据：用户ID,用户购买数量,[userid, amount]
    val userAmount = env.fromCollection(rawUserAount)

    val result = userAmount.map(new RichMapFunction[(Int,Double),String] {
      var myline:scala.collection.mutable.Map[Int,String] = mutable.Map[Int,String]()

      //重写 open 实现获取分布式缓存的方法
      override def open(parameters: Configuration): Unit = {
        //获取分布式缓存
        val file = getRuntimeContext
            .getDistributedCache
            .getFile("localFileUserInfo")

        //解析文件,获取每行,以及 I/O 流
        val str: Iterator[String] = Source
            .fromFile(file.getAbsoluteFile)
            .getLines()
        str.foreach(line => {
          val fields = line.split(",")
          myline.put(fields(0).toInt, fields(1))
        })
      }

      override def map(in: (Int, Double)): String = {
        val userName = myline.getOrElse(in._1,"")
        "用户ID:" + in._1 + "| 用户名:" + userName + "| 购买数量:" + in._2
      }
    })

    result.print()
  }
}
```

Java代码如下：

```java
//第4章/DistributeCacheDemo.java

import org.apache.commons.io.FileUtils;
import org.apache.flink.api.common.functions.RichMapFunction;
import org.apache.flink.api.java.DataSet;
import org.apache.flink.api.java.ExecutionEnvironment;
import org.apache.flink.api.java.tuple.Tuple2;
import org.apache.flink.configuration.Configuration;

import Java.io.File;
import Java.util.ArrayList;
import Java.util.HashMap;
import Java.util.List;

public class DistributeCacheDemo {

    public static void main(String[] args) throws Exception {
        //设置批处理执行环境
        final ExecutionEnvironment env =
            ExecutionEnvironment.getExecutionEnvironment();

        String cacheFile = "/flinkbook/cache/userinfo.txt";
        env.registerCachedFile(cacheFile, "localFileUserInfo", true);

        ArrayList<Tuple2<Integer,Double>> rawUserAount = new ArrayList<>();

        rawUserAount.add(new Tuple2<>(1,1000.00));
        rawUserAount.add(new Tuple2<>(2,500.20));
        rawUserAount.add(new Tuple2<>(3,800.50));

        //处理数据：用户ID,用户购买数量,[userid, amount]
        DataSet<Tuple2<Integer, Double>> userAmount =
            env.fromCollection(rawUserAount);

        DataSet<String> result = userAmount.map(
            new RichMapFunction<Tuple2<Integer, Double>, String>() {
                //保存缓存数据
                HashMap<Integer, String> allMap = new HashMap<>();
                @Override
                public void open(Configuration parameters) throws Exception {
                    super.open(parameters);
                    //获取分布式缓存的数据
                    File userInfoFile = getRuntimeContext()
                        .getDistributedCache()
                        .getFile("localFileUserInfo");
                    List<String> userInfo =
```

```java
                    FileUtils.readLines(userInfoFile,"utf-8");
                for (String value : userInfo) {
                    String[] split = value.split(",");
                    allMap.put(Integer.parseInt(split[0]), split[1]);
                }
            }

            @Override
            public String map(Tuple2<Integer, Double> value)
                        throws Exception {
                String userName = allMap.getOrDefault(value.f0,"");
                return "用户ID:" + value.f0 + "|用户名:" + userName + "|购买数量:" + value.f1;
            }
        });

    result.print();
    }
}
```

执行以上代码,输出内容如下:

```
用户ID:1|用户名:张三|购买数量:1000.0
用户ID:2|用户名:李四|购买数量:500.2
用户ID:3|用户名:王老五|购买数量:800.5
```

4.7 参数传递

5min

参数可以通过构造函数或 withParameters(Configuration)方法传递给函数。参数被序列化为函数对象的一部分,并传递给所有并行任务实例。

4.7.1 通过构造函数传参

通过构造函数(构造器)向函数对象传递参数。
Scala 代码如下:

```scala
//第4章/PassParametersDemo01.scala

import org.apache.flink.api.common.functions.FilterFunction
import org.apache.flink.api.scala._
```

```scala
/**
 * 通过构造器向函数对象传参
 */
object PassParametersDemo01 {

  def main(args: Array[String]) {
    //设置批处理执行环境
    val env = ExecutionEnvironment.getExecutionEnvironment

    //数据源
    val toFilter = env.fromElements(1, 2, 3, 4, 5)

    //FilterFunction传参:过滤出大于2的元素
    toFilter.filter(new MyFilter(2)).print()
  }

  class MyFilter(limit: Int) extends FilterFunction[Int] {
    override def filter(value: Int): Boolean = {
      value > limit
    }
  }
}
```

Java代码如下：

```java
//第4章/PassParametersDemo01.java

import org.apache.flink.api.common.functions.FilterFunction;
import org.apache.flink.api.java.DataSet;
import org.apache.flink.api.java.ExecutionEnvironment;

/**
 * 通过构造器向函数对象传参
 */
public class PassParametersDemo01 {

    public static void main(String[] args) throws Exception {
        //设置批处理执行环境
        final ExecutionEnvironment env =
                ExecutionEnvironment.getExecutionEnvironment();

        //数据源
        DataSet<Integer> toFilter = env.fromElements(1, 2, 3, 4, 5);

        //FilterFunction传参:过滤出大于2的元素
```

```java
        toFilter.filter(new MyFilter(2)).print();
    }

    private static class MyFilter implements FilterFunction<Integer> {

        private final int limit;

        public MyFilter(int limit) {
            this.limit = limit;
        }

        @Override
        public boolean filter(Integer value) throws Exception {
            return value > limit;
        }
    }
}
```

执行以上代码,输出结果如下:

```
3
4
5
```

4.7.2 通过 withParameters(Configuration)传参

通过 withParameters(Configuration)方法向函数对象传递参数。
Scala 代码如下:

```scala
//第 4 章/PassParametersDemo02.scala

import org.apache.flink.api.common.functions.RichFilterFunction
import org.apache.flink.api.scala._
import org.apache.flink.configuration.Configuration

/**
 * 通过 withParameters(Configuration)方法向函数对象传参
 */
object PassParametersDemo02 {

  def main(args: Array[String]) {
    //设置批处理执行环境
    val env = ExecutionEnvironment.getExecutionEnvironment
```

```scala
//数据源
val toFilter = env.fromElements(1, 2, 3, 4, 5)

//配置参数
val c = new Configuration()
c.setInteger("limit", 2)

toFilter
  .filter(new RichFilterFunction[Int]() {
    var limit = 0

    override def open(config: Configuration): Unit = {
      limit = config.getInteger("limit", 0)
    }

    override def filter(in: Int): Boolean = {
      in > limit
    }
  }).withParameters(c)
  .print()
```

Java代码如下：

```java
//第4章/PassParametersDemo02.java

import org.apache.flink.api.common.functions.RichFilterFunction;
import org.apache.flink.api.java.DataSet;
import org.apache.flink.api.java.ExecutionEnvironment;
import org.apache.flink.configuration.Configuration;

/**
 * 通过withParameters(Configuration)方法向函数对象传参
 */
public class PassParametersDemo02 {

    public static void main(String[] args) throws Exception {
        //设置批处理执行环境
        final ExecutionEnvironment env =
            ExecutionEnvironment.getExecutionEnvironment();

        //数据源
        DataSet<Integer> toFilter = env.fromElements(1, 2, 3, 4, 5);
```

```java
//配置参数
Configuration config = new Configuration();
config.setInteger("limit", 2);

//向 FilterFunction 传参
toFilter.filter(new RichFilterFunction< Integer >() {
    private int limit;

    @Override
    public void open(Configuration parameters) throws Exception {
        limit = parameters.getInteger("limit", 0);
    }

    @Override
    public boolean filter(Integer value) throws Exception {
        return value > limit;
    }
}).withParameters(config)
    .print();
}
}
```

执行以上代码,输出结果如下:

```
3
4
5
```

4.7.3　通过 ExecutionConfig 传递全局参数

Flink 还允许将自定义配置值传递给环境的 ExecutionConfig 接口。由于执行配置可以在所有用户函数(Rich 类型函数)中访问,因此自定义配置将在所有函数中全局可用。

通过 ExecutionConfig 传递全局参数。

Scala 代码如下:

```scala
//第 4 章/PassParametersDemo03.scala

import org.apache.flink.api.common.functions.RichFilterFunction
import org.apache.flink.api.scala._
import org.apache.flink.configuration.Configuration

object PassParametersDemo03 {
```

```scala
def main(args: Array[String]) {
  //设置批处理执行环境
  val env = ExecutionEnvironment.getExecutionEnvironment

  //数据源
  val toFilter = env.fromElements(1, 2, 3, 4, 5)

  //(1)设置自定义全局配置
  val config = new Configuration()
  config.setInteger("limit", 2)
  env.getConfig.setGlobalJobParameters(config)

  //向 FilterFunction 传参
  toFilter
    .filter(new RichFilterFunction[Int]() {
      var limit = 0

      //(2)访问全局配置中的值
      override def open(config: Configuration): Unit = {
        val globalParams = getRuntimeContext
          .getExecutionConfig
          .getGlobalJobParameters
        val globConf = globalParams.asInstanceOf[Configuration]
        limit = globConf.getInteger("limit", 0)
      }

      override def filter(in: Int): Boolean = {
        in > limit
      }
    })
    .print()
}
```

Java 代码如下：

```java
//第 4 章/PassParametersDemo03.java

import org.apache.flink.api.common.ExecutionConfig;
import org.apache.flink.api.common.functions.RichFilterFunction;
import org.apache.flink.api.java.DataSet;
import org.apache.flink.api.java.ExecutionEnvironment;
import org.apache.flink.configuration.Configuration;

public class PassParametersDemo03 {

    public static void main(String[] args) throws Exception {
        //设置批处理执行环境
```

```
final ExecutionEnvironment env =
  ExecutionEnvironment.getExecutionEnvironment();

//数据源
DataSet< Integer > toFilter = env.fromElements(1, 2, 3, 4, 5);

//(1)设置自定义全局配置
Configuration config = new Configuration();
config.setInteger("limit", 2);
env.getConfig().setGlobalJobParameters(config);

//向 FilterFunction 传参
toFilter.filter(new RichFilterFunction< Integer >() {
    private int limit;

    //(2)访问全局配置中的值
    @Override
    public void open(Configuration parameters) throws Exception {
        ExecutionConfig.GlobalJobParameters globalParams =
            getRuntimeContext()
                .getExecutionConfig()
                .getGlobalJobParameters();
        Configuration globConf = (Configuration) globalParams;
        limit = globConf.getInteger("limit", 0);
    }

    @Override
    public boolean filter(Integer value) throws Exception {
        return value > limit;
    }
})
.print();
}
}
```

执行以上代码,输出结果如下：

```
3
4
5
```

4.8 数据集中的拉链操作

某些算法中,可能需要将唯一标识符分配给数据集元素。可以使用 Flink DataSet API 的 org.apache.flink.api.java.utils 包中提供的工具类 DataSetUtils 达到此目的。该类提供了简单的实用程序方法,用于使用索引或唯一标识符对数据集中的元素执行拉链操作。

4.8.1 密集索引

DataSetUtils 类中有一个 zipWithIndex 方法,可以为元素分配连续的标签,它接收一个数据集作为输入并返回一个新的数据集,包(唯一 id,初始值)二元组。

该方法的签名如下:

```
public static < T > DataSet < Tuple2 < Long,T >> zipWithIndex(DataSet < T > input)
```

该方法为输入数据集中的所有元素分配一个唯一的 Long 值,生成的值是连续的。这种方法为元素分配连续标签的过程需要遍历两次,第 1 次用于计数,然后第 2 次标记元素。由于计数的同步,所以不能以流水线的方式进行。方法执行后返回一个由连续的 id 和初始值组成的 DataSet[(Long,T)]2 元组的新数据集。当唯一的标签足够时,它是首选方法。

例如,对数据流中的元素分配连续密集索引。

Scala 代码如下:

```scala
//第 4 章/ZipWithIndexDemo.scala

import org.apache.flink.api.scala._
import org.apache.flink.api.scala.utils.DataSetUtils

/**
 * 数据集中的元素拉链操作 - 密集索引
 */
object ZipWithIndexDemo {

  def main(args: Array[String]) {
    //设置批处理执行环境
    val env = ExecutionEnvironment.getExecutionEnvironment

    //设置两个并行分区
    env.setParallelism(2)

    //初始数据源
    val input:DataSet[String] = env.fromElements("A", "B", "C", "D", "E", "F", "G", "H")

    //分配索引
    val result: DataSet[(Long,String)] = input.zipWithIndex

    //data sink
    result.print()
  }
}
```

Java 代码如下：

```java
//第4章/ZipWithIndexDemo.java

import org.apache.flink.api.java.DataSet;
import org.apache.flink.api.java.ExecutionEnvironment;
import org.apache.flink.api.java.tuple.Tuple2;
import org.apache.flink.api.java.utils.DataSetUtils;

/**
 * 数据集中的元素拉链操作-密集索引
 */
public class ZipWithIndexDemo {

    public static void main(String[] args) throws Exception {
        //设置批处理执行环境
        final ExecutionEnvironment env =
                ExecutionEnvironment.getExecutionEnvironment();

        //设置两个并行分区
        env.setParallelism(2);

        //初始数据源
        DataSet<String> in = env.fromElements("A", "B", "C", "D", "E", "F", "G", "H");

        //分配索引
        DataSet<Tuple2<Long, String>> result = DataSetUtils.zipWithIndex(in);

        //data sink
        result.print();
    }
}
```

执行以上程序，输出的结果如下：

```
0,A
1,B
2,C
3,D
4,E
5,F
6,G
7,H
```

4.8.2 唯一索引

在很多情况下，可能不需要分配连续的标签。DataSetUtils 类还提供了另一个

zipWithUniqueId方法,这种方法以流水线的方式工作。
该方法的签名如下：

```
public static <T> DataSet<Tuple2<Long,T>> zipWithUniqueId(DataSet<T> input)
```

该方法为输入数据集中的所有元素分配一个唯一的Long值,其工作过程如下：
(1) 将一个map()函数应用于输入数据集。
(2) 每个map任务都有一个计数器c,该计数器为每个记录增加一个值。
(3) c移位n位,其中n=log2(并行任务的数量)。
(4) 要在所有任务中创建唯一的ID,需要将任务ID添加到计数器中。
(5) 对于每个记录,将收集结果计数器。

这种方法接收一个数据集作为输入,并返回一个由ID和初始值组成的tuple 2数据集。这个zipWithUniqueId()方法以流水线的方式工作,加快了标签分配的过程。

例如,对数据流中的元素分配唯一索引。
Scala代码如下：

```scala
//第4章/ZipWithUniqueIdDemo.scala

import org.apache.flink.api.scala._
import org.apache.flink.api.scala.utils.DataSetUtils

/**
 * 数据集中的元素拉链操作 - 唯一索引
 */
object ZipWithUniqueIdDemo {

  def main(args: Array[String]) {
    //设置批处理执行环境
    val env = ExecutionEnvironment.getExecutionEnvironment

    //设置两个并行分区
    env.setParallelism(2)

    //初始数据源
    val input:DataSet[String] = env.fromElements("A", "B", "C", "D", "E", "F", "G", "H")

    //分配唯一索引
    val result: DataSet[(Long,String)] = input.zipWithUniqueId

    //data sink
    result.print()
  }
}
```

Java 代码如下：

```java
//第4章/ZipWithUniqueIdDemo.java

import org.apache.flink.api.java.DataSet;
import org.apache.flink.api.java.ExecutionEnvironment;
import org.apache.flink.api.java.tuple.Tuple2;
import org.apache.flink.api.java.utils.DataSetUtils;

/**
 * 数据集中的元素拉链操作 - 唯一索引
 */
public class ZipWithUniqueIdDemo {

    public static void main(String[] args) throws Exception {
        //设置批处理执行环境
        final ExecutionEnvironment env =
                ExecutionEnvironment.getExecutionEnvironment();

        //设置两个并行分区
        env.setParallelism(2);

        //初始数据源
        DataSet<String> in = env.fromElements("A", "B", "C", "D", "E", "F", "G", "H");

        //分配唯一索引
        DataSet<Tuple2<Long, String>> result =
                        DataSetUtils.zipWithUniqueId(in);

        //data sink
        result.print();
    }
}
```

执行以上代码，输出结果如下：

```
(0,G)
(1,A)
(2,H)
(3,B)
(5,C)
(7,D)
(9,E)
(11,F)
```

4.9 Flink 批处理示例

现在有一个已经采集了的豆瓣热门电影数据集，以 CSV 格式提供。数据集是使用爬虫采集自豆瓣的热门电影，部分数据如下：

```
标题,标题链接,pic,缩略图,bd,inq,hd1,rating_score,rating_num,other
蝙蝠侠·黑暗骑士, https://movie. douban. com/subject/1851857/, 26, https://img3. doubanio. com/...,"导演：克里斯托弗·诺兰 Christopher Nolan 主演：克里斯蒂安·贝尔 Christ... 2008 / 美国 英国 / 剧情 动作 科幻 犯罪 惊悚",无尽的黑暗., / The Dark Knight,9.2,713265 人评价,/ 蝙蝠侠前传 2:黑暗骑士 / 黑暗骑士(台)
控方证人, https://movie. douban. com/subject/1296141/, 27, https://img1. doubanio. com/...,"导演：比利·怀尔德 Billy Wilder 主演：泰隆·鲍华 Tyrone Power / 玛琳... 1957 / 美国 / 剧情 犯罪 悬疑",比利·怀德满分作品., / Witness for the Prosecution,9.6,271118 人评价,/ 雄才伟略 / 情妇
活着, https://movie. douban. com/subject/1292365/, 28, https://img3. doubanio. com/...,"导演：张艺谋 Yimou Zhang 主演：葛优 You Ge / 巩俐 Li Gong / 姜武 Wu Jiang 1994 / 中国大陆 中国香港 / 剧情 历史 家庭",张艺谋最好的电影., / 人生 Lifetimes,9.2,551959 人评价,/ 人生 Lifetimes
乱世佳人, https://movie. douban. com/subject/1300267/, 29, https://img3. doubanio. com/...,"导演：维克多·弗莱明 Victor Fleming / 乔治·库克 George Cukor 主演：费... 1939 / 美国 / 剧情 历史 爱情 战争",Tomorrow is another day., / Gone with the Wind,9.3,473671 人评价,/ 飘
寻梦环游记, https://movie. douban. com/subject/20495023/, 30, https://img1. doubanio. com/...,"导演：李·昂克里奇 Lee Unkrich / 阿德里安·莫利纳 Adrian Molina 主演：... 2017 / 美国 / 喜剧 动画 奇幻 音乐",死亡不是真的逝去,遗忘才是永恒的消亡., / Coco,9.1,1033536 人评价,/ 可可夜总会(台) / 玩转极乐园(港)
末代皇帝, https://movie. douban. com/subject/1293172/, 31, https://img3. doubanio. com/...,"导演：贝纳尔多·贝托鲁奇 Bernardo Bertolucci 主演：尊龙 John Lone / 陈... 1987 / 英国 意大利 中国大陆 法国 / 剧情 传记 历史","不要跟我比惨,我比你更惨"再适合这部电影不过了., / The Last Emperor,9.2,492158 人评价,/ 末代皇帝溥仪(港)
摔跤吧！爸爸, https://movie. douban. com/subject/26387939/, 32, https://img9. doubanio. com/...,"导演：涅提·蒂瓦里 Nitesh Tiwari 主演：阿米尔·汗 Aamir Khan / 法缇玛... 2016 / 印度 / 剧情 传记 运动 家庭",你不是在为你一个人战斗,你要让千千万万的女性看到女生并不是只能相夫教子., / Dangal,9.0,1079607 人评价,/ 我和我的冠军女儿(台) / 打死不离 3 父女(港)
指环王 3:王者无敌, https://movie. douban. com/subject/1291552/, 33, https://img3. doubanio. com/...,"导演：彼得·杰克逊 Peter Jackson 主演：维果·莫腾森 Viggo Mortensen / ... 2003 / 美国 新西兰 / 剧情 动作 奇幻 冒险",史诗的终章., / The Lord of the Rings: The Return of the King,9.2,522845 人评价,/ 魔戒三部曲:王者再临(台 / 港)
少年派的奇幻漂流, https://movie. douban. com/subject/1929463/, 34, https://img3. doubanio. com/...,"导演：李安 Ang Lee 主演：苏拉·沙玛 Suraj Sharma / 伊尔凡·可汗 Irrfan... 2012 / 美国 中国台湾 英国 加拿大 / 剧情 奇幻 冒险",瑰丽壮观、无人能及的冒险之旅., / Life of Pi,9.1,1004021 人评价,/ 少年 Pi 的奇幻漂流 / 漂流少年 Pi
……
```

需要注意,数据集的第 1 行是标题行。另外,在所有字段中,只关心以下几个字段。

(1) 标题：索引为 0,表示电影名称。

(2) bd：索引为 4，表示导演姓名。
(3) rating_score：索引为 7，表示电影评分。
(4) rating_num：索引为 8，表示电影评论人数。
(5) other：索引为 9，表示电影别名。

现希望通过对该数据集的分析，回答以下问题：
(1) 数据集中采集的电影总数量是多少？
(2) 这些热门电影平均评分是多少？
(3) 找出这些热门电影中评论人数最多的十部电影。

接下来将分别以 Scala DataSet API 和 Java DataSet API 实现这个 Flink 批处理程序。

4.9.1　分析豆瓣热门电影数据集——Scala 实现

【示例 4-8】　使用 Flink Scala 批处理 API 分析豆瓣热门电影数据集。
建议按以下步骤实现：
(1) 创建项目，并将源数据集复制到项目的 input/douban/ 目录下，如图 4-4 所示。

图 4-4　将数据源文件复制到项目的 input/douban/ 目录下

(2) 新建一个 case class，代表数据集中的每行电影数据，代码如下：

```
case class Movie(
    title:String,
    director:String,
    rating:Double,
    ratingNum:String,
    other:String
)
```

(3) 回答第 1 个问题：数据集中采集的电影总数量是多少？代码如下：

```
import org.apache.flink.api.scala._

object DoubanMoviesBatch {

  //case class
  case class Movie(
    title:String,
```

```scala
      director:String,
      rating:Double,
      ratingNum:String,
      other:String
  )

  def main(args: Array[String]) {
    //设置批处理执行环境
    val env = ExecutionEnvironment.getExecutionEnvironment

    //初始数据源
    val csvFile = "input/douban/douban_movies.csv"

    val in:DataSet[Movie] = env
      .readCsvFile[Movie](
        csvFile,
        includedFields = Array(0,4,7,8,9),
        fieldDelimiter = ",",
        ignoreFirstLine = true
      )

    //问题一:统计电影总数量
    println("问题一:统计电影总数量")
    in.map(_ => 1L).reduce(_ + _).print()
  }
}
```

运行上面的代码,输出统计结果如下:

```
225
```

(4) 回答第 2 个问题:这些热门电影平均评分是多少? 代码如下:

```scala
import Java.text.DecimalFormat
import org.apache.flink.api.scala._

object DoubanMoviesBatch {

  //case class
  case class Movie(
      title:String,
      director:String,
      rating:Double,
      ratingNum:String,
      other:String
  )
```

```scala
def main(args: Array[String]) {
  //设置批处理执行环境
  val env = ExecutionEnvironment.getExecutionEnvironment

  //初始数据源
  val csvFile = "input/douban/douban_movies.csv"

  val in:DataSet[Movie] = env
    .readCsvFile[Movie](
      csvFile,
      includedFields = Array(0,4,7,8,9),
      fieldDelimiter = ",",
      ignoreFirstLine = true
    )

  //问题二:统计平均评分
  println("\n 问题二:统计平均评分");
  in.map((movie:Movie) => {
    (movie.rating, 1)
  }).reduce((t1:(Double,Int),t2:(Double,Int)) => {
    (t1._1 + t2._1, t1._2 + t2._2)
  }).map((t:(Double,Int)) => {
    val df = new DecimalFormat("#.00")
    df.format(t._1 / t._2)
  }).print()
}
```

运行上面的代码,输出统计结果如下:

```
8.83
```

即这 225 部分电影的平均评分是 8.83。

(5) 回答第 3 个问题:找出这些热门电影中评论人数最多的十部电影,代码如下:

```scala
import Java.text.DecimalFormat
import Java.util.regex.Pattern
import org.apache.flink.api.common.operators.Order
import org.apache.flink.api.scala._

object DoubanMoviesBatch {

  //case class
  case class Movie(
    title:String,
```

```scala
    director:String,
    rating:Double,
    ratingNum:String,
    other:String
)

def main(args: Array[String]) {
    //设置批处理执行环境
    val env = ExecutionEnvironment.getExecutionEnvironment

    //初始数据源
    val csvFile = "input/douban/douban_movies.csv"

    val in:DataSet[Movie] = env
      .readCsvFile[Movie](
        csvFile,
        includedFields = Array(0,4,7,8,9),
        fieldDelimiter = ",",
        ignoreFirstLine = true
      )

    //问题三:统计评论人数最多的前10部电影(Top N 问题)
    println("\n问题三:统计评论人数最多的前10部电影")
    in.map((movie:Movie) => {
      //解析出评价人数
      var ratingNum = 0L
      val pattern = Pattern.compile("\\d+")
      val matcher = pattern.matcher(movie.ratingNum)
      if(matcher.find()){
        ratingNum = matcher.group().toLong
      }
      (movie.title, ratingNum)
    }).setParallelism(1) //设置全局排序
      .sortPartition(1, Order.DESCENDING)            //按单词数量降序排序
      .first(10)
      .print()
  }
}
```

运行上面的代码,输出统计结果如下:

```
(我不是药神,1452472)
(让子弹飞,1119027)
(摔跤吧!爸爸,1079607)
(寻梦环游记,1033536)
```

```
(绿皮书,1031394)
(少年派的奇幻漂流,1004021)
(头号玩家,987671)
(你的名字,949444)
(飞屋环游记,933034)
(阿凡达,929920)
```

可以看到,电影"我不是药神"以14 521 472人次评论高居榜首,成为最受关注的电影。完整代码如下:

```scala
//第4章/DoubanMoviesBatch.scala

import Java.text.DecimalFormat
import Java.util.regex.Pattern

import org.apache.flink.api.common.operators.Order
import org.apache.flink.api.scala._

/**
 * Flink 批处理示例:豆瓣电影数据集分析
 *
 * 回答以下问题:
 *    数据集中采集的电影总数量是多少?
 *    这些热门电影平均评分是多少?
 *    找出这些热门电影中评论人数最多的10部电影
 */
object DoubanMoviesBatch {

  //case class
  case class Movie(
    title:String,
    director:String,
    rating:Double,
    ratingNum:String,
    other:String
  )

  def main(args: Array[String]) {
    //设置批处理执行环境
    val env = ExecutionEnvironment.getExecutionEnvironment

    //初始数据源
    val csvFile = "input/douban/douban_movies.csv"

    val in:DataSet[Movie] = env
```

```scala
    .readCsvFile[Movie](
      csvFile,
      includedFields = Array(0,4,7,8,9),
      fieldDelimiter = ",",
      ignoreFirstLine = true
    )

    //问题一:统计电影总数量
    println("问题一:统计电影总数量")
    in.map(_ => 1L).reduce(_ + _).print()

    //问题二:统计平均评分
    println("\n 问题二:统计平均评分");
    in.map((movie:Movie) => {
      (movie.rating, 1)
    }).reduce((t1:(Double,Int),t2:(Double,Int)) => {
      (t1._1 + t2._1, t1._2 + t2._2)
    }).map((t:(Double,Int)) => {
      val df = new DecimalFormat("#.00")
      df.format(t._1 / t._2)
    }).print()

    //问题三:统计评论人数最多的前 10 部电影(Top N 问题)
    println("\n问题三:统计评论人数最多的前 10 部电影")
    in.map((movie:Movie) => {
      //解析出评价人数
      var ratingNum = 0L
      val pattern = Pattern.compile("\\d+")
      val matcher = pattern.matcher(movie.ratingNum)
      if(matcher.find()){
        ratingNum = matcher.group().toLong
      }
      (movie.title, ratingNum)
    }).setParallelism(1) //设置全局排序
      .sortPartition(1, Order.DESCENDING)           //按单词数量降序排序
      .first(10)
      .print()
  }
}
```

4.9.2 分析豆瓣热门电影数据集——Java 实现

【示例 4-9】 使用 Flink Java 批处理 API 分析豆瓣热门电影数据集。建议按以下步骤操作:

(1) 创建项目,并将源数据集复制到项目的 input/douban/ 目录下,如图 4-5 所示。

图 4-5　将数据源文件复制到项目的 input/douban/ 目录下

(2) 新建一个 Java POJO 类,取名为 Movie.java,用来封装数据集中的每行电影数据,代码如下:

```java
public static class Movie{
    public String title;            //电影名称
    public String director;         //导演
    public double rating;           //评分
    public String ratingNum;        //评分人数
    public String other;            //其他名称

    public Movie(){}

    public Movie(String title, String director, double rating, String ratingNum, String other) {
        this.title = title;
        this.director = director;
        this.rating = rating;
        this.ratingNum = ratingNum;
        this.other = other;
    }

    @Override
    public String toString() {
        return "Movie{" +
                "title = '" + title + '\'' +
                ", rating = " + rating +
                ", ratingNum = '" + ratingNum + '\'' +
                '}';
    }
}
```

(3) 回答第 1 个问题:数据集中采集的电影总数量是多少?代码如下:

```java
import org.apache.flink.api.common.functions.MapFunction;
import org.apache.flink.api.common.functions.ReduceFunction;
import org.apache.flink.api.common.operators.Order;
```

```java
import org.apache.flink.api.java.DataSet;
import org.apache.flink.api.java.ExecutionEnvironment;
import org.apache.flink.api.java.tuple.Tuple2;
import Java.text.DecimalFormat;
import Java.util.regex.Matcher;
import Java.util.regex.Pattern;

public class DoubanMoviesBatch {

    public static void main(String[] args) throws Exception {

        //设置批处理执行环境
        final ExecutionEnvironment env =
                ExecutionEnvironment.getExecutionEnvironment();

        //初始数据源
        String csvFile = "input/douban/douban_movies.csv";
        DataSet<Movie> in = env
            .readCsvFile(csvFile)
            .ignoreFirstLine()
            .fieldDelimiter(",")
            .includeFields("1000100111")
            .pojoType(Movie.class,"title","director","rating","ratingNum","other");

        //统计电影总数量
        in.map((MapFunction<Movie, Long>) movie -> 1L)
            .reduce((ReduceFunction<Long>) (t1, t2) -> t1 + t2).print();
    }
}
```

运行上面的代码,输出统计结果如下:

225

(4) 回答第2个问题:这些热门电影平均评分是多少?代码如下:

```java
import org.apache.flink.api.common.functions.MapFunction;
import org.apache.flink.api.common.functions.ReduceFunction;
import org.apache.flink.api.common.operators.Order;
import org.apache.flink.api.java.DataSet;
import org.apache.flink.api.java.ExecutionEnvironment;
import org.apache.flink.api.java.tuple.Tuple2;
import Java.text.DecimalFormat;
import Java.util.regex.Matcher;
import Java.util.regex.Pattern;
```

```java
public class DoubanMoviesBatch {

    public static void main(String[] args) throws Exception {

        //设置批处理执行环境
        final ExecutionEnvironment env =
                ExecutionEnvironment.getExecutionEnvironment();

        //初始数据源
        String csvFile = "src/data/douban_movies.csv";
        DataSet<Movie> in = env
            .readCsvFile(csvFile)
            .ignoreFirstLine()
            .fieldDelimiter(",")
            .includeFields("1000100111")
            .pojoType(Movie.class,"title","director","rating","ratingNum","other");

        //统计平均评分
        in.map(new MapFunction<Movie, Tuple2<Double,Integer>>() {
            @Override
            public Tuple2<Double,Integer> map(Movie movie) throws Exception {
                return new Tuple2<>(movie.rating, 1);
            }
        }).reduce(new ReduceFunction<Tuple2<Double, Integer>>() {
            @Override
            public Tuple2<Double, Integer> reduce(Tuple2<Double, Integer> t1, Tuple2<Double, Integer> t2) throws Exception {
                return new Tuple2<>(t1.f0 + t2.f0, t1.f1 + t2.f1);
            }
        }).map(new MapFunction<Tuple2<Double,Integer>, Double>() {
            @Override
            public Double map(Tuple2<Double, Integer> t) throws Exception {
                DecimalFormat df = new DecimalFormat("#.00");
                return Double.valueOf(df.format(t.f0/t.f1));
            }
        }).print();
    }
}
```

运行上面的代码,输出统计结果如下:

8.83

即这225部分电影的平均评分是8.83。

(5) 回答第3个问题:找出这些热门电影中评论人数最多的10部电影,代码如下:

```java
import org.apache.flink.api.common.functions.MapFunction;
import org.apache.flink.api.common.functions.ReduceFunction;
import org.apache.flink.api.common.operators.Order;
import org.apache.flink.api.java.DataSet;
import org.apache.flink.api.java.ExecutionEnvironment;
import org.apache.flink.api.java.tuple.Tuple2;
import Java.text.DecimalFormat;
import Java.util.regex.Matcher;
import Java.util.regex.Pattern;

public class DoubanMoviesBatch {

    public static void main(String[] args) throws Exception {

        //设置批处理执行环境
        final ExecutionEnvironment env =
                ExecutionEnvironment.getExecutionEnvironment();

        //初始数据源
        String csvFile = "src/data/douban_movies.csv";
        DataSet<Movie> in = env
            .readCsvFile(csvFile)
            .ignoreFirstLine()
            .fieldDelimiter(",")
            .includeFields("1000100111")
            .pojoType(Movie.class,"title","director","rating","ratingNum","other");

        //统计评论人数最多的前10部电影
        in.map(new MapFunction<Movie, Tuple2<String,Long>>() {
            @Override
            public Tuple2<String,Long> map(Movie movie) throws Exception {
                //解析出评价人数
                long ratingNum = 0;
                Pattern pattern = Pattern.compile("\\d+");
                Matcher matcher = pattern.matcher(movie.ratingNum);
                if(matcher.find()){
                    ratingNum = Long.valueOf(matcher.group());
                }
                return new Tuple2<>(movie.title, ratingNum);
            }
        }).setParallelism(1)
        //设置全局排序
        .sortPartition(1, Order.DESCENDING)                    /按单词数量降序排序
        .first(10)
        .print();
    }
}
```

运行上面的代码,输出统计结果如下:

```
(我不是药神,1452472)
(让子弹飞,1119027)
(摔跤吧!爸爸,1079607)
(寻梦环游记,1033536)
(绿皮书,1031394)
(少年派的奇幻漂流,1004021)
(头号玩家,987671)
(你的名字,949444)
(飞屋环游记,933034)
(阿凡达,929920)
```

可以看到,电影"我不是药神"以 1 452 472 人次评论高居榜首,成为最受关注的电影,完整代码如下:

```java
//第 4 章/DoubanMoviesBatch.java

import org.apache.flink.api.common.functions.MapFunction;
import org.apache.flink.api.common.functions.ReduceFunction;
import org.apache.flink.api.common.operators.Order;
import org.apache.flink.api.java.DataSet;
import org.apache.flink.api.java.ExecutionEnvironment;
import org.apache.flink.api.java.tuple.Tuple2;

import Java.text.DecimalFormat;
import Java.util.regex.Matcher;
import Java.util.regex.Pattern;

/**
 * Flink 批处理示例:豆瓣电影数据集分析
 *
 * 回答以下问题:
 *     数据集中采集的电影总数量是多少?
 *     这些热门电影平均评分是多少?
 *     找出这些热门电影中评论人数最多的 10 部电影
 */
public class DoubanMoviesBatch {

    //POJO 类
    public static class Movie{
        public String title;            //电影名称
        public String director;         //导演
        public double rating;           //评分
        public String ratingNum;        //评分人数
        public String other;            //其他名称
```

```java
    public Movie(){}

    public Movie(String title, String director, double rating, String ratingNum, String other) {
        this.title = title;
        this.director = director;
        this.rating = rating;
        this.ratingNum = ratingNum;
        this.other = other;
    }

    @Override
    public String toString() {
        return "Movie{" +
                "title = '" + title + '\'' +
                ", rating = " + rating +
                ", ratingNum = '" + ratingNum + '\'' +
                '}';
    }
}

public static void main(String[] args) throws Exception {
    //设置批处理执行环境
    final ExecutionEnvironment env =
            ExecutionEnvironment.getExecutionEnvironment();

    //初始数据源
    String csvFile = "input/douban/douban_movies.csv";
    DataSet<Movie> in = env
        .readCsvFile(csvFile)
        .ignoreFirstLine()
        .fieldDelimiter(",")
        .includeFields("1000100111")
        .pojoType(Movie.class,"title","director","rating","ratingNum","other");

    //问题一:统计电影总数量
    System.out.println("问题一:统计电影总数量");
    in
        .map((MapFunction<Movie, Long>) movie -> 1L)
        .reduce((ReduceFunction<Long>) (t1, t2) -> t1 + t2)
        .print();

    //问题二:统计平均评分
    System.out.println("\n问题二:统计平均评分");
    in.map(new MapFunction<Movie, Tuple2<Double,Integer>>() {
        @Override
```

```java
            public Tuple2<Double,Integer> map(Movie movie) throws Exception {
                return new Tuple2<>(movie.rating, 1);
            }
        }).reduce(new ReduceFunction<Tuple2<Double,Integer>>() {
            @Override
             public Tuple2<Double, Integer> reduce(Tuple2<Double, Integer> t1, Tuple2<Double, Integer> t2) throws Exception {
                return new Tuple2<>(t1.f0 + t2.f0, t1.f1 + t2.f1);
            }
        }).map(new MapFunction<Tuple2<Double,Integer>, Double>() {
            @Override
            public Double map(Tuple2<Double, Integer> t) throws Exception {
                DecimalFormat df = new DecimalFormat("#.00");
                return Double.valueOf(df.format(t.f0/t.f1));
            }
        }).print();

        //问题三:统计评论人数最多的前10部电影
        System.out.println("\n问题三:统计评论人数最多的前10部电影");
        in.map(new MapFunction<Movie, Tuple2<String,Long>>() {
            @Override
            public Tuple2<String,Long> map(Movie movie) throws Exception {
                //解析出评价人数
                long ratingNum = 0;
                Pattern pattern = Pattern.compile("\\d+");
                Matcher matcher = pattern.matcher(movie.ratingNum);
                if(matcher.find()){
                    ratingNum = Long.valueOf(matcher.group());
                }
                return new Tuple2<>(movie.title, ratingNum);
            }
        }).setParallelism(1)
        //设置全局排序
        .sortPartition(1, Order.DESCENDING)                    //按单词数量降序排序
        .first(10)
        .print();
    }
}
```

第 5 章 使用 Table API 进行数据处理
CHAPTER 5

Apache Flink 有两个关系型 API 用于统一的流和批处理，它们分别是 Table API 和 SQL。

Table API 是一个用于 Scala 和 Java 的语言集成查询 API，它允许以一种非常直观的方式组合来自关系操作符（如选择、筛选和连接）的查询。SQL 支持基于 Apache Calcite 的操作，它实现了 SQL 标准。无论输入的是批输入（DataSet）还是流输入（DataStream），在这两个接口中指定的查询都具有相同的语义，并得出相同的结果。

Table API 和 SQL 接口与 Flink 的 DataStream 和 DataSet API 紧密集成在一起。可以轻松地在所有 API 和基于 API 的库之间切换。一旦将 DataSet/DataStream 注册为表，就可以自由地应用关系操作，如聚合、连接和选择，还可以像常规 SQL 查询一样查询表。例如，可以使用 CEP 库从 DataStream 中提取模式，然后使用表 API 分析模式，或者在对预处理数据运行 Gelly Graph 算法之前，使用 SQL 查询扫描、过滤和聚合批处理表。

5.1 依赖

根据目标编程语言的不同，用户需要将 Java 或 Scala API 添加到项目中，以便使用 Table API 和 SQL 定义管道。添加的依赖项配置内容如下：

```xml
<!-- 如果使用的是 Java 语言 -->
<dependency>
    <groupId>org.apache.flink</groupId>
    <artifactId>flink-table-api-Java-bridge_2.12</artifactId>
    <version>1.13.2</version>
    <scope>provided</scope>
</dependency>

<!-- 或者，如果使用的是 Scala 语言 -->
<dependency>
    <groupId>org.apache.flink</groupId>
```

```xml
    <artifactId>flink-table-api-scala-bridge_2.12</artifactId>
    <version>1.13.2</version>
</dependency>
```

此外，如果想在IDE中本地运行Table API和SQL程序，还必须添加以下一组模块：

```xml
<dependency>
    <groupId>org.apache.flink</groupId>
    <artifactId>flink-table-planner-blink_2.12</artifactId>
    <version>1.13.2</version>
    <scope>test</scope>
</dependency>

<!-- 注：在内部，部分表生态系统是用Scala实现的 -->
<dependency>
    <groupId>org.apache.flink</groupId>
    <artifactId>flink-streaming-scala_2.12</artifactId>
    <version>1.13.2</version>
    <scope>provided</scope>
</dependency>
```

这些依赖项将下载类路径中所需的所有JAR。下载完成后，用户就可以在代码中使用Table API了。

如果想为序列化/反序列化行或一组用户定义函数实现自定义格式或连接器，还需要添加用于SQL Client的JAR文件的依赖项，代码如下：

```xml
<dependency>
    <groupId>org.apache.flink</groupId>
    <artifactId>flink-table-common</artifactId>
    <version>1.13.2</version>
    <scope>provided</scope>
</dependency>
```

5.2 Table API 与 SQL 编程模式

Table API 和 SQL 集成在一个联合 API 中。这个 API 的核心概念是作为查询的输入和输出的 Table。Table API 和 SQL 查询可以很容易地与 DataStream 程序集成并嵌入其中。

5.2.1 TableEnvironment

TableEnvironment 是 Table API 和 SQL 集成的入口点，一个 Table 总是绑定到一个

特定的 TableEnvironment。TableEnvironment 在语言层面上是统一的,适用于所有基于 JVM 的语言(Scala 和 Java API 之间没有区别),也适用于有界和无界数据处理。它负责:

(1) 连接外部系统。
(2) 注册 catalog。
(3) 在内部 catalog 目录中注册一个 Table。
(4) 从 catalog 目录中注册和检索 Table 和其他元对象。
(5) 提供进一步的配置选项。
(6) 执行 SQL 查询。
(7) 注册用户定义函数(标量函数、表函数或聚合函数)。
(8) DataStream 和 Table 之间的转换(在 StreamTableEnvironment 中)。

在同一个查询组合中必须使用同一个 TableEnvironment 的表。通过调用 TableEnvironment 的静态 create() 方法可以创建一个 TableEnvironment。

Scala 代码如下:

```scala
import org.apache.flink.table.api.{EnvironmentSettings, TableEnvironment}

val settings = EnvironmentSettings
    .newInstance()
    .inStreamingMode()          //或.inBatchMode()
    .build()

val tEnv = TableEnvironment.create(setting)
```

Java 代码如下:

```java
import org.apache.flink.table.api.EnvironmentSettings;
import org.apache.flink.table.api.TableEnvironment;

EnvironmentSettings settings = EnvironmentSettings
    .newInstance()
    .inStreamingMode()          //或.inBatchMode()
    .build();

TableEnvironment tEnv = TableEnvironment.create(settings);
```

或者,用户可以从现有的 StreamExecutionEnvironment 创建一个 StreamTableEnvironment 来与 DataStream API 进行互操作。

Scala 代码如下:

```scala
import org.apache.flink.streaming.api.scala.StreamExecutionEnvironment
import org.apache.flink.table.api.EnvironmentSettings
```

```
import org.apache.flink.table.api.bridge.scala.StreamTableEnvironment

val env = StreamExecutionEnvironment.getExecutionEnvironment
val tEnv = StreamTableEnvironment.create(env)
```

Java 代码如下：

```
import org.apache.flink.streaming.api.environment.StreamExecutionEnvironment;
import org.apache.flink.table.api.EnvironmentSettings;
import org.apache.flink.table.api.bridge.java.StreamTableEnvironment;

StreamExecutionEnvironment env =
            StreamExecutionEnvironment.getExecutionEnvironment();
StreamTableEnvironment tEnv = StreamTableEnvironment.create(env);
```

【示例 5-1】 给出一些学生各个科目的成绩，编写 Flink 程序计算每个学生的总成绩。
Scala 代码如下：

```
//第 5 章/TableDemo00.scala

import org.apache.flink.table.api.Expressions.$
import org.apache.flink.table.api._

object TableDemo00 {

  def main(args: Array[String]) {
    //创建表环境
    val settings = EnvironmentSettings.newInstance.inBatchMode.build
    val tEnv = TableEnvironment.create(settings)

    //计算每个学生的总成绩
    tEnv
      .fromValues(
        row("张三", "math", 82),
        row("张三", "english", 92),
        row("李四", "math", 86),
        row("李四", "english", 91)
      )
      .as("name", "subject", "score")
      .groupBy( $ ("name"))
      .select( $ ("name"), $ ("score").sum.as("score_sum"))
      .execute()
      .print()
  }
}
```

Java 代码如下：

```java
//第5章/TableDemo00.java

import org.apache.flink.table.api.*;
import static org.apache.flink.table.api.Expressions.$;
import static org.apache.flink.table.api.Expressions.row;

public class TableDemo00 {

    public static void main(String[] args) throws Exception {
        //创建表环境
        EnvironmentSettings settings = EnvironmentSettings
            .newInstance()
            .inBatchMode()              //.inStreamingMode()
            .build();

        TableEnvironment tEnv = TableEnvironment.create(settings);

        //计算每个学生的总成绩
        tEnv
            .fromValues(row("张三", "math", 82),
                    row("张三","english", 92),
                    row("李四","math", 86),
                    row("李四","english", 91))
            .as("name", "subject", "score")
            .groupBy( $ ("name"))
            .select( $ ("name"), $ ("score").sum().as("score_sum"))
            .execute()
            .print();
    }
}
```

执行以上代码，输出内容如下：

```
+--------------------------------+------------------+
|                           name |        score_sum |
+--------------------------------+------------------+
|                           李四 |              177 |
|                           张三 |              174 |
+--------------------------------+------------------+
2 rows in set
```

5.2.2 Table API 与 SQL 程序的结构

所有用于批处理和流处理的 Table API 和 SQL 程序都遵循相同的模式。下面的代码示例展示了 Table API 和 SQL 程序的通用结构。

Scala 代码如下：

```scala
//创建用于批或流处理的 TableEnvironment
val tableEnv = ... //创建一个 TableEnvironment

//创建一个输入表
tableEnv.executeSql("CREATE TEMPORARY TABLE table1 ... WITH ( 'connector' = ... )")

//注册一个输出表
tableEnv.executeSql("CREATE TEMPORARY TABLE outputTable ... WITH ( 'connector' = ... )")

//从 Table API 查询创建表
val table2 = tableEnv.from("table1").select(...)

//从 SQL 查询创建一个表
val table3 = tableEnv.sqlQuery("SELECT ... FROM table1 ...")

//将 Table API 结果表发给一个 TableSink,同样适用于 SQL 结果
val tableResult = table2.executeInsert("outputTable")
tableResult...
```

Java 代码如下：

```java
//创建用于批或流处理的 TableEnvironment
TableEnvironment tableEnv = ...;            //创建一个 TableEnvironment

//创建一个输入表
tableEnv.executeSql("CREATE TEMPORARY TABLE table1 ... WITH ( 'connector' = ... )");

//注册一个输出表
tableEnv.executeSql("CREATE TEMPORARY TABLE outputTable ... WITH ( 'connector' = ... )");

//从 Table API 查询创建表
Table table2 = tableEnv.from("table1").select(...);

//从 SQL 查询创建一个表
Table table3 = tableEnv.sqlQuery("SELECT ... FROM table1 ... ");

//将 Table API 结果表发给一个 TableSink,同样适用于 SQL 结果
TableResult tableResult = table2.executeInsert("outputTable");
tableResult...
```

5.2.3 在 Catalog 中创建表

TableEnvironment 维护使用标识符创建的表的目录映射。每个标识符由三部分组成：

目录名、数据库名和对象名。如果没有指定目录或数据库,则使用当前默认值。

表可以是虚拟的(VIEW),也可以是常规的(TABLE)。VIEW可以从现有的Table对象创建,通常是Table API或SQL查询的结果。TABLE用于描述外部数据,如文件、数据库表或消息队列。

1. 临时表与永久表

表可以是临时的,并绑定到单个Flink会话的生命周期,也可以是永久的,并且跨多个Flink会话和集群可见。

永久表需要目录(catalog,如Hive Metastore)来维护关于表的元数据。一旦创建了永久表,它对于连接到目录的任何Flink会话都是可见的,并且将持续存在,直到显式地删除该表。

另一方面,临时表总是存储在内存中,并且只在创建它们的Flink会话期间存在。这些表对其他会话不可见。它们没有绑定到任何目录或数据库,但可以在其名称空间(namespace)中创建。如果删除了相应的数据库,则不会删除临时表。

可以使用与现有永久表相同的标识符注册临时表。临时表隐藏了永久表,只要临时表存在,就无法访问永久表。具有该标识符的所有查询都将对临时表执行。

这可能对实验有用。它允许首先对一个临时表运行完全相同的查询,例如只有一个数据子集,或者数据是模糊的。一旦验证了查询是正确的,就可以对实际的生产表进行查询。

2. 创建表

一个Table API对象对应于SQL术语中的VIEW(虚拟表),它封装了一个逻辑查询计划。VIEW可以在一个目录中创建。

Scala代码如下:

```scala
//获得一个TableEnvironment
import org.apache.flink.table.api.{EnvironmentSettings, TableEnvironment}

val settings = EnvironmentSettings
   .newInstance()
   .inStreamingMode()            //.inBatchMode()
   .build()

val tEnv = TableEnvironment.create(setting)

//表是一个简单的投影查询的结果
val projTable: Table = tableEnv.from("X").select(...)

//将表projTable注册为表"projectedTable"
tEnv.createTemporaryView("projectedTable", projTable)
```

Java代码如下:

```
//获得一个 TableEnvironment
import org.apache.flink.table.api.EnvironmentSettings;
import org.apache.flink.table.api.TableEnvironment;

EnvironmentSettings settings = EnvironmentSettings
    .newInstance()
    .inStreamingMode()          //.inBatchMode()
    .build();

TableEnvironment tEnv = TableEnvironment.create(setting);

//表是一个简单的投影查询的结果
Table projTable = tEnv.from("X").select(...);

//将表 projTable 注册为表 "projectedTable"
tEnv.createTemporaryView("projectedTable", projTable);
```

> **注意**：Table 对象类似于关系数据库系统中的 VIEW 对象，即定义 Table 的查询没有优化，但当另一个查询引用已注册的 Table 时将内联。如果多个查询引用同一个注册 Table，它将内联每个引用查询并执行多次，也就是说，注册 Table 的结果将不会被共享。

也可以通过连接器声明创建关系数据库中的 TABLE。连接器用于描述存储表数据的外部系统，存储系统（如 Apache Kafka 或常规文件系统）可以在这里声明。通过连接器声明创建关系数据库中的 TABLE 的模板代码。

Scala 代码如下：

```
tableEnvironment
  .connect(...)
  .withFormat(...)
  .withSchema(...)
  .inAppendMode()
  .createTemporaryTable("MyTable")
```

Java 代码如下：

```
tableEnvironment
  .connect(...)
  .withFormat(...)
  .withSchema(...)
  .inAppendMode()
  .createTemporaryTable("MyTable")
```

SQL 代码如下：

```
tableEnvironment.executeSql("CREATE [TEMPORARY] TABLE MyTable (...) WITH (...)")
```

3. 扩展表标识符

表总是使用由目录、数据库和表名组成的三部分标识符进行注册。

用户可以将其中的一个目录和一个数据库设置为"当前目录"和"当前数据库"。有了它们,上面提到的三部分标识符中的前两部分可以省略。如果没有提供它们,将引用当前目录和当前数据库。用户可以通过 Table API 或 SQL 切换当前目录和当前数据库。

标识符遵循 SQL 要求,这意味着可以使用反号字符(`)进行转义。下面使用标识符的方式进行演示。

Scala 代码如下:

```
//获得一个 TableEnvironment
import org.apache.flink.table.api.{EnvironmentSettings, TableEnvironment}

val settings = EnvironmentSettings
  .newInstance()
  .inStreamingMode()            //.inBatchMode()
  .build()

val tEnv = TableEnvironment.create(setting)

//指定当前目录和当前数据库
tEnv.useCatalog("custom_catalog")
tEnv.useDatabase("custom_database")

val table: Table = tableEnv.from("X").select(...)

//在名为 custom_catalog 的目录中注册名为 exampleView 的视图
//在名为 custom_database 的数据库中
tEnv.createTemporaryView("exampleView", table)

//在名为 custom_catalog 的目录中注册名为 exampleView 的视图
//在名为 other_database 的数据库中
tEnv.createTemporaryView("other_database.exampleView", table)

//在名为 custom_catalog 的目录中注册名为 example.View 的视图
//在名为 custom_database 的数据库中
tEnv.createTemporaryView("`example.View`", table)

//在名为 other_catalog 的目录中注册名为 exampleView 的视图
//在名为 other_database 的数据库中
tEnv.createTemporaryView("other_catalog.other_database.exampleView", table)
```

Java 代码如下：

```java
//获得一个 TableEnvironment
import org.apache.flink.table.api.EnvironmentSettings;
import org.apache.flink.table.api.TableEnvironment;

EnvironmentSettings settings = EnvironmentSettings
    .newInstance()
    .inStreamingMode()           //.inBatchMode()
    .build();

TableEnvironment tEnv = TableEnvironment.create(setting);

//指定当前目录和当前数据库
tEnv.useCatalog("custom_catalog");
tEnv.useDatabase("custom_database");

Table table = tEnv.from("X").select(...);

//在名为 custom_catalog 的目录中注册名为 exampleView 的视图
//在名为 custom_database 的数据库中
tEnv.createTemporaryView("exampleView", table);

//在名为 custom_catalog 的目录中注册名为 exampleView 的视图
//在名为 other_database 的数据库中
tEnv.createTemporaryView("other_database.exampleView", table);

//在名为 custom_catalog 的目录中注册名为 example.View 的视图
//在名为 custom_database 的数据库中
tEnv.createTemporaryView("'example.View'", table);

//在名为 other_catalog 的目录中注册名为 exampleView 的视图
//在名为 other_database 的数据库中
tEnv.createTemporaryView("other_catalog.other_database.exampleView", table);
```

5.2.4 查询表

既可以使用 Table API 查询创建的表，也可以使用 SQL 查询创建的表。

1. 使用 Table API

Table API 是用于 Scala 和 Java 的语言集成查询 API，是批处理和流数据的统一 API。
Table API 基于 Table 类，Table 类表示一个表（流或批），并提供应用关系操作的方法。这些方法返回一个新的 Table 对象，它表示在输入 Table 上应用关系操作的结果。一些关系操作由多种方法调用组成，例如 table.groupBy().select()，其中 groupBy()用于指定一个table分组,并指定()table分组上的投影。

下面的例子展示了一个简单的表API聚合查询。

Scala代码如下:

```scala
//获得一个TableEnvironment
import org.apache.flink.table.api.{EnvironmentSettings, TableEnvironment}

val settings = EnvironmentSettings
   .newInstance()
   .inStreamingMode()              //.inBatchMode()
   .build()

val tEnv = TableEnvironment.create(setting)

//注册Orders表

//扫描已注册的Orders表
val orders = tEnv.from("Orders")

//计算来自法国的所有客户的收入
val revenue = orders
  .filter($"cCountry" === "FRANCE")
  .groupBy($"cID", $"cName")
  .select($"cID", $"cName", $"revenue".sum AS "revSum")

//发出表或转换表
//执行查询
```

Scala Table API使用了以$符号开头的Scala String插值来引用Table的属性。Table API使用了Scala隐式。确保导入以下依赖:

```scala
import org.apache.flink.table.api._            //用于隐式表达式转换

//如果想与DataStream互相转换,则还要执行以下两个导入
import org.apache.flink.api.scala._
import org.apache.flink.table.api.bridge.scala._
```

Java代码如下:

```java
//获得一个TableEnvironment
import org.apache.flink.table.api.EnvironmentSettings;
import org.apache.flink.table.api.TableEnvironment;

EnvironmentSettings settings = EnvironmentSettings
```

```
    .newInstance()
    .inStreamingMode()          //.inBatchMode()
    .build();

TableEnvironment tEnv = TableEnvironment.create(setting);

//注册 Orders 表

//扫描已注册的 Orders 表
Table orders = tEnv.from("Orders");

//计算来自法国的所有客户的收入
Table revenue = orders
  .filter( $("cCountry").isEqual("FRANCE"))
  .groupBy( $("cID"), $("cName"))
  .select( $("cID"), $("cName"), $("revenue").sum().as("revSum"));

//发出表或转换表
//执行查询
```

2. 使用 SQL

Flink 的 SQL 集成基于 Apache Calcite,它实现了 SQL 标准。SQL 查询被指定为常规字符串。

下面的示例演示了如何指定查询并将结果作为 Table 返回。

Scala 代码如下:

```
//获得一个 TableEnvironment
import org.apache.flink.table.api.{EnvironmentSettings, TableEnvironment}

val settings = EnvironmentSettings
  .newInstance()
  .inStreamingMode()          //.inBatchMode()
  .build()

val tEnv = TableEnvironment.create(setting)

//注册 Orders 表

//计算来自法国的所有客户的收入
val revenue = tEnv.sqlQuery("""
  |SELECT cID, cName, SUM(revenue) AS revSum
  |FROM Orders
  |WHERE cCountry = 'FRANCE'
  |GROUP BY cID, cName
```

```
    """.stripMargin)

//向下游发送表或转换表
//执行查询
```

Java 代码如下：

```java
//获得一个TableEnvironment
import org.apache.flink.table.api.EnvironmentSettings;
import org.apache.flink.table.api.TableEnvironment;

EnvironmentSettings settings = EnvironmentSettings
    .newInstance()
    .inStreamingMode()           //.inBatchMode()
    .build();

TableEnvironment tEnv = TableEnvironment.create(setting);

//注册 Orders 表

//计算来自法国的所有客户的收入
Table revenue = tEnv.sqlQuery(
    "SELECT cID, cName, SUM(revenue) AS revSum " +
    "FROM Orders " +
    "WHERE cCountry = 'FRANCE' " +
    "GROUP BY cID, cName"
  );

//向下游发送表或转换表
//执行查询
```

下面的示例演示了如何指定将其结果插入已注册表的更新查询。

Scala 代码如下：

```scala
//获得一个TableEnvironment
import org.apache.flink.table.api.{EnvironmentSettings, TableEnvironment}

val settings = EnvironmentSettings
    .newInstance()
    .inStreamingMode()           //.inBatchMode()
    .build()

val tEnv = TableEnvironment.create(setting)

//注册 Orders 表
```

```
//注册"RevenueFrance"输出表

//计算来自法国的所有客户的收入并发送到 RevenueFrance
tEnv.sqlUpdate("""
    |INSERT INTO RevenueFrance
    |SELECT cID, cName, SUM(revenue) AS revSum
    |FROM Orders
    |WHERE cCountry = 'FRANCE'
    |GROUP BY cID, cName
  """.stripMargin)

//执行查询
```

Java 代码如下:

```
//获得一个 TableEnvironment
import org.apache.flink.table.api.EnvironmentSettings;
import org.apache.flink.table.api.TableEnvironment;

EnvironmentSettings settings = EnvironmentSettings
    .newInstance()
    .inStreamingMode()          //.inBatchMode()
    .build();

TableEnvironment tEnv = TableEnvironment.create(setting);

//注册 Orders 表
//注册"RevenueFrance"输出表

//计算来自法国的所有客户的收入并发送到 RevenueFrance
tEnv.sqlUpdate(
    "INSERT INTO RevenueFrance " +
    "SELECT cID, cName, SUM(revenue) AS revSum " +
    "FROM Orders " +
    "WHERE cCountry = 'FRANCE' " +
    "GROUP BY cID, cName"
);

//执行查询
```

3. 混合使用 Table API 和 SQL

Table API 和 SQL 查询很容易混合使用,因为它们都返回 Table 对象:
(1) 可以在 SQL 查询返回的 Table 对象上定义 Table API 查询。
(2) 可以在 Table API 查询的结果上定义 SQL 查询,方法是将结果表注册到

TableEnvironment 中,并在 SQL 查询的 FROM 子句中引用它。

5.2.5 向下游发送表

Table 是通过将其写入 TableSink 而发出的。一个 TableSink 是一个通用接口,支持各种文件格式(如 CSV、Apache Parquet、Apache Avro)、存储系统(如 JDBC、Apache HBase、Apache Cassandra、Elasticsearch)或消息系统(如 Apache Kafka、RabbitMQ)。

批处理表只能被写到一个 BatchTableSink 中,而流处理表需要一个 AppendStreamTableSink、一个 RetractStreamTableSink 或一个 UpsertStreamTableSink。

Table.executeInsert(String tableName)方法将 Table 发送到已注册的 TableSink 中。该方法通过名称从目录中查找 TableSink,并验证该 Table 的模式是否与该 TableSink 的模式相同。

下面的例子演示了如何发送一个 Table 表。

Scala 代码如下:

```scala
//获得一个 TableEnvironment
import org.apache.flink.table.api.{EnvironmentSettings, TableEnvironment}

val settings = EnvironmentSettings
  .newInstance()
  .inStreamingMode()       //.inBatchMode()
  .build()

val tEnv = TableEnvironment.create(setting)

//创建输出表
val schema = new Schema()
  .field("a", DataTypes.INT())
  .field("b", DataTypes.STRING())
  .field("c", DataTypes.LONG())

//注册 TableSink
tEnv.connect(new FileSystem().path("/path/to/file"))
  .withFormat(new Csv().fieldDelimiter('|').deriveSchema())
  .withSchema(schema)
  .createTemporaryTable("CsvSinkTable")

//使用 Table API 操作符和/或 SQL 查询计算结果表
val result: Table = ...

//将结果表发送到已注册的 TableSink
result.executeInsert("CsvSinkTable")
```

Java 代码如下:

```java
//获得一个 TableEnvironment
import org.apache.flink.table.api.EnvironmentSettings;
import org.apache.flink.table.api.TableEnvironment;

EnvironmentSettings settings = EnvironmentSettings
    .newInstance()
    .inStreamingMode()          //.inBatchMode()
    .build();

TableEnvironment tEnv = TableEnvironment.create(setting);

//创建输出表
final Schema schema = new Schema()
    .field("a", DataTypes.INT())
    .field("b", DataTypes.STRING())
    .field("c", DataTypes.LONG());

//注册 TableSink
tEnv.connect(new FileSystem().path("/path/to/file"))
    .withFormat(new Csv().fieldDelimiter('|').deriveSchema())
    .withSchema(schema)
    .createTemporaryTable("CsvSinkTable");

//使用 Table API 操作符和/或 SQL 查询计算结果表
Table result = ...

//将结果表发送到已注册的 TableSink
result.executeInsert("CsvSinkTable");
```

5.2.6　翻译并执行查询

翻译(translated)和执行查询的行为对于这两个计划器是不同的。Table API 和 SQL 查询被转换成 DataStream 程序，无论它们的输入是流还是批。查询在内部表示为逻辑查询计划，并分为两个阶段进行转换：

(1) 优化逻辑计划。

(2) 转换成 DataStream 程序。

Table API 或 SQL 查询在以下情况下被翻译：

(1) TableEnvironment.executeSql()被调用。此方法用于执行给定的语句，一旦调用此方法，SQL 查询将立即被翻译。

(2) Table.executeInsert()被调用。此方法用于将表内容插入给定的接收器路径，并且一旦调用此方法，Table API 将立即被翻译。

（3）Table.execute()被调用。此方法用于将表内容收集到本地客户机，并且一旦调用此方法，Table API 将立即被翻译。

（4）StatementSet.execute()被调用。Table（通过 StatementSet.addInsert()发到 sink)或 INSERT 语句（通过 StatementSet.addInsertSql()指定）将首先在 StatementSet 中缓冲。一旦调用 StatementSet.execute()，它们就会被转换，所有 sink 都将被优化成一个 DAG。

（5）当 Table 转换为 DataStream 时，它将被翻译。一旦转换，它就是一个常规的 DataStream 程序，并在调用 StreamExecutionEnvironment.execute()时执行。

1. 查询优化

Apache Flink 利用并扩展了 Apache Calcite 来执行复杂的查询优化。这包括一系列基于规则和成本的优化，包括：

（1）基于 Apache Calcite 的子查询解相关。

（2）投影裁剪。

（3）分区裁剪。

（4）过滤器下推。

（5）子计划去重，避免重复计算。

（6）特殊子查询重写，包括两部分：①将 IN 和 EXISTS 转换为左半连接；②将 NOT IN 和 NOT EXISTS 转换为左反连接。

（7）可选 join 重新排序。

（8）通过 table.optimizer.join-reorder-enabled 启用。

优化器不仅根据计划，还根据数据源提供的丰富统计信息和每个操作符（如 IO、CPU、网络和内存）的细粒度成本做出智能决策。

高级用户可以通过 CalciteConfig 对象提供自定义优化，CalciteConfig 对象可以通过调用 TableEnvironment#getConfig#setPlanerConfig 提供给表环境。

2. 解释表

Table API 提供了一种机制来解释用于计算 Table 的逻辑和优化的查询计划。这是通过 Table.explain()方法或 StatementSet.explain()方法完成的。Table.explain()返回 Table 的计划。StatementSet.explain()返回多个接收器的计划。它返回 3 个计划的一个字符串描述，包括：

（1）关系查询的抽象语法树，即未优化的逻辑查询计划。

（2）优化的逻辑查询计划。

（3）物理执行计划。

TableEnvironment.explainSql()和 TableEnvironment.executeSql()方法支持执行 EXPLAIN 语句，以便获取逻辑计划和优化查询计划。

例如，使用 Table.explain()方法给出 Table 的相应输出。

Scala 代码如下:

```scala
val env = StreamExecutionEnvironment.getExecutionEnvironment
val tEnv = StreamTableEnvironment.create(env)

val table1 = env.fromElements((1, "hello")).toTable(tEnv, $"count", $"word")
val table2 = env.fromElements((1, "hello")).toTable(tEnv, $"count", $"word")
val table = table1
  .where( $"word".like("F%"))
  .unionAll(table2)

println(table.explain())
```

Java 代码如下:

```java
StreamExecutionEnvironment env =
    StreamExecutionEnvironment.getExecutionEnvironment();
StreamTableEnvironment tEnv = StreamTableEnvironment.create(env);

DataStream<Tuple2<Integer, String>> stream1 =
        env.fromElements(new Tuple2<>(1, "hello"));
DataStream<Tuple2<Integer, String>> stream2 =
        env.fromElements(new Tuple2<>(1, "hello"));

//explain Table API
Table table1 = tEnv.fromDataStream(stream1, $("count"), $("word"));
Table table2 = tEnv.fromDataStream(stream2, $("count"), $("word"));
Table table = table1
  .where( $("word").like("F%"))
  .unionAll(table2);

System.out.println(table.explain());
```

上面示例的输出结果如下:

```
== Abstract Syntax Tree ==
LogicalUnion(all=[true])
  LogicalFilter(condition=[LIKE( $1, _UTF-16LE'F%')])
    FlinkLogicalDataStreamScan(id=[1], fields=[count, word])
    FlinkLogicalDataStreamScan(id=[2], fields=[count, word])

== Optimized Logical Plan ==
DataStreamUnion(all=[true], union all=[count, word])
  DataStreamCalc(select=[count, word], where=[LIKE(word, _UTF-16LE'F%')])
    DataStreamScan(id=[1], fields=[count, word])
```

```
    DataStreamScan(id = [2], fields = [count, word])

== Physical Execution Plan ==
Stage 1 : Data Source
    content : collect elements with CollectionInputFormat

Stage 2 : Data Source
    content : collect elements with CollectionInputFormat

    Stage 3 : Operator
            content : from: (count, word)
            ship_strategy : REBALANCE

            Stage 4 : Operator
                    content : where: (LIKE(word, _UTF – 16LE'F % ')), select: (count, word)
                    ship_strategy : FORWARD

                    Stage 5 : Operator
                            content : from: (count, word)
                            ship_strategy : REBALANCE
```

可以使用 StatementSet.explain()方法返回多 sink 计划的相应输出。

Scala 代码如下：

```scala
val settings = EnvironmentSettings
    .newInstance
    .useBlinkPlanner
    .inStreamingMode.build
val tEnv = TableEnvironment.create(settings)

val schema = new Schema()
    .field("count", DataTypes.INT())
    .field("word", DataTypes.STRING())

tEnv.connect(new FileSystem().path("/source/path1"))
    .withFormat(new Csv().deriveSchema())
    .withSchema(schema)
    .createTemporaryTable("MySource1")
tEnv.connect(new FileSystem().path("/source/path2"))
    .withFormat(new Csv().deriveSchema())
    .withSchema(schema)
    .createTemporaryTable("MySource2")
tEnv.connect(new FileSystem().path("/sink/path1"))
    .withFormat(new Csv().deriveSchema())
    .withSchema(schema)
```

```scala
    .createTemporaryTable("MySink1")
tEnv.connect(new FileSystem().path("/sink/path2"))
    .withFormat(new Csv().deriveSchema())
    .withSchema(schema)
    .createTemporaryTable("MySink2")

val stmtSet = tEnv.createStatementSet()

val table1 = tEnv.from("MySource1").where($ "word".like("F%"))
stmtSet.addInsert("MySink1", table1)

val table2 = table1.unionAll(tEnv.from("MySource2"))
stmtSet.addInsert("MySink2", table2)

val explanation = stmtSet.explain()
println(explanation)
```

Java 代码如下：

```java
EnvironmentSettings settings = EnvironmentSettings.newInstance()
    .useBlinkPlanner()
    .inStreamingMode()
    .build();
TableEnvironment tEnv = TableEnvironment.create(settings);

final Schema schema = new Schema()
    .field("count", DataTypes.INT())
    .field("word", DataTypes.STRING());

tEnv.connect(new FileSystem().path("/source/path1"))
    .withFormat(new Csv().deriveSchema())
    .withSchema(schema)
    .createTemporaryTable("MySource1");
tEnv.connect(new FileSystem().path("/source/path2"))
    .withFormat(new Csv().deriveSchema())
    .withSchema(schema)
    .createTemporaryTable("MySource2");
tEnv.connect(new FileSystem().path("/sink/path1"))
    .withFormat(new Csv().deriveSchema())
    .withSchema(schema)
    .createTemporaryTable("MySink1");
tEnv.connect(new FileSystem().path("/sink/path2"))
    .withFormat(new Csv().deriveSchema())
    .withSchema(schema)
    .createTemporaryTable("MySink2");
```

```java
StatementSet stmtSet = tEnv.createStatementSet();

Table table1 = tEnv.from("MySource1").where($("word").like("F%"));
stmtSet.addInsert("MySink1", table1);

Table table2 = table1.unionAll(tEnv.from("MySource2"));
stmtSet.addInsert("MySink2", table2);

String explanation = stmtSet.explain();
System.out.println(explanation);
```

上面代码的输出结果如下：

```
== Abstract Syntax Tree ==
LogicalLegacySink(name=[MySink1], fields=[count, word])
+- LogicalFilter(condition=[LIKE($1, _UTF-16LE'F%')])
   +- LogicalTableScan(table=[[default_catalog, default_database, MySource1, source:
[CsvTableSource(read fields: count, word)]]])

LogicalLegacySink(name=[MySink2], fields=[count, word])
+- LogicalUnion(all=[true])
   :- LogicalFilter(condition=[LIKE($1, _UTF-16LE'F%')])
   :  +- LogicalTableScan(table=[[default_catalog, default_database, MySource1, source:
[CsvTableSource(read fields: count, word)]]])
   +- LogicalTableScan(table=[[default_catalog, default_database, MySource2, source:
[CsvTableSource(read fields: count, word)]]])

== Optimized Logical Plan ==
Calc(select=[count, word], where=[LIKE(word, _UTF-16LE'F%')], reuse_id=[1])
+- TableSourceScan(table=[[default_catalog, default_database, MySource1, source:
[CsvTableSource(read fields: count, word)]]], fields=[count, word])

LegacySink(name=[MySink1], fields=[count, word])
+- Reused(reference_id=[1])

LegacySink(name=[MySink2], fields=[count, word])
+- Union(all=[true], union=[count, word])
   :- Reused(reference_id=[1])
   +- TableSourceScan(table=[[default_catalog, default_database, MySource2, source:
[CsvTableSource(read fields: count, word)]]], fields=[count, word])

== Physical Execution Plan ==
Stage 1 : Data Source
    content : collect elements with CollectionInputFormat
```

```
    Stage 2 : Operator
        content : CsvTableSource(read fields: count, word)
        ship_strategy : REBALANCE

        Stage 3 : Operator
            content : SourceConversion(table:Buffer(default_catalog, default_database,
MySource1, source: [CsvTableSource(read fields: count, word)]), fields:(count, word))
            ship_strategy : FORWARD

            Stage 4 : Operator
                content : Calc(where: (word LIKE _UTF-16LE'F%'), select: (count, word))
                ship_strategy : FORWARD

                Stage 5 : Operator
                    content : SinkConversionToRow
                    ship_strategy : FORWARD

                    Stage 6 : Operator
                        content : Map
                        ship_strategy : FORWARD

Stage 8 : Data Source
    content : collect elements with CollectionInputFormat

    Stage 9 : Operator
        content : CsvTableSource(read fields: count, word)
        ship_strategy : REBALANCE

        Stage 10 : Operator
            content : SourceConversion(table:Buffer(default_catalog, default_
database, MySource2, source: [CsvTableSource(read fields: count, word)]), fields:(count,
word))
            ship_strategy : FORWARD

            Stage 12 : Operator
                content : SinkConversionToRow
                ship_strategy : FORWARD

                Stage 13 : Operator
                    content : Map
                    ship_strategy : FORWARD

                    Stage 7 : Data Sink
                        content : Sink: CsvTableSink(count, word)
                        ship_strategy : FORWARD

                        Stage 14 : Data Sink
                            content : Sink: CsvTableSink(count, word)
                            ship_strategy : FORWARD
```

5.3 Table API

Table API 是一个用于流和批处理的统一的关系 API。Table API 查询可以在批或流输入上运行,而无须进行修改。Table API 是 SQL 语言的一个超级集合,是专门为使用 Apache Flink 而设计的。

5.3.1 关系运算

Table API 支持如下的关系运算符(需要注意,并不是所有的运算符都同时支持流和批运算)。

1. 扫描、投影和过滤

(1) from():类似于 SQL 查询中的 from 子句,执行注册表的扫描。
Scala 代码如下:

```
val orders = tableEnv.from("Orders")
```

Java 代码如下:

```
Table orders = tableEnv.from("Orders");
```

(2) fromValues():类似于 SQL 查询中的 values 子句,根据提供的行生成内联表。可以使用 row() 表达式来创建复合行。
Scala 代码如下:

```
val table = tEnv.fromValues(
  row(1, "ABC"),
  row(2L, "ABCDE")
)
```

Java 代码如下:

```
Table table = tEnv.fromValues(
  row(1, "ABC"),
  row(2L, "ABCDE")
);
```

以上代码片段创建的表具有以下的模式:

```
root
 |-- f0: BIGINT NOT NULL        //将原始类型 INT 和 BIGINT 泛化为 BIGINT
```

```
|-- f1: VARCHAR(5) NOT NULL     //原始类型 CHAR(3)和 CHAR(5)泛化为 VARCHAR(5)
                                //使用 VARCHAR 而不是 CHAR,因此不应用填充
```

该方法将自动从输入表达式派生类型。如果某个位置的类型不同,则该方法将尝试为所有类型找到一个通用的超类型。如果不存在公共超类型,则将引发异常。

还可以显式地指定请求的类型。它可能有助于分配更泛化的类型,例如 DECIMAL 或命名列。

Scala 代码如下:

```scala
val table = tEnv.fromValues(
    DataTypes.ROW(
        DataTypes.FIELD("id", DataTypes.DECIMAL(10, 2)),
        DataTypes.FIELD("name", DataTypes.STRING())
    ),
    row(1, "ABC"),
    row(2L, "ABCDE")
)
```

Java 代码如下:

```java
Table table = tEnv.fromValues(
    DataTypes.ROW(
        DataTypes.FIELD("id", DataTypes.DECIMAL(10, 2)),
        DataTypes.FIELD("name", DataTypes.STRING())
    ),
    row(1, "ABC"),
    row(2L, "ABCDE")
);
```

以上代码片段创建的表将具有以下的模式:

```
root
|-- id: DECIMAL(10, 2)
|-- name: STRING
```

(3) select():类似于 SQL 语句的 select 子句,执行选择操作。

Scala 代码如下:

```scala
val orders = tableEnv.from("Orders")
Table result = orders.select( $ "a", $ "c" as "d");
```

Java 代码如下:

```java
Table orders = tableEnv.from("Orders");
Table result = orders.select($("a"), $("c").as("d"));
```

可以使用星号（*）作为通配符，选择表中的所有列，代码如下：
Scala 代码如下：

```scala
Table result = orders.select($"*")
```

Java 代码如下：

```java
Table result = orders.select($("*"));
```

（4）as()：重命名字段。
Scala 代码如下：

```scala
val orders: Table = tableEnv.from("Orders").as("x", "y", "z", "t")
```

Java 代码如下：

```java
Table orders = tableEnv.from("Orders");
Table result = orders.as("x, y, z, t");
```

（5）where()/filter()：类似于 SQL 中的 where 子句。过滤不通过筛选谓词的行。
Scala 代码如下：

```scala
val orders: Table = tableEnv.from("Orders")
val result = orders.where($"a" % 2 === 0)
```

Java 代码如下：

```java
Table orders = tableEnv.from("Orders");
Table result = orders.where($("b").isEqual("red"));
```

等价地，可以使用 filter() 方法。
Scala 代码如下：

```scala
val orders: Table = tableEnv.from("Orders")
val result = orders.filter($"a" % 2 === 0)
```

Java 代码如下：

```java
Table orders = tableEnv.from("Orders");
Table result = orders.filter($("b").isEqual("red"));
```

2. 列运算符

(1) addColumns()：执行字段添加操作。如果添加的字段已经存在，则将抛出异常。Scala 代码如下：

```
val orders = tableEnv.from("Orders");
val result = orders.addColumns(concat($"c", "Sunny"))
```

Java 代码如下：

```
Table orders = tableEnv.from("Orders");
Table result = orders.addColumns(concat($("c"), "sunny"));
```

(2) addOrReplaceColumns()：执行字段添加操作。如果添加的列名与现有列名相同，则现有字段将被替换。此外，如果添加的字段有重复的字段名，则使用最后一个字段名。

Scala 代码如下：

```
val orders = tableEnv.from("Orders");
val result = orders.addOrReplaceColumns(concat($"c", "Sunny") as "desc")
```

Java 代码如下：

```
Table orders = tableEnv.from("Orders");
Table result = orders.addOrReplaceColumns(concat($("c"), "sunny").as("desc"));
```

(3) dropColumns()：执行字段删除操作。

Scala 代码如下：

```
val orders = tableEnv.from("Orders");
val result = orders.dropColumns($"b", $"c")
```

Java 代码如下：

```
Table orders = tableEnv.from("Orders");
Table result = orders.dropColumns($("b"), $("c"));
```

(4) renameColumns()：执行字段重命名操作。字段表达式应该是别名表达式，并且只能重命名现有字段。

Scala 代码如下：

```
val orders = tableEnv.from("Orders");
val result = orders.renameColumns($"b" as "b2", $"c" as "c2")
```

Java 代码如下：

```
Table orders = tableEnv.from("Orders");
Table result = orders.renameColumns( $("b").as("b2"), $("c").as("c2"));
```

3. 聚合操作

(1) groupBy()：类似于 SQL 的 group by 子句，用于按指定字段分组。

Scala 代码如下：

```
val orders: Table = tableEnv.from("Orders")
val result = orders.groupBy( $"a").select( $"a", $"b".sum().as("d"))
```

Java 代码如下：

```
Table orders = tableEnv.from("Orders");
Table result = orders.groupBy( $("a")).select( $("a"), $("b").sum().as("d"));
```

(2) groupBy()窗口聚合：在组窗口上对表进行分组和聚合，可能还有一个或多个分组键。

Scala 代码如下：

```
val orders: Table = tableEnv.from("Orders")
val result: Table = orders
    .window(Tumble over 5.minutes on $"rowtime" as "w")    //定义窗口
    .groupBy( $"a", $"w")                                   //group by key 和 window
    //访问窗口属性和聚合
    .select(
        $"a",
        $"w".start,
        $"w".end,
        $"w".rowtime,
        $"b".sum as "d")
```

Java 代码如下：

```
Table orders = tableEnv.from("Orders");
Table result = orders
    .window(Tumble.over(lit(5).minutes()).on( $("rowtime")).as("w"))  //定义窗口
    .groupBy( $("a"), $("w"))                                          //group by key 和 window
    //访问窗口属性和聚合
    .select(
        $("a"),
        $("w").start(),
```

```
    $("w").end(),
    $("w").rowtime(),
    $("b").sum().as("d")
);
```

(3) over()窗口聚合:类似于 SQL OVER 子句。基于前一行和后一行的窗口(范围),计算每行的窗口聚合。

Scala 代码如下:

```
val orders: Table = tableEnv.from("Orders")
val result: Table = orders
   //定义窗口
   .window(
     Over
       partitionBy $ "a"
       orderBy $ "rowtime"
       preceding UNBOUNDED_RANGE
       following CURRENT_RANGE
       as "w")
   //滑动聚合
   .select(
     $ "a",
     $ "b".avg over $ "w",
     $ "b".max().over( $ "w"),
     $ "b".min().over( $ "w")
   )
```

Java 代码如下:

```
Table orders = tableEnv.from("Orders");
Table result = orders
   //定义窗口
   .window(
     Over
       .partitionBy( $ ("a"))
       .orderBy( $ ("rowtime"))
       .preceding(UNBOUNDED_RANGE)
       .following(CURRENT_RANGE)
       .as("w"))
   //滑动聚合
   .select(
     $ ("a"),
     $ ("b").avg().over( $ ("w")),
     $ ("b").max().over( $ ("w")),
     $ ("b").min().over( $ ("w"))
   );
```

所有聚合必须在相同的窗口上定义,即相同的分区、排序和范围。目前,只支持 PRECEDING(UNBOUNDED 和 BOUNDED)到 CURRENT ROW 范围的窗口。目前还不支持 FOLLOWING 范围。ORDER BY 必须在单个时间属性上指定。

(4) distinct()聚合:类似于 SQL DISTINCT 聚合子句,如 COUNT(DISTINCT a)。distinct 可以应用于 GroupBy 聚合、GroupBy 窗口聚合和 Over 窗口聚合。

Scala 代码如下:

```
val orders: Table = tableEnv.from("Orders");

//group by 上的 distinct 聚合
val groupByDistinctResult = orders
  .groupBy($"a")
  .select($"a", $"b".sum.distinct as "d")

//time window group by 上的 distinct 聚合
val groupByWindowDistinctResult = orders
  .window(Tumble over 5.minutes on $"rowtime" as "w").groupBy($"a", $"w")
  .select($"a", $"b".sum.distinct as "d")

//over window 上的 distinct 聚合
val result = orders
  .window(Over
    partitionBy $"a"
    orderBy $"rowtime"
    preceding UNBOUNDED_RANGE
    as $"w")
  .select($"a", $"b".avg.distinct over $"w", $"b".max over $"w", $"b".min over $"w")
```

Java 代码如下:

```
Table orders = tableEnv.from("Orders");

//group by 上的 distinct 聚合
Table groupByDistinctResult = orders
  .groupBy($("a"))
  .select($("a"), $("b").sum().distinct().as("d"));

//time window group by 上的 distinct 聚合
Table groupByWindowDistinctResult = orders
  .window(Tumble
      .over(lit(5).minutes())
      .on($("rowtime"))
      .as("w")
  )
```

```
    .groupBy( $ ("a"), $ ("w"))
    .select( $ ("a"), $ ("b").sum().distinct().as("d"));

//over window 上的 distinct 聚合
Table result = orders
    .window(Over
        .partitionBy( $ ("a"))
        .orderBy( $ ("rowtime"))
        .preceding(UNBOUNDED_RANGE)
        .as("w"))
    .select(
        $ ("a"), $ ("b").avg().distinct().over( $ ("w")),
        $ ("b").max().over( $ ("w")),
        $ ("b").min().over( $ ("w"))
    );
```

(5) distinct()：类似于 SQL 的 distinct 子句，返回具有唯一值组合的记录。

Scala 代码如下：

```
val orders: Table = tableEnv.from("Orders")
val result = orders.distinct()
```

Java 代码如下：

```
Table orders = tableEnv.from("Orders");
Table result = orders.distinct();
```

4. join 连接

(1) 内连接：类似于 SQL 的 join 子句，连接两个表。这两个表必须具有不同的字段名，并且必须通过连接操作符或使用 where()、filter()操作符定义至少一个相等连接谓词。

Scala 代码如下：

```
val left = ds1.toTable(tableEnv, $ "a", $ "b", $ "c")
val right = ds2.toTable(tableEnv, $ "d", $ "e", $ "f")
val result = left.join(right).where( $ "a" === $ "d").select( $ "a", $ "b", $ "e")
```

Java 代码如下：

```
Table left = tableEnv.fromDataSet(ds1, "a, b, c");
Table right = tableEnv.fromDataSet(ds2, "d, e, f");
Table result = left.join(right)
    .where( $ ("a").isEqual( $ ("d")))
    .select( $ ("a"), $ ("b"), $ ("e"));
```

（2）外连接：类似于 SQL 的 left/right/full outer join 子句，连接两个表。这两个表必须具有不同的字段名，并且必须定义至少一个相等连接谓词。

Scala 代码如下：

```scala
val left = tableEnv.fromDataSet(ds1, $"a", $"b", $"c")
val right = tableEnv.fromDataSet(ds2, $"d", $"e", $"f")

val leftOuterResult = left
    .leftOuterJoin(right, $"a" === $"d")
    .select($"a", $"b", $"e")
val rightOuterResult = left
    .rightOuterJoin(right, $"a" === $"d")
    .select($"a", $"b", $"e")
val fullOuterResult = left
    .fullOuterJoin(right, $"a" === $"d")
    .select($"a", $"b", $"e")
```

Java 代码如下：

```java
Table left = tableEnv.fromDataSet(ds1, "a, b, c");
Table right = tableEnv.fromDataSet(ds2, "d, e, f");

Table leftOuterResult = left
    .leftOuterJoin(right, $("a").isEqual($("d")))
    .select($("a"), $("b"), $("e"));
Table rightOuterResult = left
    .rightOuterJoin(right, $("a").isEqual($("d")))
    .select($("a"), $("b"), $("e"));
Table fullOuterResult = left
    .fullOuterJoin(right, $("a").isEqual($("d")))
    .select($("a"), $("b"), $("e"));
```

（3）间隔连接：间隔连接是常规连接的子集，可以以流方式处理。间隔连接至少需要一个等效连接谓词和一个连接条件，该连接条件限定了两端的时间。

Scala 代码如下：

```scala
val left = ds1.toTable(tableEnv, $"a", $"b", $"c", $"ltime".rowtime)
val right = ds2.toTable(tableEnv, $"d", $"e", $"f", $"rtime".rowtime)

val result = left.join(right)
  .where($"a" === $"d" && $"ltime" >= $"rtime" - 5.minutes && $"ltime" < $"rtime" + 10.minutes)
  .select($"a", $"b", $"e", $"ltime")
```

Java 代码如下：

```java
Table left = tableEnv.fromDataSet(ds1, $("a"), $("b"), $("c"), $("ltime").rowtime());
Table right = tableEnv.fromDataSet(ds2, $("d"), $("e"), $("f"), $("rtime").rowtime
()));

Table result = left.join(right)
  .where(
    and(
        $("a").isEqual($("d")),
        $("ltime").isGreaterOrEqual($("rtime").minus(lit(5).minutes())),
        $("ltime").isLess($("rtime").plus(lit(10).minutes()))
    ))
  .select($("a"), $("b"), $("e"), $("ltime"));
```

5．集合运算

(1) union()：类似于 SQL 中的 union 子句。合并两张表并删除重复记录。这两张表必须具有相同的字段类型。这个运算只作用于批处理。

Scala 代码如下：

```scala
val left = tableEnv.from("orders1")
val right = tableEnv.from("orders2")

left.union(right)
```

Java 代码如下：

```java
Table left = tableEnv.from("orders1");
Table right = tableEnv.from("orders2");

left.union(right);
```

(2) unionAll()：类似于 SQL 中的 union all 子句。合并两张表。这两张表必须具有相同的字段类型。

Scala 代码如下：

```scala
val left = tableEnv.from("orders1")
val right = tableEnv.from("orders2")

left.unionAll(right)
```

Java 代码如下：

```
Table left = tableEnv.from("orders1");
Table right = tableEnv.from("orders2");

left.unionAll(right);
```

（3）intersect()：类似于 SQL 中的 intersect 子句。intersect 返回两张表中都存在的记录。虽然一条记录在一张或两张表中出现了不止一次，但它只会返回一次，也就是说，生成的表中没有重复的记录。这两张表必须具有相同的字段类型。这个运算只作用于批处理。

Scala 代码如下：

```
val left = tableEnv.from("orders1")
val right = tableEnv.from("orders2")

left.intersect(right)
```

Java 代码如下：

```
Table left = tableEnv.from("orders1");
Table right = tableEnv.from("orders2");

left.intersect(right);
```

（4）intersectAll()：类似于 SQL 中的 intersect all 子句。intersectAll 返回两张表中都存在的记录。如果一条记录在两张表中出现不止一次，则返回的次数与它在两张表中出现的次数相同，也就是说，生成的表可能有重复的记录。这两张表必须具有相同的字段类型。这个运算只作用于批处理。

Scala 代码如下：

```
val left = tableEnv.from("orders1")
val right = tableEnv.from("orders2")

left.intersectAll(right)
```

Java 代码如下：

```
Table left = tableEnv.from("orders1");
Table right = tableEnv.from("orders2");

left.intersectAll(right);
```

（5）minus()：类似于 SQL 中的 except 子句。minus 返回左表中存在但右表中不存在的记录。左表中的重复记录只返回一次，即删除重复记录。这两张表必须具有相同的字段类型。这个运算只作用于批处理。

Scala 代码如下：

```scala
val left = tableEnv.from("orders1")
val right = tableEnv.from("orders2")

left.minus(right)
```

Java 代码如下：

```java
Table left = tableEnv.from("orders1");
Table right = tableEnv.from("orders2");

left.minus(right);
```

（6）minusAll()：类似于 SQL 中的 except all 子句。minusAll 用于返回不存在于右表中的记录。在左表中出现 n 次在右表中出现 m 次的记录被返回($n-m$)次，也就是说，在右表中出现的重复记录会被删除。这两张表必须具有相同的字段类型。这个运算只作用于批处理。

Scala 代码如下：

```scala
val left = tableEnv.from("orders1")
val right = tableEnv.from("orders2")

left.minusAll(right)
```

Java 代码如下：

```java
Table left = tableEnv.from("orders1");
Table right = tableEnv.from("orders2");

left.minusAll(right);
```

（7）in()：类似于 SQL 中的 in 子句。如果表达式存在于给定的表的子查询中，则返回值为 true。子查询表必须包含一列。此列必须具有与表达式相同的数据类型。

Scala 代码如下：

```scala
val left = tableEnv.from("Orders1")
val right = tableEnv.from("Orders2");

val result = left.select( $ "a", $ "b", $ "c").where( $ "a".in(right))
```

Java 代码如下：

```
Table left = tableEnv.from("Orders1")
Table right = tableEnv.from("Orders2");

Table result = left.select($("a"), $("b"), $("c")).where($("a").in(right));
```

6. orderBy、offset & fetch

(1) orderBy()：类似于 SQL 中的 order by 子句。返回所有并行分区的全局排序记录。对于无界表，此操作需要对时间属性或随后的获取操作进行排序。

Scala 代码如下：

```
val result = in.orderBy($ "a".asc)
```

Java 代码如下：

```
Table result = in.orderBy($("a").asc());
```

(2) offset() & fetch()：类似于 SQL 中的 offset 和 fetch 子句。偏移操作限制来自偏移位置的(可能已排序的)结果。fetch 操作将结果限制在前 n 行(可能已经排序)。通常，这两个操作的前面有一个排序操作符。对于无界表，偏移操作需要一个 fetch 操作。

Scala 代码如下：

```
//返回排序结果的前5条记录
val result1: Table = in.orderBy($ "a".asc).fetch(5)

//跳过前3条记录并返回排序结果中所有后面的记录
val result2: Table = in.orderBy($ "a".asc).offset(3)

//跳过前10条记录并返回排序结果中的后5条记录
val result3: Table = in.orderBy($ "a".asc).offset(10).fetch(5)
```

Java 代码如下：

```
//返回排序结果的前5条记录
Table result1 = in.orderBy($("a").asc()).fetch(5);

//跳过前3条记录并返回排序结果中所有后面的记录
Table result2 = in.orderBy($("a").asc()).offset(3);

//跳过前10条记录并返回排序结果中的后5条记录
Table result3 = in.orderBy($("a").asc()).offset(10).fetch(5);
```

7. 插入操作

与 SQL 查询中的 INSERT INTO 子句类似，该方法执行插入注册输出表的操作。executeInsert()方法将立即提交一个执行插入操作的 Flink 作业。

输出表必须在 TableEnvironment 中注册，而且，注册表的模式必须与查询的模式匹配。Scala 代码如下：

```
val orders = tableEnv.from("Orders")
orders.executeInsert("OutOrders")
```

Java 代码如下：

```
Table orders = tableEnv.from("Orders");
orders.executeInsert("OutOrders");
```

5.3.2 窗口运算

Flink Table API 也支持多种窗口运算。

1. 分组窗口

分组窗口根据时间或行计数间隔将行聚合为有限组，并对每个组计算一次聚合函数。对于批处理表，窗口是按时间间隔对记录进行分组的方便快捷方式。

窗口是使用 window(w: GroupWindow)子句定义的，并且需要使用 as 子句指定的别名。为了根据窗口对表进行分组，必须在 groupBy()子句中引用窗口别名，就像常规的分组属性一样。下面的示例将演示如何在表上定义窗口聚合。

Scala 代码如下：

```
val table = input
  .window([w: GroupWindow] as $ "w")        //定义别名为 w 的窗口
  .groupBy( $ "w")                           //将表按窗口 w 分组
  .select( $ "b".sum)                        //聚合
```

Java 代码如下：

```
Table table = input
  .window([GroupWindow w].as("w"))          //定义别名为 w 的窗口
  .groupBy( $ ("w"))                         //将表按窗口 w 分组
  .select( $ ("b").sum());                   //聚合
```

在流环境中，窗口聚合只能在一个或多个属性上进行分组的情况下并行计算，例如，groupBy()子句引用一个窗口别名和至少一个附加属性。只引用窗口别名的 groupBy()子句(如上面的示例)只能由单个非并行任务计算。下面的示例显示了如何定义带有附加分组属性的窗口聚合。

Scala 代码如下：

```scala
val table = input
  .window([w: GroupWindow] as $"w")      //定义别名为w的窗口
  .groupBy($"w", $"a")                    //根据属性a和窗口w对表进行分组
  .select($"a", $"b".sum)                 //聚合
```

Java 代码如下：

```java
Table table = input
  .window([GroupWindow w].as("w"))        //定义别名为w的窗口
  .groupBy($("w"), $("a"))                //根据属性a和窗口w对表进行分组
  .select($("a"), $("b").sum());          //聚合
```

窗口属性,例如时间窗口的开始、结束或行时间戳,可以分别作为窗口别名 w.start、w.end 和 w.rowtime 的属性添加到 select()语句中。

Scala 代码如下：

```scala
val table = input
  .window([w: GroupWindow] as $"w")      //定义别名为w的窗口
  .groupBy($"w", $"a")                    //根据属性a和窗口w对表进行分组
  //聚合并增加窗口的开始、结束和行时间戳
  .select(
    $"a",
    $"w".start,
    $"w".end,
    $"w".rowtime,
    $"b".count
  )
```

Java 代码如下：

```java
Table table = input
  .window([GroupWindow w].as("w"))        //定义别名为w的窗口
  .groupBy($("w"), $("a"))                //根据属性a和窗口w对表进行分组
  //聚合并增加窗口的开始、结束和行时间戳
  .select(
    $("a"),
    $("w").start(),
    $("w").end(),
    $("w").rowtime(),
    $("b").count()
  );
```

Window 参数用于定义如何将行映射到窗口。Window 不是一个用户可以实现的接口。相反，Table API 提供了一组具有特定语义的预定义 Window 类，这些类被转换为底层的 DataStream 或 DataSet 操作。下面列出了受支持的窗口定义。

1）滚动窗口

滚动窗口将行分配给不重叠的、固定长度的连续窗口。例如，一个 5min 的滚动窗口以 5min 间隔对行进行分组。滚动窗口可以在事件时间、处理时间或行数上定义。

Scala 代码如下：

```scala
//滚动事件时间窗口
.window(Tumble over 10.minutes on $"rowtime" as $"w")

//滚动处理时间窗口（假设有一个处理时间属性 "proctime"）
.window(Tumble over 10.minutes on $"proctime" as $"w")

//滚动行计数窗口（假设有一个处理时间属性 "proctime"）
.window(Tumble over 10.rows on $"proctime" as $"w")
```

Java 代码如下：

```java
//滚动事件时间窗口
.window(Tumble.over(lit(10).minutes()).on($("rowtime")).as("w"));

//滚动处理时间窗口（假设有一个处理时间属性 "proctime"）
.window(Tumble.over(lit(10).minutes()).on($("proctime")).as("w"));

//滚动行计数窗口（假设有一个处理时间属性 "proctime"）
.window(Tumble.over(rowInterval(10)).on($("proctime")).as("w"));
```

滚动窗口是通过 Tumble 类定义的，其中包括以下方法。

(1) over：将窗口的长度定义为时间或行计数间隔。

(2) on：将 time 属性分组（时间间隔）或排序（行数）。对于批处理查询，可以是任何 Long 或 Timestamp 属性。对于流查询，这必须是一个声明的事件时间或处理时间属性。

(3) as：给窗口分配别名。别名用于在下面的 groupBy() 子句中引用窗口，并可选地在 select() 子句中选择窗口属性，如窗口开始、结束或行时间戳。

2）滑动窗口

滑动窗口具有固定的大小，并按指定的滑动间隔滑动。如果滑动间隔小于窗口大小，则滑动窗口重叠，因此，行可以分配给多个窗口。例如，一个 15min 大小的滑动窗口和 5min 的滑动间隔将每行分配给 3 个 15min 大小的不同窗口，这些窗口在 5min 的间隔内进行计算。滑动窗口可以在事件时间、处理时间或行数上定义。

Scala 代码如下：

```
//滑动事件时间窗口
.window(Slide over 10.minutes every 5.minutes on $"rowtime" as $"w")

//滑动处理时间窗口（假设有一个处理时间属性"proctime"）
.window(Slide over 10.minutes every 5.minutes on $"proctime" as $"w")

//滑动行计数窗口(假设有一个处理时间属性"proctime")
.window(Slide over 10.rows every 5.rows on $"proctime" as $"w")
```

Java 代码如下：

```
//滑动事件时间窗口
.window(Slide.over(lit(10).minutes())
        .every(lit(5).minutes())
        .on( $("rowtime"))
        .as("w"));

//滑动处理时间窗口（假设有一个处理时间属性"proctime"）
.window(Slide.over(lit(10).minutes())
        .every(lit(5).minutes())
        .on( $("proctime"))
        .as("w"));

//滑动行计数窗口(假设有一个处理时间属性"proctime")
.window(Slide.over(rowInterval(10)).every(rowInterval(5)).on( $("proctime")).as("w"));
```

滑动窗口是通过 Slide 类定义的，其中包括以下方法。

(1) over：将窗口的长度定义为时间或行计数间隔。

(2) every：定义滑动间隔，可以是时间间隔，也可以是行计数间隔。滑动间隔必须与窗口大小间隔的类型相同。

(3) on：将 time 属性分组（时间间隔）或排序（行数）。对于批处理查询，可以是任何 Long 或 Timestamp 属性。对于流查询，这必须是一个声明的事件时间或处理时间属性。

(4) as：给窗口分配别名。别名用于在下面的 groupBy()子句中引用窗口，并可选地在 select()子句中选择窗口属性，如窗口开始、结束或行时间戳。

3）会话窗口

会话窗口没有固定的大小，但它们的边界是由一个不活动的间隔定义的，即，如果在一个定义的间隔时间内没有事件出现，则会话窗口将关闭。例如，当会话窗口在 30min 不活动后观察到一行时（否则该行将被添加到现有的窗口），一个有 30min 间隔的会话窗口开始，如果 30min 内没有添加行，则关闭该会话窗口。会话窗口可以在事件时间或处理时间上工作。

Scala 代码如下：

```
//会话事件时间窗口
.window(Session withGap 10.minutes on $ "rowtime" as $ "w")

//会话处理时间窗口(假设有一个处理时间属性 "proctime")
.window(Session withGap 10.minutes on $ "proctime" as $ "w")
```

Java 代码如下：

```
//会话事件时间窗口
.window(Session.withGap(lit(10).minutes()).on( $ ("rowtime")).as("w"));

//会话处理时间窗口(假设有一个处理时间属性 "proctime")
.window(Session.withGap(lit(10).minutes()).on( $ ("proctime")).as("w"));
```

会话窗口通过 Session 类定义，其中包括以下方法。
（1）withGap：将两个窗口之间的间隔定义为时间间隔。
（2）on：将 time 属性分组（时间间隔）或排序（行数）。对于批处理查询，可以是任何 Long 或 Timestamp 属性。对于流查询，这必须是一个声明的事件时间或处理时间属性。
（3）as：给窗口分配别名。别名用于在下面的 groupBy()子句中引用窗口，并可选地在 select()子句中选择窗口属性，如窗口开始、结束或行时间戳。

2. Over 窗口

Over 窗口聚合来自标准 SQL（Over 子句），并在查询的 SELECT 子句中定义。与在 group by 子句中指定的组窗口不同，Over 窗口不会折叠行。相反，Over 窗口聚合计算每个输入行在其相邻行范围内的聚合。

Over 窗口是使用 window(w：OverWindow *)子句定义的，并通过 select() 方法中的别名引用。在表上定义 Over 窗口聚合的代码如下。

Scala 代码如下：

```
val table = input
  .window([w: OverWindow] as $ "w")           //用别名 w 定义 Over 窗口
  .select(                                     //在 Over 窗口 w 上聚合
    $ "a",
    $ "b".sum over $ "w",
    $ "c".min over $ "w"
)
```

Java 代码如下：

```
Table table = input
  .window([OverWindow w].as("w"))              //用别名 w 定义 Over 窗口
  .select(                                     //在 Over 窗口 w 上聚合
```

```
    $("a"),
    $("b").sum().over( $ ("w")),
    $("c").min().over( $ ("w"))
);
```

OverWindow 定义了计算聚合的行范围。Table API 提供了 Over 类来配置 Over 窗口的属性。Over 窗口可以在事件时间或处理时间上定义，也可以在指定时间间隔或行计数的范围上定义。支持的 Over 窗口定义作为 Over(和其他类)上的方法公开，如下所示。

(1) partitionBy：在一个或多个属性上定义输入的分区。对每个分区分别进行排序，并对每个分区分别应用聚合函数。

(2) orderBy：定义每个分区中的行顺序，从而确定聚合函数应用于行的顺序。注意，对于流查询，这必须是一个声明的事件时间或处理时间属性。目前，只支持单个排序属性。

(3) preceding：定义包含在窗口中并位于当前行之前的行的间隔。间隔可以指定为时间间隔或行计数间隔。有界的窗口被指定为间隔的大小，例如，10.minutes 时间间隔或 10.rows 行计数间隔。无界窗口使用常量指定，例如，UNBOUNDED_RANGE 作为一个时间间隔，UNBOUNDED_ROW 作为一个行计数间隔。无界窗口从分区的第 1 行开始。如果不指定 preceding 子句，则 UNBOUNDED_RANGE 和 CURRENT_RANGE 将被用作窗口的默认 preceding 和 following。

(4) following：定义包含在窗口中并跟随当前行的行的窗口间隔。间隔必须与 preceding 间隔的单位相同(时间或行计数)。目前，不支持行紧跟在当前行的窗口。相反，可以指定两个常量之一：CURRENT_ROW 用于将窗口的上界设置为当前行，CURRENT_RANGE 用于设置窗口的上界，以此来排序当前行的键，即与当前行具有相同排序键的所有行都包含在窗口中。如果省略 following 子句，则时间间隔窗口的上限定义为 CURRENT_RANGE，行数间隔窗口的上限定义为 CURRENT_ROW。

(5) as：给 Over 窗口分配别名。别名用于在下面的 select()子句中引用 Over 窗口。目前，同一个 select()调用中的所有聚合函数都必须计算同一个 Over 窗口。

1) 无界 Over 窗口

Scala 代码如下：

```
//无界事件时间 Over Window(假设有一个事件时间属性"rowtime")
.window(Over partitionBy $ "a" orderBy $ "rowtime" preceding UNBOUNDED_RANGE as "w")

//无界处理时间 Over Window(假设有一个处理时间属性"proctime")
.window(Over partitionBy $ "a" orderBy $ "proctime" preceding UNBOUNDED_RANGE as "w")

//无界事件时间行计数 Over Window(假设有一个事件时间属性"rowtime")
.window(Over partitionBy $ "a" orderBy $ "rowtime" preceding UNBOUNDED_ROW as "w")

//无界处理时间行计数 Over Window(假设有一个处理时间属性"proctime")
.window(Over partitionBy $ "a" orderBy $ "proctime" preceding UNBOUNDED_ROW as "w")
```

Java 代码如下：

```java
//无界事件时间 Over Window (假设有一个事件时间属性"rowtime")
.window(Over.partitionBy( $ ("a")).orderBy( $ ("rowtime")).preceding(UNBOUNDED_RANGE).as("w"));

//无界处理时间 Over Window (假设有一个处理时间属性"proctime")
.window(Over.partitionBy( $ ("a")).orderBy("proctime").preceding(UNBOUNDED_RANGE).as("w"));

//无界事件时间行计数 Over Window (假设有一个事件时间属性"rowtime")
.window(Over.partitionBy( $ ("a")).orderBy( $ ("rowtime")).preceding(UNBOUNDED_ROW).as("w"));

//无界处理时间行计数 Over Window (假设有一个处理时间属性"proctime")
.window(Over.partitionBy( $ ("a")).orderBy( $ ("proctime")).preceding(UNBOUNDED_ROW).as("w"));
```

2）有界 over 窗口

Scala 代码如下：

```scala
//有界事件时间 Over Window (假设有一个事件时间属性"rowtime")
.window(Over partitionBy $ "a" orderBy $ "rowtime" preceding 1.minutes as "w")

//有界处理时间 Over Window (假设有一个处理时间属性"proctime")
.window(Over partitionBy $ "a" orderBy $ "proctime" preceding 1.minutes as "w")

//有界事件时间行计数 Over Window (假设有一个事件时间属性"rowtime")
.window(Over partitionBy $ "a" orderBy $ "rowtime" preceding 10.rows as "w")

//有界处理时间行计数 Over Window (假设有一个处理时间属性"proctime")
.window(Over partitionBy $ "a" orderBy $ "proctime" preceding 10.rows as "w")
```

Java 代码如下：

```java
//有界事件时间 Over Window (假设有一个事件时间属性"rowtime")
.window(Over.partitionBy( $ ("a")).orderBy( $ ("rowtime")).preceding(lit(1).minutes()).as("w"));

//有界处理时间 Over Window (假设有一个处理时间属性"proctime")
.window(Over.partitionBy( $ ("a")).orderBy( $ ("proctime")).preceding(lit(1).minutes()).as("w"));

//有界事件时间行计数 Over Window (假设有一个事件时间属性"rowtime")
.window(Over.partitionBy( $ ("a")).orderBy( $ ("rowtime")).preceding(rowInterval(10)).as("w"));

//有界处理时间行计数 Over Window (假设有一个处理时间属性"proctime")
.window(Over.partitionBy( $ ("a")).orderBy( $ ("proctime")).preceding(rowInterval(10)).as("w"));
```

5.3.3 基于行的操作

基于行的操作生成具有多列的输出。

(1) map()：使用用户定义的标量函数或内置标量函数执行 map() 操作。如果输出类型是复合类型，则输出将被扁平化。

Scala 代码如下：

```scala
class MyMapFunction extends ScalarFunction {
  def eval(a: String): Row = {
    Row.of(a, "pre-" + a)
  }

  override def getResultType(signature: Array[Class[_]]): TypeInformation[_] =
    Types.ROW(Types.STRING, Types.STRING)
}

val func = new MyMapFunction()
val table = input.map(func($"c")).as("a", "b")
```

Java 代码如下：

```java
public class MyMapFunction extends ScalarFunction {
    public Row eval(String a) {
        return Row.of(a, "pre-" + a);
    }

    @Override
    public TypeInformation<?> getResultType(Class<?>[] signature) {
        return Types.ROW(Types.STRING, Types.STRING);
    }
}
ScalarFunction func = new MyMapFunction();
tableEnv.registerFunction("func", func);

Table table = input.map(call("func", $("c"))).as("a", "b");
```

(2) flatMap()：使用表函数执行 flatMap 操作。

Scala 代码如下：

```scala
class MyFlatMapFunction extends TableFunction[Row] {
  def eval(str: String): Unit = {
    if (str.contains("#")) {
```

```scala
      str.split("#").foreach({ s =>
        val row = new Row(2)
        row.setField(0, s)
        row.setField(1, s.length)
        collect(row)
      })
    }
  }

  override def getResultType: TypeInformation[Row] = {
    Types.ROW(Types.STRING, Types.INT)
  }
}

val func = new MyFlatMapFunction
val table = input.flatMap(func( $ "c")).as("a", "b")
```

Java 代码如下：

```java
public class MyFlatMapFunction extends TableFunction< Row > {

    public void eval(String str) {
        if (str.contains("#")) {
            String[] array = str.split("#");
            for (int i = 0; i < array.length; ++i) {
                collect(Row.of(array[i], array[i].length()));
            }
        }
    }

    @Override
    public TypeInformation< Row > getResultType() {
        return Types.ROW(Types.STRING, Types.INT);
    }
}

TableFunction func = new MyFlatMapFunction();
tableEnv.registerFunction("func", func);

Table table = input.flatMap(call("func", $("c"))).as("a", "b");
```

（3）aggregate()：使用聚合函数执行聚合操作。必须用一个 select()语句来关闭"聚合"，而这个 select 语句不支持聚合函数。如果输出类型是复合类型，则聚合的输出将被扁平化。

Scala 代码如下：

```scala
case class MyMinMaxAcc(var min: Int, var max: Int)

class MyMinMax extends AggregateFunction[Row, MyMinMaxAcc] {

  def accumulate(acc: MyMinMaxAcc, value: Int): Unit = {
    if (value < acc.min) {
      acc.min = value
    }
    if (value > acc.max) {
      acc.max = value
    }
  }

  override def createAccumulator(): MyMinMaxAcc = MyMinMaxAcc(0, 0)

  def resetAccumulator(acc: MyMinMaxAcc): Unit = {
    acc.min = 0
    acc.max = 0
  }

  override def getValue(acc: MyMinMaxAcc): Row = {
    Row.of(Integer.valueOf(acc.min), Integer.valueOf(acc.max))
  }

  override def getResultType: TypeInformation[Row] = {
    new RowTypeInfo(Types.INT, Types.INT)
  }
}

val myAggFunc = new MyMinMax
val table = input
  .groupBy($"key")
  .aggregate(myAggFunc($"a") as ("x", "y"))
  .select($"key", $"x", $"y")
```

Java 代码如下:

```java
public class MyMinMaxAcc {
    public int min = 0;
    public int max = 0;
}

public class MyMinMax extends AggregateFunction<Row, MyMinMaxAcc> {

    public void accumulate(MyMinMaxAcc acc, int value) {
```

```java
        if (value < acc.min) {
            acc.min = value;
        }
        if (value > acc.max) {
            acc.max = value;
        }
    }

    @Override
    public MyMinMaxAcc createAccumulator() {
        return new MyMinMaxAcc();
    }

    public void resetAccumulator(MyMinMaxAcc acc) {
        acc.min = 0;
        acc.max = 0;
    }

    @Override
    public Row getValue(MyMinMaxAcc acc) {
        return Row.of(acc.min, acc.max);
    }

    @Override
    public TypeInformation<Row> getResultType() {
        return new RowTypeInfo(Types.INT, Types.INT);
    }
}

AggregateFunction myAggFunc = new MyMinMax();
tableEnv.registerFunction("myAggFunc", myAggFunc);
Table table = input
    .groupBy($("key"))
    .aggregate(call("myAggFunc", $("a")).as("x", "y"))
    .select($("key"), $("x"), $("y"));
```

（4）Group Window 聚合：在组窗口（Group Window）上对表进行分组和聚合，可能有一个或多个分组键。必须用 select() 语句关闭"聚合"。该 select() 语句不支持" * "或聚合函数。

Scala 代码如下：

```scala
val myAggFunc = new MyMinMax
val table = input
    .window(Tumble over 5.minutes on $"rowtime" as "w")    //定义窗口
    .groupBy($"key", $"w")                                  //按 key 和窗口分组
    .aggregate(myAggFunc($"a") as ("x", "y"))
    .select($"key", $"x", $"y", $"w".start, $"w".end)      //访问窗口属性和聚合结果
```

Java代码如下：

```java
AggregateFunction myAggFunc = new MyMinMax();
tableEnv.registerFunction("myAggFunc", myAggFunc);

Table table = input
    //定义窗口
    .window(Tumble.over(lit(5).minutes()).on( $("rowtime")).as("w"))
    //按key和窗口分组
    .groupBy( $("key"), $("w"))
.aggregate(call("myAggFunc", $("a")).as("x", "y"))
//访问窗口属性和聚合结果
    .select( $("key"), $("x"), $("y"), $("w").start(), $("w").end());
```

(5) flatAggregate()：类似于 groupBy() 聚合。使用下面的运行表聚合操作符对分组键上的行进行分组，以按组对行进行聚合。与 AggregateFunction 的区别在于，TableAggregateFunction 可以为一个组返回 0 条或多条记录。必须用 select() 语句关闭 flatAggregate，该 select() 语句不支持聚合函数。除了可以使用 emitValue() 输出结果，还可以使用 emitUpdateWithRetract() 方法。与 emitValue() 不同，emitUpdateWithRetract() 用于发出已更新的值。该方法在 retract 模式下以增量方式输出数据，也就是说，一旦有了更新，必须在发送新的更新记录之前撤销旧记录。如果两种方法都在表聚合函数中定义，则 emitUpdateWithRetract() 方法将优先于 emitValue() 方法，因为该方法被视为比 emitValue() 方法更有效，因为它可以增量输出值。

Scala代码如下：

```scala
import Java.lang.{Integer => JInteger}
import org.apache.flink.table.api.Types
import org.apache.flink.table.functions.TableAggregateFunction

/**
 * 用于top2的累加器
 */
class Top2Accum {
    var first: JInteger = _
    var second: JInteger = _
}

/**
 * top2用户定义的表聚合函数
 */
class Top2 extends TableAggregateFunction[JTuple2[JInteger, JInteger], Top2Accum] {
```

```scala
  override def createAccumulator(): Top2Accum = {
    val acc = new Top2Accum
    acc.first = Int.MinValue
    acc.second = Int.MinValue
    acc
  }

  def accumulate(acc: Top2Accum, v: Int) {
    if (v > acc.first) {
      acc.second = acc.first
      acc.first = v
    } else if (v > acc.second) {
      acc.second = v
    }
  }

  def merge(acc: Top2Accum, its: JIterable[Top2Accum]): Unit = {
    val iter = its.iterator()
    while (iter.hasNext) {
      val top2 = iter.next()
      accumulate(acc, top2.first)
      accumulate(acc, top2.second)
    }
  }

  def emitValue(acc: Top2Accum, out: Collector[JTuple2[JInteger, JInteger]]): Unit = {
    //emit the value and rank
    if (acc.first != Int.MinValue) {
      out.collect(JTuple2.of(acc.first, 1))
    }
    if (acc.second != Int.MinValue) {
      out.collect(JTuple2.of(acc.second, 2))
    }
  }
}

val top2 = new Top2
val orders: Table = tableEnv.from("Orders")
val result = orders
  .groupBy( $ "key")
  .flatAggregate(top2( $ "a") as ( $ "v", $ "rank"))
  .select( $ "key", $ "v", $ "rank")
```

Java 代码如下：

```java
/**
 * 用于top2的累加器
 */
public class Top2Accum {
    public Integer first;
    public Integer second;
}

/**
 * top2用户定义的表聚合函数
 */
public class Top2 extends TableAggregateFunction<Tuple2<Integer, Integer>, Top2Accum> {

    @Override
    public Top2Accum createAccumulator() {
        Top2Accum acc = new Top2Accum();
        acc.first = Integer.MIN_VALUE;
        acc.second = Integer.MIN_VALUE;
        return acc;
    }


    public void accumulate(Top2Accum acc, Integer v) {
        if (v > acc.first) {
            acc.second = acc.first;
            acc.first = v;
        } else if (v > acc.second) {
            acc.second = v;
        }
    }

    public void merge(Top2Accum acc, Java.lang.Iterable<Top2Accum> iterable) {
        for (Top2Accum otherAcc : iterable) {
            accumulate(acc, otherAcc.first);
            accumulate(acc, otherAcc.second);
        }
    }

    public void emitValue(Top2Accum acc, Collector<Tuple2<Integer, Integer>> out) {
        //emit the value and rank
        if (acc.first != Integer.MIN_VALUE) {
            out.collect(Tuple2.of(acc.first, 1));
        }
        if (acc.second != Integer.MIN_VALUE) {
            out.collect(Tuple2.of(acc.second, 2));
        }
```

```
    }
}
tEnv.registerFunction("top2", new Top2());
Table orders = tableEnv.from("Orders");
Table result = orders
    .groupBy( $ ("key"))
    .flatAggregate(call("top2", $ ("a")).as("v", "rank"))
    .select( $ ("key"), $ ("v"), $ ("rank"));
```

5.4 Table API 与 DataStream API 集成

10min

在定义数据处理管道时，Table API 和 DataStream API 同样重要。

DataStream API 在一个相对低级的命令编程 API 中提供了流处理（时间、状态和数据流管理）的原语。Table API 抽象了许多内部元素，并提供了结构化和声明性的 API。这两种 API 都可以处理有界和无界流。Flink 提供了特殊的桥接功能，使与 DataStream API 的集成尽可能顺利。

5.4.1 依赖

组合 Table API 和 DataStream API 的项目需要添加桥接模块，包括到 flink-table-api-Java 或 flink-table-api-scala 和相应的特定于语言的 DataStream API 模块的传递依赖。根据所选择的编程语言，在项目中添加对应的桥接模块，内容如下：

```xml
<!-- 如果使用 Scala -->
<dependency>
  <groupId>org.apache.flink</groupId>
  <artifactId>flink-table-api-scala-bridge_2.12</artifactId>
  <version>1.13.2</version>
  <scope>provided</scope>
</dependency>

<!-- 如果使用 Java -->
<dependency>
  <groupId>org.apache.flink</groupId>
  <artifactId>flink-table-api-Java-bridge_2.12</artifactId>
  <version>1.13.2</version>
  <scope>provided</scope>
</dependency>
```

要使用 DataStream API 和 Table API 声明公共管道，需要以下导入项。

Scala 代码如下：

```scala
//Scala DataStream API
import org.apache.flink.api.scala._
import org.apache.flink.streaming.api.scala._

//桥接 Scala DataStream API 的 Table API 导入
import org.apache.flink.table.api._
import org.apache.flink.table.api.bridge.scala._
```

Java 代码如下：

```java
//Java DataStream API
import org.apache.flink.streaming.api.*;
import org.apache.flink.streaming.api.environment.*;

//桥接 Java DataStream API 的 Table API 导入
import org.apache.flink.table.api.*;
import org.apache.flink.table.api.bridge.java.*;
```

5.4.2 在 DataStream 和 Table 之间转换

Flink 提供了一个专门的 StreamTableEnvironment（在 Java 和 Scala 中），用于与 DataStream API 集成。这些环境使用额外的方法扩展常规的 TableEnvironment，并将 DataStream API 中使用的 StreamExecutionEnvironment 作为参数。

下面的代码展示了如何在两个 API 之间来回切换。Table 的列名和类型是由 DataStream 的 TypeInformation 自动派生的。由于 DataStream API 不支持在本地更改日志(changelog)处理，因此在流到表和表到流的转换期间，代码假定 append-only/insert-only 语义。

Scala 代码如下：

```scala
//第 5 章/TableDemo01.scala

import org.apache.flink.streaming.api.scala._
import org.apache.flink.table.api.bridge.scala.StreamTableEnvironment

/**
 * 在 DataStream 和 Table 之间转换
 */
object TableDemo01 {
  def main(args: Array[String]) {
    //设置流执行环境
```

```scala
val env = StreamExecutionEnvironment.getExecutionEnvironment
//创建流表执行环境
val tableEnv = StreamTableEnvironment.create(env)

//创建 DataStream
val dataStream = env.fromElements("张三","李四","王老五")

//将 insert-only DataStream 解释为一个 Table
val inputTable = tableEnv.fromDataStream(dataStream)

//将 Table 对象注册为视图并查询它
tableEnv.createTemporaryView("InputTable", inputTable)
val resultTable = tableEnv.sqlQuery("SELECT f0,CHARACTER_LENGTH(f0) FROM InputTable")

//再次将 insert-only Table 解释为一个 DataStream
val resultStream = tableEnv.toDataStream(resultTable)

//print sink
resultStream.print()

//在 DataStream API 中执行
env.execute("mix DataStream & Table API")
    }
}
```

Java 代码如下：

```java
//第 5 章/TableDemo01.java

import org.apache.flink.streaming.api.datastream.DataStream;
import org.apache.flink.streaming.api.environment.StreamExecutionEnvironment;
import org.apache.flink.table.api.Table;
import org.apache.flink.table.api.bridge.java.StreamTableEnvironment;
import org.apache.flink.types.Row;

/**
 * 在 DataStream 和 Table 之间转换
 *
 */
public class TableDemo01 {

    public static void main(String[] args) throws Exception {
        //设置流执行环境
        final StreamExecutionEnvironment env =
                StreamExecutionEnvironment.getExecutionEnvironment();
```

```java
        //创建流表执行环境
        StreamTableEnvironment tableEnv = StreamTableEnvironment.create(env);
        
        //创建 DataStream
        DataStream<String> dataStream = env.fromElements("张三","李四","王老五");
        
        //将 insert-only DataStream 解释为一个 Table
        Table inputTable = tableEnv.fromDataStream(dataStream);
        
        //将 Table 对象注册为视图并查询它
        tableEnv.createTemporaryView("InputTable", inputTable);
        Table resultTable = tableEnv.sqlQuery("SELECT f0, CHARACTER_LENGTH(f0) FROM InputTable");
        
        //再次将 insert-only Table 解释为一个 DataStream
        DataStream<Row> resultStream = tableEnv.toDataStream(resultTable);
        
        //print sink
        resultStream.print();
        
        //在 DataStream API 中执行
        env.execute();
    }
}
```

在将 Table 转换为 DataStream 或 DataSet 时,需要指定结果 DataStream 或 DataSet 的数据类型,即要将 Table 的行转换为 DataStream 的数据类型。通常最方便的转换类型是 Row。以下内容概述了不同选项的特性。

(1) Row:字段按位置映射,任意数量的字段,支持 null 值,没有类型安全的访问。

(2) POJO:字段按名称映射(POJO 字段必须命名为表字段),任意数量的字段,支持 null 值,类型安全的访问。

(3) Case Class:字段按位置映射,不支持 null 值,类型安全访问。

(4) Tuple:字段按位置映射,限制为第 22(Scala)或 25(Java)字段,不支持 null 值,类型安全访问。

(5) Atomic Type:表必须有一个单独的字段,不支持 null 值,类型安全访问。

执行上面的代码,输出结果如下:

```
7> +I[王老五, 3]
5> +I[张三, 2]
6> +I[李四, 2]
```

下面的示例展示了如何转换更新表。每个结果行表示一个更改日志(changelog)中的一个条目,该条目具有更改标志,可以通过对其调用 row.getKind()进行查询。在这个示例中,张三的第 2 个分数创建了 update before(-U)和 update after(+U)更改。

Scala 代码如下:

```scala
//第 5 章/TableDemo02.scala

import org.apache.flink.api.scala.typeutils.Types
import org.apache.flink.streaming.api.scala._
import org.apache.flink.table.api.bridge.scala.StreamTableEnvironment
import org.apache.flink.types.Row

object TableDemo02 {
  def main(args: Array[String]) {
    //设置流执行环境
    val env = StreamExecutionEnvironment.getExecutionEnvironment
    //创建流表执行环境
    val tableEnv = StreamTableEnvironment.create(env)

    //创建 DataStream
    val dataStream = env.fromElements(
      Row.of("张三", Int.box(12)),
      Row.of("李四", Int.box(10)),
      Row.of("张三", Int.box(100))
      )(Types.ROW(Types.STRING, Types.INT))

    //将 insert-only DataStream 解释为一个 Table
    val inputTable = tableEnv.fromDataStream(dataStream).as("name", "score")

    //将 Table 对象注册为视图并查询它
    tableEnv.createTemporaryView("InputTable", inputTable)
    val resultTable = tableEnv.sqlQuery("SELECT name, SUM(score) FROM InputTable GROUP BY name")

    //再次将 insert-only Table 解释为一个 DataStream
    val resultStream = tableEnv.toChangelogStream(resultTable)

    //print sink
    resultStream.print()

    //在 DataStream API 中执行
    env.execute("mix DataStream & Table API")
  }
}
```

Java 代码如下:

```java
//第 5 章/TableDemo02.java

import org.apache.flink.streaming.api.datastream.DataStream;
import org.apache.flink.streaming.api.environment.StreamExecutionEnvironment;
import org.apache.flink.table.api.Table;
import org.apache.flink.table.api.bridge.java.StreamTableEnvironment;
import org.apache.flink.types.Row;

public class TableDemo02 {

    public static void main(String[] args) throws Exception {
        //设置流执行环境
        final StreamExecutionEnvironment env =
                StreamExecutionEnvironment.getExecutionEnvironment();

        //创建流表执行环境
        StreamTableEnvironment tableEnv = StreamTableEnvironment.create(env);

        //创建 DataStream
        DataStream<Row> dataStream = env.fromElements(
                        Row.of("张三", 12),
                        Row.of("李四", 10),
                        Row.of("张三", 100));

        //将 insert-only DataStream 解释为一个 Table
        Table inputTable = tableEnv.fromDataStream(dataStream).as("name", "score");

        //将 Table 对象注册为视图并查询它
        tableEnv.createTemporaryView("InputTable", inputTable);
        Table resultTable = tableEnv.sqlQuery("SELECT name, SUM(score) FROM InputTable GROUP BY name");

        //再次将 insert-only Table 解释为一个 DataStream
        DataStream<Row> resultStream =
                tableEnv.toChangelogStream(resultTable);

        //print sink
        resultStream.print();

        //在 DataStream API 中执行
        env.execute();
    }
}
```

执行以上代码,输出结果如下:

```
5> +I[李四, 10]
2> +I[张三, 12]
2> -U[张三, 12]
2> +U[张三, 112]
```

TableEnvironment 将采用来自传入的 StreamExecutionEnvironment 的所有配置选项，但是，不能保证对 StreamExecutionEnvironment 配置的进一步更改会在实例化后传播到 StreamTableEnvironment。另外，也不支持将选项从 Table API 反向传播到 DataStream API。

建议在切换到 Table API 之前尽早设置 DataStream API 中的所有配置选项。设置配置选项的代码如下。

Scala 代码如下：

```
import Java.time.ZoneId
import org.apache.flink.api.scala._
import org.apache.flink.streaming.api.scala.StreamExecutionEnvironment
import org.apache.flink.streaming.api.CheckpointingMode
import org.apache.flink.table.api.bridge.scala._

//创建 Scala DataStream API
val env = StreamExecutionEnvironment.getExecutionEnvironment

//尽早设置各种配置

env.setMaxParallelism(256)

env.getConfig.addDefaultKryoSerializer(classOf[MyCustomType], classOf[CustomKryoSerializer])

env.getCheckpointConfig.setCheckpointingMode(CheckpointingMode.EXACTLY_ONCE)

//然后切换到 Scala Table API
val tableEnv = StreamTableEnvironment.create(env)

//尽早设置配置
tableEnv.getConfig.setLocalTimeZone(ZoneId.of("Europe/Berlin"))

//开始以这两种 API 定义管道……
```

Java 代码如下：

```
import Java.time.ZoneId;
import org.apache.flink.streaming.api.CheckpointingMode;
import org.apache.flink.streaming.api.environment.StreamExecutionEnvironment;
```

```java
import org.apache.flink.table.api.bridge.java.StreamTableEnvironment;

//创建 DataStream API
StreamExecutionEnvironment env =
        StreamExecutionEnvironment.getExecutionEnvironment();

//尽早设置各种配置

env.setMaxParallelism(256);

env.getConfig().addDefaultKryoSerializer(MyCustomType.class, CustomKryoSerializer.class);

env.getCheckpointConfig().setCheckpointingMode(CheckpointingMode.EXACTLY_ONCE);

//然后切换到 Scala Table API
StreamTableEnvironment tableEnv = StreamTableEnvironment.create(env);

//尽早设置配置
tableEnv.getConfig().setLocalTimeZone(ZoneId.of("Europe/Berlin"));

//开始以这两种 API 定义管道……
```

5.4.3 处理 insert-only 流

StreamTableEnvironment 提供了以下方法进行 DataStream API 转换。

（1）fromDataStream(DataStream)：将 insert-only 更改流和任意类型解释为表。默认情况下不传播事件时间和水印。

（2）fromDataStream(DataStream，Schema)：将 insert-only 更改流和任意类型解释为表。可选模式允许丰富列数据类型，并添加时间属性、水印策略、其他计算列或主键。

（3）createTemporaryView(String，DataStream)：给流注册一个名字以便在 SQL 中访问它。它是 createTemporaryView(String，fromDataStream(DataStream))的快捷方式。

（4）createTemporaryView(String，DataStream，Schema)：给流注册一个名字以便在 SQL 中访问它。它是 createTemporaryView（String，fromDataStream（DataStream，Schema))的快捷方式。

（5）toDataStream(DataStream)：将表转换为 insert-only 更改流。默认流记录类型为 org.apache.flink.types.Row。一个单独的 rowtime 属性列被写回 DataStream API 的记录中。水印也可以传播。

（6）toDataStream(DataStream，AbstractDataType)：将表转换为 insert-only 更改流。此方法接收一个数据类型以表示所需的流记录类型。规划器（planner）可以插入隐式强制转换并对列重新排序，以将列映射到(可能嵌套的)数据类型的字段。

(7) toDataStream(DataStream，Class)：toDataStream(DataStream，DataTypes.of(Class))的快捷方式，以快速反射地创建所需的数据类型。

从 Table API 的角度来看，从 DataStream API 转换和转换到 DataStream API 类似于从使用 SQL 中的 CREATE TABLE DDL 定义的虚拟表连接器读取或写入。

虚拟 CREATE TABLE name（schema）WITH（options）语句中的模式部分可以自动从 DataStream 的类型信息派生出来，或者完全使用 org.apache.flink.table.api.Schema 手工定义。

虚拟 DataStream 表连接器为每行公开元数据，见表 5-1。

表 5-1　虚拟 DataStream 表连接器公开的元数据

Key	Data Type	Description	R/W
rowtime	TIMESTAMP_LTZ(3) NOT NULL	流记录的时间戳	R/W

虚拟 DataStream 表源实现了 SupportsSourceWatermark，因此允许调用 SOURCE_WATERMARK()内置函数作为水印策略，以从 DataStream API 采用水印。

在不同的场景中使用 fromDataStream()的代码如下。

Scala 代码如下：

```scala
//第 5 章/TableDemo03.scala

import Java.time.Instant
import org.apache.flink.streaming.api.scala._
import org.apache.flink.table.api.Schema
import org.apache.flink.table.api.bridge.scala.StreamTableEnvironment

object TableDemo03 {

  //case class
  case class User(name: String, score: Integer, event_time: Instant)

  def main(args: Array[String]) {
    //设置流执行环境
    val env = StreamExecutionEnvironment.getExecutionEnvironment
    //创建流表执行环境
    val tableEnv = StreamTableEnvironment.create(env)

    //创建一个 DataStream
    val dataStream = env.fromElements(
      User("张三", 4, Instant.ofEpochMilli(1000)),
      User("李四", 6, Instant.ofEpochMilli(1001)),
      User("张三", 10, Instant.ofEpochMilli(1002)))
```

```scala
// === EXAMPLE 1 ===
//自动导出所有物理列
val table1 = tableEnv.fromDataStream(dataStream)
table1.printSchema()


// === EXAMPLE 2 ===
//自动导出所有物理列
//但是添加计算列(在本例中用于创建 proctime 属性列)
val table2 = tableEnv.fromDataStream(
  dataStream,
  Schema.newBuilder()
    .columnByExpression("proc_time", "PROCTIME()")
    .build())
table2.printSchema()


// === EXAMPLE 3 ===
//自动导出所有物理列
//但是添加计算列(在本例中用于创建 rowtime 属性列)
//和自定义水印策略
val table3 = tableEnv.fromDataStream(
    dataStream,
    Schema.newBuilder()
      .columnByExpression("rowtime", "CAST(event_time AS TIMESTAMP_LTZ(3))")
      .watermark("rowtime", "rowtime - INTERVAL '10' SECOND")
      .build())
table3.printSchema()


// === EXAMPLE 4 ===
//自动导出所有物理列
//但是访问流记录的时间戳以创建一个 rowtime 属性列
//也依赖于 DataStream API 中生成的水印
//假设之前已经为 dataStream 定义了水印策略(不是本例的一部分)
val table4 = tableEnv.fromDataStream(
  dataStream,
  Schema.newBuilder()
    .columnByMetadata("rowtime", "TIMESTAMP_LTZ(3)")
    .watermark("rowtime", "SOURCE_WATERMARK()")
    .build())
table4.printSchema()


// === EXAMPLE 5 ===
//手动定义物理列
//在本例中
//可以将时间戳的默认精度从 9 降低到 3
//还投影列并将'event_time'放在开头
val table5 = tableEnv.fromDataStream(
```

```
            dataStream,
            Schema.newBuilder()
                .column("event_time", "TIMESTAMP_LTZ(3)")
                .column("name", "STRING")
                .column("score", "INT")
                .watermark("event_time", "SOURCE_WATERMARK()")
                .build())
        table5.printSchema()
    }
}
```

Java 代码如下:

```
//第 5 章/TableDemo03.java

import org.apache.flink.streaming.api.datastream.DataStream;
import org.apache.flink.streaming.api.environment.StreamExecutionEnvironment;
import org.apache.flink.table.api.Schema;
import org.apache.flink.table.api.Table;
import org.apache.flink.table.api.bridge.java.StreamTableEnvironment;
import Java.time.Instant;

public class TableDemo03 {

    //POJO 类
    public static class User {
        public String name;
        public Integer score;
        public Instant event_time;

        //DataStream API 的默认构造函数
        public User() {}

        //Table API 的完全赋值构造函数
        public User(String name, Integer score, Instant event_time) {
            this.name = name;
            this.score = score;
            this.event_time = event_time;
        }
    }

    public static void main(String[] args) throws Exception {
        //设置流执行环境
        final StreamExecutionEnvironment env =
            StreamExecutionEnvironment.getExecutionEnvironment();
```

```java
//创建流表执行环境
StreamTableEnvironment tableEnv = StreamTableEnvironment.create(env);

//创建一个 DataStream
DataStream<User> dataStream = env.fromElements(
        new User("张三", 4, Instant.ofEpochMilli(1000)),
        new User("李四", 6, Instant.ofEpochMilli(1001)),
        new User("张三", 10, Instant.ofEpochMilli(1002))
);

// === EXAMPLE 1 ===
//自动导出所有物理列
Table table1 = tableEnv.fromDataStream(dataStream);
table1.printSchema();


// === EXAMPLE 2 ===
//自动导出所有物理列
//但是添加计算列(在本例中用于创建 proctime 属性列)
Table table2 = tableEnv.fromDataStream(
    dataStream,
    Schema.newBuilder().columnByExpression("proc_time", "PROCTIME()")
        .build()
);
table2.printSchema();


// === EXAMPLE 3 ===
//自动导出所有物理列
//但是添加计算列(在本例中用于创建 rowtime 属性列)
//和自定义水印策略
Table table3 = tableEnv.fromDataStream(
    dataStream,
    Schema.newBuilder()
        .columnByExpression("rowtime", "CAST(event_time AS TIMESTAMP_LTZ(3))")
        .watermark("rowtime", "rowtime - INTERVAL '10' SECOND")
        .build()
);
table3.printSchema();


// === EXAMPLE 4 ===
//自动导出所有物理列
//但是访问流记录的时间戳以创建一个 rowtime 属性列
//也依赖于 DataStream API 中生成的水印
//假设之前已经为 dataStream 定义了水印策略(不是本例的一部分)
Table table4 = tableEnv.fromDataStream(
    dataStream,
    Schema.newBuilder()
```

```
                .columnByMetadata("rowtime", "TIMESTAMP_LTZ(3)")
                .watermark("rowtime", "SOURCE_WATERMARK()")
                .build()
    );
    table4.printSchema();

    // === EXAMPLE 5 ===
    //手动定义物理列
    //在本例中
    //可以将时间戳的默认精度从 9 降低到 3
    //还投影列并将'event_time'放在开头
    Table table5 = tableEnv.fromDataStream(
        dataStream,
        Schema.newBuilder()
                .column("event_time", "TIMESTAMP_LTZ(3)")
                .column("name", "STRING")
                .column("score", "INT")
                .watermark("event_time", "SOURCE_WATERMARK()")
                .build()
    );
    table5.printSchema();
  }
}
```

DataStream 可以直接注册为一个视图(可能使用模式进行充实)。从 DataStream 创建的视图只能注册为临时视图。由于它们的内联/匿名性质,不可能在永久目录中注册它们。

在不同的场景中使用 createTemporaryView() 方法的代码如下。

Scala 代码如下：

```
//第 5 章/TableDemo04.scala

import org.apache.flink.streaming.api.scala._
import org.apache.flink.table.api.Schema
import org.apache.flink.table.api.bridge.scala.StreamTableEnvironment

object TableDemo04 {

  def main(args: Array[String]) {
    //设置流执行环境
    val env = StreamExecutionEnvironment.getExecutionEnvironment
    //创建流表执行环境
    val tableEnv = StreamTableEnvironment.create(env)

    //创建一个 DataStream
```

```scala
val dataStream: DataStream[(Long, String)] = env.fromElements((12L, "Alice"),(0L, "Bob"))

// === EXAMPLE 1 ===
//将 DataStream 注册为当前会话中的视图 MyView
//所有列都是自动派生的
tableEnv.createTemporaryView("MyView1", dataStream)
tableEnv.from("MyView1").printSchema()

// === EXAMPLE 2 ===
//将 DataStream 注册为当前会话中的视图 MyView
//提供类似于'fromDataStream'的模式来调整列
//在本例中,派生的 NOT NULL 信息已被删除
tableEnv.createTemporaryView(
  "MyView2",
  dataStream,
  Schema.newBuilder()
    .column("_1", "BIGINT")
    .column("_2", "STRING")
    .build())

tableEnv.from("MyView2").printSchema()

// === EXAMPLE 3 ===
//如果只是关于重命名列,则在创建视图之前使用 Table API
tableEnv.createTemporaryView(
  "MyView3",
  tableEnv.fromDataStream(dataStream).as("id", "name"))

tableEnv.from("MyView3").printSchema()
  }
}
```

Java 代码如下:

```java
//第 5 章/TableDemo04.java

import org.apache.flink.api.java.tuple.Tuple2;
import org.apache.flink.streaming.api.datastream.DataStream;
import org.apache.flink.streaming.api.environment.StreamExecutionEnvironment;
import org.apache.flink.table.api.Schema;
import org.apache.flink.table.api.bridge.java.StreamTableEnvironment;

public class TableDemo04 {
```

```java
public static void main(String[] args) throws Exception {
    //设置流执行环境
    final StreamExecutionEnvironment env =
            StreamExecutionEnvironment.getExecutionEnvironment();
    //创建流表执行环境
    StreamTableEnvironment tableEnv = StreamTableEnvironment.create(env);

    //创建一个 DataStream
    DataStream<Tuple2<Long, String>> dataStream = env.fromElements(
        Tuple2.of(12L, "张三"),
        Tuple2.of(0L, "李四"));

    // === EXAMPLE 1 ===
    //将 DataStream 注册为当前会话中的视图 MyView
    //所有列都是自动派生的
    tableEnv.createTemporaryView("MyView", dataStream);
    tableEnv.from("MyView").printSchema();

    // === EXAMPLE 2 ===
    //将 DataStream 注册为当前会话中的视图 MyView
    //提供类似于'fromDataStream'的模式来调整列
    //在本例中,派生的 NOT NULL 信息已被删除
    tableEnv.createTemporaryView(
        "MyView2",
        dataStream,
        Schema.newBuilder().column("f0", "BIGINT").column("f1", "STRING").build());

    tableEnv.from("MyView2").printSchema();

    // === EXAMPLE 3 ===
    //如果只是关于重命名列,则在创建视图之前使用 Table API
    tableEnv.createTemporaryView(
        "MyView3",
        tableEnv.fromDataStream(dataStream).as("id", "name")
    );
    tableEnv.from("MyView3").printSchema();
}
}
```

在不同的场景中使用 toDataStream()方法。

Scala 代码如下:

```
//第 5 章/TableDemo05.scala

import org.apache.flink.streaming.api.scala._
```

```scala
import org.apache.flink.table.api.DataTypes
import org.apache.flink.table.api.bridge.scala.StreamTableEnvironment
import org.apache.flink.types.Row

object TableDemo05 {

  case class User(name: String, score: Java.lang.Integer, event_time: Java.time.Instant)

  def main(args: Array[String]) {
    //设置流执行环境
    val env = StreamExecutionEnvironment.getExecutionEnvironment
    //创建流表执行环境
    val tableEnv = StreamTableEnvironment.create(env)

    tableEnv.executeSql(
      """
        CREATE TABLE GeneratedTable (
          name STRING,
          score INT,
          event_time TIMESTAMP_LTZ(3),
          WATERMARK FOR event_time AS event_time - INTERVAL '10' SECOND
        )
        WITH ('connector' = 'datagen')
      """
    )

    val table = tableEnv.from("GeneratedTable")

    // === EXAMPLE 1 ===
    //使用默认的 Row 实例转换
    //因为'event_time'是一个单独的rowtime属性,所以它被插入 DataStream 中
    //传播元数据和水印
    val dataStream1: DataStream[Row] = tableEnv.toDataStream(table)

    // === EXAMPLE 2 ===
    //从类'User'中提取数据类型
    //规划器对字段重新排序,并在可能的地方插入隐式强制转换,以便将内部数据结构转换为所
    //需的结构化类型
    //因为'event_time'是一个单独的rowtime属性,所以它被插入 DataStream 中
    //传播元数据和水印
    val dataStream2: DataStream[User] = tableEnv.toDataStream(table, classOf[User])

    //数据类型可以如上所示进行反射提取或显式定义
    val dataStream3: DataStream[User] =
      tableEnv.toDataStream(
        table,
```

```
        DataTypes.STRUCTURED(
            classOf[User],
            DataTypes.FIELD("name", DataTypes.STRING()),
            DataTypes.FIELD("score", DataTypes.INT()),
            DataTypes.FIELD("event_time", DataTypes.TIMESTAMP_LTZ(3))))
    }
}
```

Java 代码如下:

```java
//第 5 章/TableDemo05.java

import org.apache.flink.streaming.api.environment.StreamExecutionEnvironment;
import org.apache.flink.streaming.api.datastream.DataStream;
import org.apache.flink.table.api.DataTypes;
import org.apache.flink.table.api.Table;
import org.apache.flink.table.api.bridge.java.StreamTableEnvironment;
import org.apache.flink.types.Row;
import java.time.Instant;

public class TableDemo05 {
    //带有可变字段的 POJO
    //由于没有定义完全赋值的构造函数,所以字段顺序按字母顺序[event_time, name, score]
    public static class User {
        public String name;
        public Integer score;
        public Instant event_time;
    }

    public static void main(String[] args) throws Exception {
        //设置流执行环境
        final StreamExecutionEnvironment env =
                StreamExecutionEnvironment.getExecutionEnvironment();

        //创建流表执行环境
        StreamTableEnvironment tableEnv = StreamTableEnvironment.create(env);

        tableEnv.executeSql(
            "CREATE TABLE GeneratedTable "
            + "("
            + " name STRING,"
            + " score INT,"
            + " event_time TIMESTAMP_LTZ(3),"
            + " WATERMARK FOR event_time AS event_time - INTERVAL '10' SECOND"
```

```java
            + ")"
            + "WITH ('connector' = 'datagen')"
);

Table table = tableEnv.from("GeneratedTable");

// === EXAMPLE 1 ===
//使用默认的 Row 实例转换
//因为'event_time'是一个单独的 rowtime 属性,所以它被插入 DataStream 中
//传播元数据和水印
DataStream<Row> dataStream1 = tableEnv.toDataStream(table);

// === EXAMPLE 2 ===
//从类'User'中提取数据类型
//规划器对字段重新排序,并在可能的地方插入隐式强制转换,以便将内部数据结构转换为
//所需的结构化类型
//因为'event_time'是一个单独的 rowtime 属性,所以它被插入 DataStream 中
//传播元数据和水印
DataStream<User> dataStream2 = tableEnv.toDataStream(table, User.class);

//数据类型可以如上所示进行反射提取或显式定义
DataStream<User> dataStream3 = tableEnv
    .toDataStream(
        table,
        DataTypes.STRUCTURED(
        User.class,
        DataTypes.FIELD("name", DataTypes.STRING()),
        DataTypes.FIELD("score", DataTypes.INT()),
        DataTypes.FIELD("event_time", DataTypes.TIMESTAMP_LTZ(3)))
);
    }
}
```

注意,toDataStream()方法只支持非更新表。通常,基于时间的操作(如窗口、间隔连接或 MATCH_RECOGNIZE 子句)非常适合于紧挨着投影和过滤器等简单操作的 insert-only 管道。

带有生成更新操作的管道可以用到 toChangelogStream()方法。

5.4.4 处理变更日志流

在内部,Flink 的表运行时是一个变更日志处理器(Changelog Processor)。

StreamTableEnvironment 提供了以下方法来公开这些 Change Data Capture(CDC,变更数据捕获)功能。

（1）fromChangelogStream（DataStream）：将变更日志条目流解释为一张表。流记录类型必须是 org.apache.flink.types.Row，因为它的 RowKind 标志是在运行时计算的。默认情况下不传播事件时间和水印。该方法期望一个包含所有类型变更（在 org.apache.flink.types.RowKind 中枚举）的变更日志作为默认的 ChangelogMode。

（2）fromChangelogStream（DataStream，Schema）：允许为 DataStream 定义一个模式。

（3）fromChangelogStream（DataStream，Schema，ChangelogMode）：给出了如何将流解释为变更日志的完全控制。传递的 ChangelogMode 参数帮助规划器区分 insert-only、upsert 或 retract 行为。

（4）toChangelogStream（Table）：生成一个具有 org.apache.flink.types.Row 实例的流，并在运行时为每条记录设置 RowKind 标志。该方法支持各种类型的更新表。如果输入表包含一个 rowtime 列，则它将被传播到流记录的时间戳中。水印也将被传播。

（5）toChangelogStream（Table，Schema）：该方法可以丰富生成的列数据类型。如果有必要，规划器则可以插入隐式强制类型转换。它可以将 rowtime 写成元数据列。

（6）toChangelogStream（Table，Schema，ChangelogMode）：提供关于如何将表转换为变更日志流的完全控制。传递的 ChangelogMode 帮助规划器区分 insert-only、upsert 或 retract 行为。

从 Table API 的角度来看，和 DataStream API 的相互转换类似于从使用 SQL 中的 CREATE TABLE DDL 定义的虚拟表连接器读取或写入。这个虚拟连接器还支持读写流记录的 rowtime 元数据。虚表源实现了 SupportsSourceWatermark。

在不同的场景中使用 fromChangelogStream()。

Scala 代码如下：

```scala
//第5章/TableDemo06.scala

import org.apache.flink.api.scala.typeutils.Types
import org.apache.flink.streaming.api.scala.StreamExecutionEnvironment
import org.apache.flink.table.api.Schema
import org.apache.flink.table.api.bridge.scala.StreamTableEnvironment
import org.apache.flink.table.connector.ChangelogMode
import org.apache.flink.types.{Row, RowKind}

object TableDemo06 {

  def main(args: Array[String]) {
    //设置流执行环境
    val env = StreamExecutionEnvironment.getExecutionEnvironment
    //创建流表执行环境
    val tableEnv = StreamTableEnvironment.create(env)
```

```scala
// === EXAMPLE 1 ===
//将此流解释为 retract 流
//创建一个 Changelog DataStream
val dataStream1 = env.fromElements(
  Row.ofKind(RowKind.INSERT, "张三", Int.box(12)),
  Row.ofKind(RowKind.INSERT, "李四", Int.box(5)),
  Row.ofKind(RowKind.UPDATE_BEFORE, "张三", Int.box(12)),
  Row.ofKind(RowKind.UPDATE_AFTER, "张三", Int.box(100))
)(Types.ROW(Types.STRING, Types.INT))


//将 DataStream 解释为一个 Table
val table1 = tableEnv.fromChangelogStream(dataStream1)

//注册该 Table,并执行一个聚合
tableEnv.createTemporaryView("InputTable1", table1)
tableEnv
  .executeSql("SELECT f0 AS name, SUM(f1) AS score FROM InputTable1 GROUP BY f0")
  .print()

// === EXAMPLE 2 ===
//将流解释为 upsert 流(不需要 UPDATE_BEFORE)
//创建一个 Changelog DataStream
val dataStream2 = env.fromElements(
  Row.ofKind(RowKind.INSERT, "张三", Int.box(12)),
  Row.ofKind(RowKind.INSERT, "李四", Int.box(5)),
  Row.ofKind(RowKind.UPDATE_AFTER, "张三", Int.box(100))
)(Types.ROW(Types.STRING, Types.INT))

//将 DataStream 解释为一个 Table
val table2 = tableEnv.fromChangelogStream(
  dataStream2,
  Schema.newBuilder().primaryKey("f0").build(),
  ChangelogMode.upsert()
)

//注册表并执行聚合
tableEnv.createTemporaryView("InputTable2", table2)
tableEnv
  .executeSql("SELECT f0 AS name, SUM(f1) AS score FROM InputTable2 GROUP BY f0")
  .print()
  }
}
```

Java 代码如下:

```java
//第 5 章/TableDemo06.java

import org.apache.flink.streaming.api.datastream.DataStream;
import org.apache.flink.streaming.api.environment.StreamExecutionEnvironment;
import org.apache.flink.table.api.Schema;
import org.apache.flink.table.api.Table;
import org.apache.flink.table.api.bridge.java.StreamTableEnvironment;
import org.apache.flink.table.connector.ChangelogMode;
import org.apache.flink.types.Row;
import org.apache.flink.types.RowKind;

public class TableDemo06 {

    public static void main(String[] args) throws Exception {
        //设置流执行环境
        final StreamExecutionEnvironment env =
                StreamExecutionEnvironment.getExecutionEnvironment();
        //创建流表执行环境
        StreamTableEnvironment tableEnv = StreamTableEnvironment.create(env);

        // === EXAMPLE 1 ===
        //将此流解释为 retract 流
        //创建一个 Changelog DataStream
        DataStream<Row> dataStream1 = env.fromElements(
           Row.ofKind(RowKind.INSERT, "张三", 12),
           Row.ofKind(RowKind.INSERT, "李四", 5),
           Row.ofKind(RowKind.UPDATE_BEFORE, "张三", 12),
           Row.ofKind(RowKind.UPDATE_AFTER, "张三", 100)
        );

        //将 DataStream 解释为一个 Table
        Table table1 = tableEnv.fromChangelogStream(dataStream1);

        //注册该 Table,并执行一个聚合
        tableEnv.createTemporaryView("InputTable1", table1);
        tableEnv
            .executeSql("SELECT f0 AS name, SUM(f1) AS score FROM InputTable1 GROUP BY f0")
            .print();

        // === EXAMPLE 2 ===
        //将流解释为 upsert 流(不需要 UPDATE_BEFORE)
        //创建一个 Changelog DataStream
        DataStream<Row> dataStream2 = env.fromElements(
            Row.ofKind(RowKind.INSERT, "张三", 12),
            Row.ofKind(RowKind.INSERT, "李四", 5),
            Row.ofKind(RowKind.UPDATE_AFTER, "张三", 100)
```

```
    );

    //将 DataStream 解释为一个 Table
    Table table2 = tableEnv.fromChangelogStream(
        dataStream2,
        Schema.newBuilder().primaryKey("f0").build(),
        ChangelogMode.upsert()
    );

    //注册表并执行聚合
    tableEnv.createTemporaryView("InputTable2", table2);
    tableEnv
        .executeSql("SELECT f0 AS name, SUM(f1) AS score FROM InputTable2 GROUP BY f0")
        .print();
    }
}
```

执行以上代码,输出结果如下:

```
+----+--------------------------------+-------------+
| op |                           name |       score |
+----+--------------------------------+-------------+
| +I |                           李四 |           5 |
| +I |                           张三 |          12 |
| -D |                           张三 |          12 |
| +I |                           张三 |         100 |
+----+--------------------------------+-------------+
4 rows in set

+----+--------------------------------+-------------+
| op |                           name |       score |
+----+--------------------------------+-------------+
| +I |                           李四 |           5 |
| +I |                           张三 |          12 |
| -D |                           张三 |          12 |
| +I |                           张三 |         100 |
+----+--------------------------------+-------------+
4 rows in set
```

EXAMPLE 1 中显示的默认 ChangelogMode 应该足以满足大多数用例,因为它接受所有类型的更改。

在 EXAMPLE 2 中通过 upsert 模式将更新消息的数量减少 50% 来限制传入更改的种类,从而提高了效率。通过为 toChangelogStream() 定义一个主键和 Upsert Changelog 模式,可以减少结果消息的数量。

在不同的场景中使用 toChangelogStream()。

Scala 代码如下：

```scala
//第 5 章/TableDemo07.scala

import Java.time.Instant
import org.apache.flink.api.scala._
import org.apache.flink.streaming.api.functions.ProcessFunction
import org.apache.flink.streaming.api.scala.{DataStream, StreamExecutionEnvironment}
import org.apache.flink.table.api._
import org.apache.flink.table.api.bridge.scala.StreamTableEnvironment
import org.apache.flink.types.Row
import org.apache.flink.util.Collector

object TableDemo07 {

  def main(args: Array[String]) {
    //设置流执行环境
    val env = StreamExecutionEnvironment.getExecutionEnvironment
    //创建流表执行环境
    val tableEnv = StreamTableEnvironment.create(env)

    // === EXAMPLE 1 ===
    //以最简单和最通用的方式转换为 DataStream(no event-time)
    val simpleTable = tableEnv
      .fromValues(row("张三", 12), row("张三", 2), row("李四", 12))
      .as("name", "score")
      .groupBy( $ "name")
      .select( $ "name", $ "score".sum())

    tableEnv
      .toChangelogStream(simpleTable)
      .executeAndCollect()
      .foreach(println)

    // === EXAMPLE 2 ===
    //创建带有事件时间的表
    tableEnv.executeSql(
      """
      CREATE TABLE GeneratedTable (
        name STRING,
        score INT,
        event_time TIMESTAMP_LTZ(3),
        WATERMARK FOR event_time AS event_time - INTERVAL '10' SECOND
      )
```

```
    WITH ('connector' = 'datagen')
  """
)

val table = tableEnv.from("GeneratedTable")

//以最简单和最通用的方式转换为DataStream(with event-time)
val dataStream1: DataStream[Row] = tableEnv.toChangelogStream(table)

//由于event_time是模式中的单个时间属性,所以默认情况下它被设置为流记录的时间戳
//然而,与此同时,它仍然是Row的一部分
dataStream1.process(new ProcessFunction[Row, Unit] {
  override def processElement(
                     row: Row,
                     ctx: ProcessFunction[Row, Unit]#Context,
                     out: Collector[Unit]): Unit = {
    //prints: [name, score, event_time]
    println(row.getFieldNames(true))

    //timestamp exists twice
    assert(ctx.timestamp() == row.getFieldAs[Instant]("event_time").toEpochMilli)
  }
})
env.execute()


// === EXAMPLE 3 ===
//转换为DataStream,但将时间属性作为元数据列写出来,这意味着它不再是物理模式的一部分
val dataStream2: DataStream[Row] = tableEnv.toChangelogStream(
  table,
  Schema.newBuilder()
    .column("name", "STRING")
    .column("score", "INT")
    .columnByMetadata("rowtime", "TIMESTAMP_LTZ(3)")
    .build()
)

//流记录的时间戳由元数据定义;它不是Row的一部分
dataStream2.process(new ProcessFunction[Row, Unit] {
  override def processElement(
                     row: Row,
                     ctx: ProcessFunction[Row, Unit]#Context,
                     out: Collector[Unit]): Unit = {
    //prints: [name, score]
    println(row.getFieldNames(true))

    //timestamp exists once
```

```
            println(ctx.timestamp())
        }
    })
    env.execute()
  }
}
```

Java 代码如下：

```java
//第 5 章/TableDemo07.java

import org.apache.flink.streaming.api.datastream.DataStream;
import org.apache.flink.streaming.api.environment.StreamExecutionEnvironment;
import org.apache.flink.streaming.api.functions.ProcessFunction;
import org.apache.flink.table.api.DataTypes;
import org.apache.flink.table.api.Schema;
import org.apache.flink.table.api.Table;
import org.apache.flink.table.api.bridge.java.StreamTableEnvironment;
import org.apache.flink.table.data.StringData;
import org.apache.flink.types.Row;
import org.apache.flink.util.Collector;
import java.time.Instant;
import static org.apache.flink.table.api.Expressions.*;

public class TableDemo07 {

    public static void main(String[] args) throws Exception {
        //设置流执行环境
        final StreamExecutionEnvironment env =
                StreamExecutionEnvironment.getExecutionEnvironment();
        //创建流表执行环境
        StreamTableEnvironment tableEnv = StreamTableEnvironment.create(env);

        //创建带有事件时间的表
        tableEnv.executeSql(
          "CREATE TABLE GeneratedTable "
            + "("
            + "  name STRING,"
            + "  score INT,"
            + "  event_time TIMESTAMP_LTZ(3),"
            + "  WATERMARK FOR event_time AS event_time - INTERVAL '10' SECOND"
            + ")"
            + "WITH ('connector' = 'datagen')");

        Table table = tableEnv.from("GeneratedTable");
```

```java
// === EXAMPLE 1 ===
//以最简单和最通用的方式转换为 DataStream(no event-time)
Table simpleTable = tableEnv
    .fromValues(row("Alice", 12), row("Alice", 2), row("Bob", 12))
    .as("name", "score")
    .groupBy( $ ("name"))
    .select( $ ("name"), $ ("score").sum());

tableEnv
    .toChangelogStream(simpleTable)
    .executeAndCollect()
    .forEachRemaining(System.out::println);

// === EXAMPLE 2 ===
//以最简单和最通用的方式转换为 DataStream(with event-time)
DataStream<Row> dataStream1 = tableEnv.toChangelogStream(table);

//由于 event_time 是模式中的单个时间属性,所以默认情况下它被设置为流记录的时间戳
//然而,与此同时,它仍然是 Row 的一部分
dataStream1.process(
    new ProcessFunction<Row, Void>() {
        @Override
        public void processElement(Row row, Context ctx, Collector<Void> out) {
            //输出: [name, score, event_time]
            System.out.println(row.getFieldNames(true));

            //时间戳存在两次
            assert ctx.timestamp() == row
                .<Instant>getFieldAs("event_time")
                .toEpochMilli();
        }
    });
env.execute();


// === EXAMPLE 3 ===
//转换为 DataStream,但将时间属性作为元数据列写出来,这意味着它不再是物理模式的一部分
DataStream<Row> dataStream2 = tableEnv.toChangelogStream(
    table,
    Schema.newBuilder()
        .column("name", "STRING")
        .column("score", "INT")
        .columnByMetadata("rowtime", "TIMESTAMP_LTZ(3)")
        .build());

//流记录的时间戳由元数据定义;它不是 Row 的一部分
```

```java
dataStream2.process(new ProcessFunction<Row, Void>() {
    @Override
    public void processElement(Row row, Context ctx, Collector<Void> out) {
        //prints: [name, score]
        System.out.println(row.getFieldNames(true));

        //时间戳存在一次
        System.out.println(ctx.timestamp());
    }
});
env.execute();


// === EXAMPLE 4 ===
//对于高级用户,也可以使用更多的内部数据结构来提高效率
DataStream<Row> dataStream3 = tableEnv.toChangelogStream(
    table,
    Schema.newBuilder()
        .column("name", DataTypes.STRING().bridgedTo(StringData.class))
        .column("score", DataTypes.INT())
        .column("event_time", DataTypes.TIMESTAMP_LTZ(3).bridgedTo(Long.class))
        .build()
);
}
}
```

toChangelogStream(Table).executeandcollect()的行为等于调用 Table.execute().collect(),但是 toChangelogStream(Table)可能对测试更有用,因为它允许访问 DataStream API 中后续 ProcessFunction 中生成的水印。

5.5 Table API 实时流处理案例

下面使用 Table API 重新实现以下几个案例。

5.5.1 传感器温度实时统计

【示例 5-2】 假设一台机器上安装了传感器,用户希望从这些传感器收集数据,并编写 Flink 实时流处理程序,以便统计每个传感器每 5min 的平均温度。

注意:本案例使用 Table API 重构了 3.13.1 节的"处理 IoT 事件流"案例。

传感器收集到的事件格式如下:

```
sensor_1,1629943899014,51.087254019871054
sensor_9,1629943899014,70.44743245583899
sensor_7,1629943899014,65.53215956486392
sensor_0,1629943899014,53.210570822216546
sensor_8,1629943899014,93.12876931817556
sensor_3,1629943899014,57.55153052162809
sensor_2,1629943899014,107.61249366604993
sensor_5,1629943899014,92.02083744773739
sensor_4,1629943899014,95.7688424087137
sensor_6,1629943899014,95.04398353316257
……
```

首先,项目将依赖添加到项目的pom.xml文件中,内容如下:

```xml
<!-- 需要添加flink-csv依赖 -->
<dependency>
    <groupId>org.apache.flink</groupId>
    <artifactId>flink-csv</artifactId>
    <version>1.13.2</version>
</dependency>
```

然后,编写脚本,调用Kafka自带的生产者脚本,以每秒10条的速度将数据发送给Kafka。脚本如下:

```bash
#第5章/streamiot.sh

#!/bin/bash
BROKER=$1
if [ -z "$1" ]; then
    BROKER="localhost:9092"
fi

cat sensortemp.csv | while read line; do
    echo $line
    sleep 0.1
done | /home/hduser/bigdata/kafka_2.12-2.4.1/bin/kafka-console-producer.sh --broker-list $BROKER --topic temp
```

使用Table API的Flink流处理程序。
Scala代码如下:

```scala
//第5章/KafkaIotDemo.scala

import org.apache.flink.streaming.api.scala.StreamExecutionEnvironment
```

```scala
import org.apache.flink.table.api._
import org.apache.flink.table.api.bridge.scala.StreamTableEnvironment

object KafkaIotDemo {

  def main(args: Array[String]) {
    //设置流执行环境
    val env = StreamExecutionEnvironment.getExecutionEnvironment

    //创建表环境
    val tEnv = StreamTableEnvironment.create(env)

    //启用检查点
    env.enableCheckpointing(5000).setParallelism(1)

    //创建 Kafka 源表
    tEnv.executeSql(
      """
        |CREATE TABLE IotTemp (
        |    id string,
        |    iotime bigint,
        |    temperature double,
        |    ts AS TO_TIMESTAMP(FROM_UNIXTIME(iotime/1000, 'yyyy-MM-dd HH:mm:ss')),
        |    WATERMARK FOR ts AS ts - INTERVAL '1' SECOND
        |)
        |with(
        |    'connector' = 'kafka',
        |    'topic' = 'temp',
        |    'properties.Bootstrap.servers' = 'localhost:9092',
        |    'properties.group.id' = 'testGroup',
        |    'scan.startup.mode' = 'latest-offset',
        |    'format' = 'csv',
        |    'csv.ignore-parse-errors' = 'true',
        |    'csv.field-delimiter' = ','
        |)
      """.stripMargin)

    //创建 Kafka Sink 表
    tEnv.executeSql(
      """
        |CREATE TABLE IotAvgTemp (
        |    id string,
        |    window_end TIMESTAMP(3),
        |    avg_temp double
        |)
        |with(
```

```
              |  'connector' = 'kafka',
              |  'topic' = 'avgtemp',
              |  'properties.Bootstrap.servers' = 'localhost:9092',
              |  'format' = 'csv',
              |  'csv.field-delimiter' = ','
              |)
      """.stripMargin)

    tEnv
        //读取源表
        .from("IotTemp")
        //定义大小为 5s,滑动为 2s 的滑动窗口
        .window(Tumble over 1.seconds on $"ts" as $"w")
        //分组
        .groupBy($"id", $"w")
        //聚合
        .select( $"id",
                $"w".end.as("win_end"),
                $"temperature".avg().as("avg_temp")
        )
        .execute.print()
        //写入 Sink 表
        //.executeInsert("IotAvgTemp")
  }
}
```

Java 代码如下:

```java
//第 5 章/KafkaIotDemo.java

import org.apache.flink.streaming.api.environment.StreamExecutionEnvironment;
import org.apache.flink.table.api.Tumble;
import org.apache.flink.table.api.bridge.java.StreamTableEnvironment;
import static org.apache.flink.table.api.Expressions.*;

public class KafkaIotDemo {

    public static void main(String[] args) throws Exception {
        //设置流执行环境
        final StreamExecutionEnvironment env =
                StreamExecutionEnvironment.getExecutionEnvironment();

        //创建表环境
        StreamTableEnvironment tEnv = StreamTableEnvironment.create(env);
```

```java
//注意,一定要启用检查点
env.enableCheckpointing(1000);

env.setParallelism(1);

//创建源表
tEnv.executeSql(
    "CREATE TABLE IotTemp (" +
    " id string," +
    " iotime bigint," +
    " temperature double," +
    " ts AS TO_TIMESTAMP(FROM_UNIXTIME(iotime/1000, 'yyyy-MM-dd HH:mm:ss'))," +
    " WATERMARK FOR ts AS ts - INTERVAL '5' SECOND" +
    ")" +
    "WITH (\n" +
    " 'connector' = 'kafka',\n" +
    " 'topic' = 'temp',\n" +
    " 'properties.Bootstrap.servers' = 'localhost:9092',\n" +
    " 'properties.group.id' = 'testGroup',\n" +
    " 'scan.startup.mode' = 'latest-offset',\n" +
    " 'format' = 'csv',\n" +
    " 'csv.ignore-parse-errors' = 'true',\n" +
    " 'csv.field-delimiter' = ',\n" +
    ")"
);

//创建 Sink 表
tEnv.executeSql(
    "CREATE TABLE IotAvgTemp (" +
    " id string," +
    " window_end TIMESTAMP(3)," +
    " avg_temp double" +
    ")" +
    "WITH (" +
    " 'connector' = 'kafka'," +
    " 'topic' = 'avgtemp'," +
    " 'properties.Bootstrap.servers' = 'localhost:9092'," +
    " 'format' = 'csv'," +
    " 'csv.field-delimiter' = ','" +
    ")"
);

tEnv
    //读取源表
    .from("IotTemp")
    //定义滚动窗口
```

```
            .window(Tumble.over(lit(1).seconds()).on($("ts")).as("w"))
            //分组
            .groupBy($("id"),$("w"))
            //聚合
            .select($("id"),$("w").end().as("win_end"),$("temperature").avg().as("avg_temp"))
            //.execute().collect().forEachRemaining(System.out::println);
            .execute().print();
            //写入Sink表
            //.executeInsert("IotAvgTemp");
    }
}
```

接下来,建议按以下步骤执行:

(1) 启动 ZooKeeper 服务和 Kafka 服务。打开一个终端窗口,启动 ZooKeeper(不要关闭),命令如下:

```
$ ./bin/zookeeper-server-start.sh config/zookeeper.properties
```

(2) 打开另一个终端窗口,启动 Kafka 服务(不要关闭),命令如下:

```
$ ./bin/kafka-server-start.sh config/server.properties
```

(3) 打开第 3 个终端窗口,在 Kafka 中创建一个名为 temp 的主题(Topic),命令如下:

```
$ ./bin/kafka-topics.sh --create --Bootstrap-server localhost:9092 --replication-factor 1 --partitions 1 --topic temp
```

(4) 查看已经创建的主题,命令如下:

```
$ ./bin/kafka-topics.sh --list --Bootstrap-server localhost:9092
```

(5) 运行上面编写的流处理程序(相当于 Kafka 的消费者程序)。
(6) 在第 3 个终端窗口,执行传感器生成脚本,命令如下:

```
$ ./streamiot.sh
```

(7) 观察第 5 步运行流程序的控制台,应该看到的输出内容如图 5-1 所示。

5.5.2 车辆超速实时监测

3min

使用 Table API 和 SQL 实现物流公司运输车辆超速实时监测程序。

```
+----+----------+-------------------------+--------------------+
| op |       id |                 win_end |           avg_temp |
+----+----------+-------------------------+--------------------+
| +I | sensor_8 | 2021-08-26 10:11:47.000 |   91.17537874481508 |
| +I | sensor_2 | 2021-08-26 10:11:47.000 |  111.99315044277473 |
| +I | sensor_4 | 2021-08-26 10:11:47.000 |  100.14499019450446 |
| +I | sensor_6 | 2021-08-26 10:11:47.000 |   93.87981149048566 |
| +I | sensor_3 | 2021-08-26 10:11:47.000 |   61.75141384143203 |
| +I | sensor_5 | 2021-08-26 10:11:47.000 |   89.54459167996558 |
| +I | sensor_8 | 2021-08-26 10:11:48.000 |   90.86763958004099 |
| +I | sensor_3 | 2021-08-26 10:11:48.000 |   61.79318995473199 |
| +I | sensor_2 | 2021-08-26 10:11:48.000 |  111.22972954270946 |
| +I | sensor_6 | 2021-08-26 10:11:48.000 |   94.09083634082995 |
| +I | sensor_7 | 2021-08-26 10:11:48.000 |   67.15824053415402 |
| +I | sensor_5 | 2021-08-26 10:11:48.000 |   90.44831606878837 |
| +I | sensor_4 | 2021-08-26 10:11:48.000 |  100.97907052029765 |
| +I | sensor_1 | 2021-08-26 10:11:48.000 |   60.257149639424455 |
| +I | sensor_9 | 2021-08-26 10:11:48.000 |   72.35319925509846 |
| +I | sensor_0 | 2021-08-26 10:11:48.000 |   49.46435741106236 |

Process finished with exit code 1
```

图 5-1 流程序输出内容

注意：本案例使用 Table API 重构了 3.13.2 节的"车辆超速实时监测"案例。

【**示例 5-3**】 使用 Table API 和 SQL 编写 Flink 流处理程序，监听 Kafka cars 主题的车速信息（CSV 格式），将超速信息（平均速度>100km/h）发送到 Kafka fastcars 主题（JSON 格式）。

首先，在项目的 pom.xml 文件中添加依赖项，内容如下：

```xml
<dependency>
    <groupId>org.apache.flink</groupId>
    <artifactId>flink-json</artifactId>
    <version>${flink.version}</version>
</dependency>
```

接下来，编辑 Flink 流处理管道程序。

Scala 代码如下：

```scala
//第 5 章/FastCarsDetectWithTableAPI.scala

import org.apache.flink.streaming.api.scala.StreamExecutionEnvironment
import org.apache.flink.table.api._
import org.apache.flink.table.api.bridge.scala.StreamTableEnvironment

/**
 * 物流公司车辆超速检测程序
```

```scala
 *
 * 使用 Table API
 */
object FastCarsDetectWithTableAPI {

  def main(args: Array[String]) {
    //设置流执行环境
    val env = StreamExecutionEnvironment.getExecutionEnvironment

    //创建表环境
    val tEnv = StreamTableEnvironment.create(env)

    //启用检查点
    env.enableCheckpointing(50000).setParallelism(1)

    //创建 Kafka 源表
    tEnv.executeSql(
      """
        |CREATE TABLE cars_tb (
        |   carId string,
        |   speed int,
        |   acceleration double,
        |   intime bigint,
        |   ts AS TO_TIMESTAMP(FROM_UNIXTIME(intime/1000, 'yyyy-MM-dd HH:mm:ss')),
        |   WATERMARK FOR ts AS ts - INTERVAL '2' SECOND
        |)
        |with(
        |   'connector' = 'kafka',
        |   'topic' = 'cars',
        |   'properties.Bootstrap.servers' = 'localhost:9092',
        |   'properties.group.id' = 'testGroup',
        |   'scan.startup.mode' = 'latest-offset',
        |   'format' = 'csv',
        |   'csv.ignore-parse-errors' = 'true',
        |   'csv.field-delimiter' = ','
        |)
      """.stripMargin)

    //创建 Kafka Sink 表
    tEnv.executeSql(
      """
        |CREATE TABLE fastcars_tb (
        |   carId string,
        |   window_start TIMESTAMP(3),
        |   window_end TIMESTAMP(3),
        |   avgSpeed double
```

```
      |)
      |with(
      |    'connector' = 'kafka',
      |    'topic' = 'fastcars',
      |    'properties.Bootstrap.servers' = 'localhost:9092',
      |    'format' = 'json'
      |)
    """.stripMargin)

  tEnv
    //读取源表
    .from("cars_tb")
    //定义大小为 5s,滑动为 2s 的滑动窗口
    .window(Slide over 5.seconds every 2.seconds on $"ts" as $"w")
    //分组
    .groupBy($"carId", $"w")
    //聚合
    .select( $"carId",
             $"w".start.as("win_start"),
             $"w".end.as("win_end"),
             $"speed".avg().as("avgSpeed")
    )
    //过滤超速信息
    .where( $"avgSpeed" > 100)
    //写入 Sink 表
    .executeInsert("fastcars_tb")
  }
}
```

Java 代码如下：

```
//第 5 章/FastCarsDetectWithTableAPI.java

import org.apache.flink.streaming.api.environment.StreamExecutionEnvironment;
import org.apache.flink.table.api.Slide;
import org.apache.flink.table.api.bridge.java.StreamTableEnvironment;
import static org.apache.flink.table.api.Expressions.$;
import static org.apache.flink.table.api.Expressions.lit;

/**
 * 物流公司车辆超速检测程序
 *
 * 使用 Table API
 */
public class FastCarsDetectWithTableAPI {
```

```java
public static void main(String[] args) throws Exception {
    //设置流执行环境
    final StreamExecutionEnvironment env =
            StreamExecutionEnvironment.getExecutionEnvironment();

    //创建表环境
    StreamTableEnvironment tEnv = StreamTableEnvironment.create(env);

    //注意,一定要启用检查点
    env.enableCheckpointing(1000);

    env.setParallelism(1);

    //创建 Kafka 源表
    tEnv.executeSql(
        "CREATE TABLE cars_tb (" +
        " carId string," +
        " speed int," +
        " acceleration double," +
        " intime bigint," +
        " ts AS TO_TIMESTAMP(FROM_UNIXTIME(intime/1000, 'yyyy-MM-dd HH:mm:ss'))," +
        " WATERMARK FOR ts AS ts - INTERVAL '2' SECOND" +
        ")" +
        "WITH (\n" +
        " 'connector' = 'kafka',\n" +
        " 'topic' = 'cars',\n" +
        " 'properties.Bootstrap.servers' = 'localhost:9092',\n" +
        " 'properties.group.id' = 'testGroup',\n" +
        " 'scan.startup.mode' = 'latest-offset',\n" +
        " 'format' = 'csv',\n" +
        " 'csv.ignore-parse-errors' = 'true',\n" +
        " 'csv.field-delimiter' = ',',\n" +
        ")"
    );

    //创建 Kafka Sink 表
    tEnv.executeSql(
        "CREATE TABLE fastcars_tb (" +
        " carId string," +
        " window_start TIMESTAMP(3)," +
        " window_end TIMESTAMP(3)," +
        " avgSpeed double" +
        ")" +
        "WITH (" +
        " 'connector' = 'kafka',\n" +
        " 'topic' = 'fastcars',\n" +
        " 'properties.Bootstrap.servers' = 'localhost:9092',\n" +
        " 'format' = 'json'\n" +
```

```
                ")"
            );

            tEnv
                //读取源表
                .from("cars_tb")
                //定义大小为 5s,滑动为 2s 的滑动窗口
                .window(Slide.over(lit(5).seconds())
                            .every(lit(2).seconds())
                            .on($("ts")).as("w"))
                //分组
                .groupBy($("carId"), $("w"))
                //聚合
                .select($("carId"),
                        $("w").start().as("win_start"),
                        $("w").end().as("win_end"),
                        $("speed").avg().as("avgSpeed")
                )
                //过滤超速信息
                .where($("avgSpeed").isGreater(100))
                //写入 Sink 表
                .executeInsert("fastcars_tb");
    }

}.
```

接下来,请按以下步骤执行。

(1) 启动 ZooKeeper,命令如下:

```
$ ./bin/zookeeper-server-start.sh config/zookeeper.properties
```

(2) 启动 Kafka,命令如下:

```
$ ./bin/kafka-server-start.sh config/server.properties
```

(3) 创建两个主题,命令如下:

```
$ ./bin/kafka-topics.sh --zookeeper localhost:2181 --list
$ ./bin/kafka-topics.sh --zookeeper localhost:2181 --replication-factor 1 --partitions 1 --create --topic cars
$ ./bin/kafka-topics.sh --zookeeper localhost:2181 --replication-factor 1 --partitions 1 --create --topic fastcars
```

(4) 先在一个新的终端窗口中,执行消费者脚本,来拉取 fastcars 主题数据,命令如下:

```
$ ./bin/kafka-console-consumer.sh --Bootstrap-server localhost:9092 --topic fastcars
```

(5) 执行前面开发的流计算程序。
(6) 执行模拟数据源程序,命令如下:

```
$ cd cars
$ Java -jar fastcars_fat.jar localhost:9092 cars
```

(7) 可以看到持续不断地生成模拟车辆运行速度数据,数据如下:

```
Writing car6,115,38.44529,1631158426779
Writing car1,29,10.961765,1631158428433
Writing car9,98,9.312117,1631158429434
Writing car8,17,60.017864,1631158430435
Writing car8,18,87.79231,1631158431436
Writing car8,70,84.12387,1631158432437
Writing car4,114,46.86547,1631158433438
Writing car0,55,15.12608,1631158434439
Writing car8,10,10.488677,1631158435440
……
```

(8) 回到消息者脚本执行窗口,查看超速数据。可以监测到类似如下的超速信息(JSON 格式):

```
{"carId":"car6","window_start":"2021-09-09 11:33:42","window_end":"2021-09-09 11:33:47","avgSpeed":115.0}
{"carId":"car6","window_start":"2021-09-09 11:33:44","window_end":"2021-09-09 11:33:49","avgSpeed":115.0}
{"carId":"car6","window_start":"2021-09-09 11:33:46","window_end":"2021-09-09 11:33:51","avgSpeed":115.0}
{"carId":"car4","window_start":"2021-09-09 11:33:50","window_end":"2021-09-09 11:33:55","avgSpeed":114.0}
{"carId":"car4","window_start":"2021-09-09 11:33:52","window_end":"2021-09-09 11:33:57","avgSpeed":114.0}
……
```

95min

5.5.3 电商用户行为实时分析

【示例 5-4】 综合运用 Flink 的各种 API,基于 EventTime 实现分析电商用户行为。
电商平台中的用户行为频繁且较复杂,系统上线运行一段时间后,可以收集到大量的用户行为数据,进而利用大数据技术进行深入挖掘和分析,得到感兴趣的商业指标并增强对风险的控制。

电商用户行为数据多样，整体可以分为用户行为习惯数据和业务行为数据两大类。用户行为习惯数据包括用户的登录方式、上线的时间点及时长、单击和浏览页面、页面停留时间及页面跳转等，可以从中进行流量统计和热门商品的统计，也可以深入挖掘用户的特征；这些数据往往可以从 Web 服务器日志中直接读取，而业务行为数据就是用户在电商平台中针对每个业务（通常是某个具体商品）所进行的操作，一般会在业务系统中相应的位置埋点，然后收集日志进行分析。业务行为数据又可以简单分为两类：一类是能够明显地表现出用户兴趣的行为，例如对商品的收藏、喜欢、评分和评价，可以从中对数据进行深入分析，得到用户画像，进而对用户给出个性化的推荐商品列表，这个过程往往会用到机器学习相关的算法；另一类则是常规的业务操作，但需要着重关注一些异常状况以做好风控，例如登录和订单支付。

本项目限于数据，只实现实时热门商品统计。

1. 数据源说明

本案例使用阿里天池的一份淘宝用户行为数据集，格式为 CSV。本数据集包含了 2017 年 11 月 25 日至 2017 年 12 月 3 日之间约一百万随机用户的所有行为（行为包括单击、购买、加购、喜欢）。数据集的每行表示一条用户行为，由用户 ID、商品 ID、商品类目 ID、行为类型和时间戳组成，并以逗号分隔。

注意：从阿里天池上下载的数据集 UserBehavior.csv 解压缩后为 3.41GB，这里截取其中一部分（485 730 行）用于开发测试，命名为 UserBehavior_part.csv。

关于数据集中每列的详细描述，见表 5-2。

表 5-2 淘宝用户行为数据集说明

列名称	含义	说明
User ID	用户 ID	整数类型，序列化后的用户 ID
Item ID	商品 ID	整数类型，序列化后的商品 ID
Category ID	商品类目 ID	整数类型，序列化后的商品所属类目 ID
Behavior Type	行为类型	字符串，枚举类型，包括'pv'、'buy'、'cart'、'fav'
Timestamp	时间戳	长整数，行为发生的时间戳

其中用户行为类型共有 4 种，它们分别如下。

(1) pv：商品详情页 pv，等价于单击。

(2) buy：购买商品。

(3) cart：将商品加入购物车。

(4) fav：收藏商品。

部分用户行为数据示例如下：

```
543462,1715,1464116,pv,1511658000
662867,2244074,1575622,pv,1511658000
561558,3611281,965809,pv,1511658000
894923,3076029,1879194,pv,1511658000
834377,4541270,3738615,pv,1511658000
...
```

2. 预备知识

Flink SQL 提供了一些与日期处理有关的函数。

1) TO_TIMESTAMP

将 BIGINT 类型的日期或者 VARCHAR 类型的日期转换成 TIMESTAMP 类型,其语法如下:

```
TIMESTAMP TO_TIMESTAMP(BIGINT time)                    //time:毫秒
TIMESTAMP TO_TIMESTAMP(VARCHAR date)                   //date:yyyy-MM-dd HH:mm:ss
TIMESTAMP TO_TIMESTAMP(VARCHAR date, VARCHAR format)
```

2) FROM_UNIXTIME

返回值为 VARCHAR 类型的日期值,默认日期格式为 yyyy-MM-dd HH:mm:ss,若指定日期格式,则按指定格式输出。如果任一输入参数是 NULL,则返回 NULL,其语法如下:

```
VARCHAR FROM_UNIXTIME(BIGINT unixtime[, VARCHAR format])
```

说明:

(1) 参数 unixtime 为长整型,是以秒为单位的时间戳。

(2) 参数 format 可选,为日期格式,默认格式为 yyyy-MM-dd HH:mm:ss,表示返回 VARCHAR 类型的符合指定格式的日期,如果有参数为 NULL 或解析错误,则返回 NULL。

3. 任务实现

使用 Table API 和 SQL,实现如下描述的数据实时处理管道:

(1) 将用户行为数据采集到 Kafka。

(2) 使用 Table API 读取 Kafka 并写入 MySQL。

(3) 用 Grafana 实时可视化显示。

首先,向项目添加依赖。因为要读取的 Kafka 的用户行为事件是 CSV 格式的,所以在项目的 pom.xml 文件中添加的依赖如下:

```
<!-- 需要添加 flink-csv 依赖 -->
<dependency>
    <groupId>org.apache.flink</groupId>
```

```xml
        <artifactId>flink-csv</artifactId>
        <version>1.13.2</version>
</dependency>
```

创建 Kafka 源表的 SQL 语句如下：

```sql
CREATE TABLE user_behavior (
    user_id BIGINT,
    item_id BIGINT,
    category_id BIGINT,
    behavior STRING,
    ts TIMESTAMP(3),
    proctime AS PROCTIME(), -- 使用计算列生成处理时间属性
    WATERMARK FOR ts AS ts - INTERVAL '5' SECOND -- 定义 ts 列上的水印,将 ts 标记为事件时间属性
) WITH (
    'connector' = 'kafka',                                  -- 使用 Kafka 连接器
    'topic' = 'user_behavior',                              -- Kafka 主题
    'scan.startup.mode' = 'earliest-offset',                -- 从头开始读取
    'properties.Bootstrap.servers' = 'kafka:9094',          -- Kafka Broker 地址
    'format' = 'json'                                       -- 数据格式
);
```

在上面的 SQL 语句中,按照数据的格式声明了 5 个字段,除此之外,还通过计算列语法和 PROCTIME() 内置函数声明了一个产生处理时间的虚拟列。另外通过 WATERMARK 语法,在 ts 字段上声明了 watermark 策略(容忍 5s 乱序),ts 字段因此也成了事件时间列。

接下来,编写流处理代码。

Scala 代码实现:

```scala
//第 5 章/UserBehaviorDemo.scala

import org.apache.flink.streaming.api.scala.StreamExecutionEnvironment
import org.apache.flink.table.api._
import org.apache.flink.table.api.bridge.scala.StreamTableEnvironment

object UserBehaviorDemo {

  def main(args: Array[String]) {
    //设置流执行环境
    val env = StreamExecutionEnvironment.getExecutionEnvironment

    //创建表环境
    val tEnv = StreamTableEnvironment.create(env)
```

```scala
//启用检查点,并将并行度设置为1
env.enableCheckpointing(5000).setParallelism(1)

//创建Kafka源表
tEnv.executeSql(
    """
      |CREATE TABLE user_behavior (
      |   user_id bigint,
      |   item_id bigint,
      |   category_id bigint,
      |   behavior string,
      |   behavior_time bigint,
      |   ts AS TO_TIMESTAMP(FROM_UNIXTIME(behavior_time, 'yyyy-MM-dd HH:mm:ss')),
      |   proctime AS PROCTIME(),
      |   WATERMARK FOR ts AS ts - INTERVAL '5' SECONDS
      |)
      |with(
      |   'connector' = 'kafka',
      |   'topic' = 'user_behavior',
      |   'properties.Bootstrap.servers' = 'localhost:9092',
      |   'properties.group.id' = 'testGroup',
      |   'scan.startup.mode' = 'latest-offset',
      |   'format' = 'csv',
      |   'csv.ignore-parse-errors' = 'true',
      |   'csv.field-delimiter' = ','
      |)
    """.stripMargin)

//创建MySQL Sink表
tEnv.executeSql(
    """
      |CREATE TABLE buy_cnt_per_hour (
      |   hour_of_day TIMESTAMP(3),
      |   buy_cnt BIGINT
      |)
      |with(
      |   'connector' = 'JDBC',
      |   'url' = 'JDBC:mysql://localhost:3306/xueai8?useSSL=false',
      |   'table-name' = 'buy_cnt_per_hour',
      |   'driver' = 'com.mysql.JDBC.Driver',
      |   'username' = 'root',
      |   'password' = 'admin'
      |)
    """.stripMargin)

tEnv
```

```
            //读取源表
            .from("user_behavior")
            //定义大小为 5s,滑动为 2s 的滑动窗口
            .window(Tumble over 1.minute on $ "ts" as $ "w")
            //分组
            .groupBy( $ "w")
            //聚合
            .select( $ "w".start.as("hour_of_day"), $ "item_id".count.as("buy_cnt"))
            //.execute.print()
            //写入 Sink 表
            .executeInsert("buy_cnt_per_hour")
    }
}
```

Java 代码如下：

```java
//第 5 章/UserBehaviorDemo.java

import org.apache.flink.streaming.api.environment.StreamExecutionEnvironment;
import org.apache.flink.table.api.Tumble;
import org.apache.flink.table.api.bridge.java.StreamTableEnvironment;
import org.apache.flink.table.expressions.TimeIntervalUnit;
import org.apache.flink.util.TimeUtils;

import static org.apache.flink.table.api.Expressions.$ ;
import static org.apache.flink.table.api.Expressions.lit;

public class UserBehaviorDemo {

    public static void main(String[] args) throws Exception {
        //设置流执行环境
        final StreamExecutionEnvironment env =
                StreamExecutionEnvironment.getExecutionEnvironment();

        //创建表环境
        StreamTableEnvironment tEnv = StreamTableEnvironment.create(env);

        //启用检查点,并将并行度设置为 1
        env.enableCheckpointing(1000).setParallelism(1);

        //创建 Kafka 源表
        tEnv.executeSql(
          "CREATE TABLE user_behavior (" +
          " user_id bigint," +
          " item_id bigint," +
```

```
    " category_id bigint," +
    " behavior string," +
    " behavior_time bigint," +
    " ts AS TO_TIMESTAMP(FROM_UNIXTIME(behavior_time, 'yyyy-MM-dd HH:mm:ss'))," +
    " proctime AS PROCTIME()," +
    " WATERMARK FOR ts AS ts - INTERVAL '5' SECOND" +
    ")" +
    "WITH (" +
    " 'connector' = 'kafka'," +
    " 'topic' = 'user_behavior'," +
    " 'properties.Bootstrap.servers' = '192.168.190.133:9092'," +
    " 'properties.group.id' = 'testGroup'," +
    " 'scan.startup.mode' = 'latest-offset'," +
    " 'format' = 'csv'," +
    " 'csv.ignore-parse-errors' = 'true'," +
    " 'csv.field-delimiter' = ','" +
    ")"
);

//创建 MySQL Sink 表
tEnv.executeSql("CREATE TABLE buy_cnt_per_hour (" +
    "  hour_of_day TIMESTAMP(3)," +
    "  buy_cnt BIGINT" +
    ") WITH (" +
    "  'connector' = 'JDBC'," +
    "  'url' = 'JDBC:mysql://localhost:3306/xueai8?useSSL=false'," +
    "  'table-name' = 'buy_cnt_per_hour'," +
    "  'driver'   = 'com.mysql.JDBC.Driver'," +
    "  'username' = 'root'," +
    "  'password' = 'admin'" +
    ")"
);

//统计每分钟的成交量
/*
    统计每小时的成交量就是每小时共有多少"buy"的用户行为
    因此需要用到 TUMBLE 窗口函数,按照一小时切窗
    然后每个窗口分别统计"buy"的个数,这可以通过先过滤出"buy"的数据,然后通过COUNT
    (*) 实现
    SELECT HOUR(TUMBLE_START(ts, INTERVAL '1' HOUR)), COUNT(*)
    FROM user_behavior
    WHERE behavior = 'buy'
    GROUP BY TUMBLE(ts, INTERVAL '1' HOUR);
*/
/* 在MySQL 数据库中创建相应的结果表
    CREATE TABLE xueai8.buy_cnt_per_hour (
```

```
            hour_of_day timestamp,
            buy_cnt bigint
        )
      */

        tEnv
            //读取源表
            .from("user_behavior")
            //过滤出购买行为
            .filter($("behavior").isEqual("buy"))
            //定义滚动窗口
            .window(Tumble.over(lit(1).minute()).on($("ts")).as("w"))
            //分组
            .groupBy($("w"))
            //聚合
            .select($("w").start().as("hour_of_day"),$("item_id").count().as("buy_cnt"))
            //使用.extract(TimeIntervalUnit.MINUTE)抽取日期中的指定部分
            .select($("w").start().extract(TimeIntervalUnit.MINUTE).as("hour_of_day"),
$("item_id").count().as("buy_cnt"))
            .execute().collect().forEachRemaining(System.out::println);
            //写入 Sink 表
            .executeInsert("buy_cnt_per_hour");
    }
}
```

4. 执行过程

建议按以下步骤执行程序：

（1）启动 Kafka。

首先打开一个终端，运行 ZooKeeper，命令如下：

```
$ ./bin/zookeeper-server-start.sh ./config/zookeeper.properties
```

再打开一个终端，运行 Kafka 服务器，命令如下：

```
$ ./bin/kafka-server-start.sh ./config/server.properties
```

再打开一个终端，创建名为 user_behavior 的 Kafka 主题，命令如下：

```
$ ./bin/kafka-topics.sh --create --zookeeper localhost:2181 --replication-factor 1 --partitions 1 --topic user_behavior
```

查看已经存在的主题，命令如下：

```
$ ./bin/kafka-topics.sh -- list -- zookeeper localhost:2181
```

(2) 在MySQL上创建接收表,SQL语句如下:

```
CREATE TABLE buy_cnt_per_hour (
    hour_of_day timestamp,
    buy_cnt bigint
);
```

(3) 运行可视化终端Grafana。

Grafana是一款用Go语言开发的开源数据可视化工具,可以做数据监控和数据统计,带有告警功能。

首先安装Grafana,安装步骤如下:

① 下载安装包,然后解压缩即可。

② 在命令行启动Grafana服务器,命令如下:

```
E:\BigData\Grafana\grafana-7.5.0\bin>grafana-server.exe
```

③ 打开浏览器,访问地址为http://localhost:3000/。

④ 在Grafana中要先配置好数据源,指向MySQL中的接收表,如图5-2所示。

图5-2 在Grafana中配置数据源

⑤ 在Grafana中创建dashboard,查询获得的数据,使用的SQL语句如下:

```sql
SELECT
    UNIX_TIMESTAMP(hour_of_day) as time_sec,
    buy_cnt as value
FROM xueai8.buy_cnt_per_hour
```

为了更好地看到动态变化的效果,在一个仪表板上设置显示的时间范围,如图 5-3 所示。

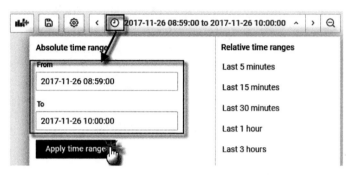

图 5-3　在 Grafana 的一个仪表板上设置显示的时间范围

(4) 运行 Flink 流程序。

(5) 执行数据生产者脚本 streamuserbehavior.sh。它会调用 Kafka 自带的生产者脚本,以每秒 10 条的速度将数据发送给 Kafka 的 user_behavior 主题。编辑生产数据的脚本文件 streamuserbehavior.sh,代码如下:

```bash
#第5章/streamuserbehavior.sh

#!/bin/bash
BROKER=$1
if [ -z "$1" ]; then
      BROKER="localhost:9092"
fi

cat UserBehavior_part.csv | while read line; do
      echo $line
      sleep 0.1
done | ~/bigdata/kafka_2.11-2.4.1/bin/kafka-console-producer.sh --broker-list $BROKER --topic user_behavior
```

执行该脚本,使用的命令如下:

```
$ ./streamuserbehavior.sh
```

(6) 观察 Grafana 中数据的实时呈现效果,如图 5-4 所示。

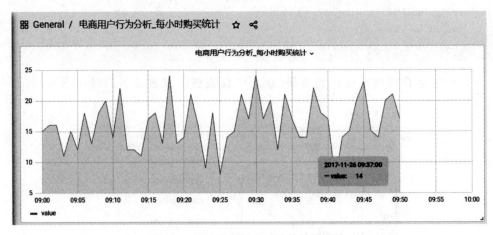

图 5-4　用户购买量实时统计结果显示

第 6 章 Flink on YARN

CHAPTER 6

Flink 已内置支持应用程序可以在 YARN 上执行,即任何使用 Flink API 构建的应用程序都可以在 YARN 上轻松执行。如果用户已经拥有 YARN 集群,则不需要设置或安装任何东西。在本章中,我们将看到如何利用现有的 Hadoop/YARN 集群并行执行 Flink 任务。

6.1 Flink on YARN session

从 Flink 1.8 开始,Hadoop 不再包含在 Flink 的安装包中,所以需要单独下载并复制到 Flink 的 lib 目录下。

6.1.1 下载 Flink 集成 Hadoop 依赖包

如果使用的是 Hadoop 2,则可从 Flink 官网下载 flink-shaded-hadoop2-uber-2.7.5-1.10.0.jar,如图 6-1 所示。

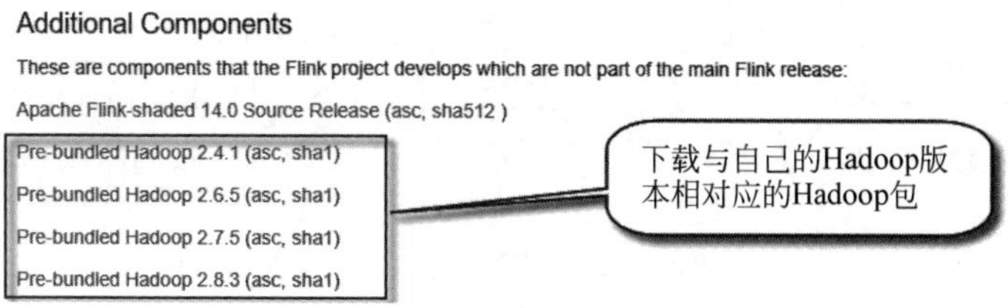

图 6-1 从 Flink 官网下载兼容 Hadoop 2 的 JAR 包

如果使用的是 Hadoop 3,则可从 Maven 库下载,如图 6-2 所示。

本书中使用 Hadoop-3.2.2,因此应下载最新的 flink-shaded-hadoop-3-uber-3.1.1.7.2.9.0-173-9.0.jar 包。将下载的 Hadoop 包复制到各个节点安装的 Flink 的 lib 目录下。

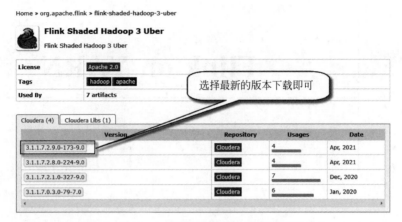

图 6-2 从 Maven 库下载兼容 Hadoop 3 的 JAR 包

6.1.2 运行 Flink on YARN session

Apache Hadoop YARN 是一个资源提供器,在许多数据处理框架中很受欢迎。Flink 服务提交给 YARN 的 ResourceManager,后者由 YARN NodeManager 管理的机器生成 container 资源容器。Flink 将其 JobManager 和 TaskManager 实例部署到这些 container 资源容器中。

Flink on YARN 的执行过程,如图 6-3 所示。

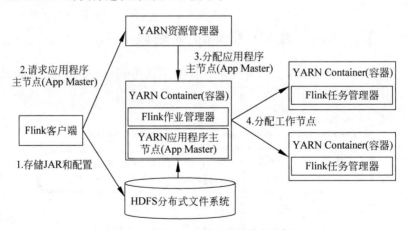

图 6-3 Flink on YARN 的执行过程

图 6-3 显示了 Flink on YARN 的内部工作情况。它经过以下步骤:
(1) 检查是否设置了 Hadoop 和 YARN 配置目录。
(2) 如果是,则联系 HDFS 并将 JAR 和配置存储在 HDFS 上。
(3) 联系 Node Manager 分配 Application Master。

(4) 一旦分配了 Application Master,启动 Flink 作业管理器。

(5) 稍后,根据给定的配置参数启动 Flink 任务管理器。

Flink 可以根据运行在 Job Manager 上的作业所需的处理槽的数量动态地分配和解除已分配的 Task Manager 资源。

确保设置了 HADOOP_CLASSPATH 环境变量。如果没有,则使用如下的命令设置:

```
$ export HADOOP_CLASSPATH = 'hadoop classpath'
```

一旦确定设置了 HADOOP_CLASSPATH 环境变量,就可以在 YARN 上启动一个 Flink 会话(Flink on YARN session),并提交 Flink 作业程序。

Flink on YARN session 是一个会话,它在各个节点上启动所有必需的 Flink 服务(Job Manager 和 Task Manage),以便用户可以开始执行 Flink 作业。

要启动 Flink on YARN session,执行的命令如下(确保已经启动了 Hadoop 集群):

```
#(0)export HADOOP_CLASSPATH
$ export HADOOP_CLASSPATH = 'hadoop classpath'

#(1)启动 Flink on YARN session
$ cd ~/bigdata/flink-1.13.2
$ ./bin/yarn-session.sh --detached
```

启动 Flink on YARN session 时提供了--detached 参数,表示客户端将在提交被接受后停止。现在,可以通过命令输出最后一行打印的 URL 或通过 YARN Resource Manager Web UI 访问 Flink Web 界面。例如,在命令输出最后一行打印的 URL 信息,如图 6-4 所示。

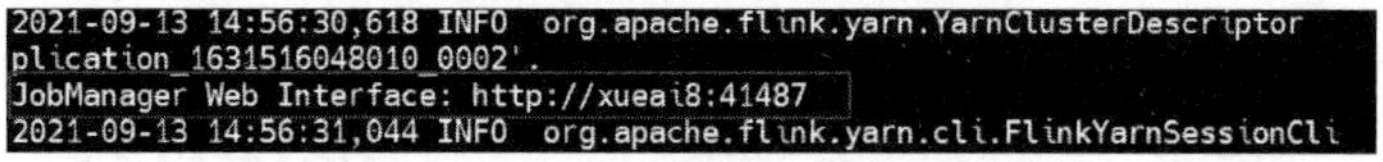

```
2021-09-13 14:56:30,618 INFO  org.apache.flink.yarn.YarnClusterDescriptor
plication_1631516048010_0002'.
JobManager Web Interface: http://xueai8:41487
2021-09-13 14:56:31,044 INFO  org.apache.flink.yarn.cli.FlinkYarnSessionCli
```

图 6-4 Flink Web 界面信息

其中的端口号(这里为 41487)每次启动可能都会变化。打开浏览器,访问 Flink Web UI 界面,如图 6-5 所示。

也可以通过 YARN Resource Manager Web UI 访问 Flink Web UI 界面,如图 6-6 所示。

这种模式下会启动 Flink on YARN session,并且会启动 Flink 的两个必需服务:Job Manager 和 Task Manager,然后就可以向集群提交作业。同一个 session 中可以提交多个 Flink 作业。

在启动过程中,有可能会遇到以下启动错误信息:

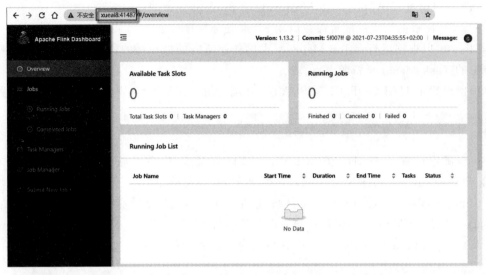

图 6-5 访问 Flink Web UI 界面

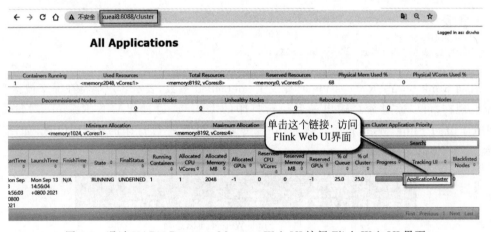

图 6-6 通过 YARN Resource Manager Web UI 访问 Flink Web UI 界面

```
2021-06-04 18:18:12,867 ERROR org.apache.flink.yarn.cli.FlinkYarnSessionCli      [] - 
Error while running the Flink session.
Java.lang.NoSuchMethodError: org.apache.commons.cli.Option.builder(LJava/lang/String;)
Lorg/apache/commons/cli/Option$Builder;
        at org.apache.flink.yarn.cli.FlinkYarnSessionCli.<init>(FlinkYarnSessionCli.java:
230) ~[flink-dist_2.12-1.13.1.jar:1.13.1]
        at org.apache.flink.yarn.cli.FlinkYarnSessionCli.<init>(FlinkYarnSessionCli.java:
156) ~[flink-dist_2.12-1.13.1.jar:1.13.1]
        at org.apache.flink.yarn.cli.FlinkYarnSessionCli.main(FlinkYarnSessionCli.java:851)
[flink-dist_2.12-1.13.1.jar:1.13.1]
```

```
The program finished with the following exception:

Java.lang.NoSuchMethodError: org.apache.commons.cli.Option.builder(LJava/lang/String;)
Lorg/apache/commons/cli/Option$Builder;
        at org.apache.flink.yarn.cli.FlinkYarnSessionCli.<init>(FlinkYarnSessionCli.java:
230)
        at org.apache.flink.yarn.cli.FlinkYarnSessionCli.<init>(FlinkYarnSessionCli.java:
156)
        ...
```

如果遇到这样的启动错误信息,则表示缺少 commons-cli-1.4.jar 包。解决方法就是从网上下载 commons-cli-1.4.jar 包,然后将其复制到 Flink 的 lib 目录下,重新启动即可。

必须确保设置了 YARN_CONF_DIR 和 HADOOP_CONF_DIR 环境变量,以便 Flink 能够找到所需的配置。也可以在启动 Flink on YARN session 时,详细说明任务管理器的数量、每个任务管理器的内存和要使用的插槽,命令如下:

```
$ ./bin/yarn-session.sh -n 2 -tm 1024 -s 4
```

以上参数的说明如下。

(1) -n,--container <arg>:要分配的 YARN container 数量(=Task Managers 数量)。

(2) -tm,--taskManagerMemory <arg>:每个 TaskManager Container 的内存(单位为 MB)。

(3) -s,--slots <arg>:每个 TaskManager 的 slot 的数量。

所以上面的命令启动了两个 TaskManager,每个 TaskManager 内存为 1GB 且占用 4 个资源槽。在启动 Flink on YARN session 时会加载 conf/flink-config.yaml 配置文件,用户可以根据自己的需求去修改里面的相关参数。

6.1.3 提交 Flink 作业

现在有了一个 Flink on YARN session,用户准备向其提交一个 Flink 作业。

1. 提交批处理作业

提交 Flink 批处理作业示例程序,命令如下:

```
$ ./bin/flink run ./examples/batch/WordCount.jar -- input hdfs://localhost:8020/data/flink/wc.txt
```

这将调用在 YARN 集群上执行的 Flink 作业。

Flink 应用程序主 UI 的作业执行的屏幕截图如图 6-7 所示。

Flink 批处理作业执行计划的截图如图 6-8 所示。

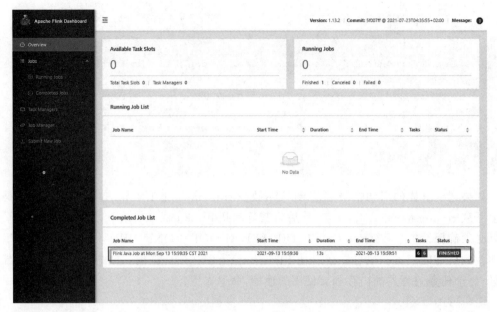

图 6-7　批处理作业执行截图

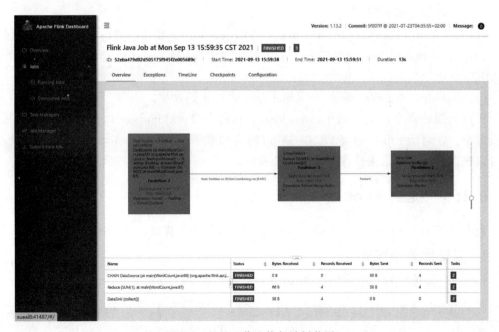

图 6-8　批处理作业执行计划截图

执行结果如图 6-9 所示。

如果不想将结果输到终端，而是保存在文件中，则可以使用--output 参数指定保存结果

图 6-9 批处理作业执行结果

的地方,命令如下:

```
$ ./bin/flink run ./examples/batch/WordCount.jar -- input hdfs://localhost:8020/data/flink/wc.txt
-- output hdfs://localhost:8020/data/flink/out
```

需要注意的是,上面的--input 和--output 参数并不是 Flink 内部的参数,而是在 WordCount 程序中定义的。另外在指定路径时一定要记得加上模式,例如上面的 hdfs://, 否则程序会在本地寻找文件。

2. 提交流处理作业

提交 Flink 流处理作业示例程序,命令如下:

```
$ ./bin/flink run ./examples/streaming/TopSpeedWindowing.jar
```

Flink 应用程序主 UI 的作业执行的屏幕截图如图 6-10 所示。

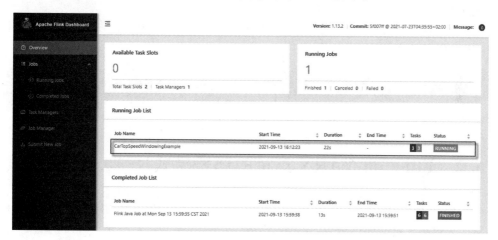

图 6-10 流处理作业执行截图

Flink 流处理作业执行计划的截图如图 6-11 所示。

6.1.4 停止 Flink on YARN session

一旦处理完成,用户可以通过两种方法停止 Flink on YARN session。

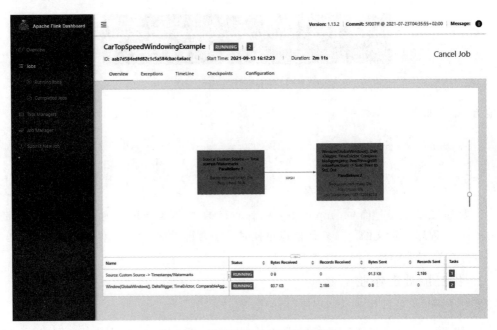

图 6-11　流处理作业执行计划截图

第一种方法是在开始该会话的控制台上简单地按快捷键 Ctrl＋C,这将发送终止信号并停止 YARN session。

第二种方法是通过命令来停止该会话(需要根据 yarn-session.sh 命令的输出信息替换应用程序 id),命令如下:

```
$ echo "stop" | ./bin/yarn-session.sh -id application_xxxxx_xxx stop
```

这样可以立即看到 Flink YARN 应用程序被杀掉。

6.2　Flink on YARN 支持的部署模式

对于生产使用,建议以 Per-Job 或 Application 模式部署 Flink 应用程序,因为这些模式为应用程序提供了更好的隔离。

6.2.1　Application 模式

Application 模式将在 YARN 上启动一个 Flink 集群,其中应用程序 JAR 的 main()方法将在 YARN 中的 JobManager 上执行。应用程序一完成,该集群就会关闭。可以通过命令 yarn application -kill <ApplicationId>或取消 Flink 作业手动停止集群,命令如下:

```
$ ./bin/flink run -application -t yarn-application ./examples/streaming/
TopSpeedWindowing.jar
```

一旦部署了 Application 模式集群,就可以与它进行交互,以执行取消或获取保存点等操作,命令如下:

```
#列出集群上正在运行的作业
$ ./bin/flink list -t yarn-application -Dyarn.application.id=application_XXXX_YY

#取消正在运行的作业
$ ./bin/flink cancel -t yarn-application -Dyarn.application.id=application_XXXX_YY
<jobId>
```

需要注意,取消 Application 集群上的作业将停止该集群。

要释放 Application 模式的全部潜力,可以考虑与 yarn.provided.lib.dirs 配置选项一起使用它,并将应用程序 JAR 包预先上传到集群中所有节点都可以访问的位置。在本例中,命令如下:

```
$ ./bin/flink run-application -t yarn-application \
 -Dyarn.provided.lib.dirs="hdfs://myhdfs/my-remote-flink-dist-dir" \
 hdfs://myhdfs/jars/my-application.jar
```

上述操作将使作业提交变得更加轻量级,因为所需的 Flink JAR 和应用程序 JAR 将由指定的远程位置提取,而不是由客户端发送到集群。

6.2.2 Per-Job 集群模式

Per-Job 集群模式将在 YARN 上启动一个 Flink 集群,然后在本地运行应用程序 JAR,最后将 JobGraph 提交给 YARN 上的 JobManager。如果传递 --detached 参数,则客户端将在提交被接受后停止。一旦作业停止,YARN 集群就会停止。命令如下:

```
$ ./bin/flink run -t yarn-per-job --detached ./examples/streaming/TopSpeedWindowing.jar
```

一旦部署了 Per-Job 集群,就可以与它进行交互,以执行取消或获取保存点等操作,交互命令如下:

```
#列出集群上正在运行的作业
$ ./bin/flink list -t yarn-per-job -Dyarn.application.id=application_XXXX_YY

#取消正在运行的作业
$ ./bin/flink cancel -t yarn-per-job -Dyarn.application.id=application_XXXX_YY
<jobId>
```

需要注意，取消 Per-Job 集群上的作业将停止集群。

6.2.3 session 模式

6.1 节已经描述了 session 模式。session 模式有以下两种操作模式。

(1) attached mode(默认)：yarn-session.sh 客户端将 Flink 集群提交给 YARN，但是客户端继续运行，跟踪集群的状态。如果集群失败，则客户端将显示错误。如果客户端被终止，则会发出关闭集群的信号。

(2) detached mode(-d 或--detached)：yarn-session.sh 客户端将 Flink 集群提交给 YARN，然后客户端返回。需要调用另一个客户端或 YARN 工具来停止 Flink 集群。

session 模式将在/tmp/.yarn-properties -< username >目录下创建一个隐藏的 YARN 属性文件，它将在提交作业时由命令行接口提取，用于集群发现。

也可以在提交 Flink 作业时，在命令行接口中手动指定目标 YARN 集群，命令如下：

```
$ ./bin/flink run -t yarn-session \
  -Dyarn.application.id=application_XXXX_YY \
  ./examples/streaming/TopSpeedWindowing.jar
```

可以重新附加到 Flink on YARN session，使用的命令如下：

```
$ ./bin/yarn-session.sh -id application_XXXX_YY
```

除了通过 conf/flink-conf.yaml 文件传递配置外，也可以使用-Dkey=value 参数在提交时将任何配置传递给./bin/yarn-session.sh 客户端。

Flink on YARN session 客户端也有一些常用来设置的“快捷参数”，可以通过./bin/yarn-session.sh -h 命令列出这些快捷参数，如图 6-12 所示。

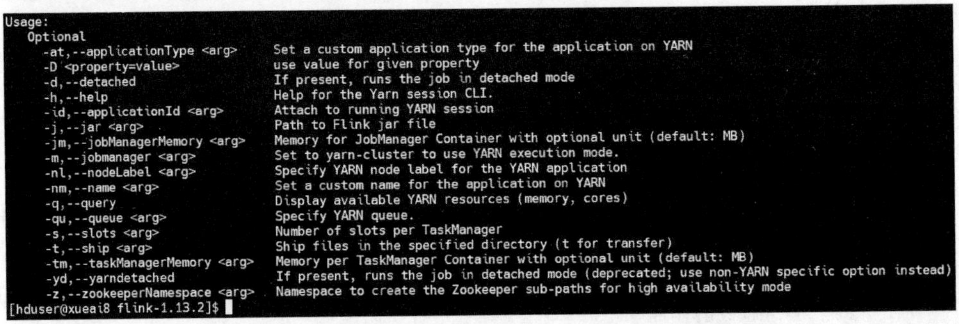
```
Usage:
   Optional
     -at,--applicationType <arg>     Set a custom application type for the application on YARN
     -D <property=value>             use value for given property
     -d,--detached                   If present, runs the job in detached mode
     -h,--help                       Help for the Yarn session CLI.
     -id,--applicationId <arg>       Attach to running YARN session
     -j,--jar <arg>                  Path to Flink jar file
     -jm,--jobManagerMemory <arg>    Memory for JobManager Container with optional unit (default: MB)
     -m,--jobmanager <arg>           Set to yarn-cluster to use YARN execution mode.
     -nl,--nodeLabel <arg>           Specify YARN node label for the YARN application
     -nm,--name <arg>                Set a custom name for the application on YARN
     -q,--query                      Display available YARN resources (memory, cores)
     -qu,--queue <arg>               Specify YARN queue.
     -s,--slots <arg>                Number of slots per TaskManager
     -t,--ship <arg>                 Ship files in the specified directory (t for transfer)
     -tm,--taskManagerMemory <arg>   Memory per TaskManager Container with optional unit (default: MB)
     -yd,--yarndetached              If present, runs the job in detached mode (deprecated; use non-YARN specific option instead)
     -z,--zookeeperNamespace <arg>   Namespace to create the Zookeeper sub-paths for high availability mode
[hduser@xueai8 flink-1.13.2]$
```

图 6-12　Flink on YARN session 客户端快捷参数

第 7 章 基于 Flink 构建流批一体数仓

基于 Flink 构建流批一体的实时数仓是目前数据仓库领域比较火的实践方案。随着 Flink 的不断迭代,其提供的一系列技术特性使用户构建流批一体的应用变得越来越方便。本章将以 Flink 1.13 版本为例,介绍这些特性的基本使用方式,主要包括以下内容:

(1) Flink 集成 Hive。
(2) Hive Catalog 与 Hive Dialect。
(3) Flink 读写 Hive。
(4) Flink upsert-kafka 连接器。
(5) Flink CDC 的 connector。

7.1 Flink 集成 Hive 数仓

使用 Hive 构建数据仓库已经成为比较普遍的一种解决方案。目前,一些比较常见的大数据处理引擎,无一例外地兼容 Hive。Flink 从 1.9 版本开始支持集成 Hive,在 Flink 1.10 版本对 Hive 的集成也达到了生产级别的要求。值得注意的是,不同版本的 Flink 对于 Hive 的集成有所差异,本书将以 Flink 1.13 版本为例,阐述 Flink 集成 Hive 的过程和步骤。

7.1.1 Flink 集成 Hive 的方式

Flink 提供了与 Hive 的双重集成。

(1) 持久化元数据。第一重集成是 Flink 利用 Hive 的 Metastore 作为持久化的 Catalog 目录,可通过 HiveCatalog 将不同会话中的 Flink 元数据存储到 Hive Metastore 中。例如,可以使用 HiveCatalog 将 Kafka 源表或 Elasticsearch 源表存储在 Hive Metastore 中,这样该表的元数据信息会被持久化到 Hive 的 Metastore 对应的元数据库中,在后续的 SQL 查询中,就可以重复使用它们。

(2) 利用 Flink 来读写 Hive 的表。第二重集成是提供了 Flink 作为读取和写入 Hive

表的替代引擎。HiveCatalog 的设计提供了与 Hive 良好的兼容性，用户可以"开箱即用"地访问其已有的 Hive 表。不需要修改现有的 Hive Metastore，也不需要更改表的数据位置或分区。

7.1.2 Flink 集成 Hive 的步骤

首先，需要了解 Flink 支持的 Hive 版本信息。Flink 支持的 Hive 版本如图 7-1 所示。

大版本	V1	V2	V3	V4	V5	V6	V7
1.0	1.0.0	1.0.1					
1.1	1.1.0	1.1.1					
1.2	1.2.0	1.2.1	1.2.2				
2.0	2.0.0	2.0.1					
2.1	2.1.0	2.1.1					
2.2	2.2.0						
2.3	2.3.0	2.3.1	2.3.2	2.3.3	2.3.4	2.3.5	2.3.6
3.1	3.1.0	3.1.1	3.1.2				

图 7-1 Flink 支持的 Hive 版本

值得注意的是，对于不同的 Hive 版本，可能在功能方面有所差异，这些差异取决于所使用的 Hive 版本，而不取决于 Flink。一些版本的功能差异如下：

(1) Hive 内置函数在使用 Hive 1.2.0 及更高版本时支持。

(2) 列约束，也就是 PRIMARY KEY 和 NOT NULL，在使用 Hive 3.1.0 及更高版本时支持。

(3) 更改表的统计信息，在使用 Hive 1.2.0 及更高版本时支持。

(4) DATE 列统计信息，在使用 Hive 1.2.0 及更高版本时支持。

(5) 使用 Hive 2.0.x 版本时不支持写入 ORC 表。

Apache Hive 是基于 Hadoop 构建的，所以还需要 Hadoop 的依赖，这需要先配置好 HADOOP_CLASSPATH。配置 HADOOP_CLASSPATH，需要在 /etc/profile 文件中配置环境变量，内容如下：

```
export HADOOP_CLASSPATH = '$HADOOP_HOME/bin/hadoop classpath'
```

这一点非常重要，否则在使用 Flink SQL CLI 查询 Hive 中的表时，会报如下错误：

```
Java.lang.ClassNotFoundException: org.apache.hadoop.mapred.JobConf
```

另外，集成 Hive 需要额外添加一些依赖 JAR 包，并将其放置在 Flink 安装目录下的 lib 文件夹下，这样才能通过 Table API 或 SQL Client 与 Hive 进行交互。

Flink 官网提供了以下两种方式添加 Hive 的依赖项。

(1) 第一种是使用 Flink 提供的 Hive JAR 包(根据使用的 Metastore 的版本来选择对应的 Hive JAR 包)。

(2) 如果使用的 Hive 版本与 Flink 提供的 Hive JAR 包兼容的版本不一致,则可以选择第二种方式,即分别添加每个所需的 JAR 包。

建议优先使用 Flink 提供的 Hive JAR 包,这种方式比较简单方便。可用的 JAR 包及其适用的 Hive 版本如图 7-2 所示。

Metastore version	Maven dependency	SQL Client JAR
1.0.0 - 1.2.2	flink-sql-connector-hive-1.2.2	Download
2.0.0 - 2.2.0	flink-sql-connector-hive-2.2.0	Download
2.3.0 - 2.3.6	flink-sql-connector-hive-2.3.6	Download
3.0.0 - 3.1.2	flink-sql-connector-hive-3.1.2	Download

图 7-2 可用的 JAR 包及其适用的 Hive 版本

用户可以根据使用的 Hive 版本,下载对应的 JAR 包即可。例如本书使用的 Hive 版本为 Hive 3.1.2,所以只需下载 flink-sql-connector-hive-3.1.2,并将其放置到 Flink 安装目录的 lib 文件夹下。Flink 的 Hive 连接器包含了 flink-hadoop 兼容和 flink-orc JAR 包。

上面列举的 JAR 包是用户在使用 Flink SQL CLI 所需要的 JAR 包,除此之外,根据不同的 Hive 版本,还需要添加如下 3 个 JAR 包(以 Hive 3.1.2 为例):

(1) flink-connector-hive_2.12-1.13.2.jar。

(2) hive-exec-3.1.2.jar。

(3) antlr-runtime-3.5.2.jar。

其中,hive-exec-3.1.2.jar 包存在于 Hive 安装路径下的 lib 文件夹。最终,Flink 的 lib 下的 JAR 包如图 7-3 所示。

/home/hduser/bigdata/flink-1.13.2/lib/

Name	Size (KB)	Last modified	Owner	Group	Access
..					
antlr-runtime-3.5.2.jar	163	2021-09-14...	hduser	hduser	-rw-rw-r--
commons-cli-1.4.jar	52	2021-08-09...	hduser	hduser	-rw-rw-r--
flink-connector-hive_2.12-1.13.2.jar	7 604	2021-09-13...	hduser	hduser	-rw-rw-r--
flink-csv-1.13.2.jar	90	2021-07-23...	hduser	hduser	-rw-r--r--
flink-dist_2.12-1.13.2.jar	103 645	2021-07-23...	hduser	hduser	-rw-r--r--
flink-json-1.13.2.jar	144	2021-07-23...	hduser	hduser	-rw-r--r--
flink-shaded-hadoop-3-uber-3.1.1.7.2.9.0-173-9.0.jar	58 207	2021-08-09...	hduser	hduser	-rw-rw-r--
flink-shaded-zookeeper-3.4.14.jar	7 529	2021-04-10...	hduser	hduser	-rw-r--r--
flink-sql-connector-hive-3.1.2_2.12-1.13.2.jar	47 700	2021-09-14...	hduser	hduser	-rw-rw-r--
flink-table-blink_2.12-1.13.2.jar	37 644	2021-07-23...	hduser	hduser	-rw-r--r--
flink-table_2.12-1.13.2.jar	34 198	2021-07-23...	hduser	hduser	-rw-r--r--
hive-exec-3.1.2.jar	39 671	2019-08-23...	hduser	hduser	-rw-r--r--
log4j-1.2-api-2.12.1.jar	65	2020-03-25...	hduser	hduser	-rw-r--r--
log4j-api-2.12.1.jar	270	2020-03-25...	hduser	hduser	-rw-r--r--
log4j-core-2.12.1.jar	1 635	2020-03-25...	hduser	hduser	-rw-r--r--
log4j-slf4j-impl-2.12.1.jar	22	2020-03-25...	hduser	hduser	-rw-r--r--

图 7-3 需要添加到 Flink lib 目录下的 JAR 包

如果使用 Maven 构建自己的程序，则需要在 pom.xml 文件中包含以下依赖项：

```xml
<!-- Flink Dependency -->
<dependency>
    <groupId>org.apache.flink</groupId>
    <artifactId>flink-connector-hive_2.12</artifactId>
    <version>1.13.2</version>
    <scope>provided</scope>
</dependency>

<dependency>
    <groupId>org.apache.flink</groupId>
    <artifactId>flink-table-api-Java-bridge_2.12</artifactId>
    <version>1.13.2</version>
    <scope>provided</scope>
</dependency>

<!-- Hive Dependency -->
<dependency>
    <groupId>org.apache.hive</groupId>
    <artifactId>hive-exec</artifactId>
    <version>3.1.2</version>
    <scope>provided</scope>
</dependency>
```

将上面的 3 个 JAR 包添加至 Flink 的 lib 目录下后，就可以使用 Flink 操作 Hive 的数据表了。

7.1.3 Flink 连接 Hive 模板代码

Flink 使用 HiveCatalog 可以通过批或者流的方式来处理 Hive 中的表。这就意味着 Flink 既可以作为 Hive 的一个批处理引擎，也可以通过流处理的方式来读写 Hive 中的表，从而支持实时数仓的应用和流批一体的落地。

通过表环境或 YAML 配置使用 catalog 接口和 HiveCatalog 连接到现有的 Hive 安装，以及连接到 Hive 的模板。

Scala 代码如下：

```scala
val settings = EnvironmentSettings.newInstance()
    .useBlinkPlanner()
    .inStreamingMode()
    .build()
val tableEnv = TableEnvironment.create(settings)

val name = "myhive"
```

```
val defaultDatabase = "default"
val hiveConfDir = "/home/hduser/bigdata/hive-3.1.2/conf"

val hive = new HiveCatalog(name, defaultDatabase, hiveConfDir)
tableEnv.registerCatalog("myhive", hive)

//将 HiveCatalog 设置为会话的当前目录
tableEnv.useCatalog("myhive")
```

Java 代码如下：

```
EnvironmentSettings settings = EnvironmentSettings.newInstance()
    .useBlinkPlanner()
    .inStreamingMode()
    .build();
TableEnvironment tableEnv = TableEnvironment.create(settings);

String name = "myhive";
String defaultDatabase = "default";
String hiveConfDir = "/home/hduser/bigdata/hive-3.1.2/conf";

HiveCatalog hive = new HiveCatalog(name, defaultDatabase, hiveConfDir);
tableEnv.registerCatalog("myhive", hive);

//将 HiveCatalog 设置为会话的当前目录
tableEnv.useCatalog("myhive");
```

7.2 批流一体数仓构建实例

Flink 1.13.2 的 Hive Streaming 功能极大地提高了 Hive 的实时性能，对于 ETL 操作非常有用，也可以以一定的灵活性满足流连续查询的需求。

【示例 7-1】 基于 7.1 节的内容基础，本节以电商用户行为实时分析为例，构建一个批流一体数据仓库案例。

实现架构如图 7-4 所示。

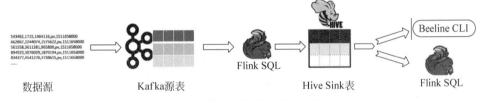

图 7-4 批流一体数据仓库架构

7.2.1 数据集说明

本案例准备了一份淘宝用户行为数据集,保存为 CSV 文件。本数据集包含了 2017 年 11 月 25 日至 2017 年 12 月 3 日之间约一百万随机用户的所有行为(行为包括单击、购买、加购、喜欢)。数据集的每行表示一条用户行为,由用户 ID、商品 ID、商品类目 ID、行为类型和时间戳组成,并以逗号分隔。

注意:从天池上下载的数据集 UserBehavior.csv 解压缩后为 3.41GB,这里截取其中一部分(485 730 行)用于开发测试,命名为 UserBehavior_part.csv。

关于数据集中每列的详细描述,见表 7-1。

表 7-1 数据集字段说明

列名称	含义	说明
User ID	用户 ID	整数类型,序列化后的用户 ID
Item ID	商品 ID	整数类型,序列化后的商品 ID
Category ID	商品类目 ID	整数类型,序列化后的商品所属类目 ID
Behavior Type	行为类型	字符串,枚举类型,包括('pv','buy','cart','fav')
Timestamp	时间戳	长整数,行为发生的时间戳

注意到,用户行为类型共有 4 种,它们分别如下。
(1) pv:商品详情页 pv,等价于单击。
(2) buy:商品购买。
(3) cart:将商品加入购物车。
(4) fav:收藏商品。
部分用户行为数据示例如下:

```
543462,1715,1464116,pv,1511658000
662867,2244074,1575622,pv,1511658000
561558,3611281,965809,pv,1511658000
894923,3076029,1879194,pv,1511658000
834377,4541270,3738615,pv,1511658000
……
```

为了模拟源数据流,这里需要编写好生产数据的脚本 streamuserbehavior.sh,它会调用 Kafka 自带的生产者脚本,以每秒 10 条的速度将数据发送给 Kafka 的 user_behavior 主题。脚本 streamuserbehavior.sh 的代码如下:

```
#第7章/streamuserbehavior.sh

#!/bin/bash
BROKER=$1
if [ -z "$1" ]; then
      BROKER="localhost:9092"
fi

cat UserBehavior_part.csv | while read line; do
      echo $line
      sleep 0.1
done | ~/bigdata/kafka_2.11-2.4.1/bin/kafka-console-producer.sh --broker-list
$BROKER --topic user_behavior
```

本案例中需要对日期时间进行处理，Flink SQL 提供了如下一些与日期处理有关的函数。

1) TO_TIMESTAMP

将 BIGINT 类型的日期或者 VARCHAR 类型的日期转换成 TIMESTAMP 类型，其语法格式如下：

```
TIMESTAMP TO_TIMESTAMP(BIGINT time)                    //time:毫秒
TIMESTAMP TO_TIMESTAMP(VARCHAR date)                   //date:yyyy-MM-dd HH:mm:ss
TIMESTAMP TO_TIMESTAMP(VARCHAR date, VARCHAR format)
```

2) FROM_UNIXTIME

返回值为 VARCHAR 类型的日期值，默认日期格式：yyyy-MM-dd HH:mm:ss，若指定日期格式，则按指定格式输出。如果任一输入参数是 NULL，则返回 NULL，其语法格式如下：

```
VARCHAR FROM_UNIXTIME(BIGINT unixtime[, VARCHAR format])
```

其参数说明如下：

(1) 参数 unixtime 为长整型，是以秒为单位的时间戳。

(2) 参数 format 可选，为日期格式，默认格式为 yyyy-MM-dd HH:mm:ss，表示返回 VARCHAR 类型的符合指定格式的日期，如果有参数为 NULL 或解析错误，则返回 NULL。

7.2.2 创建 Flink 项目

首先使用 IntelliJ IDEA，创建一个 Flink Maven 项目。修改 pom.xml 文件，添加如下依赖项。

Scala Maven 依赖：

```xml
<dependency>
    <groupId>org.apache.flink</groupId>
    <artifactId>flink-scala_2.12</artifactId>
    <version>1.13.2</version>
    <scope>provided</scope>
</dependency>
<dependency>
    <groupId>org.apache.flink</groupId>
    <artifactId>flink-streaming-scala_2.12</artifactId>
    <version>1.13.2</version>
    <scope>provided</scope>
</dependency>

<!-- 连接器依赖 -->
<dependency>
    <groupId>org.apache.flink</groupId>
    <artifactId>flink-connector-base</artifactId>
    <version>1.13.2</version>
</dependency>

<!-- Kafka 连接器依赖 -->
<dependency>
    <groupId>org.apache.flink</groupId>
    <artifactId>flink-connector-kafka_2.12</artifactId>
    <version>1.13.2</version>
</dependency>
<dependency>
    <groupId>org.apache.flink</groupId>
    <artifactId>flink-sql-connector-kafka_2.12</artifactId>
    <version>1.13.2</version>
</dependency>
<dependency>
    <groupId>org.apache.kafka</groupId>
    <artifactId>kafka-clients</artifactId>
    <version>2.4.1</version>
</dependency>

<!-- Hive 连接器依赖 -->
<dependency>
    <groupId>org.apache.flink</groupId>
    <artifactId>flink-connector-hive_2.12</artifactId>
    <version>1.13.2</version>
</dependency>
<dependency>
```

```xml
    <groupId>org.apache.flink</groupId>
    <artifactId>flink-sql-connector-hive-3.1.2_2.12</artifactId>
    <version>1.13.2</version>
</dependency>

<!-- Hive 依赖 -->
<dependency>
    <groupId>org.apache.hive</groupId>
    <artifactId>hive-exec</artifactId>
    <version>3.1.2</version>
</dependency>

<!-- 表依赖 -->
<dependency>
    <groupId>org.apache.flink</groupId>
    <artifactId>flink-table-api-scala-bridge_2.12</artifactId>
    <version>1.13.2</version>
</dependency>
<dependency>
    <groupId>org.apache.flink</groupId>
    <artifactId>flink-table-planner-blink_2.12</artifactId>
    <version>1.13.2</version>
</dependency>
<dependency>
    <groupId>org.apache.flink</groupId>
    <artifactId>flink-table-common</artifactId>
    <version>1.13.2</version>
</dependency>

<dependency>
    <groupId>org.apache.flink</groupId>
    <artifactId>flink-csv</artifactId>
    <version>1.13.2</version>
</dependency>
<dependency>
    <groupId>org.apache.flink</groupId>
    <artifactId>flink-json</artifactId>
    <version>1.13.2</version>
</dependency>
```

Java Maven 依赖：

```xml
<dependency>
    <groupId>org.apache.flink</groupId>
    <artifactId>flink-Java</artifactId>
```

```xml
            <version>1.13.2</version>
            <scope>provided</scope>
        </dependency>
        <dependency>
            <groupId>org.apache.flink</groupId>
            <artifactId>flink-streaming-Java_2.12</artifactId>
            <version>1.13.2</version>
            <scope>provided</scope>
        </dependency>

        <!-- 连接器依赖 -->
        <dependency>
            <groupId>org.apache.flink</groupId>
            <artifactId>flink-connector-base</artifactId>
            <version>1.13.2</version>
        </dependency>

        <!-- Kafka连接器依赖 -->
        <dependency>
            <groupId>org.apache.flink</groupId>
            <artifactId>flink-connector-kafka_2.12</artifactId>
            <version>1.13.2</version>
        </dependency>
        <dependency>
            <groupId>org.apache.flink</groupId>
            <artifactId>flink-sql-connector-kafka_2.12</artifactId>
            <version>1.13.2</version>
        </dependency>
        <dependency>
            <groupId>org.apache.kafka</groupId>
            <artifactId>kafka-clients</artifactId>
            <version>2.4.1</version>
        </dependency>

        <!-- Hive连接器依赖 -->
        <dependency>
            <groupId>org.apache.flink</groupId>
            <artifactId>flink-connector-hive_2.12</artifactId>
            <version>1.13.2</version>
        </dependency>

        <!-- Hive依赖 -->
        <dependency>
            <groupId>org.apache.hive</groupId>
            <artifactId>hive-exec</artifactId>
            <version>3.1.2</version>
```

```xml
</dependency>
<dependency>
    <groupId>org.apache.flink</groupId>
    <artifactId>flink-sql-connector-hive-3.1.2_2.12</artifactId>
    <version>1.13.2</version>
</dependency>

<!-- 表依赖 -->
<dependency>
    <groupId>org.apache.flink</groupId>
    <artifactId>flink-table-api-Java-bridge_2.12</artifactId>
    <version>1.13.2</version>
</dependency>
<dependency>
    <groupId>org.apache.flink</groupId>
    <artifactId>flink-table-planner-blink_2.12</artifactId>
    <version>1.13.2</version>
</dependency>
<dependency>
    <groupId>org.apache.flink</groupId>
    <artifactId>flink-table-common</artifactId>
    <version>1.13.2</version>
</dependency>

<dependency>
    <groupId>org.apache.flink</groupId>
    <artifactId>flink-csv</artifactId>
    <version>1.13.2</version>
</dependency>
<dependency>
    <groupId>org.apache.flink</groupId>
    <artifactId>flink-json</artifactId>
    <version>1.13.2</version>
</dependency>

<!-- 兼容Hadoop -->
<dependency>
    <groupId>org.apache.flink</groupId>
    <artifactId>flink-shaded-hadoop-3</artifactId>
    <version>3.1.1.7.2.9.0-173-9.0</version>
    <scope>provided</scope>
</dependency>
```

注意：如果 flink-shaded-hadoop-3-xxx-xx-xx 无法通过 Maven 正确加载，应手工将其添加到项目依赖中。

Flink 集成 Hive 时当不支持嵌入式 Metastore 时，需要对 Hive 的配置做出修改，在

hive-site.xml 文件中配置 hive.metastore.uris，并运行一个 Hive Metastore。

配置 hive.metastore.uris 的内容如下：

```xml
<property>
    <name>hive.metastore.uris</name>
    <value>thrift://localhost:9083</value>
</property>
```

因此，在项目的 resources 目录下，创建 hive-site.xml 文件，并编辑内容如下：

```xml
<?xml version="1.0" encoding="UTF-8" standalone="no"?>
<?xml-stylesheet type="text/xsl" href="configuration.xsl"?>
<configuration>
    <property>
        <name>hive.metastore.warehouse.dir</name>
        <value>hdfs://localhost:8020/user/hive/warehouse</value>
    </property>
    <property>
        <name>spark.sql.hive.metastore.version</name>
        <value>3.1.2</value>
    </property>
    <property>
        <name>hive.metastore.uris</name>
        <value>thrift://localhost:9083</value>
    </property>

    <!-- 配置 Hive Metastore:MySQL 连接信息 -->
    <property>
        <name>Javax.jdo.option.ConnectionURL</name>
        <value>JDBC:mysql://localhost:3306/hive</value>
    </property>
    <property>
        <name>Javax.jdo.option.ConnectionDriverName</name>
        <value>com.mysql.JDBC.Driver</value>
    </property>
    <property>
        <name>Javax.jdo.option.ConnectionUserName</name>
        <value>root</value>
    </property>
    <property>
        <name>Javax.jdo.option.ConnectionPassword</name>
        <value>123456</value>
    </property>
</configuration>
```

注意：应将其中的 IP 地址和数据库连接信息改为读者自己的连接信息。

运行 hive metastore，命令如下：

```
$ ./bin/hive -- service metastore
```

7.2.3 创建执行环境

程序的第 1 步是创建执行环境。
Scala 代码如下：

```scala
import org.apache.flink.streaming.api.scala.StreamExecutionEnvironment
import org.apache.flink.table.api.EnvironmentSettings
import org.apache.flink.table.api.TableEnvironment
import org.apache.flink.table.api.bridge.scala.StreamTableEnvironment

val env = StreamExecutionEnvironment.getExecutionEnvironment
env.enableCheckpointing(20000).setParallelism(2)

val settings = EnvironmentSettings
  .newInstance()
  .inStreamingMode()
  .build()

val tEnv = StreamTableEnvironment.create(env, settings)

tEnv.getConfig
  .getConfiguration
  .set(ExecutionCheckpointingOptions.CHECKPOINTING_MODE,    CheckpointingMode.EXACTLY_ONCE)

tEnv.getConfig
  .getConfiguration
  .set(ExecutionCheckpointingOptions.CHECKPOINTING_INTERVAL, Duration.ofSeconds(20))
```

Java 代码如下：

```java
import org.apache.flink.table.api.EnvironmentSettings;
import org.apache.flink.table.api.TableEnvironment;
import org.apache.flink.streaming.api.environment.StreamExecutionEnvironment;
import org.apache.flink.table.api.bridge.java.StreamTableEnvironment;
```

```
StreamExecutionEnvironment env = 
            StreamExecutionEnvironment.getExecutionEnvironment();
env.enableCheckpointing(1000).setParallelism(3);

EnvironmentSettings settings = EnvironmentSettings
    .newInstance()
    .inStreamingMode()
    .build();

StreamTableEnvironment tEnv = StreamTableEnvironment.create(env, settings);

tEnv.getConfig()
    .getConfiguration()
    .set(ExecutionCheckpointingOptions.CHECKPOINTING_MODE, CheckpointingMode.EXACTLY_ONCE);

tEnv.getConfig()
    .getConfiguration()
    .set(ExecutionCheckpointingOptions.CHECKPOINTING_INTERVAL, Duration.ofSeconds(20));
```

在 Flink 1.13.2 的 Table/SQL API 中，FileSystem 连接器是由 StreamingFileSink 组件的一个增强版本（名为 StreamingFileWriter）实现的。只有 checkpoint 检查点成功，StreamingFileSink 写入的文件才会从挂起（pending）状态变为完成（finished）状态，从而允许它们在下游安全地被读取，所以在上面的代码中必须开启检查点并设置一个合理的间隔。

7.2.4 注册 HiveCatalog

接下来，需要在 Hive Metastore 中注册 HiveCatalog。

Scala 代码如下：

```
val catalogName    = "myhive"
val defaultDatabase = "default"
val hiveConfDir    = "/home/hduser/bigdata/hive-3.1.2/conf"

val catalog = new HiveCatalog(catalogName, defaultDatabase, hiveConfDir)
tEnv.registerCatalog(catalogName, catalog)

//将 HiveCatalog 设置为会话的当前目录
tEnv.useCatalog(catalogName)
```

Java 代码如下：

```
String catalogName     = "myhive";
String defaultDatabase = "default";
```

```
    String hiveConfDir     = "/home/hduser/bigdata/hive-3.1.2/conf";

    HiveCatalog catalog = new HiveCatalog(catalogName, defaultDatabase, hiveConfDir);
    tEnv.registerCatalog(catalogName, catalog);

    //将 HiveCatalog 设置为会话的当前目录
    tEnv.useCatalog(catalogName);
```

7.2.5 创建 Kafka 流表

Kafka 主题存储 JSON 格式的日志信息,并在构建表时计算事件时间和水印列。为此,需要创建 Kafka 流源表。

Scala 代码如下:

```
//创建 Kafka 源表
tEnv.executeSql(
    """
      |CREATE TABLE stream_db.user_behavior_kafka (
      | user_id bigint,
      | item_id bigint,
      | category_id bigint,
      | behavior string,
      | ts bigint,
      | procTime AS PROCTIME(),
      | eventTime AS TO_TIMESTAMP(FROM_UNIXTIME(ts, 'yyyy-MM-dd HH:mm:ss')),
      | WATERMARK FOR eventTime AS eventTime - INTERVAL '5' SECONDS
      |)
      |with(
      | 'connector' = 'kafka',
      | 'topic' = 'user_behavior',
      | 'properties.Bootstrap.servers' = 'xueai8:9092',
      | 'properties.group.id' = 'testGroup',
      | 'scan.startup.mode' = 'latest-offset',
      | 'format' = 'csv',
      | 'csv.ignore-parse-errors' = 'true',
      | 'csv.field-delimiter' = ','
      |)
    """.stripMargin
)
```

Java 代码如下:

```
//创建 Kafka 源表
tEnv.executeSql(
```

```
        "CREATE TABLE stream_db.user_behavior_kafka (" +
        " user_id bigint," +
        " item_id bigint," +
        " category_id bigint," +
        " behavior string," +
        " ts bigint," +
        " procTime AS PROCTIME()," +
        " eventTime AS TO_TIMESTAMP(FROM_UNIXTIME(ts, 'yyyy-MM-dd HH:mm:ss'))," +
        " WATERMARK FOR eventTime AS eventTime - INTERVAL '5' SECONDS" +
        " )" +
        " with(" +
        " 'connector' = 'kafka'," +
        " 'topic' = 'user_behavior'," +
        " 'properties.Bootstrap.servers' = 'xueai8:9092'," +
        " 'properties.group.id' = 'testGroup'," +
        " 'scan.startup.mode' = 'latest-offset'," +
        " 'format' = 'csv'," +
        " 'csv.ignore-parse-errors' = 'true'," +
        " 'csv.field-delimiter' = ','" +
        " )"
);
```

HiveCatalog 之前已经注册过,所以用户可以在 Hive 中看到创建的 Kafka 流表的元数据。在 Hive CLI 中查看,执行的命令如下:

```
hive> DESCRIBE FORMATTED stream_db.user_behavior_kafka;
```

可以看到元数据信息(注意,该表没有事实上的列),信息如下:

```
# col_name              data_type               comment

# Detailed Table Information
Database:               stream_db
OwnerType:              USER
Owner:                  null
CreateTime:             Wed Sep 15 10:58:44 CST 2021
LastAccessTime:         UNKNOWN
Retention:              0
Location:               hdfs://xueai8:8020/user/hive/warehouse/stream_db.db/user_behavior_kafka
Table Type:             MANAGED_TABLE
Table Parameters:
        flink.connector         kafka
        flink.csv.field-delimiter       ,
        flink.csv.ignore-parse-errors true
```

```
    flink.format          csv
    flink.properties.Bootstrap.servers     localhost:9092
    flink.properties.group.id        testGroup
    flink.scan.startup.mode latest-offset
    flink.schema.0.data-type         BIGINT
    flink.schema.0.name      user_id
    flink.schema.1.data-type         BIGINT
    flink.schema.1.name      item_id
    flink.schema.2.data-type         BIGINT
    flink.schema.2.name      category_id
    flink.schema.3.data-type         VARCHAR(2147483647)
    flink.schema.3.name      behavior
    flink.schema.4.data-type         BIGINT
    flink.schema.4.name      ts
    flink.schema.5.data-type         TIMESTAMP(3) WITH LOCAL TIME ZONE NOT NULL
    flink.schema.5.expr      PROCTIME()
    flink.schema.5.name      procTime
    flink.schema.6.data-type         TIMESTAMP(3)
    flink.schema.6.expr      TO_TIMESTAMP(FROM_UNIXTIME('ts', 'yyyy-MM-dd HH:mm:ss'))
    flink.schema.6.name      eventTime
    flink.schema.watermark.0.rowtime        eventTime
    flink.schema.watermark.0.strategy.data-type    TIMESTAMP(3)
    flink.schema.watermark.0.strategy.expr 'eventTime' - INTERVAL '5' SECOND
    flink.topic           user_behavior
    transient_lastDdlTime 1631674724

# Storage Information
SerDe Library:              org.apache.hadoop.hive.serde2.lazy.LazySimpleSerDe
InputFormat:                org.apache.hadoop.mapred.TextInputFormat
OutputFormat:               org.apache.hadoop.hive.ql.io.IgnoreKeyTextOutputFormat
Compressed:                 No
Num Buckets:                -1
Bucket Columns:             []
Sort Columns:               []
Storage Desc Params:
    serialization.format  1
Time taken: 1.282 seconds, Fetched: 51 row(s)
```

7.2.6 创建 Hive 表

Flink SQL 提供了符合 HiveQL 语法标准的 DDL,以此来指定 SqlDialect。

为了观察结果,下面的表使用了 days/hours/minutes 的三级分区,在实践中可能没有那么细粒度(10min 甚至 1h 的分区可能更合适)。

Scala 代码如下:

```scala
tableEnv.getConfig.setSqlDialect(SqlDialect.HIVE)  //切换到 HiveQL

tableEnv.executeSql("CREATE DATABASE IF NOT EXISTS hive_tmp")
tableEnv.executeSql("DROP TABLE IF EXISTS hive_tmp.analytics_access_log_hive")

tEnv.executeSql(
    """
      |CREATE TABLE hive_db.user_behavior_hive (
      |    user_id bigint,
      |    item_id bigint,
      |    category_id bigint,
      |    behavior string,
      |    ts bigint
      |) PARTITIONED BY (
      |    ts_date STRING,
      |    ts_hour STRING,
      |    ts_minute STRING
      |) STORED AS PARQUET
      |TBLPROPERTIES (
      |    'sink.partition-commit.trigger' = 'partition-time',
      |    'sink.partition-commit.delay' = '1 min',
      |    'sink.partition-commit.policy.kind' = 'metastore,success-file',
      |    'partition.time-extractor.timestamp-pattern' = '$ts_date $ts_hour:$ts_minute:00'
      |)
    """.stripMargin
)
```

Java 代码如下：

```java
tEnv.getConfig().setSqlDialect(SqlDialect.HIVE);   //切换为 HiveQL

tEnv.executeSql("CREATE DATABASE IF NOT EXISTS hive_db");
tEnv.executeSql("DROP TABLE IF EXISTS hive_db.user_behavior_hive");

tEnv.executeSql(
    "CREATE TABLE hive_db.user_behavior_hive (" +
    "   user_id bigint," +
    "   item_id bigint," +
    "   category_id bigint," +
    "   behavior string," +
    "   ts bigint" +
    ") PARTITIONED BY (" +
    "   ts_date STRING," +
    "   ts_hour STRING," +
```

```
"    ts_minute STRING" +
") STORED AS PARQUET\n" +
" TBLPROPERTIES (" +
"    'sink.partition-commit.trigger' = 'partition-time'," +
"    'sink.partition-commit.delay' = '1 min'," +
"    'sink.partition-commit.policy.kind' = 'metastore,success-file'," +
"    'partition.time-extractor.timestamp-pattern' = '$ts_date $ts_hour:$ts_minute:00'" +
")"
);
```

Hive 表的参数复用了 SQL FileSystem Connector 的参数,与分区提交密切相关。下面简单介绍上述 4 个参数。

(1) sink.partition-commit.trigger:触发分区提交的时间特征。默认值为处理时间,所以这里使用分区时间 partition-time,它是根据分区时间戳提交的,分区时间戳是分区中数据的事件时间。

(2) partition.time-extractor.timestamp-pattern:分区时间戳的抽取格式。需要写成 yyyy-MM-dd HH:mm:ss,并替换为 Hive 表中对应的分区字段。显然,Hive 表的分区字段值来自流表中定义的事件时间。

(3) sink.partition-commit.delay:触发分区提交中的时延。当时间特征被设置为 partition-time 时,分区将不会实际提交,直到水印时间戳大于分区创建时间加上这个时延。该参数值最好与分区粒度相同,例如当 Hive 表按 1h 进行分区时,可设置为 1h,当按 10min 进行分区时,可设置为 10min。

(4) sink.partition-commit.policy.kind:分区提交策略可以理解为使分区在下游可见的附加操作。Metastore 表示更新 Hive Metastore 中的表元数据,success-file 表示 create_in partitionSUCCESS 标签文件。先更新 Metastore (添加分区),然后写入成功文件 '_SUCCESS'。

在 Hive 中可以查看元数据(程序执行之后才会创建),命令如下:

```
hive> DESCRIBE FORMATTED hive_db.user_behavior_hive;
```

可以看到输出的元数据信息,信息如下:

```
# col_name            data_type               comment
user_id               bigint
item_id               bigint
category_id           bigint
behavior              string
ts                    bigint
```

```
# Partition Information
# col_name              data_type           comment
ts_date                 string
ts_hour                 string
ts_minute               string

# Detailed Table Information
Database:               hive_db
OwnerType:              USER
Owner:                  null
CreateTime:             Wed Sep 15 10:58:46 CST 2021
LastAccessTime:         UNKNOWN
Retention:              0
Location:
hdfs://xueai8:8020/user/hive/warehouse/hive_db.db/user_behavior_hive
Table Type:             MANAGED_TABLE
Table Parameters:
        COLUMN_STATS_ACCURATE   {\"BASIC_STATS\":\"true\"}
        bucketing_version       2
        numFiles                0
        numPartitions           0
        numRows                 0
        partition.time-extractor.timestamp-pattern        $ts_date $ts_hour:$ts_minute:00
        rawDataSize             0
        sink.partition-commit.delay       1 min
        sink.partition-commit.policy.kind     metastore,success-file
        sink.partition-commit.trigger     partition-time
        totalSize               0
        transient_lastDdlTime   1631674726

# Storage Information
SerDe Library: org.apache.hadoop.hive.ql.io.parquet.serde.ParquetHiveSerDe
InputFormat:   org.apache.hadoop.hive.ql.io.parquet.MapredParquetInputFormat
OutputFormat:  org.apache.hadoop.hive.ql.io.parquet.MapredParquetOutputFormat
Compressed:             No
Num Buckets:            -1
Bucket Columns:         []
Sort Columns:           []
Storage Desc Params:
        serialization.format    1
Time taken: 0.558 seconds, Fetched: 46 row(s)
```

7.2.7 流写 Hive 表

接下来，实现读取 Kafka 主题并实时写入 Hive 数据仓库的功能。注意，在本例中，流表中的事件时间被转换为 Hive 的分区。

Scala 代码如下：

```
tEnv.getConfig.setSqlDialect(SqlDialect.DEFAULT)              //切换为标准 SQL

tEnv.executeSql(
    """
        |INSERT INTO hive_db.user_behavior_hive
        |SELECT
        |  user_id,item_id,category_id,behavior,ts,
        |  DATE_FORMAT(eventTime,'yyyy-MM-dd') as ts_date,
        |  DATE_FORMAT(eventTime,'HH') as ts_hour,
        |  DATE_FORMAT(eventTime,'mm') as ts_minute
        |FROM stream_db.user_behavior_kafka
        |WHERE item_id > 0
    """.stripMargin
)
```

Java 代码如下：

```
tEnv.getConfig().setSqlDialect(SqlDialect.DEFAULT); //切换为标准 SQL

tEnv.executeSql(
        "INSERT INTO hive_db.user_behavior_hive\n" +
        "SELECT\n" +
        "   user_id,item_id,category_id,behavior,ts," +
        "   DATE_FORMAT(eventTime,'yyyy-MM-dd') as ts_date," +
        "   DATE_FORMAT(eventTime,'HH') as ts_hour," +
        "   DATE_FORMAT(eventTime,'mm') as ts_minute\n" +
        "FROM stream_db.user_behavior_kafka\n" +
        "WHERE item_id > 0"
);
```

查看流接收器的结果，打开浏览器，访问地址 http://192.168.190.133:9870/。可以看到按小时分区存储，如图 7-5 所示。

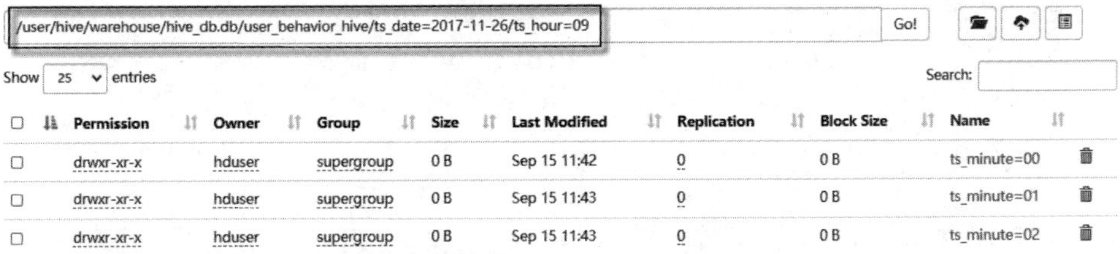
/user/hive/warehouse/hive_db.db/user_behavior_hive/ts_date=2017-11-26/ts_hour=09
Go!
Show 25 entries
Search:
Permission
Owner
Group
Size
Last Modified
Replication
Block Size
Name
drwxr-xr-x hduser supergroup 0 B Sep 15 11:42 0 0 B ts_minute=00
drwxr-xr-x hduser supergroup 0 B Sep 15 11:43 0 0 B ts_minute=01
drwxr-xr-x hduser supergroup 0 B Sep 15 11:43 0 0 B ts_minute=02

图 7-5 以小时为分区的 Hive 存储

可以看到按分钟分区存储，如图 7-6 所示。

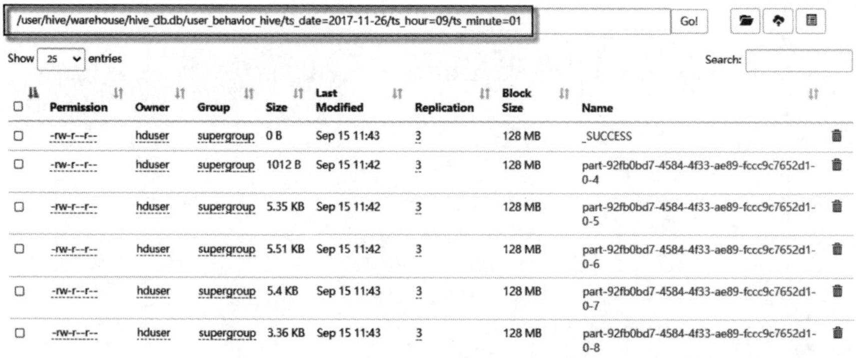

/user/hive/warehouse/hive_db.db/user_behavior_hive/ts_date=2017-11-26/ts_hour=09/ts_minute=01 Go!

Show 25 entries Search:

	Permission	Owner	Group	Size	Last Modified	Replication	Block Size	Name
☐	-rw-r--r--	hduser	supergroup	0 B	Sep 15 11:43	3	128 MB	_SUCCESS
☐	-rw-r--r--	hduser	supergroup	1012 B	Sep 15 11:42	3	128 MB	part-92fb0bd7-4584-4f33-ae89-fccc9c7652d1-0-4
☐	-rw-r--r--	hduser	supergroup	5.35 KB	Sep 15 11:42	3	128 MB	part-92fb0bd7-4584-4f33-ae89-fccc9c7652d1-0-5
☐	-rw-r--r--	hduser	supergroup	5.51 KB	Sep 15 11:42	3	128 MB	part-92fb0bd7-4584-4f33-ae89-fccc9c7652d1-0-6
☐	-rw-r--r--	hduser	supergroup	5.4 KB	Sep 15 11:43	3	128 MB	part-92fb0bd7-4584-4f33-ae89-fccc9c7652d1-0-7
☐	-rw-r--r--	hduser	supergroup	3.36 KB	Sep 15 11:43	3	128 MB	part-92fb0bd7-4584-4f33-ae89-fccc9c7652d1-0-8

图 7-6 以分钟为分区的 Hive 存储

因为并行度是 2,所以每次写都会生成两个文件。当写入分区中的所有数据时,同时生成一个__success 文件。如果它是一个正在写入的分区,则将看到.inprogress 文件。

检查 Hive 查询以确保数据是正确的(需要确保已经启动了 YARN 集群),执行以下 Hive SQL 语句:

```
hive> SELECT from_UNIXtime(min(cast(ts AS BIGINT))), from_UNIXtime(max(cast(ts AS BIGINT)))
    > FROM hive_db.user_behavior_hive
    > WHERE ts_date = '2017-11-26' AND ts_hour = '09' AND ts_minute = '01';
```

查询结果如下:

```
2017-11-26 01:01:00     2017-11-26 01:01:59
Time taken: 135.823 seconds, Fetched: 1 row(s)
```

执行过程如图 7-7 所示。

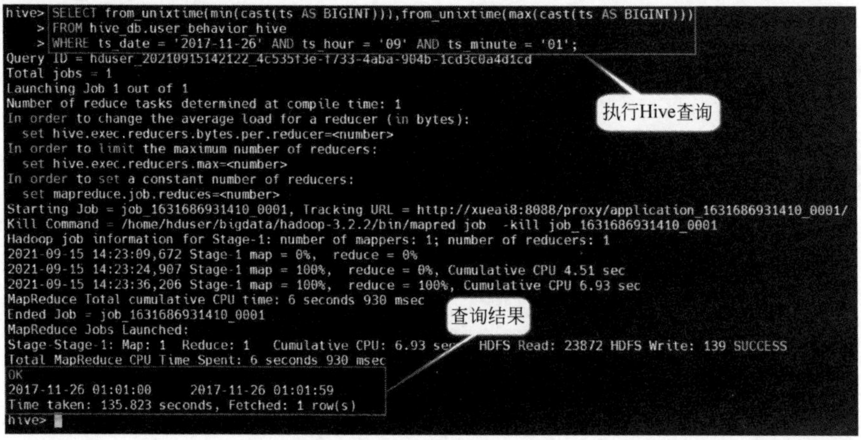

```
hive> SELECT from_unixtime(min(cast(ts AS BIGINT))),from_unixtime(max(cast(ts AS BIGINT)))
    > FROM hive_db.user_behavior_hive
    > WHERE ts_date = '2017-11-26' AND ts_hour = '09' AND ts_minute = '01';
Query ID = hduser_20210915142122_4c535f3e-f733-4aba-904b-1cd3c0a4d1cd
Total jobs = 1
Launching Job 1 out of 1
Number of reduce tasks determined at compile time: 1
In order to change the average load for a reducer (in bytes):
  set hive.exec.reducers.bytes.per.reducer=<number>
In order to limit the maximum number of reducers:
  set hive.exec.reducers.max=<number>
In order to set a constant number of reducers:
  set mapreduce.job.reduces=<number>
Starting Job = job_1631686931410_0001, Tracking URL = http://xueai8:8088/proxy/application_1631686931410_0001/
Kill Command = /home/hduser/bigdata/hadoop-3.2.2/bin/mapred job  -kill job_1631686931410_0001
Hadoop job information for Stage-1: number of mappers: 1; number of reducers: 1
2021-09-15 14:23:09,672 Stage-1 map = 0%,  reduce = 0%
2021-09-15 14:23:24,907 Stage-1 map = 100%,  reduce = 0%, Cumulative CPU 4.51 sec
2021-09-15 14:23:36,206 Stage-1 map = 100%,  reduce = 100%, Cumulative CPU 6.93 sec
MapReduce Total cumulative CPU time: 6 seconds 930 msec
Ended Job = job_1631686931410_0001
MapReduce Jobs Launched:
Stage-Stage-1: Map: 1  Reduce: 1   Cumulative CPU: 6.93 sec   HDFS Read: 23872 HDFS Write: 139 SUCCESS
Total MapReduce CPU Time Spent: 6 seconds 930 msec
OK
2017-11-26 01:01:00     2017-11-26 01:01:59
Time taken: 135.823 seconds, Fetched: 1 row(s)
hive>
```


图 7-7 Hive 查询执行过程

7.2.8 动态读取 Hive 流表

为了使 Hive 表成为流源，需要启用动态表选项，并通过表提示为 Hive 数据流指定参数。例如，使用 Hive 简单计算商品 PV 示例。

Scala 代码如下：

```scala
tableEnv.getConfig
  .getConfiguration
  .setBoolean(TableConfigOptions.TABLE_DYNAMIC_TABLE_OPTIONS_ENABLED, true)

//分钟级查询
val result = tEnv.sqlQuery(
    """
      |SELECT behavior,count(1) AS behavior_cnt
      |FROM hive_db.user_behavior_hive
      |/* + OPTIONS(
      |  'streaming-source.enable' = 'true',
      |  'streaming-source.monitor-interval' = '1 min',
      |  'streaming-source.consume-start-offset' = '2017-11-26 09:01:00'
      |) */
      |WHERE behavior = 'buy'
      |GROUP BY behavior
    """.stripMargin
)

tEnv
    .toChangelogStream(result)
    .executeAndCollect()
    .foreach(println)
```

Java 代码如下：

```java
tEnv.getConfig()
  .getConfiguration()
  .setBoolean(TableConfigOptions.TABLE_DYNAMIC_TABLE_OPTIONS_ENABLED, true);

//分钟级查询
Table result = tEnv.sqlQuery(
    "SELECT behavior,count(1) AS behavior_cnt\n" +
    "FROM hive_db.user_behavior_hive\n" +
    "/* + OPTIONS(\n" +
    "  'streaming-source.enable' = 'true',\n" +
    "  'streaming-source.monitor-interval' = '1 min',\n" +
    "  'streaming-source.consume-start-offset' = '2017-11-26 09:01:00'\n" +
```

```
        ") * /\n" +
        "WHERE behavior = 'buy'\n" +
        "GROUP BY behavior"
);

tEnv
    .toChangelogStream(result)
    .executeAndCollect()
    .forEachRemaining(System.out::println);
```

在上面的代码中,3个表提示参数的含义解释如下。

(1) streaming-source.enable:设置为true,表示Hive表可以作为Source表使用。

(2) streaming-source.monitor-interval:感知Hive表中新增数据的周期,应设置为1min以上。对于分区表,监视新分区的生成以增量读取数据。

(3) streaming-source.consume-start-offset:开始消费的时间戳,也需要写成yyyy-MM-dd HH:mm:ss。

最后,由于SQL语句中有ORDER BY和LIMIT逻辑,需要调用toRetractStream()方法将其转换为一个undo流以输出结果。

7.2.9 完整示例代码

到目前为止,已经完成了如下两个流程序。

(1) FlinkHiveJob:读取Kafka主题并实时写入Hive数仓的Flink流程序。

(2) HiveAnalysisJob:将上一个作业写入的Hive表作为流源表,进行每分钟聚合统计。

下面是本案例两个程序的完整代码。

1. FlinkHiveJob

FlinkHiveJob类的完整实现代码如下。

Scala代码如下:

```
//第 7 章/FlinkHiveJob.scala

import java.time.Duration
import org.apache.flink.streaming.api.CheckpointingMode
import org.apache.flink.streaming.api.environment.ExecutionCheckpointingOptions
import org.apache.flink.streaming.api.scala._
import org.apache.flink.table.api.{EnvironmentSettings, SqlDialect}
import org.apache.flink.table.api.bridge.scala.StreamTableEnvironment
import org.apache.flink.table.catalog.hive.HiveCatalog

object FlinkHiveJob {
```

```scala
def main(args: Array[String]) {
  //设置流执行环境
  val env = StreamExecutionEnvironment.getExecutionEnvironment

  env.setParallelism(2)

  val settings = EnvironmentSettings
    .newInstance()
    .inStreamingMode()
    .build()

  val tEnv = StreamTableEnvironment.create(env, settings)

  tEnv.getConfig
    .getConfiguration
    .set(ExecutionCheckpointingOptions.CHECKPOINTING_MODE, CheckpointingMode.EXACTLY_ONCE)

  tEnv.getConfig
    .getConfiguration
    .set(ExecutionCheckpointingOptions.CHECKPOINTING_INTERVAL, Duration.ofSeconds(20))

  //在 Hive Metastore 中注册 HiveCatalog
  val catalogName = "myhive"
  val defaultDatabase = "default"
  val hiveConfDir = "./src/main/resources"

  val catalog = new HiveCatalog(catalogName, defaultDatabase, hiveConfDir)
  tEnv.registerCatalog(catalogName, catalog)

  //将 HiveCatalog 设置为会话的当前目录
  tEnv.useCatalog(catalogName)

  //创建 Kafka 流表
  tEnv.executeSql("CREATE DATABASE IF NOT EXISTS stream_db")
  tEnv.executeSql("DROP TABLE IF EXISTS stream_db.user_behavior_kafka")

  //创建 Kafka 源表
  tEnv.executeSql(
    """
      |CREATE TABLE stream_db.user_behavior_kafka (
      |  user_id bigint,
      |  item_id bigint,
      |  category_id bigint,
      |  behavior string,
      |  ts bigint,
```

```
        |    procTime AS PROCTIME(),
        |    eventTime AS TO_TIMESTAMP(FROM_UNIXTIME(ts, 'yyyy-MM-dd HH:mm:ss')),
        |    WATERMARK FOR eventTime AS eventTime - INTERVAL '5' SECONDS
        |)
        |with(
        |    'connector' = 'kafka',
        |    'topic' = 'user_behavior',
        |    'properties.Bootstrap.servers' = 'localhost:9092',
        |    'properties.group.id' = 'testGroup',
        |    'scan.startup.mode' = 'latest-offset',
        |    'format' = 'csv',
        |    'csv.ignore-parse-errors' = 'true',
        |    'csv.field-delimiter' = ','
        |)
      """.stripMargin)

    //在Hive中可以查看元数据
    //hive> DESCRIBE FORMATTED stream_db.user_behavior_kafka;

    //创建Hive Sink表
    tEnv.getConfig.setSqlDialect(SqlDialect.HIVE)

    tEnv.executeSql("CREATE DATABASE IF NOT EXISTS hive_db")
    tEnv.executeSql("DROP TABLE IF EXISTS hive_db.user_behavior_hive")

    tEnv.executeSql(
      """
        |CREATE TABLE hive_db.user_behavior_hive (
        |    user_id bigint,
        |    item_id bigint,
        |    category_id bigint,
        |    behavior string,
        |    ts bigint
        |) PARTITIONED BY (
        |    ts_date STRING,
        |    ts_hour STRING,
        |    ts_minute STRING
        |) STORED AS PARQUET
        |TBLPROPERTIES (
        |    'sink.partition-commit.trigger' = 'partition-time',
        |    'sink.partition-commit.delay' = '1 min',
        |    'sink.partition-commit.policy.kind' = 'metastore,success-file',
        |    'partition.time-extractor.timestamp-pattern' = '$ts_date $ts_hour:$ts_minute:00'
        |)
      """.stripMargin)
```

```
    )

    //在Hive中可以查看元数据
    //hive> DESCRIBE FORMATTED hive_db.user_behavior_hive;

    //注意,流表中的事件时间被转换为Hive的分区
    tEnv.getConfig.setSqlDialect(SqlDialect.DEFAULT) //切换为标准SQL

    tEnv.executeSql(
      """
        |INSERT INTO hive_db.user_behavior_hive
        |SELECT
        |   user_id,item_id,category_id,behavior,ts,
        |   DATE_FORMAT(eventTime,'yyyy-MM-dd') as ts_date,
        |   DATE_FORMAT(eventTime,'HH') as ts_hour,
        |   DATE_FORMAT(eventTime,'mm') as ts_minute
        |FROM stream_db.user_behavior_kafka
        |WHERE item_id > 0
      """.stripMargin
    )
  }
}
```

Java 代码如下：

```java
//第7章/FlinkHiveJob.java

import org.apache.flink.streaming.api.CheckpointingMode;
import org.apache.flink.streaming.api.environment.ExecutionCheckpointingOptions;
import org.apache.flink.streaming.api.environment.StreamExecutionEnvironment;
import org.apache.flink.table.api.EnvironmentSettings;
import org.apache.flink.table.api.SqlDialect;
import org.apache.flink.table.api.bridge.java.StreamTableEnvironment;
import org.apache.flink.table.catalog.hive.HiveCatalog;
import java.time.Duration;

public class FlinkHiveJob {

    public static void main(String[] args) throws Exception {
        //设置流执行环境
        final StreamExecutionEnvironment env =
                StreamExecutionEnvironment.getExecutionEnvironment();

        env.setParallelism(2);
```

```java
        EnvironmentSettings settings = EnvironmentSettings
            .newInstance()
            .inStreamingMode()
            .build();

        StreamTableEnvironment tEnv = StreamTableEnvironment.create(env, settings);

        tEnv.getConfig()
            .getConfiguration()
            .set(ExecutionCheckpointingOptions.CHECKPOINTING_MODE,
CheckpointingMode.EXACTLY_ONCE);

        tEnv.getConfig()
            .getConfiguration()
            .set(ExecutionCheckpointingOptions.CHECKPOINTING_INTERVAL,
Duration.ofSeconds(20));

        //注册HiveCatalog
        String catalogName = "myhive";
        String defaultDatabase = "default";
        String hiveConfDir = "./src/main/resources";

        HiveCatalog catalog = new HiveCatalog(catalogName, defaultDatabase, hiveConfDir);
        tEnv.registerCatalog(catalogName, catalog);

        //将HiveCatalog设置为会话的当前目录
        tEnv.useCatalog(catalogName);

        //创建Kafka流表
        tEnv.executeSql("CREATE DATABASE IF NOT EXISTS stream_db");
        tEnv.executeSql("DROP TABLE IF EXISTS stream_db.user_behavior_kafka");

        //创建Kafka源表
        tEnv.executeSql(
            "CREATE TABLE stream_db.user_behavior_kafka (" +
            "    user_id bigint," +
            "    item_id bigint," +
            "    category_id bigint," +
            "    behavior string," +
            "    ts bigint," +
            "    procTime AS PROCTIME()," +
            "    eventTime AS TO_TIMESTAMP(FROM_UNIXTIME(ts, 'yyyy-MM-dd HH:mm:ss'))," +
            "    WATERMARK FOR eventTime AS eventTime - INTERVAL '5' SECONDS" +
            " )" +
            " with(" +
            "    'connector' = 'kafka'," +
```

```
            "  'topic' = 'user_behavior'," +
            "  'properties.Bootstrap.servers' = 'localhost:9092'," +
            "  'properties.group.id' = 'testGroup'," +
            "  'scan.startup.mode' = 'latest-offset'," +
            "  'format' = 'csv'," +
            "  'csv.ignore-parse-errors' = 'true'," +
            "  'csv.field-delimiter' = ','" +
            ")"
);

//在 Hive 中可以查看元数据
//hive> DESCRIBE FORMATTED stream_db.user_behavior_kafka;

//创建 Hive Sink 表
tEnv.getConfig().setSqlDialect(SqlDialect.HIVE);          //切换为 HiveQL 方言

tEnv.executeSql("CREATE DATABASE IF NOT EXISTS hive_db");
tEnv.executeSql("DROP TABLE IF EXISTS hive_db.user_behavior_hive");

tEnv.executeSql(
    "CREATE TABLE hive_db.user_behavior_hive (" +
    "  user_id bigint," +
    "  item_id bigint," +
    "  category_id bigint," +
    "  behavior string," +
    "  ts bigint" +
    ") PARTITIONED BY (" +
    "  ts_date STRING," +
    "  ts_hour STRING," +
    "  ts_minute STRING" +
    ") STORED AS PARQUET\n" +
    " TBLPROPERTIES (" +
    "  'sink.partition-commit.trigger' = 'partition-time'," +
    "  'sink.partition-commit.delay' = '1 min'," +
    "  'sink.partition-commit.policy.kind' = 'metastore,success-file'," +
    "  'partition.time-extractor.timestamp-pattern' = '$ts_date $ts_hour:$ts_minute:00'" +
    ")"
);

//在 Hive 中可以查看元数据
//hive> DESCRIBE FORMATTED hive_db.user_behavior_hive;

//注意,流表中的事件时间被转换为 Hive 的分区
tEnv.getConfig().setSqlDialect(SqlDialect.DEFAULT);       //切换为标准 SQL
```

```
        tEnv.executeSql(
            "INSERT INTO hive_db.user_behavior_hive\n" +
            "SELECT\n" +
            "    user_id,item_id,category_id,behavior,ts," +
            "    DATE_FORMAT(eventTime,'yyyy-MM-dd') as ts_date," +
            "    DATE_FORMAT(eventTime,'HH') as ts_hour," +
            "    DATE_FORMAT(eventTime,'mm') as ts_minute\n" +
            "FROM stream_db.user_behavior_kafka\n" +
            "WHERE item_id > 0"
        );
    }
}
```

2. HiveAnalysisJob

HiveAnalysisJob 类的完整实现代码如下：

Scala 代码如下：

```
//第7章/HiveAnalysisJob.scala

import org.apache.flink.streaming.api.scala._
import org.apache.flink.table.api.EnvironmentSettings
import org.apache.flink.table.api.bridge.scala.StreamTableEnvironment
import org.apache.flink.table.api.config.TableConfigOptions
import org.apache.flink.table.catalog.hive.HiveCatalog

object HiveAnalysisJob {
  def main(args: Array[String]) {
    //设置流执行环境
    val env = StreamExecutionEnvironment.getExecutionEnvironment

    env.setParallelism(2)

    val settings = EnvironmentSettings
      .newInstance()
      .inStreamingMode()
      .build()

    //val tEnv = TableEnvironment.create(setting)
    //val tEnv = StreamTableEnvironment.create(env)
    val tEnv = StreamTableEnvironment.create(env, settings)

    //启用动态表选项
    tEnv.getConfig
      .getConfiguration
```

```scala
      .setBoolean(TableConfigOptions.TABLE_DYNAMIC_TABLE_OPTIONS_ENABLED, true)

    //在 Hive Metastore 中注册 HiveCatalog
    val catalogName = "myhive"
    val defaultDatabase = "default"
    val hiveConfDir = "./src/main/resources"

    val catalog = new HiveCatalog(catalogName, defaultDatabase, hiveConfDir)
    tEnv.registerCatalog(catalogName, catalog)

    //将 HiveCatalog 设置为会话的当前目录
    tEnv.useCatalog(catalogName)

    //分钟级查询
    val result = tEnv.sqlQuery(
      """
        |SELECT behavior,count(1) AS behavior_cnt
        |FROM hive_db.user_behavior_hive
        |/* + OPTIONS(
        |  'streaming-source.enable' = 'true',
        |  'streaming-source.monitor-interval' = '1 min',
        |  'streaming-source.consume-start-offset' = '2017-11-26 09:01:00'
        |) */
        |WHERE behavior = 'buy'
        |GROUP BY behavior
      """.stripMargin
    )

    tEnv
      .toChangelogStream(result)
      .executeAndCollect()
      .foreach(println)
  }
}
```

Java 代码如下：

```
//第 7 章/HiveAnalysisJob.java

import org.apache.flink.streaming.api.environment.StreamExecutionEnvironment;
import org.apache.flink.table.api.EnvironmentSettings;
import org.apache.flink.table.api.Table;
import org.apache.flink.table.api.bridge.java.StreamTableEnvironment;
import org.apache.flink.table.api.config.TableConfigOptions;
import org.apache.flink.table.catalog.hive.HiveCatalog;
```

```java
public class HiveAnalysisJob {

    public static void main(String[] args) throws Exception {
        //设置流执行环境
        final StreamExecutionEnvironment env =
                StreamExecutionEnvironment.getExecutionEnvironment();

        env.setParallelism(2);

        EnvironmentSettings settings = EnvironmentSettings
            .newInstance()
            .inStreamingMode()
            .build();

        StreamTableEnvironment tEnv = StreamTableEnvironment.create(env, settings);

        //启用动态表选项
        tEnv.getConfig()
            .getConfiguration()
            .setBoolean(TableConfigOptions.TABLE_DYNAMIC_TABLE_OPTIONS_ENABLED, true);

        //注册HiveCatalog
        String catalogName = "myhive";
        String defaultDatabase = "default";
        String hiveConfDir = "./src/main/resources";

        HiveCatalog catalog = new HiveCatalog(catalogName, defaultDatabase, hiveConfDir);
        tEnv.registerCatalog(catalogName, catalog);

        //将HiveCatalog设置为会话的当前目录
        tEnv.useCatalog(catalogName);

        //分钟级查询
        Table result = tEnv.sqlQuery(
            "SELECT behavior,count(1) AS behavior_cnt\n" +
            "FROM hive_db.user_behavior_hive\n" +
            "/* + OPTIONS(\n" +
            "    'streaming-source.enable' = 'true',\n" +
            "    'streaming-source.monitor-interval' = '1 min',\n" +
            "    'streaming-source.consume-start-offset' = '2017-11-26 09:01:00'\n" +
            ") */\n" +
            "WHERE behavior = 'buy'\n" +
            "GROUP BY behavior"
        );

        tEnv
```

```
        .toChangelogStream(result)
        .executeAndCollect()
        .forEachRemaining(System.out::println);
    }
}
```

7.2.10 执行步骤

至此,所有的代码已经开发完毕。接下来,建议按以下步骤执行。
(1) 启动 Kafka,创建 user_behavior topic。假设用户现在位于 Kafka 的安装目录下。
打开一个终端,运行 ZooKeeper,命令如下:

```
$ ./bin/zookeeper-server-start.sh ./config/zookeeper.properties
```

另打开一个终端,运行 Kafka 服务器,命令如下:

```
$ ./bin/kafka-server-start.sh ./config/server.properties
```

再打开一个终端,创建 Kafka 主题 user_behavior,命令如下:

```
$ ./bin/kafka-topics.sh --create --zookeeper localhost:2181 --replication-factor 1 --partitions 1 --topic user_behavior
```

查看已经存在的 Topics,命令如下:

```
$ ./bin/kafka-topics.sh --list --zookeeper localhost:2181
```

运行 Kafka Producer Console(不做)。再打开一个终端,执行 Kafka 自带的生产者脚本,从命令行向 Kafka 指定 Topic 发送消息,命令如下:

```
$ ./bin/kafka-console-producer.sh --broker-list localhost:9092 --topic user_behavior
```

运行 Kafka Consumer Console(不做)。再打开一个终端,执行 Kafka 自带的消费者脚本,消费 user_behavior 主题的消息,命令如下:

```
$ ./bin/kafka-console-consumer.sh --Bootstrap-server localhost:9092 --topic user_behavior
```

(2) 启动 Hive Metastore 服务。新打开一个终端窗口,执行的命令如下:

```
$ hive --service metastore
```

(3) 启动 Flink 集群。新打开一个终端窗口,执行的命令如下:

```
$ cd ~/bigdata/flink-1.13.2
$ export HADOOP_CLASSPATH='hadoop classpath'
$ ./bin/start-cluster.sh
```

(4) 先运行 HiveAnalysisJob 程序,执行连续聚合查询。

(5) 然后运行前面已经开发好的 Flink 流 ETL 程序 FlinkHiveJob,读取 Kafka topic 实时用户购买行为数据,实时入仓。

(6) 执行数据生产者脚本,模拟产生实时用户购买行为数据。执行前面已经创建的脚本 streamuserbehavior.sh,命令如下:

```
$ ./streamuserbehavior.sh
```

(7) 查看 HiveAnalysisJob 程序统计结果,结果如图 7-8 所示。

图 7-8 Hive 查询执行过程

7.3 纯 SQL 构建批流一体数仓

在 7.2 节中,使用 Flink Table API 集成 Hive 实现了批流一体数据仓库的构建。实际上,对于数据分析人员,可以直接使用 Flink SQL CLI 来集成 Hive,实现纯 SQL 无代码构建批流一体数据仓库。

7.3.1 使用 Flink SQL 客户端

Flink 的 Table & SQL API 可以处理使用 SQL 语言编写的查询语句,但是这些查询需要嵌入用 Java 或 Scala 编写的表程序中。此外,这些程序在提交到集群之前需要用构建工具打包。这或多或少限制了 Java/Scala 程序员使用 Flink。

Flink SQL 客户端的目的是提供一种简单的方式来编写、调试和将表程序提交到 Flink 集群上,而无须写一行 Java 或 Scala 代码。Flink 支持的 SQL 语言,包括数据定义语言(DDL)、数据操作语言(DML)和查询语言。Flink 的 SQL 支持基于 Apache Calcite 的操作,它实现了 SQL 标准。

Flink SQL 目前支持的 SQL 语句包括:

(1) SELECT (Queries)。

(2) CREATE TABLE,DATABASE,FUNCTION。

(3) DROP TABLE,DATABASE,FUNCTION。

(4) ALTER TABLE,DATABASE,FUNCTION。

(5) INSERT。

Flilnk SQL 客户端命令行界面(CLI)能够在命令行中检索和可视化分布式应用中实时产生的结果。Flink SQL Client 绑定在常规的 Flink 发行版中,因此可以开箱即用。它只需一个可以执行表程序的正在运行的 Flink 集群。使用 Flink SQL 客户端的步骤如下。

(1) 先启动 Flink 集群,命令如下:

```
$ cd ~/bigdata/flink-1.13.2
$ export HADOOP_CLASSPATH=`hadoop classpath`
$ ./bin/start-cluster.sh
```

(2) 再启动 SQL 客户端脚本,命令如下:

```
$ ./bin/sql-client.sh
```

启动界面如图 7-9 所示。

图 7-9 Flink SQL 客户端启动界面

启动 CLI 后,可以使用 HELP 命令列出所有可用的 SQL 语句,如图 7-10 所示。

CLI 支持 3 种模式来维护和可视化结果。

(1) 表格模式(Table Model)将结果物化到内存中,并以常规的分页表表示形式将其可

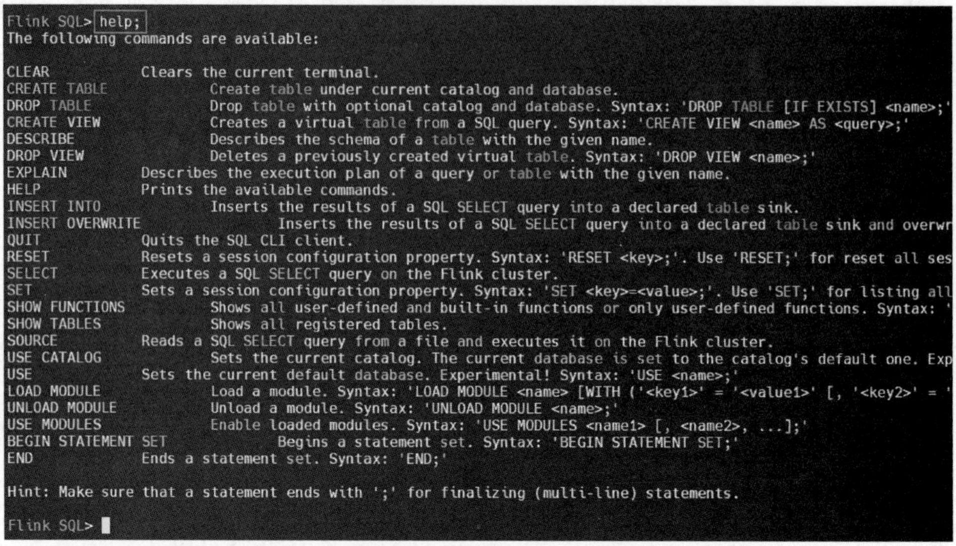

图 7-10　使用 HELP 命令列出所有可用的 SQL 语句

视化。可以通过在 CLI 中执行以下命令来启用它（默认使用表格模式）：

```
Flink SQL> SET sql-client.execution.result-mode=table;
```

（2）变更日志模式（Changelog Model）不物化结果，而是可视化由插入（+）和撤销（-）组成的连续查询产生的结果流。

```
Flink SQL> SET sql-client.execution.result-mode=changelog;
```

（3）表模式（Tableau Model）更像是一种传统的方式，它将以 Tableau 格式直接在屏幕上显示结果。显示内容将受到查询执行类型（execution.type）的影响。

```
Flink SQL> SET sql-client.execution.result-mode=tableau;
```

需要注意，当在流查询中使用 Tableau 模式时，结果将在控制台上连续打印。如果此查询的输入数据是有界的，则在 Flink 处理完所有输入数据后，作业将终止，打印也将自动停止。否则如果想终止一个正在运行的查询，只需按快捷键 Ctrl+C，在本例中，作业和打印将停止。

在构建 SQL 查询原型期间，这 3 种模式都很有用。在所有这些模式中，结果都存储在 SQL 客户端的 Java 堆内存中。为了保持 CLI 界面的响应，Changelog 模式只显示最新的 1000 个更改。表模式允许在仅受可用主存和配置的最大行数（max-table-result-rows）限制的更大结果中进行导航。

注意：在批处理环境中执行的查询只能使用 Table 或 Tableau 模式检索。

为了验证设置和集群连接，可以输入第 1 个 SQL 查询，并按 Enter 键执行它。SQL 查询语句如下：

```
Flink SQL> select 'Hello World';
```

这个查询不需要表源，只产生一个 row 结果。CLI 将从集群中检索结果并对其进行可视化。按 Q 键可以关闭结果视图。

执行过程如图 7-11 所示。

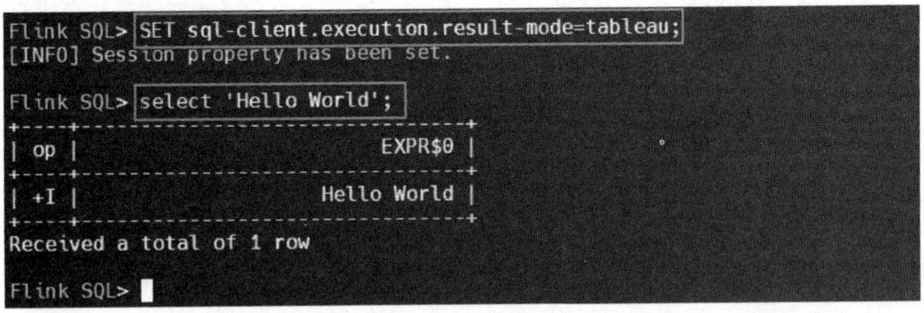

图 7-11　验证设置和集群连接的 SQL 查询执行过程

要退出 Flink SQL CLI 命令行，可执行的命令如下：

```
Flink SQL> exit;
```

7.3.2　集成 Flink SQL CLI 和 Hive

用户可以使用 Flink SQL CLI 来直接连接 Hive，而无须使用 Table API。

1. 依赖添加

将所有 Hive 依赖项（下面的 JAR 列表）添加到 Flink 分布式安装的 /lib 目录下，包括：

（1）mariadb-Java-client-2.1.2.jar。
（2）hive-exec-3.1.2.jar。
（3）flink-connector-hive_2.12-1.13.2.jar。
（4）flink-sql-connector-hive-3.1.2_2.12-1.13.2.jar。
（5）flink-shaded-hadoop-3-uber-3.1.1.7.2.9.0-173-9.0.jar。

如图 7-12 所示。

2. 编辑 hive-site.xml 文件

将 Hive 安装配置文件 hive-site.xml 复制到 Flink 安装的 conf/ 目录下，或者在该位置

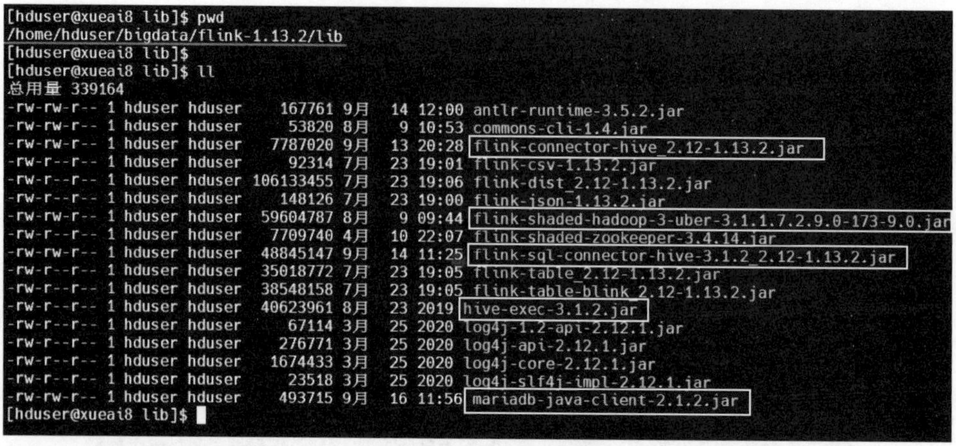

```
[hduser@xueai8 lib]$ pwd
/home/hduser/bigdata/flink-1.13.2/lib
[hduser@xueai8 lib]$
[hduser@xueai8 lib]$ ll
总用量 339164
-rw-rw-r-- 1 hduser hduser     167761 9月  14 12:00 antlr-runtime-3.5.2.jar
-rw-rw-r-- 1 hduser hduser      53820 8月   9 10:53 commons-cli-1.4.jar
-rw-rw-r-- 1 hduser hduser    7787020 9月  13 20:28 flink-connector-hive_2.12-1.13.2.jar
-rw-r--r-- 1 hduser hduser      92314 7月  23 19:01 flink-csv-1.13.2.jar
-rw-r--r-- 1 hduser hduser  106133455 7月  23 19:06 flink-dist_2.12-1.13.2.jar
-rw-r--r-- 1 hduser hduser     148126 7月  23 19:00 flink-json-1.13.2.jar
-rw-rw-r-- 1 hduser hduser   59604787 8月   9 09:44 flink-shaded-hadoop-3-uber-3.1.1.7.2.9.0-173-9.0.jar
-rw-r--r-- 1 hduser hduser    7709740 4月  10 22:07 flink-shaded-zookeeper-3.4.14.jar
-rw-rw-r-- 1 hduser hduser   48845147 9月  14 11:25 flink-sql-connector-hive-3.1.2_2.12-1.13.2.jar
-rw-r--r-- 1 hduser hduser   35018772 7月  23 19:05 flink-table_2.12-1.13.2.jar
-rw-r--r-- 1 hduser hduser   38548158 7月  23 19:05 flink-table-blink_2.12-1.13.2.jar
-rw-r--r-- 1 hduser hduser   40623961 8月  23 2019 hive-exec-3.1.2.jar
-rw-r--r-- 1 hduser hduser      67114 3月  25 2020 log4j-1.2-api-2.12.1.jar
-rw-r--r-- 1 hduser hduser     276771 3月  25 2020 log4j-api-2.12.1.jar
-rw-r--r-- 1 hduser hduser    1674433 3月  25 2020 log4j-core-2.12.1.jar
-rw-r--r-- 1 hduser hduser      23518 3月  25 2020 log4j-slf4j-impl-2.12.1.jar
-rw-r--r-- 1 hduser hduser     493715 9月  16 11:56 mariadb-java-client-2.1.2.jar
[hduser@xueai8 lib]$
```

图 7-12 添加 Hive 依赖项

新创建一个 hive-site.xml 配置文件。编辑其内容如下：

```xml
<?xml version = "1.0" encoding = "UTF-8" standalone = "no"?>
<?xml-stylesheet type = "text/xsl" href = "configuration.xsl"?>
<configuration>
    <property>
        <name>hive.metastore.warehouse.dir</name>
        <value>hdfs://localhost:8020/user/hive/warehouse</value>
    </property>
    <property>
        <name>spark.sql.hive.metastore.version</name>
        <value>3.1.2</value>
    </property>
    <property>
        <name>hive.metastore.uris</name>
        <value>thrift://localhost:9083</value>
    </property>

    <!-- 配置 Hive Metastore:MySQL 连接信息 -->
    <property>
        <name>Javax.jdo.option.ConnectionURL</name>
        <value>JDBC:mysql://localhost:3306/hive</value>
    </property>
    <property>
        <name>Javax.jdo.option.ConnectionDriverName</name>
        <value>com.mysql.JDBC.Driver</value>
    </property>
    <property>
        <name>Javax.jdo.option.ConnectionUserName</name>
```

```
            <value>root</value>
        </property>
        <property>
            <name>javax.jdo.option.ConnectionPassword</name>
            <value>admin</value>
        </property>
</configuration>
```

注意：应将其中的 IP 地址和数据库连接信息改为读者自己的连接信息。

当 Flink 集成的 Hive 不支持 Embedded Metastore 时，需要对 Hive 的配置进行修改，在 hive-site.xml 文件中配置 hive.metastore.uris，并运行一个 Hive Metastore。

配置 hive.metastore.uris 的内容如下：

```
<property>
    <name>hive.metastore.uris</name>
    <value>thrift://localhost:9083</value>
</property>
```

运行 Hive Metastore 的命令如下：

```
$ ./bin/hive -- service metastore
```

7.3.3　注册 HiveCatalog

HiveCatalog 会连接 Hive Metastore 并桥接 Flink 和 Hive 之间的元数据。目前，这个 HiveCatalog 可以提供读取 Hive 元数据的能力，包括数据库、表、表分区、简单的数据类型、表和列的统计信息。

Flink 可以通过 HiveCatalog 读取 Hive 的 MetaData。Flink 还实现了 HiveTableSource，使 Flink 作业可以直接读取 Hive 中普通表和分区表的数据，以及对分区进行裁剪。

Flink 使用 HiveCatalog 可以通过批或者流的方式来处理 Hive 中的表。这就意味着 Flink 既可以作为 Hive 的一个批处理引擎，也可以通过流处理的方式来读写 Hive 中的表，从而支持实时数仓的应用和流批一体的落地。

从 Flink 1.13 开始，Flink 推荐使用 SQL 脚本来初始化 session 会话。

Flink SQL 查询需要一个执行它的配置环境。Flink SQL Client 支持 -i 启动选项，在启动 Flink SQL Client 时执行一个初始化 SQL 文件设置环境。所谓的初始化 SQL 文件可以使用 Flink DDL 来定义可用的 catalog 目录、表源和表 sink、用户定义函数及执行和部署所需的其他属性。

这时,启动 Flink SQL Client 的语法如下:

```
> ./sql-client.sh -i init1.sql,init2.sql
```

其中多个初始化 SQL 文件使用逗号进行分隔。例如,在 Flink 配置目录 conf 下,创建一个用于 Flink SQL Client 初始化的 init.sql 文件,编辑其内容如下:

```
-- 定义可用的 catalogs

CREATE CATALOG MyHiveCatalog
  WITH (
    'type' = 'hive',
    'default-database' = 'default',
    'hive-conf-dir' = '/home/hduser/bigdata/flink-1.13.2/conf'
  );

USE CATALOG MyHiveCatalog;
```

在上面这个初始化脚本文件中,连接 Hive catalog 目录,使用 MyHiveCatalog 作为当前目录,并指定在 Tableau 模式下运行探索性查询。

如果要读写 Hive 表,因为 Hive 表对应的数据仓库地址在 HDFS 上,所以要先启动 HDFS,命令如下:

```
$ start-dfs.sh
```

然后启动 Flink 集群,命令如下:

```
$ export HADOOP_CLASSPATH = '$HADOOP_HOME/bin/hadoop classpath'
$ ./bin/start-cluster.sh
```

接下来启动 FlinkSQL CLI,命令如下:

```
$ ./bin/sql-client.sh -i conf/init.sql
```

接下来,可以查看注册的 catalog,命令如下:

```
> show catalogs;
```

可以看到如图 7-13 所示的输出内容。

接下来,切换到 MyHiveCatalog,命令如下:

```
> use catalog MyHiveCatalog;
```

查看 MyHiveCatalog 目录下包含哪些数据库，命令如下：

```
> show databases;
```

查看默认当前数据库（default）中包含哪些数据表，命令如下：

```
> show tables;
```

执行过程如图 7-14 所示。

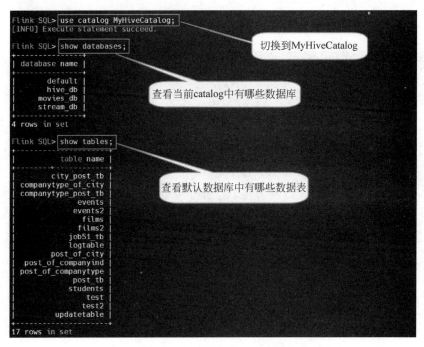

图 7-14　查看 MyHiveCatalog

在 Flink SQL Client 中执行如下 SQL 语句，创建并操作 Hive 表：

```
-- 使用 Hive 的 DDL 语法
SET table.sql-dialect=hive;
```

```
-- 创建表
CREATE TABLE test_hive (
    name string,
    age int
) STORED AS ORC;

-- 查看数据表
show tables;

-- 插入数据
INSERT INTO test_hive SELECT '张三', 20;

-- 将结果显示模式设置为 Tableau
SET sql-client.execution.result-mode=tableau;

-- 执行查询
select * from test_hive;
```

执行结果如图 7-15 所示。

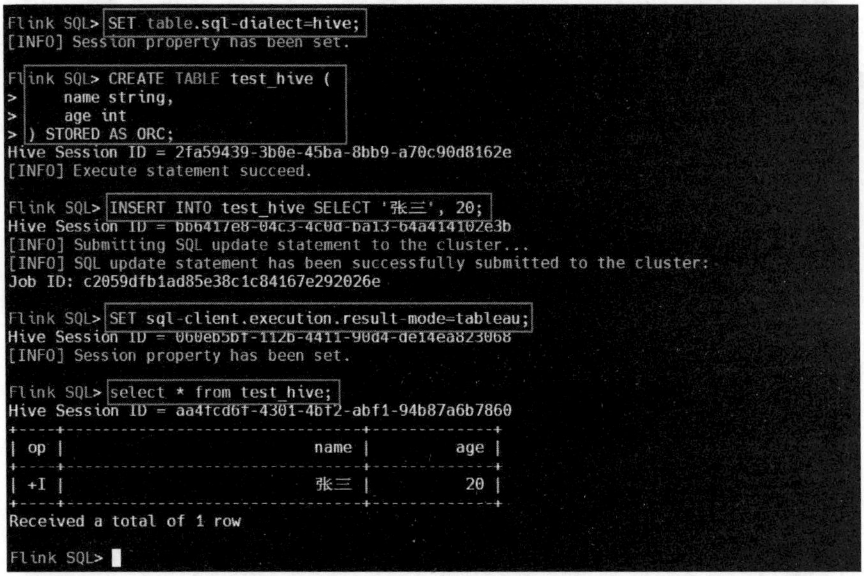

图 7-15 创建并操作 Hive 表

如果要删除 test_hive 表,则可执行下面的 SQL 语句:

```
> drop table test_hive;
```

7.3.4 使用 SQL Client 提交作业

当执行查询或插入语句时,Flink SQL Client 支持使用进入交互模式或使用-f 选项提交 SQL 语句。用户在交互式命令行中提交作业,或者使用-f 选项来执行 SQL 文件。在这两种模式中,Flink SQL Client 都支持解析和执行 Flink 支持的所有类型的 SQL 语句。

1. 交互式命令行

在交互式命令行中,Flink SQL Client 可读取用户输入并在获取分号(;)时执行语句。

如果语句执行成功,则 Flink SQL Client 将打印成功消息。当获取错误时,SQL Client 也将打印错误消息。默认情况下,错误信息只包含错误原因。为了输出完整的异常堆栈进行调试,应通过以下命令将 sql-client.verbose 设置为 true。

```
> SET 'sql-client.verbose' = 'true'
```

2. 执行 SQL 文件

SQL Client 支持执行带有-f 选项的 SQL 脚本文件。SQL Client 将逐个执行 SQL 脚本文件中的语句,并为每个已执行语句打印执行消息。一旦一条语句失败,SQL Client 就会退出,所有剩余的语句都不会被执行。

下面是这样一个文件的示例,代码如下:

```sql
CREATE TEMPORARY TABLE users (
    user_id BIGINT,
    user_name STRING,
    user_level STRING,
    region STRING,
    PRIMARY KEY (user_id) NOT ENFORCED
) WITH (
    'connector' = 'upsert-kafka',
    'topic' = 'users',
    'properties.Bootstrap.servers' = '...',
    'key.format' = 'csv',
    'value.format' = 'avro'
);

-- 设置同步模式
SET 'table.dml-sync' = 'true';

-- 设置作业名称
SET 'pipeline.name' = 'SqlJob';

-- 设置作业提交的队列
SET 'yarn.application.queue' = 'root';
```

```
-- 设置作业并行性
SET 'parallism.default' = '100';

-- 从指定保存点路径恢复
SET 'execution.savepoint.path' = '/tmp/flink-savepoints/savepoint-cca7bc-bb1e257f0dab';

-- 定义了查询之后,可以将它作为一个长期运行的分离的Flink作业提交给集群
-- 为此,需要使用INSERT INTO语句指定存储结果的目标系统
INSERT INTO pageviews_enriched
SELECT *
FROM pageviews AS p
LEFT JOIN users FOR SYSTEM_TIME AS OF p.proctime AS u
ON p.user_id = u.user_id;
```

在上面的配置中,包含了:

(1) 定义从CSV文件中读取的时态表源users。
(2) 设置属性,例如作业名称。
(3) 设置保存点路径。
(4) 提交一个从指定保存点路径加载保存点的SQL作业。

3. 执行一组SQL语句

Flink SQL客户端作为单个Flink作业执行每个INSERT INTO语句,然而,有时不是最佳的,因为管道的某些部分可以重用。Flink SQL Client支持STATEMENT SET语法来执行一组SQL语句。这是Table API中StatementSet的等价特性。STATEMENT SET语法包含一个或多个INSERT INTO语句。STATEMENT SET块中的所有语句都进行了整体优化,并作为单个Flink作业执行。联合优化和执行允许重用公共的中间结果,因此可以显著地提高执行多个查询的效率。

STATEMENT SET语法如下:

```
BEGIN STATEMENT SET;
    -- 一个或多个INSERT INTO语句
    { INSERT INTO|OVERWRITE < select_statement >; } +
END;
```

STATEMENT SET中包含的语句必须用分号(;)分隔。
在交互式模式下,执行STATEMENT SET的示例代码如下:

```
Flink SQL> CREATE TABLE pageviews (
> user_id BIGINT,
> page_id BIGINT,
> viewtime TIMESTAMP,
> proctime AS PROCTIME()
```

```
> ) WITH (
> 'connector' = 'kafka',
> 'topic' = 'pageviews',
> 'properties.Bootstrap.servers' = '...',
> 'format' = 'avro'
> );
[INFO] Execute statement succeed.

Flink SQL> CREATE TABLE pageview (
> page_id BIGINT,
> cnt BIGINT
> ) WITH (
> 'connector' = 'JDBC',
> 'url' = 'JDBC:mysql://localhost:3306/mydatabase',
> 'table-name' = 'pageview'
> );
[INFO] Execute statement succeed.

Flink SQL> CREATE TABLE uniqueview (
> page_id BIGINT,
> cnt BIGINT
> ) WITH (
> 'connector' = 'JDBC',
> 'url' = 'JDBC:mysql://localhost:3306/mydatabase',
> 'table-name' = 'uniqueview'
> );
[INFO] Execute statement succeed.

Flink SQL> BEGIN STATEMENT SET;
[INFO] Begin a statement set.

Flink SQL> INSERT INTO pageviews
> SELECT page_id, count(1)
> FROM pageviews
> GROUP BY page_id;
[INFO] Add SQL update statement to the statement set.

Flink SQL> INSERT INTO uniqueview
> SELECT page_id, count(distinct user_id)
> FROM pageviews
> GROUP BY page_id;
[INFO] Add SQL update statement to the statement set.

Flink SQL> END;
[INFO] Submitting SQL update statement to the cluster...
[INFO] SQL update statement has been successfully submitted to the cluster:
Job ID: 6b1af540c0c0bb3fcfcad50ac037c862
```

在 SQL 文件中，执行 STATEMENT SET 的示例代码如下：

```sql
CREATE TABLE pageviews (
  user_id BIGINT,
  page_id BIGINT,
  viewtime TIMESTAMP,
  proctime AS PROCTIME()
) WITH (
  'connector' = 'kafka',
  'topic' = 'pageviews',
  'properties.Bootstrap.servers' = '...',
  'format' = 'avro'
);

CREATE TABLE pageview (
  page_id BIGINT,
  cnt BIGINT
) WITH (
  'connector' = 'JDBC',
  'url' = 'JDBC:mysql://localhost:3306/mydatabase',
  'table-name' = 'pageview'
);

CREATE TABLE uniqueview (
  page_id BIGINT,
  cnt BIGINT
) WITH (
  'connector' = 'JDBC',
  'url' = 'JDBC:mysql://localhost:3306/mydatabase',
  'table-name' = 'uniqueview'
);

BEGIN STATEMENT SET;

INSERT INTO pageviews
SELECT page_id, count(1)
FROM pageviews
GROUP BY page_id;

INSERT INTO uniqueview
SELECT page_id, count(distinct user_id)
FROM pageviews
GROUP BY page_id;

END;
```

默认情况下，SQL Client 异步执行 DML 语句。这意味着，SQL Client 将向 Flink 集群提交 DML 语句的作业，而不是等待作业完成，因此，SQL Client 可以同时提交多个作业。这对于流作业非常有用，这些作业通常是长时间运行的。

SQL Client 确保一条语句可成功地提交到集群。一旦提交了语句，CLI 将显示关于 Flink 作业的信息，代码如下：

```
Flink SQL> INSERT INTO MyTableSink SELECT * FROM MyTableSource;
[INFO] Table update statement has been successfully submitted to the cluster:
Cluster ID: StandaloneClusterId
Job ID: 6f922fe5cba87406ff23ae4a7bb79044
```

SQL Client 在提交后不会跟踪正在运行的 Flink 作业的状态。CLI 进程可以在提交后关闭，而不影响分离查询。Flink 的 Restart Strategy（重启策略）解决了容错问题。可以使用 Flink 的 Web 界面、命令行或 REST API 取消查询。

然而，对于批处理用户，更常见的情况是，下一个 DML 语句需要等待，直到前一个 DML 语句完成。为了同步执行 DML 语句，可以设置表。在 SQL Client 中将 table.dml-sync 选项设置为 true，代码如下：

```
Flink SQL> SET 'table.dml-sync' = 'true';
[INFO] Session property has been set.

Flink SQL> INSERT INTO MyTableSink SELECT * FROM MyTableSource;
[INFO] Submitting SQL update statement to the cluster...
[INFO] Execute statement in sync mode. Please wait for the execution finish...
[INFO] Complete execution of the SQL update statement.
```

如果想终止作业，则只需按快捷键 Ctrl+C 取消。

7.3.5　构建批流一体数仓完整过程

至此，已经掌握了如何通过 Flink SQL Client 由 HiveCatalog 来连接和访问 Hive。接下来将重构 7.2 节的项目，通过 Flink SQL Client 集成 Hive 实现纯 SQL 无代码构建批流一体数据仓库。实现架构如图 7-16 所示。

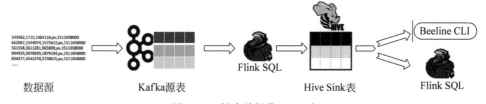

图 7-16　创建并操作 Hive 表

因为这里 Flink SQL Client 要访问 Kafka 源表，所以需要将 Flink Kafka 相关的连接器也添加到Flink安装的lib目录下。包括以下依赖JAR包：

(1) kafka-clients-2.4.1.jar。

(2) flink-connector-base-1.13.2.jar。

(3) flink-connector-kafka_2.12-1.13.2.jar。

(4) flink-sql-connector-kafka_2.12-1.13.2.jar。

对于Flink，需要在其配置文件flink-conf.yaml中添加如下配置：

```
execution.checkpointing.interval: 10s   #checkpoint间隔时间
execution.checkpointing.tolerable-failed-checkpoints: 10   #checkpoint失败容忍次数
```

接下来，建议按以下步骤执行：

第1步：环境准备。

首先准备好运行环境，包括Hadoop集群、Hive Metastore、Kafka集群、Flink on YARN session，步骤如下。

(1) 启动HDFS集群。在终端窗口中，执行的命令如下：

```
$ start-dfs.sh
$ start-yarn.sh
```

(2) 启动Hive Metastore服务。在终端窗口中，执行的命令如下（保持运行）：

```
$ hive --service metastore
```

(3) 启动Kafka集群。按下面的注释说明执行：

```
#切换到Kafka安装目录
$ cd ~/bigdata/kafka_2.12-2.4.1

#打开一个终端，运行ZooKeeper
$ ./bin/zookeeper-server-start.sh ./config/zookeeper.properties

#另打开一个终端，运行Kafka服务器
$ ./bin/kafka-server-start.sh ./config/server.properties

#再打开一个终端，创建名为user_behavior的Kafka topic
$ ./bin/kafka-topics.sh --create --zookeeper localhost:2181 --replication-factor 1 --partitions 1 --topic user_behavior

#查看已经存在的Topics
$ ./bin/kafka-topics.sh --list --zookeeper localhost:2181
```

(4) 启动Flink集群（注意，这里使用Flink on YARN session模式）。在终端窗口中，执行的命令如下：

```
$ cd ~/bigdata/flink-1.13.2

$ export HADOOP_CLASSPATH=`hadoop classpath`
$ ./bin/yarn-session.sh --detached
```

第 2 步：创建 Flink SQL 执行的脚本文件。

按以下步骤执行：

（1）首先创建以下的目录，用来存放执行的脚本文件，命令如下：

```
$ mkdir -p ~/data/flink/sql/
$ cd ~/data/flink/sql/
```

（2）编辑 Flink SQL Client 初始化脚本 conf/init.sql，将显示模式设置为 Tableau，设置内容如下：

```
...
SET sql-client.execution.result-mode = tableau;
...
```

（3）创建第 1 个 SQL 脚本文件 kafka2hive.sql，实现读取 Kafka user_behavior 主题，并实时写入 Hive 的 user_behavior_hive 表中。编辑脚本文件 kafka2hive.sql 的内容如下：

```
-- 第7章/kafka2hive.sql

-- 创建 Kafka 源表
CREATE DATABASE IF NOT EXISTS stream_db;

DROP TABLE IF EXISTS stream_db.user_behavior_kafka;

CREATE TABLE stream_db.user_behavior_kafka (
    user_id bigint,
    item_id bigint,
    category_id bigint,
    behavior string,
    ts bigint,
    procTime AS PROCTIME(),
    eventTime AS TO_TIMESTAMP(FROM_UNIXTIME(ts, 'yyyy-MM-dd HH:mm:ss')),
    WATERMARK FOR eventTime AS eventTime - INTERVAL '5' SECONDS
)
with(
    'connector' = 'kafka',
    'topic' = 'user_behavior',
    'properties.Bootstrap.servers' = 'xueai8:9092',
```

```sql
    'properties.group.id' = 'testGroup',
    'scan.startup.mode' = 'latest-offset',
    'format' = 'csv',
    'csv.ignore-parse-errors' = 'true',
    'csv.field-delimiter' = ','
);

-- 创建 Hive Sink 表
SET table.sql-dialect = hive;

CREATE DATABASE IF NOT EXISTS hive_db;
DROP TABLE IF EXISTS hive_db.user_behavior_hive;

CREATE TABLE hive_db.user_behavior_hive (
    user_id bigint,
    item_id bigint,
    category_id bigint,
    behavior string,
    ts bigint
) PARTITIONED BY (
    ts_date STRING,
    ts_hour STRING,
    ts_minute STRING
) STORED AS PARQUET
TBLPROPERTIES (
    'sink.partition-commit.trigger' = 'partition-time',
    'sink.partition-commit.delay' = '1 min',
    'sink.partition-commit.policy.kind' = 'metastore,success-file',
    'partition.time-extractor.timestamp-pattern' = '$ts_date $ts_hour:$ts_minute:00'
);

-- 执行 ETL
SET table.sql-dialect = default;

INSERT INTO hive_db.user_behavior_hive
SELECT
    user_id,item_id,category_id,behavior,ts,
    DATE_FORMAT(eventTime,'yyyy-MM-dd') as ts_date,
    DATE_FORMAT(eventTime,'HH') as ts_hour,
    DATE_FORMAT(eventTime,'mm') as ts_minute
FROM stream_db.user_behavior_kafka
WHERE item_id > 0;
```

(4) 创建第 2 个 SQL 脚本文件 analysis_hive.sql,将 Hive 的 user_behavior_hive 表作为流表,将分析周期缩短到分钟级。编辑脚本文件 analysis_hive.sql 的内容如下:

第7章 基于Flink构建流批一体数仓

```sql
-- 第7章/analysis_hive.sql

set table.dynamic-table-options.enabled=true;
set execution.runtime-mode=streaming;

SELECT behavior,count(1) AS behavior_cnt
FROM hive_db.user_behavior_hive
/*+ OPTIONS(
    'streaming-source.enable' = 'true',
    'streaming-source.monitor-interval' = '1 min',
    'streaming-source.consume-start-offset' = '2017-11-26 09:01:00'
) */
WHERE behavior = 'buy'
GROUP BY behavior;
```

第3步：提交 Flink 作业 1。

在一个新的终端窗口中，启动 Flink SQL Client，连接 Hive，并通过-f 选项指定执行 analysis_hive.sql 脚本文件，命令如下：

```
$ cd ~/bigdata/flink-1.13.2
$ ./bin/sql-client.sh -i conf/init.sql -f ~/data/flink/sql/analysis_hive.sql
```

第4步：提交 Flink 作业 2。

在一个新的终端窗口中，启动 Flink SQL Client，连接 Hive，并通过-f 选项指定执行 kafka2hive.sql 脚本文件，命令如下：

```
$ cd ~/bigdata/flink-1.13.2
$ ./bin/sql-client.sh -i conf/init.sql -f ~/data/flink/sql/kafka2hive.sql
```

第5步：执行数据源数据生成脚本。

执行数据生产者脚本，模拟产生实时用户购买行为数据。执行前面已经创建的脚本 streamuserbehavior.sh，命令如下：

```
$ cd ~/data/flink/userbehavior/
$ ./streamuserbehavior.sh
```

第6步：观察作业 1 的统计结果。

查看作业 1 的执行窗口，可以看到输出内容，如图 7-17 所示。

第7步：观察作业 2 的执行过程。

打开浏览器，访问 Flink Web UI 界面，可以观察到 Flink 作业的执行情况，如图 7-18 所示。

单击图中标示的 job，查看作业执行详情，如图 7-19 所示。

图 7-17　观察作业 1 统计结果

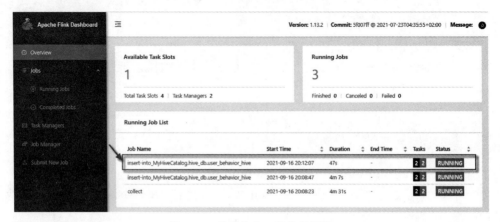

图 7-18　观察作业 2 的执行情况

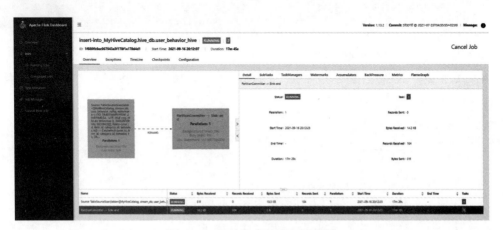

图 7-19　查看作业 2 的执行详情

第 8 步：使用 Hive CLI 进行批量查询(可选)。

当然，也可以使用 Hive CLI 进行批量查询，SQL 语句如下：

```
hive> select behavior,count(*) as cnt
    > from user_behavior_hive
    > group by behavior;
```

可以得到类似下面的统计结果：

```
buy   115
cart  300
fav   172
pv    5004
```

第 9 步：使用 Flink SQL Client 进行批量查询（可选）。

或者，使用 Flink SQL Client 连接 Hive 进行批量查询。这时，创建另一个 SQL 脚本文件 analysis_hive_batch.sql，用于将 Hive 的 user_behavior_hive 表作为源表，以批处理的方式对其进行聚合统计。编辑脚本文件 analysis_hive_batch.sql 的代码如下：

```sql
#第 7 章/analysis_hive_batch.sql

set table.dynamic-table-options.enabled=true;

set execution.runtime-mode=batch;

SELECT behavior,count(1) AS behavior_cnt
FROM hive_db.user_behavior_hive
GROUP BY behavior;
```

注意，因为 checkpoint 不支持 batch 批处理，所以需要先在 Flink 配置文件 flink-conf.yaml 中注释掉检查点配置，如图 7-20 所示。

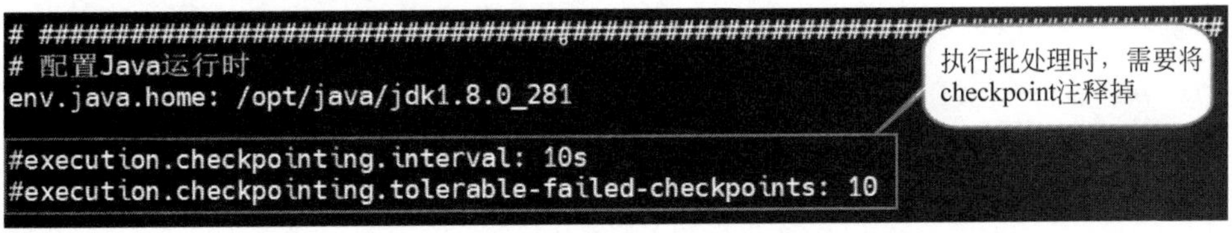

```
# ##################################################################
# 配置Java运行时
env.java.home: /opt/java/jdk1.8.0_281

#execution.checkpointing.interval: 10s
#execution.checkpointing.tolerable-failed-checkpoints: 10
```

执行批处理时，需要将 checkpoint 注释掉

图 7-20　在配置文件 flink-conf.yaml 中注释掉检查点配置

然后，通过 Flink SQL Client 连接 Hive 执行批量聚合统计，代码如下：

```
$ ./bin/sql-client.sh -i conf/init.sql -f ~/data/flink/sql/analysis_hive_batch.sql
```

执行过程如图 7-21 所示。

图 7-21 执行批量聚合统计

第 8 章 基于 Flink 和 Iceberg 数据湖构建实时数仓

CHAPTER 8

随着大数据技术的发展,下一代大数据技术的变革技术(数据湖技术)越来越受到企业的关注。本章将介绍如何基于 Flink 流计算和 Iceberg 数据湖框架来构建实时数仓。

8.1 现代数据湖概述

9min

随着数据量增长到前所未有的新水平,处理这种增长的新工具和技术也出现了,其中一个演变的领域是数据湖。传统的数据仓库工具用于从数据驱动业务智能。业界随后认识到,数据仓库通过执行写模式限制了智能的潜力。显然,在收集数据时不能考虑所收集数据集的所有维度。这意味着,强制执行模式或删除很少的条目(因为它们看起来似乎毫无用处),从长远来看可能会损害业务智能。此外,数据仓库技术无法跟上数据增长的步伐。由于数据仓库一般基于数据库和结构化数据格式,它被证明不足以应对当前数据驱动领域所面临的挑战。

这导致了数据湖的出现,这些数据湖针对非结构化和半结构化数据进行了优化,可以轻松地扩展到 PB 级规模,并允许更好地集成各种工具,以帮助企业最大限度地利用其数据。

8.1.1 什么是数据湖

数据湖的概念是什么?一般来讲,用户将企业产生的数据维护在一个平台上,称为"数据湖",如图 8-1 所示。

图 8-1 中这个湖的数据源是多种多样的,有些可能是结构化数据,有些可能是非结构化数据,有些甚至可能是二进制数据。有一些人站在湖口,用设备检测水质,对应数据湖上的流处理操作;有多台水泵从数据湖抽水,对应数据湖的批处理操作;也有一群人在船头或岸上钓鱼,对应的是数据科学家从中提取数据价值(数据挖掘)。

综上所述,数据湖有以下 4 个主要特征:
(1) 原始数据的存储,原始数据来源丰富。
(2) 支持多种计算模型。

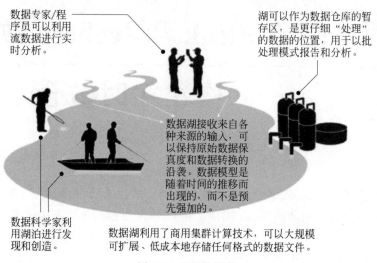

图 8-1 理解数据湖

(3) 具有完善的数据管理能力。能够访问多个数据源，实现不同数据之间的连接，支持模式管理和权限管理。

(4) 灵活的底层存储。一般采用 DS3、OSS 和 HDFS 作为廉价的分布式文件系统，采用特定的文件格式和缓存来满足相应场景的数据分析需求。

数据湖的这 4 个主要特征见表 8-1。

表 8-1 数据湖的主要特征

特 征	描 述
存储原始数据	结构化数据；半结构化数据；非结构化数据；二进制数据（图片等）
多种计算模型	批处理；流计算；交互式分析；机器学习
完善的数据管理	多种数据源接入；数据连接；Schema 管理；权限管理
灵活的底层存储	S3/OSS/HDFS；Parquet/Avro/Orc；数据缓存加速

通常在不同的上下文环境中术语"数据湖"具有不同含义，但在所有数据湖定义中有几个重要属性是一致的：

(1) 支持非结构化和半结构化数据。

(2) 可扩展到 PB 级或更高。

(3) 使用类 SQL 的接口与存储的数据交互。

(4) 能够尽可能无缝地连接各种分析工具。

(5) 最后，现代数据湖通常是分离存储和分析工具的组合。

8.1.2 数据湖架构

那么,开源数据湖的架构是什么样的呢?数据湖的架构如图 8-2 所示。

图 8-2 数据湖架构

这个数据湖架构主要分为以下 4 个层次:

(1)底部是分布式文件系统。云上的用户将更多地使用 S3、OSS 等对象存储。非云用户通常使用自己的 HDFS。

(2)第 2 层是数据加速层。数据湖架构是一个完全分离存储和计算的架构。如果所有数据访问都通过远程读取文件系统上的数据,则性能和成本将非常高。如果能将一些频繁访问的热点数据缓存到本地计算节点中,则实现冷热分离就很自然了。一方面可以获得良好的本地读取性能,另一方面可以节省远程访问的带宽。在这一层,通常选择云上开源的 Alluxio 或者 JindoFS。

(3)第 3 层是表格式层,主要是将一批数据文件封装成具有业务意义的表,并提供如 ACID、快照、模式和分区等表级语义。表格式通常与 Delta、Iceberg 和 Hudi 等开源项目相对应。

(4)顶层是不同计算场景的计算引擎。开源计算引擎一般包括 Spark、Flink、Hive、Presto、Hive MR 等。这些计算引擎可以同时访问同一数据湖的表。

在图 8-2 所示的架构图中,第 3 层表格式层是因可用性挑战和数据一致性要求而导致的一个全新的软件项目类别。这些项目位于存储平台和分析平台之间,并以本地方式处理对象存储平台,同时向最终用户提供强大的 ACID 保证。

8.1.3 开源数据湖框架

目前市场上有三大开源数据湖框架,分别是 Delta Lake、Apache Iceberg 和 Hudi。另外,Apache Hive 也正在努力向数据湖演化。

1. Delta Lake

Delta Lake 是一个开源平台，它为 Apache Spark 带来了 ACID 事务。Delta Lake 由 Databricks 开发，它运行在现有的存储平台（如 S3、HDFS、Azure）上，并完全兼容 Apache Spark API。特别提供了以下特性。

（1）Spark 上的 ACID 事务：可序列化的隔离级别确保读取器不会看到不一致的数据。

（2）可扩展元数据处理：利用 Spark 的分布式处理能力，轻松处理具有数十亿文件的 PB 级表的所有元数据。

（3）流和批统一：Delta Lake 中的表既是一个批表，又是一个流源（source）和流接收器（sink）。流数据采集、批量历史回填、交互式查询都是开箱即用的。

（4）模式强制：自动处理模式变化，以防止在摄入期间插入坏记录。

（5）时间旅行：数据版本控制支持回滚、完整的历史审计跟踪和可重现的机器学习实验。

（6）upsert 和 delete：支持合并、更新和删除操作，以支持复杂的用例，如更改数据捕获、缓慢变化维度（Slowly Changing Dimension，SCD）操作、流式 Upsert 等。

注意：缓慢变化维度就是数据仓库维度表中，那些随时间变化比较不明显，但仍然会发生变化的维度。

2. Apache Iceberg

Apache Iceberg 是一种用于巨大分析数据集的开放表格式。Iceberg 向 Presto 和 Spark 添加了使用高性能格式的表，其工作方式与 SQL 表类似。Iceberg 的重点是避免不愉快的意外，帮助模式演化和避免无意的数据删除。用户不需要了解分区来获得快速查询。

（1）模式演进支持添加、删除、更新或重命名，并且没有副作用。

（2）隐藏分区防止用户错误导致不正确的结果或极慢的查询。

（3）分区布局演变可以随着数据量或查询模式的改变而更新表的布局。

（4）时间旅行可以实现使用完全相同的表快照的可重复查询，或者让用户方便地检查更改。

（5）版本回滚允许用户通过将表重置为良好状态来快速纠正问题。

3. Apache Hudi

Apache Hudi（简称 Hudi）是 Uber 创建的一个数据湖框架。Hudi 于 2019 年 1 月加入 Apache 孵化器进行孵化，并于 2020 年 5 月晋升为 Apache 顶级项目。它是最流行的数据湖框架之一。

Hudi 支持以下特征：

（1）Upsert 支持快速、可插入的索引。

（2）支持回滚的原子发布数据。

(3) 写入器和查询之间的快照隔离。
(4) 保存点数据恢复。
(5) 使用统计管理文件大小，以及布局。
(6) 异步压缩行和列数据。
(7) 跟踪沿袭 (lineage) 的时间线元数据。
(8) 利用聚类优化数据湖布局。

8.2 基于 Flink＋Iceberg 构建企业数据湖

随着大数据存储和处理需求的多样化，如何构建一个统一的数据湖存储，并在其上进行多种形式的数据分析成了企业构建大数据生态的一个重要方向。Netflix 发起的 Apache Iceberg 项目具备 ACID 能力的表格式中间件成为大数据、数据湖领域炙手可热的方向。

8.2.1 Apache Iceberg 的优势

Apache Iceberg 作为一个数据湖解决方案，它支持 ACID 事务、修改和删除（基于 Spark 2.4.5＋以上）、独立于计算引擎、支持表结构和分区方式动态变更等特性，同时，为了支持流式数据的写入，引入 Flink 作为流式处理框架，并将 Iceberg 作为 Flink sink 的终表解决方案。

Apache Iceberg 的主要优势如下。

(1) 写入：支持事务，写入即可见，并提供 upset/merge into 的能力。

(2) 读取：支持以流的方式读取增量数据。Flink Table Source 及 Spark 结构化流都支持；不惧怕 Schema 的变化。

(3) 计算：通过抽象不绑定任何引擎。提供原生的 Java Native API，从生态上来讲，目前支持 Spark、Flink、Presto、Hive。

(4) 存储：对底层存储进行了抽象，不绑定于任何底层存储；支持隐藏分区和分区进化，方便业务进行数据分区策略；支持 Parquet、ORC、Avro 等格式来兼容行存储和列存储。

当然了，Iceberg 并不支持行级 update，仅支持 insert into/overwrite，Iceberg 团队在未来将致力于解决这个问题。

Apache Iceberg 同时支持 Apache Flink 的 DataStream API 和 Table API，以将记录写入 Iceberg 表。目前，只集成了 Iceberg 和 Apache Flink 1.11.x，两者兼容性关系如图 8-3 所示。

目前已知在用的 Iceberg 的大厂如下。

(1) 国外：Netflix、Apple、Linkined、Adobe、Dremio 等。

(2) 国内：腾讯、网易、阿里云等。

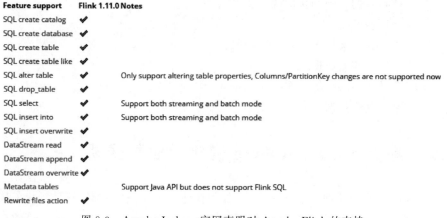

图 8-3　Apache Iceberg 官网声明对 Apache Flink 的支持

8.2.2　Apache Iceberg 经典业务场景

那么，Flink 可以与数据湖结合哪些经典应用场景呢？在这里，当讨论业务场景时，默认选择 Apache Iceberg 作为数据湖模型。

1. 场景 1：构建实时数据管道

首先，Flink+Iceberg 最经典的场景是构建一个实时数据管道，如图 8-4 所示。

图 8-4　实时数据管道

业务端产生的大量日志数据被导入消息队列中，如 Kafka。在使用 Flink 流计算引擎执行 ETL 之后，它被导入 Apache Iceberg 原始表中。一些业务场景需要直接运行分析作业来分析原始表的数据，而另一些场景则需要进一步净化数据，然后，可以启动一个新的 Flink 作业来使用来自 Apache Iceberg 表的增量数据，并在处理后将其写入经过处理的 Iceberg 表。此时，可能有业务需要进一步聚合数据，因此可以继续在 Iceberg 表上启动增量 Flink 作业，并将聚合的数据结果写入聚合表。

通过 Flink+Iceberg 构建的实时数据管道，可以借助 Flink 实现数据的精确一次性语义地入湖和出湖，并且新定稿的数据可以检查点周期内可见。

当然，这个场景也可以通过 Flink+Hive 实现，但是在 Flink+Hive 实现方案中，写入 Hive 的数据更多的是用于数据仓库的数据分析，而不是增量拉取。一般情况下，Hive 按分区增量写时间大于 15min。长时间高频率的 Flink 写入会导致分区膨胀。

而 Iceberg 允许 1min 甚至 30s 的增量写入，这可以极大地提高端到端数据的实时性能。上层的分析作业可以看到更新的数据，而下游的增量作业可以读取更新的数据。

2. 场景2：实时分析RDBMS的变量日志

第2个经典场景是可以使用Flink+Iceberg来分析来自关系型数据库（如MySQL）的binlog，如图8-5所示。

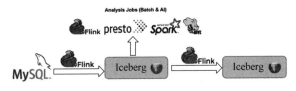

图8-5　实时数据管道

一方面，Apache Flink本身支持CDC数据解析。通过Flink CDC连接器拉出binlog数据后，自动转换为insert、delete和update，Flink运行时可以识别这些消息，供用户进行进一步的实时计算。

另一方面，Apache Iceberg完美地实现了相等删除功能，即用户定义要删除的记录，直接写入Apache Iceberg表中，这样便可删除相应的行，实现了数据湖的流删除。在未来的Iceberg版本中，用户将不需要设计任何额外的业务领域，只需编写几行代码就可以将binlog流传输到Apache Iceberg。

此外，当CDC数据成功输入Iceberg之后，还可以通过常见的计算引擎，如Presto、Spark、Hive等，实时读取Iceberg表中的最新数据。

3. 场景3：近实时场景的流批统一

第3个经典场景是近实时场景的流批统一。

在常用的Lambda架构中，有实时处理模块和离线处理模块。实时模块一般由Flink、Kafka和HBase构建，离线模块一般由Parquet和Spark等组件构建，如图8-6所示。

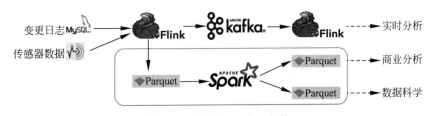

图8-6　常用的Lambda架构

这种架构涉及的计算组件和存储组件很多，系统维护和业务开发成本非常高。

在实际应用中，有很多场景的实时性要求不是很苛刻，例如，可以放宽到分钟级的水平。这种场景称为近实时场景。可以使用Flink+Iceberg来优化整个架构，如图8-7所示。

在这个优化后的架构中，实时数据通过Flink写入Iceberg表。近实时链路仍然可以通过Flink计算增量数据。离线链路还可以通过Flink批计算读取快照进行全局分析，并得到相应的分析结果，供用户在不同场景下进行读取和分析。经过这样的改进，我们将计算引擎

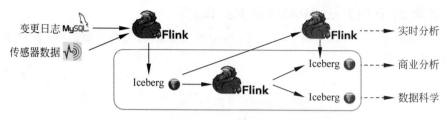

图 8-7　近实时场景的流批统一

统一到 Flink 中,将存储组件统一到 Iceberg 中,大大降低了整个系统的维护和开发成本。

4. 场景 4:使用 Iceberg 全量数据和 Kafka 增量数据引导新的 Flink 作业

第 4 个场景使用 Apache Iceberg 全量数据和 Kafka 增量数据引导新的 Flink 作业。

例如,用户现有的流作业是在线运行的。突然有一天,一个客户提出他们遇到了一个新的计算场景,需要设计一个新的 Flink 作业。他们浏览了去年的历史数据,并接收了正在生成的 Kafka 增量数据。那么这时我们应该怎么做呢?

用户仍然可以使用通用的 Lambda 架构。离线链路通过 Kafka→Flink→Iceberg 同步写入数据湖。因为 Kafka 的成本较高,所以可以保留最近 7 天的数据。Iceberg 存储成本低,可以存储全部历史数据(根据检查点划分多个数据间隔)。当启动一个新的 Flink 作业时,只需拉出 Iceberg 中的数据,然后在运行后顺利地接收 Kafka 的数据,架构如图 8-8 所示。

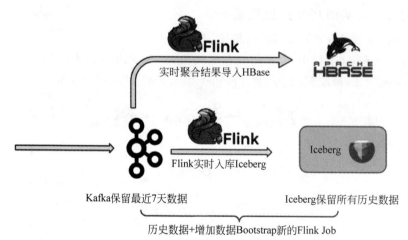

图 8-8　使用 Iceberg 全量数据和 Kafka 增量数据引导新的 Flink 作业

5. 场景 5:数据订正

第 5 个场景和第 4 个场景有点相似。还是在 Lambda 架构中,实时处理链路由于事件丢失或到达顺序的问题可能导致流计算的结果不一定完全准确,此时一般需要充分的历史数据来纠正实时计算的结果,而 Apache Iceberg 可以很好地发挥这个作用,通过它可以以高性价比管理历史数据，如图 8-9 所示。

图 8-9 使用 Apache Iceberg 实现数据订正

8.2.3 应用 Apache Iceberg 的准备工作

在使用 Flink 集成 Iceberg 之前，有几项准备工作要做。

1. Flink 集成 Iceberg

首先要从 Apache Iceberg 官网下载 Iceberg 的 JAR 包，如图 8-10 所示。

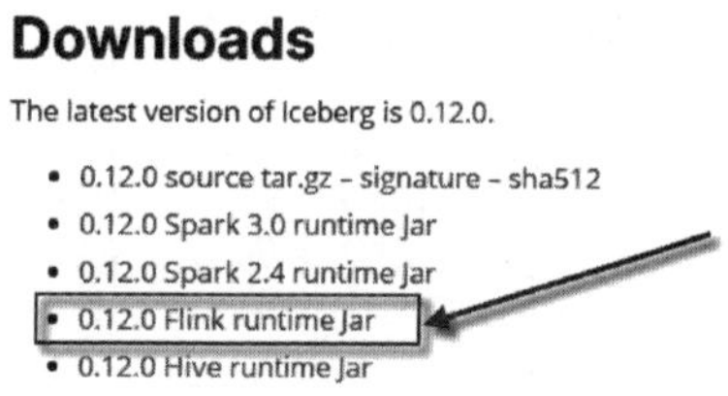

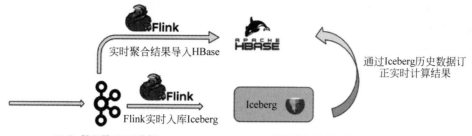

图 8-10 下载 Apache Iceberg JAR 包

单击以上链接，下载 iceberg-flink-runtime-0.12.0.jar 包。Apache Iceberg 使用嵌入式程序的方式工作，因此只需将下载的 iceberg-flink-runtime-0.12.0.jar 包复制到 Flink 安装的 lib 目录下，如图 8-11 所示。

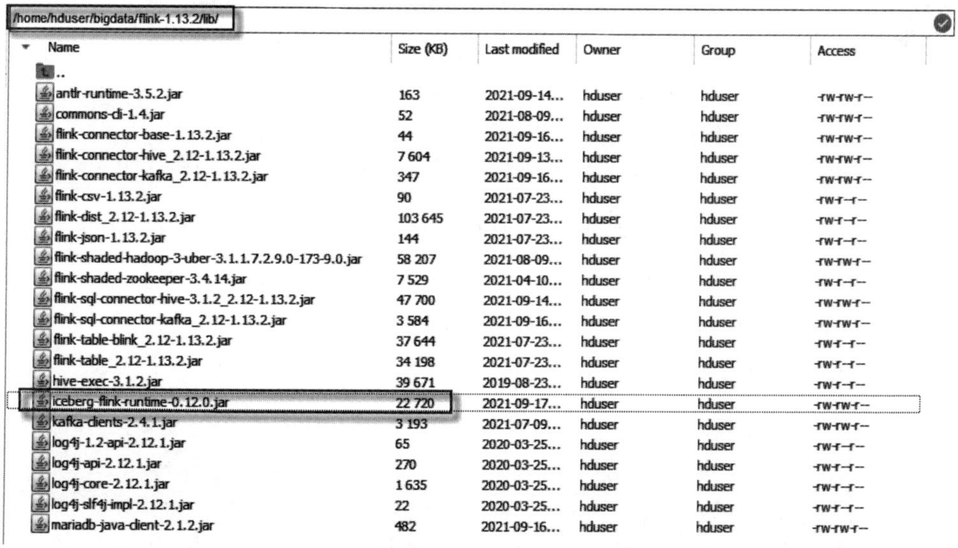

/home/hduser/bigdata/flink-1.13.2/lib/

Name	Size (KB)	Last modified	Owner	Group	Access
..					
antlr-runtime-3.5.2.jar	163	2021-09-14...	hduser	hduser	-rw-rw-r--
commons-cli-1.4.jar	52	2021-08-09...	hduser	hduser	-rw-rw-r--
flink-connector-base-1.13.2.jar	44	2021-09-16...	hduser	hduser	-rw-rw-r--
flink-connector-hive_2.12-1.13.2.jar	7 604	2021-09-13...	hduser	hduser	-rw-rw-r--
flink-connector-kafka_2.12-1.13.2.jar	347	2021-09-16...	hduser	hduser	-rw-rw-r--
flink-csv-1.13.2.jar	90	2021-07-23...	hduser	hduser	-rw-r--r--
flink-dist_2.12-1.13.2.jar	103 645	2021-07-23...	hduser	hduser	-rw-r--r--
flink-json-1.13.2.jar	144	2021-07-23...	hduser	hduser	-rw-r--r--
flink-shaded-hadoop-3-uber-3.1.1.7.2.9.0-173-9.0.jar	58 207	2021-08-09...	hduser	hduser	-rw-rw-r--
flink-shaded-zookeeper-3.4.14.jar	7 529	2021-04-10...	hduser	hduser	-rw-r--r--
flink-sql-connector-hive-3.1.2_2.12-1.13.2.jar	47 700	2021-09-14...	hduser	hduser	-rw-rw-r--
flink-sql-connector-kafka_2.12-1.13.2.jar	3 584	2021-09-16...	hduser	hduser	-rw-rw-r--
flink-table-blink_2.12-1.13.2.jar	37 644	2021-07-23...	hduser	hduser	-rw-r--r--
flink-table_2.12-1.13.2.jar	34 198	2021-07-23...	hduser	hduser	-rw-r--r--
hive-exec-3.1.2.jar	39 671	2019-08-23...	hduser	hduser	-rw-r--r--
iceberg-flink-runtime-0.12.0.jar	22 720	2021-09-17...	hduser	hduser	-rw-rw-r--
kafka-clients-2.4.1.jar	3 193	2021-07-09...	hduser	hduser	-rw-rw-r--
log4j-1.2-api-2.12.1.jar	65	2020-03-25...	hduser	hduser	-rw-r--r--
log4j-api-2.12.1.jar	270	2020-03-25...	hduser	hduser	-rw-r--r--
log4j-core-2.12.1.jar	1 635	2020-03-25...	hduser	hduser	-rw-r--r--
log4j-slf4j-impl-2.12.1.jar	22	2020-03-25...	hduser	hduser	-rw-r--r--
mariadb-java-client-2.1.2.jar	482	2021-09-16...	hduser	hduser	-rw-rw-r--

图 8-11 将下载的 iceberg-flink-runtime-0.12.0.jar 包并复制到 Flink 安装的 lib 目录下

在 Flink 中使用 Apache Iceberg，不需要特定的安装。

2. Flink 连接 Hive

为了在 Flink 中创建 Iceberg 表，官方推荐使用 Flink SQL Client。

默认情况下，Apache Iceberg 包含了用于 Hadoop Catalog 的 Hadoop JAR。如果想使用 Hive Catalog，则需要在打开 Flink SQL 客户端时加载 Hive JAR。

关于 Flink 集成 Hive，可参考 9.3.2 节"集成 Flink SQL CLI 和 Hive"。确保将所有 Hive 依赖项添加到 Flink 分布式安装的/lib 目录下。这些依赖项包括：

(1) mariadb-Java-client-2.1.2.jar。

(2) hive-exec-3.1.2.jar。

(3) flink-connector-hive_2.12-1.13.2.jar。

(4) flink-sql-connector-hive-3.1.2_2.12-1.13.2.jar。

(5) flink-shaded-hadoop-3-uber-3.1.1.7.2.9.0-173-9.0.jar。

以上这些依赖，也可以在启动 Flink SQL Client 时通过-j 选项动态指定，命令如下：

```
$ export HADOOP_CLASSPATH = 'hadoop classpath'

$ ./bin/sql-client.sh embedded \
    -j <flink-runtime-directory>/iceberg-flink-runtime-0.12.0.jar \
    -j <hive-bundlded-jar-directory>/flink-sql-connector-hive-3.1.2_2.12-1.13.2.jar \
    shell
```

Apache Iceberg 的官方案例是通过 Flink SQL Client 实现的，这个客户端比较大，而且不是很好用（这是主要原因），所以建议使用 Zeppelin 而不是 SQL Client。

在 Zeppelin 中使用 Apache Iceberg 时，首先要指定依赖包，命令如下：

```
%flink.conf
flink.execution.packages \ org.apache.flink:flink-connector-kafka_2.12:1.13.2
flink.execution.jars /home/dijie/iceberg/iceberg-flink-runtime-0.12.0.jar
```

Flink 集成 Apache Iceberg 时，通常需要先创建 Catalog。建立一个新的 Catalog 目录和指定 Iceberg 目录，命令如下：

```
%flink.ssql
CREATE CATALOG iceberg_catalog WITH (
  'type' = 'iceberg',
  'catalog-type' = 'hive',
  'uri' = 'thrift://localhost:9083',
  'clients' = '5',
  'property-version' = '1',
  'warehouse' = 'hdfs://localhost:8020/user/hive/warehouse'
);
```

查看 Catalog 是否创建成功，执行的命令如下：

```
% flink.ssql
show catalogs;
```

3. 配置 Flink 检查点

最后，还需要配置 Flink 的 checkpoint，因为目前 Flink 向 Iceberg 提交信息是在每次 checkpoint 时触发的。在 Flink 中配置 checkpoint 的方法如下。

在 flink-conf.yaml 添加如下配置：

```
#checkpoint 间隔时间
execution.checkpointing.interval: 10s


#checkpoint 失败容忍次数
execution.checkpointing.tolerable-failed-checkpoints: 10
```

8.2.4 创建和使用 Catalog

Apache Iceberg 支持 3 种 Catalog，分别是 Hive Catalog、Hadoop Catalog 和自定义 Catalog。

1. Hive Catalog

使用 Hive Catalog，可以将表结构持久存储到 Hive 的 Metastore 中。例如，创建一个名为 hive_catalog 的 Iceberg Hive Catalog，代码如下：

```
CREATE CATALOG hive_catalog WITH (
  'type'='iceberg',
  'catalog-type'='hive',
  'uri'='thrift://localhost:9083',
  'clients'='5',
  'property-version'='1',
  'warehouse'='hdfs://localhost:8020/user/hive/warehouse'
);
```

各个参数的解释如下。

(1) type：对于 Iceberg 表格式，只使用 Iceberg（必选）。

(2) catalog-type：Iceberg 当前支持 Hive 或 Hadoop Catalog 类型（必选）。

(3) uri：Hive Metastore 的 Thrift URI（必选）。

(4) clients：Hive Metastore 客户端池大小，默认值为 2（可选）。

(5) property-version：版本号，用于描述属性版本。如果属性格式改变，则此属性可用于向后兼容。当前的属性版本是1(可选)。

（6）warehouse：Hive Warehouse 位置，如果既没有设置 hive-conf-dir 来指定包含 hive-site.xml 配置文件的位置，也没有将正确的 hive-site.xml 添加到 classpath，则用户应该指定此路径。

（7）hive-conf-dir：包含 hive-site.xml 配置文件的目录路径，该配置文件将用于提供定制的 Hive 配置值。当创建 Iceberg Catalog 时如果在 hive-conf-dir 和 warehouse 都设置了，则来自< hive-conf-dir >/hive-site.xml（或 classpath 的 Hive 配置）的 hive.metastore.warehouse.dir 值将被 warehouse 的值覆盖。

（8）cache-enabled：是否启用 Catalog 缓存，默认值为 true。

创建之后，可以使用 show catalogs 命令查看当前所有的 Catalogs。

2. Hadoop Catalog

Apache Iceberg 还支持 HDFS 中基于目录的 Catalog，可以使用 'catalog-type' = 'hadoop' 来配置，代码如下：

```
CREATE CATALOG hadoop_catalog WITH (
    'type' = 'iceberg',
    'catalog-type' = 'hadoop',
    'warehouse' = 'hdfs://localhost:8020/user/hive/warehouse',
    'property-version' = '1'
);
```

其中 warehouse 是用于存放元数据文件和数据文件的 HDFS 目录（必选）。

3. 自定义 Catalog

Flink 还支持通过指定 catalog-impl 属性来加载自定义的 Iceberg Catalog 实现。当设置 catalog-impl 时，将忽略 catalog-type 的值，代码如下：

```
CREATE CATALOG my_catalog WITH (
    'type' = 'iceberg',
    'catalog-impl' = 'com.my.custom.CatalogImpl',
    'my-additional-catalog-config' = 'my-value'
);
```

4. 通过 YAML 配置创建 Catalog

可以在启动 Flink SQL Client 之前，在 conf/sql-client-defaults.yaml 配置文件中注册 Catalog，内容如下：

```
catalogs:
  - name: my_catalog
    type: iceberg
```

```
catalog-type: hadoop
warehouse: hdfs://localhost:8020/warehouse/path
```

8.2.5　Iceberg DDL 命令

Apache Iceberg 支持如下常用的 DDL 命令。

1）CREATE DATABASE
默认情况下，Iceberg 将使用 Flink 中的 default 数据库。如果不想在 default 数据库下创建表，则可以创建一个单独的数据库，例如，创建一个名为 iceberg_db 的数据库，代码如下：

```
create database iceberg_db;
use iceberg_db;
```

2）CREATE TABLE
表创建命令。使用方法如下：

```
CREATE TABLE hive_catalog.default.sample (
    id BIGINT COMMENT 'unique id',
    data STRING
);
```

创建表的 create 命令现在支持最常用的 Flink create 子句，包括：
（1）PARTITION BY (column1, column2, ...) 配置分区，Flink 还不支持隐藏分区。
（2）COMMENT 'table document' 设置一个表描述。
（3）WITH ('key'='value', ...) 设置将存储在 Apache Iceberg 表属性中的表配置。
目前不支持计算列、主键、水印定义等。

3）PARTITIONED BY
要创建分区表，使用 PARTITIONED BY，代码如下：

```
CREATE TABLE hive_catalog.default.sample (
    id BIGINT COMMENT 'unique id',
    data STRING
) PARTITIONED BY (data);
```

Apache Iceberg 支持隐藏分区，但 Apache Flink 不支持通过列上的函数进行分区，所以现在无法在 Flink DDL 中支持隐藏分区。

4）CREATE TABLE LIKE
要创建与另一个表具有相同模式、分区和表属性的表，可以使用 CREATE TABLE

LIKE，代码如下：

```
CREATE TABLE hive_catalog.default.sample (
    id BIGINT COMMENT 'unique id',
    data STRING
);

CREATE TABLE hive_catalog.default.sample_like LIKE hive_catalog.default.sample;
```

5）ALTER TABLE

目前，Iceberg 只支持在 Flink 1.11 中修改表属性。修改表属性的代码如下：

```
ALTER TABLE hive_catalog.default.sample SET ('write.format.default' = 'avro')
```

6）ALTER TABLE…RENAME TO

对表重命名，代码如下：

```
ALTER TABLE hive_catalog.default.sample RENAME TO hive_catalog.default.new_sample;
```

7）DROP TABLE

删除表 default.sample，代码如下：

```
DROP TABLE hive_catalog.default.sample;
```

8.2.6 Iceberg SQL 查询

Apache Iceberg 现在支持 Flink 中的流和批读取。用户可以执行以下 SQL 命令，将执行类型从 streaming 模式切换到 batch 模式，反之亦然，代码如下：

```
-- 以流模式为当前会话上下文执行 Flink 作业
SET execution.type = streaming

-- 以批处理模式为当前会话上下文执行 Flink 作业
SET execution.type = batch
```

1）Flink 批量读取

如果想通过提交 Flink 批处理作业检查 Iceberg 表中的所有行，则执行语句如下：

```
SET execution.type = batch;
SELECT * FROM hive_catalog.default.sample;
```

2) Flink 流式读取

Apache Iceberg 支持 Flink 流作业中从历史快照 id 开始的增量数据处理,执行语句如下:

```
-- 设置当前会话以流模式提交 Flink 作业
SET execution.type = streaming ;

-- 启用此开关,因为流式读取 SQL 将在 Flink SQL 提示选项中提供一些作业选项
SET table.dynamic-table-options.enabled=true;

-- 从 Iceberg 当前快照读取所有记录,然后从该快照开始读取增量数据
SELECT * FROM sample /* + OPTIONS('streaming' = 'true', 'monitor-interval' = '1s') */ ;

-- 从 id 为"3821550127947089987"的快照开始读取所有增量数据(将排除该快照的记录)
SELECT * FROM sample /* + OPTIONS('streaming' = 'true', 'monitor-interval' = '1s', 'start-snapshot-id' = '3821550127947089987') */ ;
```

这些是可以设置在 Flink SQL 提示选项中用于流作业的选项。

(1) monitor-interval:连续监控新提交数据文件的时间间隔(默认值:"1s")。

(2) start-snapshot-id:流作业开始的快照 id。

8.2.7　Iceberg SQL 写入

目前,Apache Iceberg 在 Flink 1.11 中支持 INSERT INTO 和 INSERT OVERWRITE 语句。

1) INSERT INTO

要使用 Flink 流作业将新数据追加到表中,可使用 INSERT INTO 语句,代码如下:

```
INSERT INTO hive_catalog.default.sample VALUES (1, 'a');
INSERT INTO hive_catalog.default.sample SELECT id, data from other_kafka_table;

insert into sample values (1,'test1');
insert into sample values (2,'test2');

insert into sample_partition PARTITION(data = 'city') SELECT 86;
```

2) INSERT OVERWRITE

若要用查询结果替换表中的数据,则可在批作业中使用 INSERT OVERWRITE(Flink 流作业不支持 INSERT OVERWRITE)语句。对于 Iceberg 表来讲,覆盖是原子操作。

有由 SELECT 查询产生的行的分区将被替换,代码如下:

```
INSERT OVERWRITE sample VALUES (1, 'a');
```

Apache Iceberg 还支持通过 select 查询值覆盖给定的分区，代码如下：

```
INSERT OVERWRITE hive_catalog.default.sample PARTITION(data = 'a') SELECT 6;
```

对于 Iceberg 分区表，当所有的分区列在 partition 子句中设置了一个值时，它用于插入一个静态分区，否则如果部分分区列（所有分区列的前缀部分）在 partition 子句中设置了一个值，则它用于将查询结果写入一个动态分区。对于一个未分区的 Iceberg 表，它的数据将被 INSERT OVERWRITE 完全覆盖。

8.2.8 使用 DataStream 读取

Iceberg 目前在 Java API 中支持流和批读取。

1) 批读取

例如，从 Iceberg 表中读取所有记录，然后在 Flink 批作业中将其打印到标准输出控制台，代码如下：

```
StreamExecutionEnvironment env =
    StreamExecutionEnvironment.createLocalEnvironment();

TableLoader tableLoader =
    TableLoader.fromHadooptable("hdfs://localhost:8020/warehouse/path");

DataStream<RowData> batch = FlinkSource.forRowData()
    .env(env)
    .tableLoader(loader)
    .streaming(false)
    .build();

//将所有记录打印到标准输出
batch.print();

//提交并执行此批读取作业
env.execute("Test Iceberg Batch Read");
```

2) 流读取

例如，读取从 snapshot-id '3821550127947089987' 开始的增量记录，并在 Flink 流作业中打印到标准输出控制台，代码如下：

```
StreamExecutionEnvironment env = StreamExecutionEnvironment.createLocalEnvironment();

TableLoader tableLoader = TableLoader.fromHadooptable("hdfs://localhost:8020/warehouse/path");
```

```
DataStream<RowData> stream = FlinkSource.forRowData()
    .env(env)
    .tableLoader(loader)
    .streaming(true)
    .startSnapshotId(3821550127947089987)
    .build();

//将所有记录打印到标准输出
stream.print();

//提交并执行此流读取作业
env.execute("Test Iceberg Batch Read");
```

8.2.9 使用 DataStream 写入

Apache Iceberg 支持从不同的 DataStream 写入 Iceberg 表。

1）追加数据

它支持将 DataStream<RowData>和 DataStream<Row>写入 Iceberg Sink 表中。模板代码如下：

```
StreamExecutionEnvironment env = ...;

DataStream<RowData> input = ... ;
Configuration hadoopConf = new Configuration();

TableLoader tableLoader = TableLoader
    .fromHadooptable("hdfs://localhost:8020/warehouse/path",hadoopConf);

FlinkSink.forRowData(input)
    .tableLoader(tableLoader)
    .build();

env.execute("Test Iceberg DataStream");
```

Iceberg API 还允许用户将泛型化的 DataStream<T>写入 Iceberg 表。

2）覆盖数据

为了动态地覆盖现有 Iceberg 表中的数据，可以在 FlinkSink builder 中设置 overwrite 标志。模板代码如下：

```
StreamExecutionEnvironment env = ...;

DataStream<RowData> input = ... ;
Configuration hadoopConf = new Configuration();
```

```
TableLoader tableLoader = TableLoader
.fromHadooptable("hdfs://localhost:8020/warehouse/path",hadoopConf);

FlinkSink.forRowData(input)
    .tableLoader(tableLoader)
    .overwrite(true)
    .build();

env.execute("Test Iceberg DataStream");
```

8.2.10 重写文件操作

通过提交 Flink 批作业，Iceberg 提供了将小文件改写为大文件的 API。这个 Flink 操作的行为与 Spark 的 rewriteDataFiles 相同，代码如下：

```
import org.apache.iceberg.flink.actions.Actions;

TableLoader tableLoader = TableLoader.fromHadooptable("hdfs://localhost:8020/warehouse/path");
Table table = tableLoader.loadTable();
RewriteDataFilesActionResult result = Actions.forTable(table)
        .rewriteDataFiles()
        .execute();
```

8.2.11 未来改进

在当前的 Flink Iceberg 集成工作中，有一些特性还不支持，包括：
(1) 不支持创建带有隐藏分区的 Iceberg 表。
(2) 不支持创建包含计算列的 Iceberg 表。
(3) 不支持创建带水印的 Iceberg 表。
(4) 不支持添加列、删除列、重命名列、更改列。
这些特性将在未来的版本中逐步得到支持和改进。

8.3 基于 Flink+Iceberg 构建准实时数仓

Apache Flink 是大数据领域流行的流批处理计算引擎。数据湖是云时代的一种新技术架构。这导致了基于 Iceberg、Hudi 和 Delta Lake 的解决方案的兴起。Apache Iceberg 目前支持 Flink 通过 DataStream API/Table API 将数据写入 Iceberg 表，并为 Apache Flink 1.11.x 提供集成支持。

【示例 8-1】 使用 Apache Flink 和 Apache Iceberg 构建实时数据仓库。

本节将解释并演示如何使用 Apache Flink 和 Apache Iceberg 构建实时数据仓库。

8.3.1 实时数仓构建

接下来，集成 Flink+Iceberg+Presto 实现基于数据湖的准实时数仓案例。整体设计架构如图 8-12 所示。

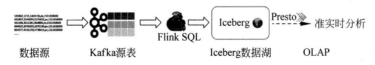

图 8-12 集成 Flink+Iceberg+Presto 实现基于数据湖的准实时数仓架构

继续使用淘宝用户行为数据集 UserBehavior_part.csv 来实时生成用户行为数据，使用 Kafka 作为流数据入口，使用 Flink SQL Client 执行实时入库操作，使用 Presto 执行分钟级的 OLAP 联机分析处理。

因为这里 Flink SQL Client 要访问 Kafka 源表及使用 Iceberg 表格式，所以需要将相关的 JAR 包添加到 Flink 安装路径的 lib 目录下，包括：

(1) iceberg-flink-runtime-0.12.0.jar。

(2) kafka-clients-2.4.1.jar。

(3) flink-connector-base-1.13.2.jar。

(4) flink-connector-kafka_2.12-1.13.2.jar。

(5) flink-sql-connector-kafka_2.12-1.13.2.jar。

如图 8-13 所示。

/home/hduser/bigdata/flink-1.13.2/lib/

Name	Size (KB)	Last modified	Owner	Group	Access
..					
antlr-runtime-3.5.2.jar	163	2021-09-14...	hduser	hduser	-rw-rw-r--
commons-cli-1.4.jar	52	2021-08-09...	hduser	hduser	-rw-rw-r--
flink-clients_2.12-1.13.2.jar	198	2021-09-18...	hduser	hduser	-rw-rw-r--
flink-connector-base-1.13.2.jar	44	2021-09-16...	hduser	hduser	-rw-rw-r--
flink-connector-hive_2.12-1.13.2.jar	7 604	2021-09-13...	hduser	hduser	-rw-rw-r--
flink-connector-kafka_2.12-1.13.2.jar	347	2021-09-16...	hduser	hduser	-rw-rw-r--
flink-csv-1.13.2.jar	90	2021-07-23...	hduser	hduser	-rw-r--r--
flink-dist_2.12-1.13.2.jar	103 645	2021-07-23...	hduser	hduser	-rw-r--r--
flink-json-1.13.2.jar	144	2021-07-23...	hduser	hduser	-rw-r--r--
flink-shaded-hadoop-3-uber-3.1.1.7.2.9.0-...	58 207	2021-08-09...	hduser	hduser	-rw-rw-r--
flink-shaded-zookeeper-3.4.14.jar	7 529	2021-04-10...	hduser	hduser	-rw-r--r--
flink-sql-connector-hive-3.1.2_2.12-1.13.2.jar	47 700	2021-09-14...	hduser	hduser	-rw-rw-r--
flink-sql-connector-kafka_2.12-1.13.2.jar	3 584	2021-09-16...	hduser	hduser	-rw-rw-r--
flink-streaming-java_2.12-1.13.2.jar	1 393	2021-09-18...	hduser	hduser	-rw-rw-r--
flink-table-api-java-bridge_2.12-1.13.2.jar	122	2021-09-18...	hduser	hduser	-rw-rw-r--
flink-table-api-scala-bridge_2.12-1.13.2.jar	51	2021-09-18...	hduser	hduser	-rw-rw-r--
flink-table-blink_2.12-1.13.2.jar	37 644	2021-07-23...	hduser	hduser	-rw-r--r--
flink-table_2.12-1.13.2.jar	34 198	2021-07-23...	hduser	hduser	-rw-r--r--
hive-exec-3.1.2.jar	39 671	2019-08-23...	hduser	hduser	-rw-r--r--
iceberg-flink-runtime-0.12.0.jar	22 720	2021-09-17...	hduser	hduser	-rw-rw-r--
kafka-clients-2.4.1.jar	3 193	2021-07-09...	hduser	hduser	-rw-rw-r--
log4j-1.2-api-2.12.1.jar	65	2020-03-25...	hduser	hduser	-rw-r--r--
log4j-api-2.12.1.jar	270	2020-03-25...	hduser	hduser	-rw-r--r--
log4j-core-2.12.1.jar	1 635	2020-03-25...	hduser	hduser	-rw-r--r--
log4j-slf4j-impl-2.12.1.jar	22	2020-03-25...	hduser	hduser	-rw-r--r--
mariadb-java-client-2.1.2.jar	482	2021-09-16...	hduser	hduser	-rw-rw-r--

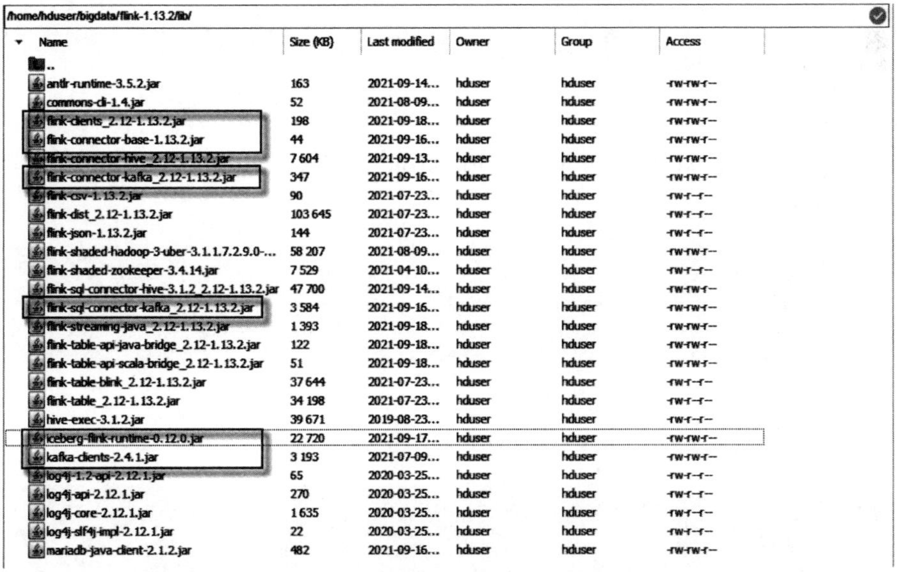

图 8-13 将相关的 JAR 包添加到 Flink 安装路径的 lib 目录下

接下来，建议按以下步骤执行：

第1步：环境准备。

首先准备好运行环境，包括 Hadoop 集群、Hive Metastore 服务、Kafka 集群、Flink on YARN session。建议按以下步骤操作：

（1）启动 HDFS 集群。在终端窗口中，执行的命令如下：

```
$ start-dfs.sh
$ start-yarn.sh
```

（2）启动 Hive Metastore 服务。在终端窗口中，执行的命令如下（保持运行）：

```
$ hive --service metastore
```

（3）启动 Kafka 集群。建议按下面的注释说明执行。

```
#切换到 Kafka 安装目录
$ cd ~/bigdata/kafka_2.12-2.4.1

#打开一个终端，运行 ZooKeeper
$ ./bin/zookeeper-server-start.sh ./config/zookeeper.properties

#另打开一个终端，运行 Kafka 服务器
$ ./bin/kafka-server-start.sh ./config/server.properties

#再打开一个终端，创建名为 user_behavior 的 Kafka topic
$ ./bin/kafka-topics.sh --create --zookeeper localhost:2181 --replication-factor 1 --partitions 1 --topic user_behavior

#查看已经存在的 Topics
$ ./bin/kafka-topics.sh --list --zookeeper localhost:2181
```

（4）启动 Flink 集群（注意，这里使用 Flink on YARN session 模式）。在终端窗口中，执行的命令如下：

```
$ cd ~/bigdata/flink-1.13.2
$ export HADOOP_CLASSPATH='hadoop classpath'
$ ./bin/yarn-session.sh --detached
```

第2步：启动 Flink SQL 客户端。

启动 Flink SQL Client。在终端窗口中，执行的命令如下：

```
$ export HADOOP_CLASSPATH='hadoop classpath'
$ cd ~/bigdata/flink-1.13.2
$ ./bin/sql-client.sh embedded --session application_1623048264828_0001
```

注意，记得将应用程序 id 修改为读者当前正在运行的应用程序 id。

第 3 步：创建 Kafka 流表（源表）。

在默认 catalog 下，创建 Kafka 源表。在 Flink SQL Client 命令行，执行以下语句：

```sql
-- 查看 catalogs
show catalogs;

-- 创建 Kafka 源表
DROP TABLE IF EXISTS user_behavior_kafka;

CREATE TABLE user_behavior_kafka (
    user_id bigint,
    item_id bigint,
    category_id bigint,
    behavior string,
    ts bigint,
    procTime AS PROCTIME(),
    eventTime AS TO_TIMESTAMP(FROM_UNIXTIME(ts, 'yyyy-MM-dd HH:mm:ss')),
    WATERMARK FOR eventTime AS eventTime - INTERVAL '5' SECONDS
)
with(
    'connector' = 'kafka',
    'topic' = 'user_behavior',
    'properties.Bootstrap.servers' = 'xueai8:9092',
    'properties.group.id' = 'testGroup',
    'scan.startup.mode' = 'latest-offset',
    'format' = 'csv',
    'csv.ignore-parse-errors' = 'true',
    'csv.field-delimiter' = ','
);

show catalogs;
```

第 4 步：创建 Iceberg Hive Catalog 及 Iceberg Sink 表。

Flink 会实时读取 Kafka 的用户购买行为记录并写入 Iceberg Sink 表中，因此，下面先定义一个可用的 Iceberg Catalog，然后在该 Catalog 中定义数据库和接收表。继续在 Flink SQL Client 命令行执行以下语句：

```sql
-- 定义可用的 Iceberg Catalogs
CREATE CATALOG iceberg_hive_catalog WITH (
  'type' = 'iceberg',
  'catalog-type' = 'hive',
  'uri' = 'thrift://localhost:9083',
```

```sql
    'clients' = '5',
    'property-version' = '1',
    'hive-conf-dir' = '/home/hduser/bigdata/flink-1.13.2/conf'
);

-- 切换到 Iceberg Catalog
USE CATALOG iceberg_hive_catalog;

-- 创建 iceberg_db 数据库
CREATE DATABASE IF NOT EXISTS iceberg_db;

-- 查看当前 Catalog 中有哪些数据库
show databases;

-- 切换数据库
use iceberg_db;

-- 查看当前 Catalog 中的当前数据库中有哪些表
show tables;

-- 创建 Iceberg Sink 表
DROP TABLE IF EXISTS iceberg_db.user_behavior_hive;

CREATE TABLE iceberg_db.user_behavior_hive (
        user_id bigint,
        item_id bigint,
        category_id bigint,
        behavior string,
        ts bigint,
        ts_date string,
        ts_hour string,
        ts_minute string
) PARTITIONED BY (ts_date,ts_hour,ts_minute)
WITH ('connector' = 'iceberg','write.format.default' = 'ORC');

-- 执行实时 ETL
INSERT INTO iceberg_db.user_behavior_hive
SELECT
    user_id,item_id,category_id,behavior,ts,
    FROM_UNIXTIME(ts, 'yyyy-MM-dd') as ts_date,
    FROM_UNIXTIME(ts, 'HH') as ts_hour,
    FROM_UNIXTIME(ts, 'mm') as ts_minute
FROM default_catalog.default_database.user_behavior_kafka
WHERE item_id > 0;
```

8.3.2 执行 OLAP 联机分析

接下来，使用 Presto 执行 OLAP 联机分析处理，对 iceberg_db.user_behavior_hive 表执行聚合操作，每 10min 统计一次用户的各种行为。

Apache Presto 是由 Facebook 开发的一个分布式并行查询执行引擎，对低时延和交互式查询分析进行了优化。Presto 可以轻松地运行查询，而且不需要停机时间，甚至可以从 GB 级扩展到 PB 级。

建议按以下步骤配置和使用 Presto。

第 1 步：配置 Iceberg 连接器。

Apache Iceberg 连接器允许查询 Iceberg 表中存储的数据。

要配置 Apache Iceberg 连接器，需要在 Presto 安装目录下的 etc/catalog/子目录中创建一个 catalog 目录属性文件 iceberg.properties（etc/catalog/iceberg.properties），命令如下：

```
$ cd ~/bigdata/presto-server-0.261/etc/catalog
$ nano iceberg.properties
```

该属性文件需要包含以下内容：

```
connector.name=iceberg
hive.metastore.uri=thrift://localhost:9083
```

第 2 步：运行 Presto 服务器。

在前台运行 Presto 服务器。在终端窗口中执行的命令如下：

```
$ cd ~/bigdata/presto-server-0.261
$ bin/launcher run
```

第 3 步：运行 Presto CLI 客户端。

通过 Presto CLI 连接 Presto 服务器。在终端窗口中执行的命令如下：

```
$ cd ~/bigdata/presto-server-0.261
$ ./bin/presto-cli-0.261-executable.jar --server localhost:8989
```

然后在 Presto CLI 中执行如下 SQL 语句：

```
presto> show catalogs;
presto> show schemas from iceberg;
presto> show tables from iceberg.iceberg_db;
```

执行过程和结果如图 8-14 所示。

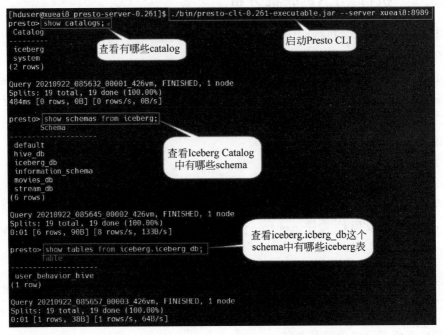

图 8-14　在 Presto CLI 中执行 SQL 语句过程

第 4 步：执行聚合操作。

接下来，执行聚合查询，统计各种用户购买行为。在 Presto CLI 中执行 Presto SQL 查询，语句如下：

```
presto> select behavior, count(1) as cnt
     -> from iceberg.iceberg_db.user_behavior_hive
     -> group by behavior;
```

执行过程和执行结果如图 8-15 所示。

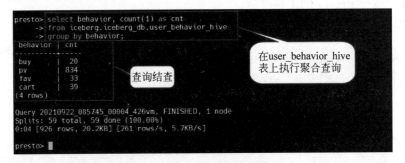

图 8-15　统计各种用户购买行为

第 5 步：退出 Presto CLI。

要退出 Presto CLI，执行的命令如下：

```
presto> exit;
```

第 6 步：关闭 Presto 服务器。

执行完所有命令后，可以使用以下命令停止 Presto 服务器：

```
$ bin/launcher stop
```

图 书 推 荐

书 名	作 者
HarmonyOS 应用开发实战（JavaScript 版）	徐礼文
HarmonyOS 原子化服务卡片原理与实战	李洋
鸿蒙操作系统开发入门经典	徐礼文
鸿蒙应用程序开发	董昱
鸿蒙操作系统应用开发实践	陈美汝、郑森文、武延军、吴敬征
HarmonyOS 移动应用开发	刘安战、余雨萍、李勇军 等
HarmonyOS App 开发从 0 到 1	张诏添、李凯杰
HarmonyOS 从入门到精通 40 例	戈帅
JavaScript 基础语法详解	张旭乾
华为方舟编译器之美——基于开源代码的架构分析与实现	史宁宁
Android Runtime 源码解析	史宁宁
鲲鹏架构入门与实战	张磊
鲲鹏开发套件应用快速入门	张磊
华为 HCIA 路由与交换技术实战	江礼教
深度探索 Go 语言——对象模型与 runtime 的原理、特性及应用	封幼林
深度探索 Flutter——企业应用开发实战	赵龙
Flutter 组件精讲与实战	赵龙
Flutter 组件详解与实战	［加］王浩然（Bradley Wang）
Flutter 跨平台移动开发实战	董运成
Dart 语言实战——基于 Flutter 框架的程序开发（第 2 版）	亢少军
Dart 语言实战——基于 Angular 框架的 Web 开发	刘仕文
IntelliJ IDEA 软件开发与应用	乔国辉
Vue＋Spring Boot 前后端分离开发实战	贾志杰
Vue.js 快速入门与深入实战	杨世文
Vue.js 企业开发实战	千锋教育高教产品研发部
Python 从入门到全栈开发	钱超
Python 全栈开发——基础入门	夏正东
Python 全栈开发——高阶编程	夏正东
Python 全栈开发——数据分析	夏正东
Python 游戏编程项目开发实战	李志远
Python 人工智能——原理、实践及应用	杨博雄 主编，于营、肖衡、潘玉霞、高华玲、梁志勇 副主编
Python 深度学习	王志立
Python 预测分析与机器学习	王沁晨
Python 异步编程实战——基于 AIO 的全栈开发技术	陈少佳
Python 数据分析实战——从 Excel 轻松入门 Pandas	曾贤志
Python 数据分析从 0 到 1	邓立文、俞心宇、牛瑶
Python Web 数据分析可视化——基于 Django 框架的开发实战	韩伟、赵盼

续表

书 名	作 者
Python 玩转数学问题——轻松学习 NumPy、SciPy 和 Matplotlib	张骞
Pandas 通关实战	黄福星
深入浅出 Power Query M 语言	黄福星
FFmpeg 入门详解——音视频原理及应用	梅会东
云原生开发实践	高尚衡
云计算管理配置与实战	杨昌家
虚拟化 KVM 极速入门	陈涛
虚拟化 KVM 进阶实践	陈涛
边缘计算	方娟、陆帅冰
物联网——嵌入式开发实战	连志安
动手学推荐系统——基于 PyTorch 的算法实现（微课视频版）	於方仁
人工智能算法——原理、技巧及应用	韩龙、张娜、汝洪芳
跟我一起学机器学习	王成、黄晓辉
TensorFlow 计算机视觉原理与实战	欧阳鹏程、任浩然
分布式机器学习实战	陈敬雷
计算机视觉——基于 OpenCV 与 TensorFlow 的深度学习方法	余海林、翟中华
深度学习——理论、方法与 PyTorch 实践	翟中华、孟翔宇
深度学习原理与 PyTorch 实战	张伟振
AR Foundation 增强现实开发实战（ARCore 版）	汪祥春
ARKit 原生开发入门精粹——RealityKit＋Swift＋SwiftUI	汪祥春
HoloLens 2 开发入门精要——基于 Unity 和 MRTK	汪祥春
巧学易用单片机——从零基础入门到项目实战	王良升
Altium Designer 20 PCB 设计实战（视频微课版）	白军杰
Cadence 高速 PCB 设计——基于手机高阶板的案例分析与实现	李卫国、张彬、林超文
Octave 程序设计	于红博
ANSYS 19.0 实例详解	李大勇、周宝
AutoCAD 2022 快速入门、进阶与精通	邵为龙
SolidWorks 2020 快速入门与深入实战	邵为龙
SolidWorks 2021 快速入门与深入实战	邵为龙
UG NX 1926 快速入门与深入实战	邵为龙
西门子 S7-200 SMART PLC 编程及应用（视频微课版）	徐宁、赵丽君
三菱 FX3U PLC 编程及应用（视频微课版）	吴文灵
全栈 UI 自动化测试实战	胡胜强、单镜石、李睿
pytest 框架与自动化测试应用	房荔枝、梁丽丽
软件测试与面试通识	于晶、张丹
智慧教育技术与应用	［澳］朱佳（Jia Zhu）
智慧建造——物联网在建筑设计与管理中的实践	［美］周晨光（Timothy Chou）著；段晨东、柯吉译
敏捷测试从零开始	陈霁、王富、武夏

续表

书　名	作　者
深入理解微电子电路设计——电子元器件原理及应用(原书第5版)	[美]理查德·C.耶格(Richard C. Jaeger)、[美]特拉维斯·N.布莱洛克(Travis N. Blalock)著；宋廷强 译
深入理解微电子电路设计——数字电子技术及应用(原书第5版)	[美]理查德·C.耶格(Richard C. Jaeger)、[美]特拉维斯·N.布莱洛克(Travis N. Blalock)著；宋廷强 译
深入理解微电子电路设计——模拟电子技术及应用(原书第5版)	[美]理查德·C.耶格(Richard C. Jaeger)、[美]特拉维斯·N.布莱洛克(Travis N. Blalock)著；宋廷强 译